The Editors

Dr. Rajan Kumar Gupta (1963) has worked on Ecophysiology of Antarctic Cyanobacteria for his Ph.D. degree with Late Prof. A.K. Kashyap of Centre of Advanced Study in Botany, Banaras Hindu University, Varanasi. For the past twenty years he has been working on various aspects of Antarctic microflora. Dr. Gupta was deputed by Govt. of India for his participation as Biological Scientist in Antarctica twice. He has participated in XI and XIV Indian Scientific Expeditions to Antarctica during 1991-92 and 1994-95. He has visited several countries like Mauritius, Japan, Nepal, Thailand, South Africa, Belgium, Singapore, Srilanka etc. for presentation of his work on different aspects of algae. Dr. Gupta has worked on various aspects of cyanobacteria, *i.e.*, morphology, ecology and nitrogen fixation, biotechnological applications and published more than 70 technical papers in various national and overseas journals and more than 40 chapters in various books. Dr. Gupta has published 3 Botany Practical Books and 1 book on Paryavaran Adhyan Environmental Studies) and 7 reference (research) books entitled "Glimpses of Cyanobacteria", "Advances in Applied Phycology", "Soil Microflora", "Microbial Biotechnology and Ecology Vol. 1 & Vol. 2" and "Diversity: An Overview and Diversity of Lower Plants". Seven students have been awarded the D.Phil degree and four are working under his supervision for their D.Phil degree of Various Universities of Uttarakhand. Dr. Gupta is a recipient of Research Award from University Grants Commission, New Delhi. Dr. Gupta is member of number of organizations in India and abroad. He is the Fellow of the Society for Environment and Ecoplanning and International Botanical Society and chaired various sessions in the conferences in India and abroad. He is in the editorial and advisory board of various journals. Presently, Dr. Gupta is teaching Microbiology and Biotechnology in Department of Botany, Dr. P.D.B.H. Government Post Graduate College, Kotdwar, Pauri Garhwal, Uttarakhand.

Dr. Nasim Akhtar (1963) obtained his M.Sc. and Ph.D. degree under the able guidance of one of the eminent Plant Tissue Culturist of India (Late) Prof. V. S. Jaiswal at the Centre of Advanced Study in Botany, Banaras Hindu University, Varanasi. Subsequently he has worked in the area of Transgenic Plant Research and Plant Molecular Biology for his Post Doctoral Research at Indian Institute of Technology, Kharagpur. Dr. Nasim Akhtar is presently working in the capacity of Associate Professor at Department of Biotechnology, GITAM Institute of Technology, GITAM University, Gandhi Nagar Campus, Rushikonda, Visakhapatnam, (A.P.), India.

He is actively engaged in the research on various aspects of Plant Biotechnology for the last 23 years particularly on induction of somatic embryogenesis in plants. Presently he is working on production of anticancer drug production from *Vinca rosea*. He is looking for the miniaturization of optimization process for secondary metabolite production from callus and cell suspension culture.

He has published four research papers on the regulation of somatic embryogenesis in guava, and written several review chapters on somatic embryogenesis in guava published by Kluwer Academic Publisher, Netherland and Daya Publications Pvt. Ltd, New Delhi. A Protocol for high efficiency somatic embryogenesis in guava has been published in Springer Series: Methods in Molecular Biology.

Dr. Deepak Vyas (1964) M.Sc. 1987 from Dr. Hari Singh Gaur University, Sagar and Ph.D. 1992 from Banaras Hindu University, Varanasi, is Associate Professor in Department of Botany, Dr. H.S. Gaur University, Sagar. Dr. Vyas has 16 years of teaching and 23 years of Research Experience. 15 students has obtained their Ph.D. degree under guidance and 3 are working for their Ph.D.

He is UGC Research Awardee and reciepient of International award on Ozone depletion. Dr. He published about 119 papers in National and International Journals. He has edited 2 books "Microflora" and "Microbial Biotechnology and Ecology", member of editorial board of many and various organisation, participated and organised various symposia/seminars/workshops and botanical excursions and delivered a number of lectures in different and institutions. Have worked in University administration in various capacities, as in university examination and result processing, central evaluation, coordinator central mission cell, joint proctor, as incharge University Botanical Garden since 2011 etc. At Vyas is working on AM biotechnology and mushroom biology. He is also providing ices on mushroom cultivation and marketing for mushroom growers of the region.

Biotechnology
An Overview

h
a
A
hi
Me

his g
Dr. V
Vyas
"Soil
journals
conferen
universiti
coordinat
university
present Dr
extension se

Biotechnology
An Overview

Rajan Kumar Gupta
M.Sc., Ph.D. F.I.C.E.R.
Associate Professor, Department of Botany,
Dr. P.D.B.H. Government Post Graduate College, Kotdwar, Pauri Garhwal,
Uttarakhand, India
(Affiliated to HNB Garhwal University, Srinagar, UK, India)

Nasim Akhtar
M.Sc., Ph.D.
Associate Professor, Department of Biotechnology
GITAM Institute of Technology, GITAM University,
Visakhapatanam – 520 045 Andhra Pradesh, India

Deepak Vyas
M.Sc., Ph.D.
Associate Professor, Department of Botany,
Dr. HS Gaur University (Central University)
Sagar, Madhya Pradesh, India

2015
Daya Publishing House®
A Division of
Astral International Pvt. Ltd.
New Delhi – 110 002

Cataloging in Publication Data--DK
Courtesy: D.K. Agencies (P) Ltd. <docinfo@dkagencies.com>

Biotechnology : an overview / Rajan Kumar Gupta, Nasim Akhtar,
Deepak Vyas.
    p.   cm.

Contributed articles.
Includes bibliographical references and index.
ISBN 978-93-5130-303-9 (International Edition)

1. Biotechnology. 2. Biotechnology--India.   I. Gupta, Rajan Kumar, 1963-
II. Akhtar, Nasim, 1963- III. Vyas, Deepak, 1964-

DDC 660.6   23

Published by	:	**Daya Publishing House®** *A Division of* **Astral International Pvt. Ltd.** – ISO 9001:2008 Certified Company – House No. 96, Gali No. 6, Block-C, 30ft Road, Tomar Colony, Burari New Delhi-110 084 E-mail: info@astralint.com Website: www.astralint.com
Sales Office	:	4760-61/23, Ansari Road, Darya Ganj New Delhi-110 002 Ph. 011-23245578, 23244987
Laser Typesetting	:	**Classic Computer Services** Delhi - 110 035
Printed at	:	**Thomson Press India Limited**

Acknowledgements

The editors are thankful to all the academicians and scientists whose contributions have enriched this volume. We also express our deep sense of gratitude to our parents whose blessings have always prompted us to pursue academic activities deeply. It is quite possible that in a work of this nature, some mistakes might have crept in text inadvertently and for these we owe undiluted responsibility.

We are grateful to all authors for their contribution to present book. The Editor Rajan Kumar Gupta is thankful to University Grants Commission, New Delhi and to Uttarakhand Council for Science and Technology, Dehradun, UK for providing financial assistance in the form of project. Rajan Kumar Gupta wish to place on record his special thanks to his wife Mrs Alka and two daughters Akriti and Ayushi and his research team for their cooperation in all his academic and scientific endeavours. Nasim Akhtar and Deepak Vyas gratefully acknowledge their institutions and family without their support this work was not possible. Finally, we will always remain debtor to all our well wishers for their blessings without which this book would not have come to light.

Rajan Kumar Gupta
Nasim Akhtar
Deepak Vyas

Preface

We are living in revolutionary era of information technology, industrial technology and biotechnology that has significantly transformed all spheres of life and the lifestyle. The selective breeding of dogs dates back over 14,000 years ago and about 10,000 years ago, human being extended his control over other life forms to include the domestication of animals. Much before the beginning of civilization our ancestors were fermenting grains and fruits to create alcoholic beverages. This applied aspect of micro-organisms, like that of animals, was completely trivial based on taste, smell, and vision, and not on any knowledge of genetic mechanisms for selective breeding. Biological inventions towards the entire twenty century have changed the sustainability in the present century. The discovery of DNA structure in 1953, genetically modified tomato in 1995, cloning of Dolly the sheep in 1996, and the sequencing of the entire human genome at the end of millennium testify awareness that the biotech miracle is inescapable today.

Biotechnology is a diverse field dealing with the application of biological discoveries to transform industry, agriculture, and medicine in a new dimension. This biotechnological application is much hyped created by overzealous promotion of biotechnology companies gravitated by press and media to cure diseases, develop new drugs, and feed the world's hunger through genetically modified organisms, resulting in confusion over what is real and what is fanciful speculation. Biotechnology is variably portrayed as either the next dot-com ride for those with excess capital to invest or as simply not worth following as an investment vehicle. There is no debate that biotech is a global business phenomenon. Worldwide successes and failures of biotech companies are tracked even for a weak signal that could boom like dot-com boom of 1990s.

As the investment in biotechnology varies considerably from one country to the next by virtue of corporate and government funding, variations in public acceptance of biotechnology products, and the country's political environment. The successful investors and business executives in the biotechnology space understand and capitalize on the global interdependencies in the industry. The success of applying biotechnology to commercialize products still depends on substantial investment in R&D undertaken by private and public investors, researchers, and entrepreneurs.

Being multidisciplinary nature, biotechnology does not necessarily offer the single, best route; rather it can effectively be used as one of the tools or integrated into other processes. The strict rules on selection of programs, organisms, methods, technologies and biosafety as well as new legislation on restricted use of genome modifications of vertebrates, higher plants, genetically modified food, patenting of transgenic animals or sequenced parts of genomes, biotechnology has emerged with a very high standard high-tech and safe technology. Definitely the application of biotechnology invariably has reduced either operating costs or capital costs or both. This has led to a more sustainable process with lowered ecological disturbances by reducing some or many energy use, water use, waste-water or greenhouse gas production.

The ethical and social problems arising in agriculture and medicine are still controversial. The acceptance of "Biotechnology" in medicine, agriculture, food and pharma production has become a political matter all over the globe.

The aim of this volume on Biotechnology: An Overview is to keep the reader informed on the latest progress made in microbiology, plant and animal cell culture to the extent of industrial technology. The volume is consisting of 28 different chapters contributed by the author(s) having vast experience in teaching and research in the wide arena of biotechnology.

The present book comprised of 28 chapters on different aspect of Biotechnology. Chapter 1ˢᵗ on quorum sensing is totally based on new frontier of Biotechnology and Microbial Communication System. Chapter 2ⁿᵈ will provide latest information on Bio Fuel Technology. Chapters 3, 12, 17, 19, 21 are based on Environmental Biotechnology, which provide information not only on Ecosystem but also give an idea, how to commercialize the technique to improve the soil quality including phytoremediation. Chapters 4 and 25 gives an overview of microbial diversity and ultimately how these microbes are utilized commercially for the benefits of the human being. Chapter 5 Explains new concept of Bioinformatics related to plant world.

Chapters 6, 26, 27 and 28: As we are aware of commercial aspect of plants and their role in controlling various environmental problems at one end and on other end they are the sources of food, medicine etc. many elements which are the source of nutrients for the growth of the plants more in Nano Size. Therefore, Nanotechnology though being a new science but in one way or other contributes not only in terms of information but also ultimately improvise commercial aspect of various plants, Whether they are used as agriculture crops, horticultural crops, floriculture or as wild plants. Similarly in Chapter 7 which is related to latest hot talk about GM crops related food. Since the present population of world is suffering from shortage of food and to overcome this problem, GM crops may provide some solutions for that.

Chapters 8, 10 and 11 gives latest information about Biofertilizer Technology. Biofertilizer not only reduces the cost of chemical fertilizer but also enhances the soil quality. These days' numbers of bio inoculants are available on commercial scale to enrich the soil health as well as plant health. Chapter 25 gives an account on Plant which is greatest sources of medicine and drug. Taxol an important drug of cancer from *Taxus bacata*. Fungi are Achlorophylls Plants are store house of many medicines and industrially important enzymes. An application of new biotechnological tools or molecular marker; certainly help us to explore new vistas of science. Chapters 16, 18 and 20 deals with plant Cell Factory, Bioreactor and, Microbial Mega Cell in advancement of biological processes provides commercial aspect of plants and microbial system. Chapters 22, 23 and 24 deals with risk associated with Transgenic Plants, Biosensors and their commercial aspect and molecular mechanism, which enhance drought tolerances in plants and ultimately, such plant valued for commercial purpose.

Rajan Kumar Gupta

Nasim Akhtar

Deepak Vyas

Contents

Biotechnology: An Overview (2015)   Pages 1–7
Editors: Rajan Kumar Gupta, Nasim Akhtar and Deepak Vyas
Published by: DAYA PUBLISHING HOUSE, NEW DELHI

Chapter 1

Quorum Sensing: Microbial Communication System

Santosh Kumar Karn*

*Ambala College of Engineering and Applied Research,
Ambala – 133 001, Haryana*

ABSTRACT

Quorum sensing (autoinduction) is a term that describes an environmental sensing system that allows bacteria to monitor their own population density. Autoinduction relies upon the interaction of a small diffusible signal molecule (the autoinducer) with a transcriptional activator protein. These signal molecules diffuse from bacterial cells and accumulate in the environment as a function of cell growth. Once a threshold concentration is reached, these signals serve as co-inducers to regulate the transcription of a set of target genes. Gram-positive and Gram-negative bacteria use quorum sensing communication circuits to regulate a diverse array of physiological activities like symbiosis, virulence, competence, conjugation, antibiotic production, motility, sporulation, and biofilm formation. In Gram-negative bacteria, most autoinducer belong to the family of N-acyl homoserine lactones (AHL). Gram-positive bacteria use processed oligo-peptide to communicate. Cell-to-cell communication has played an important role in many diverse community based function it also helps the establishment of a population in changing environment according to the requirement. This sensing can be put to numerous uses such as in controlling plant and animal diseases, regulation of the production of useful/toxic fermentation products, etc. in the future.

Keywords: Quorum sensing, Cell-to-cell communication, Autoinducer, N-acylhomoserine lactones, Oligo-peptides.

* Author: E-mail: santoshkarn@gmail.com

Introduction

Quorum sensing is a process by which micro-organisms communicate and interact with each other through the use of pheromone like molecules (Bassler, 1999). This type of communication system is used by multiple species of microbe to essentially count their neighbours and once at a critical density, co-ordinate a variety of different group activities through control of gene expression at population level (Bassler, 1999). In this way, bacteria are able to essentially act as a multicelullar organism.

These bacterial communication systems were considered anomalous, and in general bacteria as a whole were not believed to use cell-cell communication. The exchange of chemical signals between cells/organisms was assumed to be a trait highly characteristic of eukaryotes. The recent advance in the field of cell-cell communication in bacteria has now shown that many bacteria probably communicate using secreted chemical molecules to coordinate the behaviour of the group. Furthermore, we now know that different classes of chemical signal are employed by the some bacteria, use more than one chemical signal and/or more than one type of signal to communicate, that complex hierarchical regulatory circuit have evolved to integrate and process the sensory information, and that the signals can be used to differentiate between species in consortia. It seems clear now that the ability to communicate both within and between species is critical for bacterial survival and interaction in natural habitats.

The signal molecule used for communication was dubbed as autoinducer, owing to its origin inside the bacterial cell. The desired response can be arrived at by attainment of quorum employing the autoinducer and the process was labeled as autoinduction. In other words, the whole circuit relies on the intracellular production and export of a low-molecular mass signalling molecule, the extracellular concentration of which grows with the population density of the producing organism. The signaling molecule can be sensed and reimported into these cells, thus allowing the whole population to respond to changing environment/requirement once a critical concentration (cell density) has been achieved.

Several classes of microbial-derived signaling molecules have now been identified. Broadly, these can be divided into two main categories (i) amino acids and short peptide derivatives commonly used by Gram-positive bacteria (Lazazzera and Grossman, 1998) and (ii) fatty acid derivatives called homoserine lactones (HSL) frequently used by Gram-negative bacteria (Whitehead *et al.*, 2001). Whatever may be the nature of the signal molecule, the whole network functions by its reentry into the cell either via diffusion or an active transport (Whitehead *et al.*, 2001). The signaling mechanism involves subsequent interaction of the signal with intracellular effectors that will induce the pathway for the concerned phenotype.

In the past decade quorum sensing circuits have been identified in over twenty five species of Gram-negative bacteria (Parsek and Greenberg, 2000). In every case except those of *Vibrio harveyi* and *Monocentris xanthus* the quorum sensing circuits identified in Gram negative bacteria resemble the canonical quorum sensing circuit of the symbiotic bacterium *V. fischeri*. Specifically, these Gram-negative bacterial quorum sensing circuits contain, at a minimum, homologues of two *V. fischeri* regulatory proteins called LuxI and LuxR. The LuxI-like proteins are responsible for the biosynthesis of a specific acylated homoserine lactone signaling molecule HSL known as anautoinducer. The autoinducer concentration increases with increasing cell-population density. The LuxR-like proteins bind cognate HSL autoinducers that have achieved a critical threshold concentration, and the LuxR-autoinducer complexes also activate target gene transcription (Engebrecht *et al.*, 1983). Using this quorum sensing mechanism Gram-negative bacteria can efficiently couple gene expression to fluctuations in cell-population density. Twenty five species of bacteria that mediate quorum sensing by means of a LuxI/LuxR-type circuit, the *V. fischeri, Pseudomonas aeruginosa, Agrobacterium tumefaciens,* and *Erwinia carotovora systems* are understood clearly.

Quorum-Sensing in Gram-Negative Bacteria (Bioluminescence: The Lux System)

The first incidence of such a biological phenomenon came to light with the discovery of luminescence produced by certain marine bacteria such as *V. fischeri* and *V. harveyi*. These bacteria, when free-living in sea water at low cell density are non-luminescent. However, when grown to high cell densities in the laboratory, *V. fischeri* culture bioluminisces with a blue-green light. Interestingly this bacterium commonly form symbiotic relationship with some fishes (such as the Japanese pinecone fish *Monocentris japonica*) and squid species (such as *Euprymna scolopes*) (Visick and McFall-Ngai 2000). These marine animals carry a specialized organ called the light organ, in which bacteria like *V. fischeri* are housed. *E. scolopes* may express bio luminescent appearance in dark environments due to the maintenance of a high-density *V. fischeri* population (10^{10}–10^{11} cells ml^{-1}) in the light organ. This bioluminescent phenotype is exploited by the squid in order to perform a behavioural phenomenon called counter-illumination. At night, the squid camouflages itself from predators residing below it by controlling the intensity of light that it projects downwards, thus eliminating a visible shadow created by moonlight. This is a case of perfect symbiosis, as in return *E. scolopes* provides the *V. fischeri* population with nutrients. The presence of luminescent-competent *V. fischeri* cells in the light organ of juvenile squid is crucial for the correct development of this organ (Visick and McFall-Ngai 2000). Further studies on *V. fischeri* revealed that the bacterium grows very fast, directly entering the exponential phase, but the luminescence increases only at about mid-log phase of its growth (Hastings and Greenberg 1999). The sudden increase in luminescence was attributed to the transcriptional regulation of the enzyme, luciferase, which in turn corresponded to a threshold density of cells. This whole circuit is based on the bacterial assessment of its population density by means of release of chemical signaling molecules or autoinducers. The autoinducer then establishes a communication between the cells that gets reflected in the expression of a particular gene, in this case, the luciferase gene (*lux*). For a long time, bioluminescene expressed by *V. fischeri* remained a model system to study density dependent. For a long time, bioluminescene expressed by *V. fischeri* remained a model system to study density dependent expression of a gene function (Hastings and Greenberg, 1999).

Quorum-Sensing Gram-Positive Bacteria

A number of Gram-positive bacteria are also reported to quorum sensing system. The nature of the signal molecules used in these systems differs from those of Gram negative organisms (Dunny and Leonard 1997). Quorum sensing is used to regulate the development of bacterial competence in *Bacillus subtilis* and *Streptococcus pneumoniae*, conjugation in *Enterococcus faecalis*, and virulence in *Staphylococcus aureus* (Dunny and Leonard 1997). In *Pneuomococci*, five genes have been implicated in the peptide-mediated regulatory circuit like, com ABCDE for competence development (Lee and Morrison 1999). The peptide signal required for development of the competent state is called competence stimulating peptide (CSP). CSP is a 17-amino acid peptide that is produced from a 41-amino acid precursor peptide called Com C27. The ComAB-ATP binding cassette (ABC) transporter processes and secretes CSP extracellularly (Pestova *et al.*, 1996). Com C expression is normally maintained at a basal level, allowing production of peptide in proportion to cell numbers. Com D acts as a membrane bound receptor/kinase and acts through a response regulator, Com E, to transmit a signal reflecting the extracellular abundance of CSP to responder genes. High levels of CSP induce auto phosphorylation of Com D, which leads to subsequent transfer of the phosphoryl group to ComE (Pestova *et al.*, 1996). Phospho-Com E activates transcription of the com X gene. Com X is an alternative sigma factor that initiates the transcription of competence-specific operon involved in DNA uptake and recombination by recognizing a com box (also called as cin-box) consensus sequence (TACGAATA) in their promoter regions (Welch *et al.*, 2000). Due to the highly diverse nature of phenotypes governed by cell-cell signaling cascade, it is difficult to describe all of them.

Quorum Sensing in Prokaryote and Eukaryote

Although quorum sensing signal molecules have largely been considered effectors of prokaryotic gene expression, they can also affect the behavior of eukaryotic cells. AHL are known to have immunomodulatory effects. They also induce relaxation of blood vessels. Apparently some bacteria have the power to influence the host immune responses to their benefit, and stimulate the delivery of nutrients for their survival by increasing the blood supply. But signal molecules may also benefit the host. 'Probiotic' bacteria are thought to be beneficial to the host organism and are added as dietary supplements in health promoting food. Cultures of *Bacillus subtilis*, for example, have been used to treat dysentery and other intestinal problems. Recently it was revealed that *B. subtilis* produces a quorum sensing signal molecule, the competence and sporulation stimulating factor (CSF), which induces the synthesis of the heat shock protein Hsp27 in the intestine. This protects intestinal cells against oxidative damage and loss of barrier function. The marine alga *Ulva* releases zoospores into the water. These attach to a suitable surface and differentiate into new algae. The zoospores are known to settle preferentially on to sites of concentrated AHL biosynthesis.

Implications

As described, there is a widespread occurrence of cell-cell signalling among different bacterial species. Also, many organisms utilize the same species of molecule to regulate different phenotypes. Thus, one would predict that some form of interspecies communication would be likely in environments where different autoinducer producing bacterial species inhabit a common place. Results from numerous studies have shown that various LuxR homologues can interact with non-cognate acyl HSL molecules (Welch *et al.*, 2000; Wood *et al.*, 1997). Depending on the LuxR homologue and the acyl HSL, such interactions can result in the activation of the specific transcriptional regulator. Conversely when assayed in the presence of the cognate acyl HSL, other species of acyl HSL have been found to essentially block activation of the LuxR homologue, presumably by competing for the ligand-binding site on the protein. It seems likely that in the environment, one bacterial community could produce acyl HSLs not to inhibit the quorum-sensing phenotypes expressed by another community but to regulate physiological processes of its own. One reason why bacteria like *Xanthomonas* sp. use non-acyl HSL based quorum sensing system could be to gain a competitive advantage over their neighbouring bacteria by avoiding such interference and crosstalk. It is important to remember that a bacterial species could also respond to the presence of foreign acyl HSL by utilizing the signaling molecules to up or down regulate competitively advantageous phenotypes. For example, this phenomenon could be used for expression of competitor inhibitory antibiotics. A study led by (Wood *et al.*, 1997) demonstrated that phenazine biosynthesis can be stimulated in one population of *Pseudomonas aureofaciens* by acyl HSL produced by a distinct population of the same organism. Similarly, TraR of *Agrobacterium tumefaciens* can also respond to signals (cognate and non cognate) produced by other micro-organism that occupy its habitat (Cha *et al.*, 1998). Thus the development of specific signaling molecules used to sense not only cons specific bacteria but also certain non con specific bacteria present in certain specific niches, is a clever mechanism for community level regulation of gene expression. Many of the recognized acyl HSL producing micro-organisms are known for their capacity to associate with higher organisms either in a pathogenic or symbiotic relationship. Higher organisms have also evolved mechanisms that enable them to detect and respond to acyl HSL messaging systems in order to prevent or limit infection (Teplitski *et al.*, 2000). For example, the macroalga *Delisea pulchra* produces compounds, commonly known as furanones, which have the ability to specifically interfere with acyl HSL-mediated quorum-sensing system (Rasmussen *et al.*, 2000). While serving as prokaryotic cell-to-cell signals, some acyl HSLs may act as virulence factors. In particular one of the *P. aeruginosa* produced molecules OdDHL could act as a potential modulatory agent of the mammalian immune systems (Telford *et al.*, 1998). Interspecies communication through the use of autoinducers has been used as a possible mechanism by which the pathogenicity of certain virulent bacteria such as *Burkholderia cepacia* is enhanced (McKenny *et al.*, 1995). Thus, when determining the role of autoinducers in nature, the synergistic effects of the surrounding environment, including other bacteria and/or host must be considered.

Quorum Quenching

The most important implication of all this for human beings is that once we understand how bacteria talk, we can find ways to block their communication. It is hardly surprising that there has been a widespread surge of interest in quorum quenching *i.e.* blocking quorum sensing in bacteria. Already, many higher organisms including plants and animals have been found to produce AHL inactivating enzymes. Their purpose is presumably to inhibit quorum sensing and defend the organism against bacterial infection. In mammals enzymes that inactivate AHL have been found in serum and airway epithelia. Such natural quorum-quenching mechanisms may be used to develop a new generation of antimicrobials. Already, the crystal structures of several quorum-quenching enzymes have been found and their catalytic mechanisms elucidated. Thus, quorum sensing is emerging as an area that has immense research potential to understand how micro-organisms communicate among themselves as well as with other organisms.

References

1. Bassler, B.L, 1999. How bacteria talk to each other: Regulation of gene expression by quorum sensing. *Curr. Opin. Microbiol.*, 2: 582–587.

2. Cha, C., Gao, P., Chen, Y.C., Shaw, P.D. and Farrand, S.K., 1998. Production of acyl-homoserine lactone quorum-sensing signals by gram-negative plant-associated bacteria. *Mol. Plant-Microb. Interact.*, 11: 1119–1129.

3. Dunny, G.M. and Leonard, B.A., 1997. Cell-cell communication in gram-positive bacteria. *Annu. Rev. Microbiol.*, 51: 527–564.

4. Engebrecht, J. Nealson, K. and Silverman, M., 1983. Bacterial bioluminescence: isolation and genetic analysis of functions from *Vibrio fischeri. Cell*, 32: 773–781.

5. Hastings, J.W. and Greenberg, E.P., 1999. Quorum sensing: the explanation of a curious phenomenon reveals a common characteristic of bacteria. *J. Bacteriol.*, 181: 2667–2668.

6. Lazazzera, B.A. and Grossman, A.D., 1998. The ins and outs of peptide signaling. *Trends Microbiol.*, 7: 288–294.

7. Lee, M.S. and Morrison, D.A., 1999. Identification of a new regulator in *Streptococcus pneumoniae* linking quorum sensing to competence for genetic transformation. *J. Bacteriol.*, 181: 5004–5016.

8. McKenny, D., Brown, K.E. and Allison, D.G., 1995. Influence of *Pseudomonas aeruginosa* exproducts on virulence of *Burkholderia cepacia*: Evidence for interspecies communication. *J. Bacteriol.*, 177: 6989–6992.

9. Parsek, M.R. and Greenberg, E.P., 2000. Acylhomoserine lactone quorum sensing in gram-negative bacteria: a signaling mechanism involved in associations with higher organisms. *Proc. Natl. Acad. Sci. USA*, 97: 8789–8793.

10. Pestova, E.V., Havarstein, L.S. and Morrison, D.A., 1996. Regulation of competence for genetic transformation in *Streptococcus pneumonia* by an auto-induced peptide pheromone and a two-component regulatory system. *Mol. Microbiol.*, 21: 853–862.

11. Rasmussen, T.B., Manefield, M. and Andersen, J.B., 2000. How *Delisea pulchra* furanones affect quorum sensing and swarming motility in *Serratia liquefaciens* MG1. *Microbiology*, 146: 3237–3244.

12. Telford, G., Wheeler, D., Williams, P., Tomkins, P.T., Appleby, P., Sewell, H., Stewart, G.S.A.B., Bycroft, B.W. and Pritchard, D.I., 1998 The *Pseudomonas aeruginosa* quorum-sensing signal molecule *N*-(3-oxododecanoyl)-L-homoserine lactone has immunomodulatory activity. *Infect. Immun.*, 66: 36–42.

13. Teplitski, M., Robinson, J.B. and Bauer, W.D., 2000. Plants secrete substances that mimic bacterial N-acyl homoserine lactone signal activities and affect population density-dependent behaviours in associated bacteria. *Mol. Plant-Microbe Interact.*, 13: 637–648.

14. Visick, K.L. and McFall-Ngai, M.J., 2000. An exclusive contract: Specificity in the *Vibrio fischeri-Euprymna scolopes* partnership. *J. Bacteriol.*, 182: 1779–1787.

15. Welch, M., Todd, D.E., Whitehead, N.A., McGowan, S.J., Bycroft, B.W. and Salmond, G.P.C., 2000. N-acyl homoserine lactone binding to the CarR receptor determines quorum-sensing specificity in *Erwinia*. *EMBO J.*, 19: 631–641.

16. Whitehead, N.A., Barnard, A.M.L., Slater, H., Simpson, N.J.L., and Salmond, G.P.C., 2001. Quorum-sensing in gram-negative bacteria. *FEMS Microbiol. Rev.*, 25: 365–404.

17. Wood, D.W., Gong, F., Daykin, M.M., Williams, P. and Pierson, L.S., 1997. N-Acyl-homoserine-lactone-mediated regulation of phenazine gene expression by *Pseudomonas aureofaciens* in the wheat rhizosphere. *J. Bacteriol.*, 179: 7663–7670.

Biotechnology: An Overview (2015) Pages 9–20
Editors: Rajan Kumar Gupta, Nasim Akhtar and Deepak Vyas
Published by: DAYA PUBLISHING HOUSE, NEW DELHI

Chapter 2

Micro-algae: A Potent Candidate for Carbon Sequestration and Biofuel Production

Shiv Shanker Pandey, Vivek Ambastha and
Budhi Sagar Tiwari*

School of Life Sciences, Jawaharlal Nehru University,
New Delhi – 110 067

ABSTRACT

Severe reduction in fossil based fuels due to urbanization and industrial development have resulted a great need of developing sustainable source of clean energy. Biofuels are becoming worldwide leader in the development of renewable energy resources. Algae are among the most potentially significant sources of sustainable biofuels in the future of renewable energy. There are many advantages for production of algae-based biofules. Algae have a rapid growth rate, can be cultivated in brackish coastal water and seawater besides being housed in other fresh water bodies and land which is not suitable for agriculture. Algae can sequester carbon dioxide emmiting from industries and also can use nitrogen, phosphate, silicon and sulfate nutrients from human or animal waste. It is worthwhile to mention that algal biofuel production is thought to help in stabilizing the concentration of carbon dioxide in the atmosphere and decrease global warming impacts. Secondly, algae based biodiesel is non toxic, without sulfur, highly biodegradable and relatively harmless to the environment if spilled. Algae are capable of producing in excess of 30 times more oil per acre than corn and soybean crops. Currently, algal biofuel production has not been commercialized due to high costs associated with production, harvesting and oil extraction but the technology is progressing. Extensive research has been conducted to determine the utilization of microalgae as an alternative energy source and algal oil production commercially viable. This chapter reviews the current status of utilization of microalgae for carbon sequestration and biofuel production. The microalgal species mostly used for biodiesel production are presented and their main advantages are compared with other available biodiesel feedstocks.

Keywords: Micro-algae, Carbon sequestration, Carbon-concentrating mechanism, Bioethanol, Biofuel, Biodiesel, Hydrogen production.

---

* *Corresponding Author:* E-mail: budhi@rediffmail.com; bstiwari@mail.jnu.ac.in

Introduction

The use of fossil fuels is now widely accepted as unsustainable due to depleting resources and the accumulation of greenhouse gases in the environment that have already exceeded the "dangerously high" threshold of 450 ppm CO_2. To achieve environmental and economic sustainability, fuel production processes are required that are not only renewable, but also capable of sequestering atmospheric CO_2. Currently, nearly all renewable energy sources (*e.g.* wind, hydroelectric, solar, geothermal, tidal) target the electricity market, while fuels make up a much larger share of the global energy demand (~66 per cent). Higher oil prices and increased interest in energy security have stimulated new public and private investment in algal biofuels research. Biofuels are therefore rapidly being developed.

Biofuels are solid, liquid or gaseous fuels derived from organic matter. They are generally divided into primary and secondary biofuels (Figure 2.1). While primary biofuels such as fuelwood are used in an unprocessed form primarily for heating, cooking or electricity production, secondary biofuels such as bioethanol and biodiesel

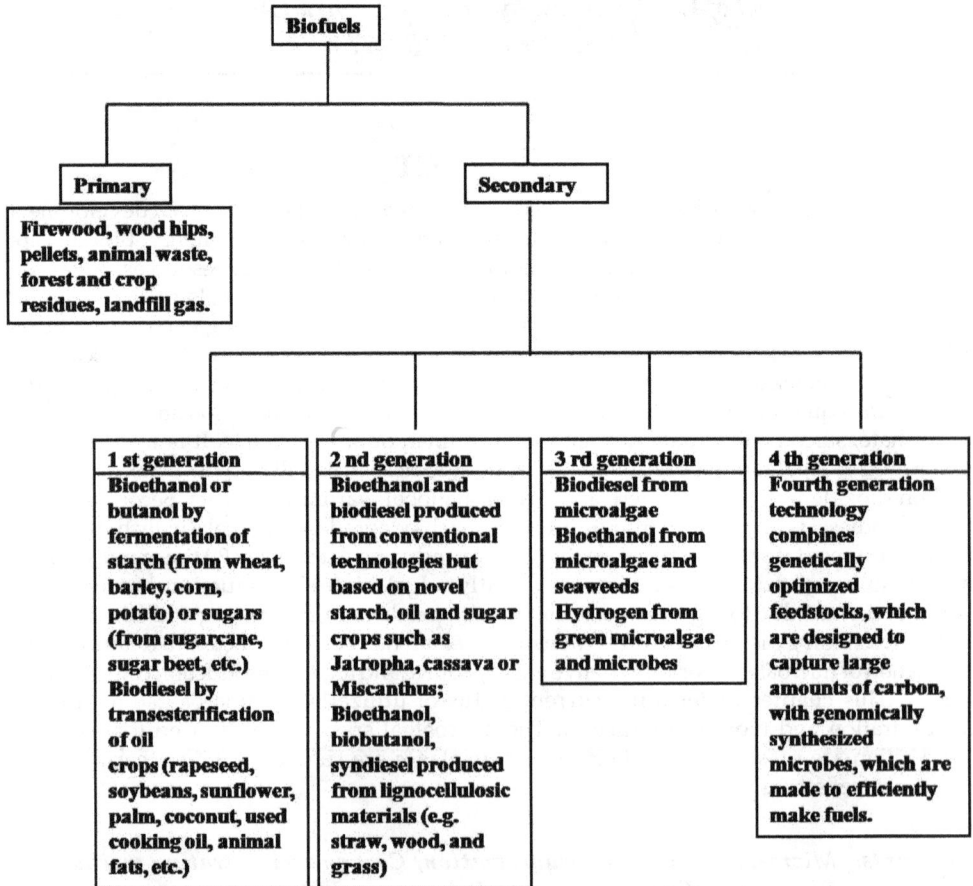

Figure 2.1. Different generations of biofuels.

are produced by processing biomass and are able to be used in vehicles and various industrial processes. The secondary biofuels can be categorized into four generations on the basis of different parameters, such as the type of processing technology, type of feedstock and their level of development. All types of processes of biofuels production have their own limitations. First and second-generation biofuels account for 99 per cent of today's global biofuel production. 3rd-generation biofuel is basically advanced algae-based biodiesel while 4th-generation biofuels are created using petroleum-like hydroprocessing or advanced biochemistry.

There are many advantages of microalgae for using as a biofuel source. Microalgae are single-cell, photosynthetic organisms, grow rapidly, can be cultivated to have a high oil mainly triacylglyceride and high energy content (Table 2.1), can be grown on non arable land with brackish or salt water and most abundant form of plant life, responsible for more than half of the world's primary production of oxygen. While terrestrial plants in temperate climates can achieve a photoconversion efficiency of only below 1 per cent, microalgae can convert up to 5 per cent of the solar energy into chemical energy (Rösch *et al.*, 2012). These characteristics make them an attractive source of biomass for renewable biofuels (Savage *et al.*, 2012). Some algal strains are capable of doubling their mass several times per day. In some cases, more than half of that mass consists of lipids or triacylglycerides- the same material found in vegetable oils. These bio-oils can be used to produce such advanced biofuels as biodiesel, green diesel, green gasoline, and green jet fuel.

Table 2.1. Comparison of lipid sources for biodiesel.

Crop	Oil Yield (L/ha)	Land Area Needed (M ha)[a]	Percent of existing US cropping area
Corn	172	1540	846
Soybean	446	594	326
Canola	1190	223	122
Jatropha	1892	140	77
Coconut	2698	99	54
Oil palm	5950	45	24
Microalgae[b]	136,900	2	1.1
Microalgae[c]	58,700	4.5	2.5

a: For meeting 50 per cent of all transport fuel needs of the United States.

b: 70 per cent oil (by wt) in biomass.

c: 30 per cent oil (by wt) in biomass.

Carbon Sequestration by Microalgae

Carbon dioxide sequestration and sustainable fuel production are the two main problems at global level. One of the most promising approaches for the production of pollution-free energy is the use of micro-organisms in photobioreactor, consuming carbon dioxide and producing biomass and biofuels. Because of fast growth rate of algae and their ability to grow in a wide range of environment, possibly these organisms can capture most of the CO_2 emitted by power plants, vehicles and industrial sources and fix carbon in to biofuels. Using algae for reducing CO_2 in the atmosphere is known as algae-based Carbon Capture Technology. Very high ability of CO_2 fixatation by microalgae is due to presence of carbon-concentrating mechanism (CCM), a metabolic system that allows the cells to enrich the amount of CO_2 at the site of Rubisco up to 1000-fold over that in the surrounding medium. Besides this biomass of few strains of algae are rich in oil, that is used as a feedstock for biofuel production. Algae-based CO_2 capture has triggered significant research interest in governments, academia and industry.

Biofuel Production by Microalgae

The largest proportion of biofuels is produced from micro-algae and higher plants which use photosynthesis to convert solar energy into chemical energy. Figure 2.2 shows the relation of the photosynthetic pathway and glycolytic pathway to biofuel production in micro-algae. Algae of different families have ability to produce and accumulate a large fraction of their dry matter as oil as listed in Table 2.2 (Chisti *et al.*, 2007, Meng *et al.*, 2009 and Gouveia *et al.*, 2009).

Table 2.2. Constituents of algae (per cent of Dry Matter) (Becker *et al.*, 2007).

Alga	Protein	Carbohydrates	Lipids
Anabaena cylindrica	43–56	25–30	4–7
Aphanizomenon flos-aquae	62	23	3
Chlamydomonas rheinhardii	48	17	21
Chlorella pyrenoidosa	57	26	2
Chlorella vulgaris	51–58	12–17	14–22
Dunaliella salina	57	32	6
Euglena gracilis	39–61	14–18	14–20
Porphyridium cruentum	28–39	40–57	9–14
Scenedesmus obliquus	50–56	10–17	12–14
Spirogyra sp.	6–20	33–64	11–21
Arthrospira maxima	60–71	13–16	6–7
Spirulina platensis	46–63	8–14	4–9
Synechococcus sp.	63	15	11

Increase in carbon dioxide concentration besides methane and oxides of nitrogen in the atmosphere is leading to climate change. These greenhouse gases (GHGs) cause depletion of ozone layer protecting the atmosphere against UV radiation, thereby warming the atmosphere. The average concentration of CO_2 increased from 315 ppm in 1960 to 380 ppm in 2007 (IPCC, 2007). There has been a 35 per cent increase in CO_2 emission worldwide since 1990 (Kaladharan *et al.*, 2009). Carbon fixation by photoautotrophic algae has the potential to diminish the release of CO_2 into the atmosphere and in helping to alleviate the trend toward global warming. Primary

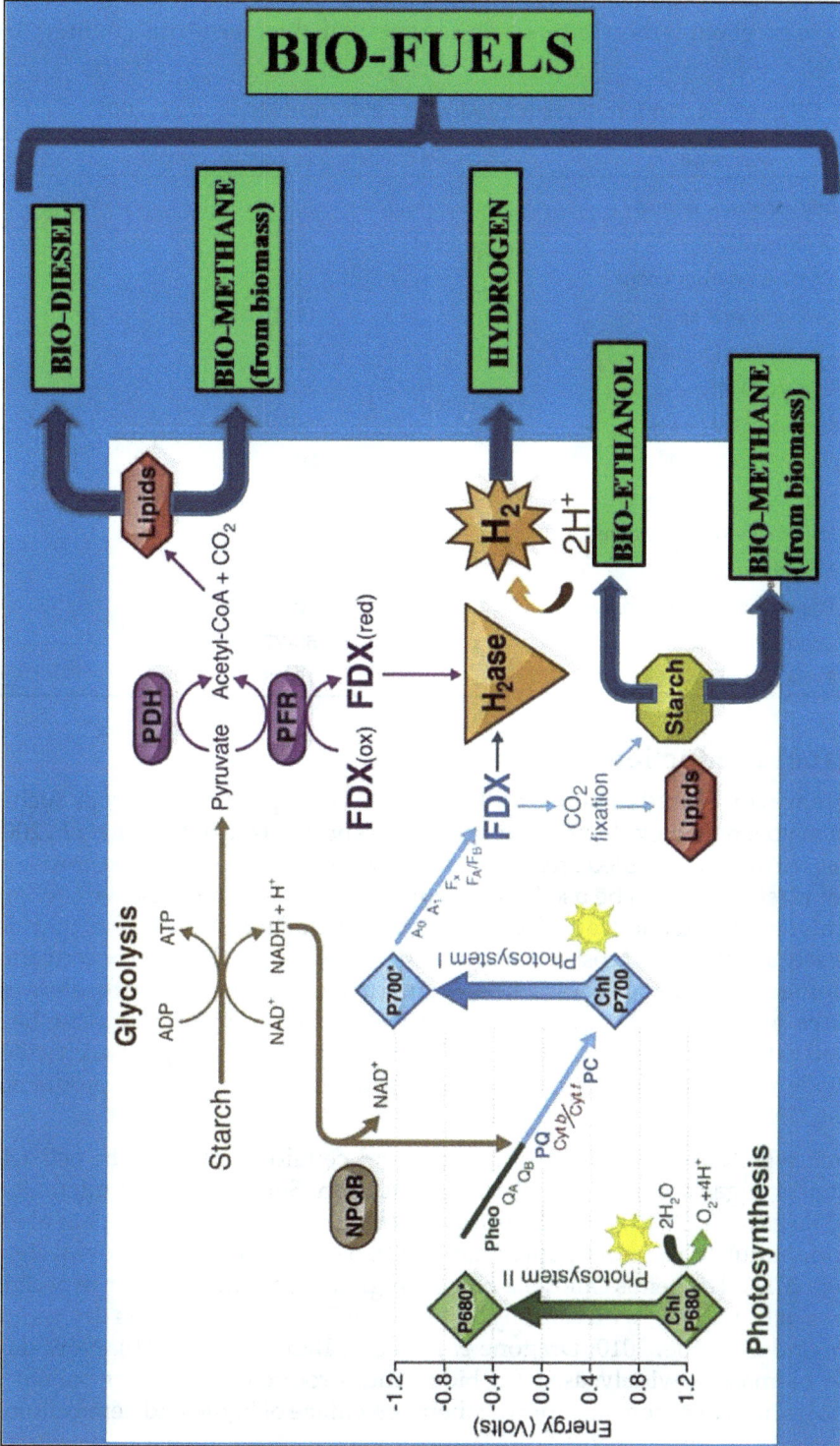

Figure 2.2. The process of photosynthesis converts solar energy into chemical energy is a key to all biofuel production systems in plants.

producers of coastal and marine ecosystems such as phytoplankton, seaweed and seagrass are excellent carbon sequestering agents than their terrestrial counterparts (Zou, 2005).

Table 2.3. Oil content of microalgae (Chisti *et al.*, 2007 and Meng *et al.*, 2009).

Microalga	Oil Content (per cent dry weight)
Botryococcus braunii	25-75
Chlorella sp.	28-32
Crypthecodinium cohnii	20
Cylindrotheca sp.	16-37
Dunaliella primolecta	23
Isochrysis sp.	25-33
Monallanthus salina	>20
Nannochloris sp.	20-35
Nannochloropsis sp.	31-68
Neochloris oleoabundans	35-54
Nitzschia sp.	45-47
Phaeodactylum tricornutum	20-30
Schizochytrium sp.	55-77
Tetraselmis sueica	15-23

Bioethanol Production

Algae are considered as the only alternative to current bioethanol crops such as corn and soybean as they do not require arable land (Chisti, 2007; Hu *et al.*, 2008; Singh *et al.*, 2010c). Water filled areas that are not suitable for growing food crops and industrial waste water can be used for the cultivation of algal biomass without any compromise with land and water resource for the production of bioethanol that also not adversely affect the food cost (Singh *et al.*, 2011). In addition, algae can be converted directly into energy, such as biodiesel, bioethanol and biomethanol and therefore can be a source of renewable energy. The unicellular marine microalgae have been considered to be an abounding resource for carotenoids, lipids, and polysaccharides, and are widely investigated in the fields of food supplements and bio-fuel production (Liau *et al.*, 2010).

Starch which is the storage component and cellulose which is the cell wall component of algae are used for ethanol production. Some species of microalgae such as *Chlorella, Chlamydomonas, Dunaliella, Scenedesmus,* and *Tetraselmis* have been shown to accumulate a large amount of carbohydrates (>40 per cent of the dry weight) (John *et al.*, 2011). The genus Chlorella possess high carbohydrate content, especially the species of *C. vulgaris*, with carbohydrates being 37–55 per cent of its dry weight (Brennan and Owende, 2010; Dragone *et al.*, 2011; Illman *et al.*, 2000). Now days Chlorella biomass is widely used for bioethanol production. Another benefit in bioethanol production from algae is that their percentage of lignin and hemicellulose is lower as compared to other lignocellulosic plants (Harun *et al.*, 2010). It was also found that the Nitrogen starvation triggered the accumulation of carbohydrates in the microalga, achieving a carbohydrate content of 51.3 per cent after 4-day starvation (Ho *et al.*, 2012-II) indicating their utility even in harsh environment.

Hydrogen Production

Hydrogen is regarded as a promising energy source of the future, since it is non-toxic, extremely environment friendly, easily converted to electricity and burns cleanly. The combustion of hydrogen results in the formation of steam and liquid water. In this respect, the use of hydrogen is completely safe from environmental standpoint. The hydrogen demand is rapidly increasing world-wide. Currently, hydrogen is produced by fossil fuel-based processes which emit large amounts of carbon dioxide, sulphur dioxide and nitrogen oxides. Carbon dioxide emissions are the most important cause of global warming which rises the average temperature on earth. The emissions of carbon dioxide have dramatically increased within the last 50 years and are still increasing. While the CO_2 concentration was 320 ppm in 1960, it increased by ca. 20 per cent up to 380 ppm in 2010. Moreover, pollutants such as nitrous oxides create smog and haze. Thus, there is a great need of new energy sources such as hydrogen which should be pollution free, environment friendly and sustainable. Biological hydrogen production has thus recently received renewed attention owing to urban air pollution and global warming concerns.

Hydrogen can be produced from a variety of feedstocks; from fossil resources such as natural gas and coal, and from renewable resources such as biomass and water with input from renewable energy sources (*e.g.* sunlight, wind, wave or hydro-power). A variety of process technologies can be used, including chemical, biological, electrolytic, photolytic and thermo-chemical. Each technology is in a different stage of development, and each offers unique opportunities, benefits and challenges. Local availability of feedstock, maturity of technology, market applications and demand, policy issues and costs will influence the choice and timing of the various options for hydrogen production.

Chlamydomonas reinhardtii is used as a model organism for biological hydrogen production. Figure 2.3 shows the process of hydrogen production in green micro algae *Chlamydomonas reinhardtii*. [Fe]-hydrogenase is the key enzyme that catalyzes H_2 production using electrons from ferredoxin. *Chlamydomonas reinhardtii* contains two Fe-hydrogenases (HydA1 and HydA2). HydA1 is located in the chloroplast and linked to the photosynthetic electron transport chain via the ferredoxin PetF (Forestier *et al.*, 2003, Happe *et al.*, 1993, 2002). Two copies of Fe-hydrogenase genes were also identified in the green alga *Scenedesmus obliquus* (Florin *et al.*, 2001 and Wünschiers *et al.*, 2001). Major hassle for hydrogen production is that Hydrogenase is highly sensitive to oxygen. Generation of oxygen during photolysis of water inhibit the hydrogenase activity and stop hydrogen production. But it was found that under Sulphur deprived condition, photosynthetic O_2 evolution in algae became inactivated, resulting in the light-induced establishment of anaerobic conditions and hydrogen photoproduction for several days (Kosourov *et al.*, 2007). Hydrogen production in

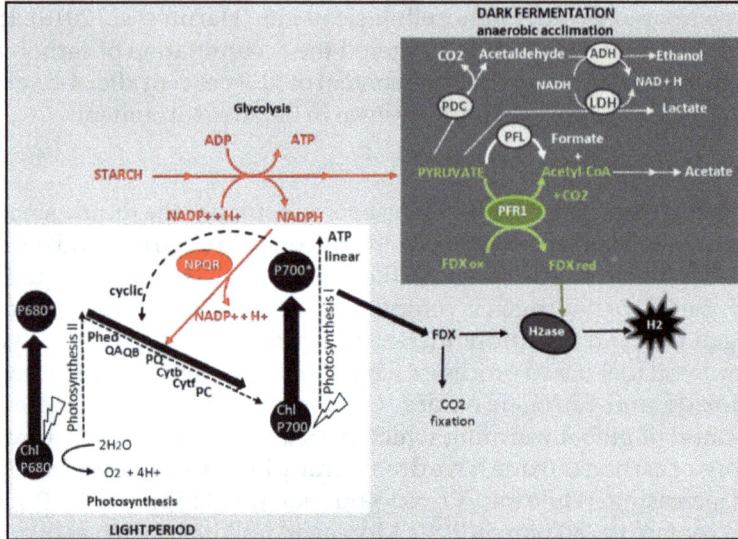

Figure 2.3. Simplified illustration of the physiological pathways used for H$_2$ production in *Chlamydomonas* (Stern *et al.*, 2009). The two photoproduction pathways involving PSII and PSI under the light period are showed in black. Electrons excited to higher energy (low potential) by PSI are able to reduce ferredoxin (FDX), the physiological electron donor to hydrogenase. Both the PSII-dependent and NPQR-dependent (red) pathways require reduction of the PQ pool and PSI activity. In the case of the PSII-independent pathway (reactions in red), reducing power formed by the catabolism of organic substrates is used for reduction of the PQ pool. During dark fermentation the oxidation of pyruvate catalysed by PFR (green reactions) is used to reduce ferredoxin and putatively mediates the observed production of H$_2$ in the dark. White reactions show the parallel main fermentative products from pyruvate, competing with H$_2$. Dashed arrows show linear and cyclic electron flow.

sulphur deprived condition is currently studied extensively worldwide. Stress factor such as nutrients starvation, light intensity, salt concentration and temperature also affect the hydrogen production (Skjanes *et al.*, 2012).

Genetically Engineered Microalgae for Efficient Biofuel Generation

Genetic engineering of microalgae is considered as efficient approach for enhaced hydrogen production. Genes involved in carbohydrate and lipid metabolisms are the major target for genitic modification. Genetic strategies can be applied for increasing glucan storage and decreasing starch degradation to increase biofuel production in microalgae. The rate limiting step of starch synthesis ADP-glucose pyrophosphorylase (AGPase) is most appropriate enzyme for increasing strach biosynthesis (Radakovits *et al.*, 2010). Overexpression of oxygen tolerent hydrogenase and Pyruvate ferridoxin oxidoreductase (PFR1) will be good approach for hydrogen production. Pyruvate ferredoxin oxidoreductase (PFR1), which is responsible for oxidation of pyruvate in the chloroplast yielding acetyl-CoA, reduces ferredoxin and CO$_2$ and mediates the production of H$_2$ in the dark. Under mixotrophic growth conditions *PFR1* becomes upregulated (Gomes de Oliveira Dal'Molin *et al.*, 2011). Second approach to enhance the biofuel production through genetic engineering is to make algal cell more efficient to utilize the sun light *i.e.*, to increase the photosynthetic efficiency or light utilization efficiency. Major problem in mass culture condition is that fully pigmented *Chlamydomonas* cells do not permit complete transmittance of light that results in lower utilization of the solar energy for hydrogen production and most of the solar energy wastefully dissipates in the form of heat. Therefore genetic engineering is applied to reduce the size of chlorophyll antenna size and also to reduce the chlorophyll content. It was found that truncated Chl antenna cells permit greater transmittance of light and overall better solar utilization by the culture (Polle *et al.*, 2002 and Melis *et al.*, 2009). To reduce the light-harvesting chlorophyll antenna size *Chlamydomonas tla* (truncated light-harvesting chlorophyll antenna size) mutant were developed (Tetali *et al.*, 2007, Kirst *et al.*, 2012 a, b). There are three types of *tla* mutants- *tla1*, *tla2* and *tla3*. These mutants showed lighter-green phenotype and lower chlorophyll content than that of wild-type strains. In tla2 and tla3 mutant Chl antenna size of the photosystems was only about 65 per cent and 40 per cent of that in the wild type respectively. The mutant contains a correspondingly lower amount of light-harvesting proteins and has lower steady-state levels of Lhcb mRNA (Melis *et al.*, 2003). The *tla* mutants require higher light intensity for the saturation of photosynthesis. Moreover, the mutant showed greater solar conversion efficiencies and a higher photosynthetic productivity than the wild type under mass culture conditions (Melis *et al.*, 2003, Tetali *et al.*, 2007, Kirst *et al.*, 2012 a, b). These mutant can be use to enhance the hydrogen and other biofuel production.

Future Perspectives

Biofuels production from algae is not economical by using the technology available today. Lowering this cost will require coordinated R and D across a wide range of technical sectors. The engineering of sustainable microalgal systems is very important to lower the cast of biofuel production from microalgae. Several genomics and metabolomics based technologies are already in use and in progress and can be use to obtain best performing microalgal strain. Screening of natural habitates to isolate a valuable strain will be a good approach. The natural isolates can be also improved for high growth rate, starch biosynthesis, lipid biosynthesis, oil content and nutritional values by implying metabolomics, genetics, proteomics and lipidomics technology. Combination of system biology approach together with genetic engineering, metabolic and protein engineering can provide a way to identify regulatory and rate limiting steps and differentially expressed genes and metabolites that can be used to maximize the biosynthetic pathway related to biofuel production in a micro-algal cell.

Acknowledgement

Acknowledgement is due to DBT, India for financial support as Ramlingaswami fellowship grant to BST.

References

1. Becker., 2007. Micro-algae as a source of protein. *Biotechnol. Adv.*, 25: 207–210.

2. Brennan, L. and Owende, P., 2010. Biofuels from microalgae: A review of technologies for production, processing, and extractions of biofuels and co-products. *Renew. Sust. Energy Rev.*, 14(2): 557–577.

3. Chisti, Y., 2007. Biodiesel from microalgae. *Biotechnol Adv.*, 25: 294–306.

4. Dextera, J. and Pengcheng, Fu, 2009. Metabolic engineering of cyanobacteria for ethanol production. *Energy Environ. Sci.*, 2: 857–864.

5. Dragone, G., Fernandes, B.D., Abreu, A.P., Vicente, A.A. and Teixeira, J.A., 2011. Nutrient limitation as a strategy for increasing starch accumulation in microalgae. *Appl. Energy*, 88(10): 3331–3335.

6. Florin, L., Tsokoglou, A. and Happe, T., 2001. A novel type of iron hydrogenase in the green alga*Scenedesmus obliquus* is linked to the photosynthetic electron transport chain. *J. Biol. Chem.*, 276: 6125–6132.

7. Forestier, M., King, P., Zhang, L., Posewitz, M., Schwarzer, S., Happe, T., Ghirardi, M.L. and Seibert, M., 2003. Expression of two [Fe]-hydrogenases in *Chlamydomonas reinhardtii* under anaerobic conditions. *Eur. J. Biochem.* 270: 2750–2758.

8. Gomes de Oliveira Dal'Molin, Lake-Ee Quek, Robin W. Palfreyman, Lars K. Nielsen, 2011. AlgaGEM: A genome-scale metabolic reconstruction of algae based on the *Chlamydomonas reinhardtii* genome. *BMC Genomics*, 12(Suppl 4): S5.

9. Gouveia, L., Marques, A.E., da Silva, T.L. and Reis, A., 2009. *Neochloris oleabundans* UTEX# 1185: A suitable renewable lipid source for biofuel production. *J. Ind. Microbiol. Biotechnol.*, 36: 821–826.

10. Happe, T. and Kaminski, A., 2002. Differential regulation of the Fe-hydrogenase during anaerobic adaptation in the green alga *Chlamydomonas reinhardtii*. *Eur. J. Biochem.*, 269: 1022–1032.

11. Happe, T. and Naber, J.D., 1993. Isolation, characterization and N-terminal amino acid sequence of hydrogenase from the green alga *Chlamydomonas reinhardtii*. *Eur. J. Biochem.*, 214: 475–481.

12. Harun, R., Danquah, M.K. and Forde, G.M., 2010. Microalgal biomass as a fermentation feedstock for bioethanol production. *J. Chem. Technol. Biotechnol.*, 85: 199–203.

13. Ho, S.-H. *et al.*, 2012a. Bioethanol production using carbohydrate-rich microalgae biomass as feedstock. *Bioresour. Technol.*, http://dx.doi.org/10.1016/j.biortech. 2012.10.015

14. Ho, S.-H. *et al.*, 2012b. Characterization and optimization of carbohydrate production from an indigenous microalga *Chlorella vulgaris* FSP–E. *Bioresour. Technol.*, http://dx.doi.org/10.1016/j.biortech.2012.10.100

15. Hu, Q., Sommerfeld, M., Jarvis, E., Ghirardi, M. Posewitz, M., Seibert, M. and Darzins, A. 2008. Microalgal triacylglycerols as feedstocks for biofuel production: perspectives and advances. *Plant J.*, 54: 621–639.

16. Illman, A.M., Scragg, A.H. and Shales, S.W., 2000. Increase in *Chlorella* strains caloric values when grown in low nitrogen medium. *Enzyme Microb. Technol.*, 27(8): 631–635.

17. John, R.P., Anisha, G.S., Nampoothiri, K.M. and Pandey, A., 2011. Micro and macroalgal biomass: A renewable source for bioethanol. *Bioresour. Technol.*, 102(1): 186–193.

18. Kaladharan, P., Veena, S. and Vivekanandan, 2009. Carbon sequestration by a few marine algae: observation and projection sequestration. p. 107 – 110.

19. Kosourov, S., Patrusheva, E., Ghirardi, M.L., Seibert, M. and Tsygankov, A., 2007. A comparison of hydrogen photoproduction by sulfur deprived Chlamydomonas reinhardtii under different growth conditions. *Journal of Biotechnology*, 128(4): 776–787.

20. Liau, B.C., Shen, C.T., Liang, F.P., Hong, S.E., Hsu, S.L., Jong, T.T. and Chang, C.M., 2010. Supercritical fluids extraction and anti-solvent purification of carotenoids from microalgae and associated bioactivity. *J. Supercrit. Fluids*, 55: 169–175.

21. Meng, X., Yang, J., Xu, X., Zhang, L., Nie, Q. and Xian, M., 2009. Biodiesel production from oleaginous micro-organisms. *Renew. Energy*, 34: 1–5.

22. Nguyen, Thi Hong Minh and Vu, Van Hanh, 2012. Bioethanol production from marine algae biomass: Prospect and troubles. *J. Viet. Env.*, 3(1): 25–29

23. Radakovits, R., Jinkerson, R.E., Darzins, A. and Posewitz, M.C., 2010. Genetic engineering of algae for enhanced biofuel production. *Eukaryotic Cell*, p. 486–501.

24. Rösch, C., Skarka, J. and Wegerer, N., 2012. Materials flow modeling of nutrient recycling in biodiesel production from microalgae. *Bioresour. Technol.*, 107: 191–199.

25. Savage, P.E., 2012. Algae under pressure and in hot water. *Science*, 338: 1039; DOI: 10.1126/science.1224310.

26. Singh, A., Nigam, P.S. and Murphy, J.D., 2010. Mechanism and challenges in commercialisation of algal biofuels. *Bioresour. Technol.* doi: 10.1016/j.biortech .2010.06.057.

27. Singh, A., Nigam, P.S. and Murphy, J.D., 2011. Renewable fuels from algae: An answer to debatable land based fuels. *Bioresource Technol.*, 102: 10–16.

28. Skjanes, K., Rebours, C. and Lindblad, P., 2012. Potential for green microalgae to produce hydrogen, pharmaceuticals and other high value products in a combined process. *Critical Reviews in Biotechnology*. DOI: 10.3109/07388551.2012.681625

29. Stern, D.B., Harris, E.H. and Witman, G., 2009. *The Chlamydomonas Sourcebook: Organellar and Metabolic Processes, Volume 2*. San Diego, CA, USA, Academic Press, p. 1040

30. Wünschiers, R., Stangier, K., Senger, H. and Schulz, R., 2001. Molecular evidence for a Fe–hydrogenase in the green alga *Scenedesmus obliquus*. *Curr. Microbiol.*, 42: 353–360.

31. Zou, D., 2005. Effects of elevated atmospheric CO_2 on growth, photosysnthesis and nitrogen metabolism in the economic brown seaweed, *Hizikia fusiforme* (Sargassaceae, Phaeophyta). *Aquaculture*, 250: 726–735.

Biotechnology: An Overview (2015)　　　　　　　　Pages 21–29
Editors: Rajan Kumar Gupta, Nasim Akhtar and Deepak Vyas
Published by: DAYA PUBLISHING HOUSE, NEW DELHI

Chapter 3

A Brief Introduction to Ecosystem: Its Structure and Functions

Fouzia Ishaq¹, Amir Khan² and M. Salman Khan³*

¹*Limnological Research Lab, Department of Zoology and Environmental
Science, Gurukula Kangri University, Haridwar, Uttarakhand*
²*Glocal School of Life Sciences, The Glocal Univesrity,
Mirzapur Pole, Saharanpur, , Uttar Pradesh*
³*Departmentment of Biotechnology, Integral University,
Lucknow – 226 026, Uttar Pradesh*

ABSTRACT

The ecosystem concept is fundamental to examination of human impacts on life on earth. It provides a way of looking at the functional interactions between life and environment which helps us to understand the behaviour of ecological systems, and predict their response to human or natural environmental changes. Ecosystems are found throughout the biosphere. The ecosystem concept provides a convenient means of structuring and understanding the highly complex system which is our world. Plants and animals live where they have water, food, and shelter. An ecosystem contains all the plants, animals, and nonliving things in an environment. The different parts of an ecosystem work together. Change to ecosystems may be caused by human actions. One of the issues that give rise to the greatest concern among scientists concerned with the environment, and among the public at large, is the effects that humans are having upon ecosystems and their functioning.

Keywords: Ecosystem, Ecosystem functioning, Human impacts, Environment.

* *Corresponding Author:* E-mail: amiramu@gmail.com

Introduction

An ecosystem has been defined in two ways; 1) It is an energy-driven complex of a community of organisms and its controlling environment. 2) An ecosystem is a community of living organisms together with the physical processes that occur within an environment.

These two definitions, nearly 25 years apart, provide consistent statements on the key attributes of ecosystems. These key attributes are directly related to the concepts of functional ecology. In particular, interactions between the physical environment and organisms and between organisms and other organisms direct the evolutionary trends of competition, tolerance of stress, and tolerance of disturbance. These interactions are central to the functional processes specified in the definitions of ecosystems (Pullin, 2002).

Energy Flow in Ecosystems

All organisms need energy to grow, move, repair, and reproduce. Most organisms get their energy from sunlight. This can happen either directly or indirectly. Lettuce and most other plants, get energy directly from sunlight through photosynthesis. During photosynthesis, plant leaves produce glucose. The plants use the chemical energy in glucose to carry out life functions (DeAngelis, 1980). In an ecosystem, plants are called producers because they use energy from sunlight to make, or produce, their own food. A rabbit, however, cannot get energy directly from sunlight. But as the rabbit eats the lettuce, it indirectly gets energy from the Sun that is stored in the leaves. Organisms that get energy by eating other organisms are called consumers. The fungus cannot make its own food from sunlight, but it doesn't eat other organisms either. It gets it by breaking down the remains of organisms that were once alive, such as trees that have fallen down. Organisms such as the fungus are called decomposers. They release materials from dead plants and animals back into the environment, where other consumers can use them. Without decomposers, nothing would decay. Most living things on Earth depend on the Sun's energy either directly or indirectly. The leaves of a berry bush use energy from the Sun to make food known as glucose. Plants use the chemical energy in glucose as energy for their life functions. Plants are producers; organisms that can make their own food (Ulanowicz and Kay, 1991). Animals cannot use sunlight to make their own food. Animals are consumers; organisms that get energy by eating other organisms. When a bear eats berries, it gets the energy stored in them. The bear uses energy from the Sun indirectly. Toadstools cannot make their own food. But they cannot eat other organisms either. When organisms die and fall to the ground, their bodies decay. A decomposer is an organism that gets energy by breaking down the remains of dead organisms. Toadstools are decomposers. Decomposers return the materials from the dead organism's body back into the environment. Decomposers help provide materials that other organisms can use. Without decomposers, nothing would ever decay. Dead organisms would just pile up forever (Jonsson and Malmqvist, 2000).

Food Chain

As you know, organisms either use energy from sunlight to produce their own food or they eat other organisms that have energy. A food chain shows one possible path of how organisms within an ecosystem get their food. Because the original source of energy is sunlight, a food chain begins with plant life and ends with an animal. Notice that the arrows in a food chain always point toward the organism that receives the energy (Post *et al.*, 2000) *e.g.*, in the food chain, the food chain that connects the path of energy from wheat, to the mouse, to the snake, and on to the owl. In an ecosystem, some organisms produce food, while others consume food. This is how energy travels in an ecosystem. A food chain shows a path of energy through an ecosystem. Follow the food chain from the microscopic organisms to the common mussel, then to the herring gull (Krause *et al.*, 2002).

Food Web

Every chain has a producer that makes its own food and consumers that eat other organisms. Most organisms are part of more than one food chain and eat more than one kind of food (Dunne *et al.*, 2002). Because organisms in an ecosystem often belong to more than one food chain, the food chains become interconnected, or mixed. These interconnected food chains form a food web. Study the food web is given here. Wheat, clover, and dandelions are the producers at the bottom of this food web. The owl and the hawk are the consumers at the top because no animals in this ecosystem eat them (Berryman, 1993). Many different food chains exist in an ecosystem. Food chains have producers and consumers. Consumers often eat other consumers. Organisms may be part of several food chains. A food web is made up of several food chains that are interconnected.

Energy Pyramids

A food chain shows the path that energy takes from producers to consumers. However, it does not give any information about how much energy moves from organism to organism. Not all of the energy that plants receive from sunlight is available to be passed on to animals that eat the plant. This is because the plant uses some energy to stay alive. The same is true for animals. They use energy to grow, move, and reproduce (Polis and Strong, 1996). They pass on only the energy that is left over.

An energy pyramid shows how energy moves through an ecosystem. In an energy pyramid, the greatest amount of energy is available from the trees and bushes on the bottom level. Giraffes eat these plants then use most of the energy they get to carry out life processes. When a lion eats a giraffe, there is little energy stored in the giraffe's body to pass on to the lion. Because of this, an ecosystem needs many giraffes to support a small number of lions. A food chain shows how energy travels from producers to the top consumer. But a food chain does not show how much energy moves from one organism to another. Not all of the energy that a green plant takes in from the Sun moves to other organisms. The plant uses some of the energy for its own life processes. Some energy is lost as heat. This repeats throughout a food chain. A snake uses energy to slide along the ground. A wood mouse uses energy to dig itself a hole. Organisms must use energy to grow, move, and reproduce. So, only part of the energy can move to the next level of the food chain. An energy pyramid is a model that shows how energy moves through an ecosystem. The pyramid gets smaller as it nears the top. There is more energy at lower the levels (Levine, 1980). There is less energy towards the top of the pyramid because most of it has been used by organisms for life processes or has been given off as heat. Only energy stored in the tissues of an organism can pass from one level to the next.

Competition between Organisms for Resources

All plants and animals need food, water, and space. Within an ecosystem, these resources are limited, so there is always a competition for them. Animals with different needs can live side by side with little competition. This is because the birds eat different foods. These birds do not need to compete for food in this ecosystem. Competition occurs only when organisms of an ecosystem have the same needs. Sometimes competition is between members of the same species, such as two herons. If there is a drought and the marsh becomes dry, the herons that can survive with less food and water have a better chance of survival than those who need more. Sometimes competition is between different species. Suppose a stork came to this marsh to find food. Since storks and herons eat the same kind of fish and frogs, the two species would compete for the same resources (Downing and Leibold, 2002).

Ecosystems Change

All ecosystems sustain natural changes over time. People also cause changes to ecosystems. Sometimes ecosystems change. First, one part of the ecosystem changes. Then the other parts change too. Long ago, many wolves lived in Yellowstone National Park. The wolves ate elk and other animals. People wanted to get rid of the wolves. They killed many of them. When the wolves were gone, there were not enough animals to eat the elk. Yellowstone's elk population grew out of control. There were too many elk and not enough food. Many of the elk died. Finally, people divided to bring wolves back to the park (Knops *et al.*, 2002).

Natural Changes

In the summer of 1988, raging forest fires burned throughout Yellowstone National Park. The fires, which were started by lightning, charred one-third of this national park. The park follows a "natural burn" rule. This means that fires started accidentally by humans are put out, but fires started by a natural event, such as lightning, are allowed to burn unless they threaten people's lives and property. Park managers know that natural disasters, such as forest fires, volcanic eruptions, and floods, are an important part of ecology. They change ecosystems by killing old plants and allowing new ones to grow. Succession is a series of changes that occur in an ecosystem (Jorgensen *et al.*, 2000). This is how succession worked in Yellowstone after the fires. Plants called pioneer species began to grow on the damaged land. Pioneer plants can grow under difficult conditions. The following spring, about two dozen different kinds of plants began to grow out of the ashes. Many of these plants had existed before the fires, but only as roots. The forest floor had been so thick that they could not compete for the resources necessary to grow stems and leaves. As the pioneer plants died each season, their bodies decomposed and built up the soil. After enough soil formed, other organisms were able to live in the ecosystem. Seeds took root and formed new plants in the rich soil. Some of these seeds may have survived the fire because they were buried deep in the ground. Others blew in from unburned areas. For the most part, park rangers did not replant Yellowstone. Yellowstone's forests replanted themselves.

Competition

Competition is the struggle between organisms to survive when resources are limited. Like all organisms, the animals on the African savannah need food, water, and shelter. The animals that survive get these resources. Organisms that have different needs can live together without competing. Zebras eat the tall, coarse grass. After the zebras eat, the wildebeests eat the shorter grass left behind. Competition takes place between organisms that have similar needs. Resource, such as food, water, and shelter are limited in an ecosystem (Dierssen, 2000). Organisms can survive when their adaptations are best suited to their conditions. Organisms with adaptations not well suited to their conditions will not survive. Some competition takes place between members of the same species. For instance, lack of rainfall can make water scarce. Only some zebras that compete for water will survive. The successful ones will be those who can live on less water. Competition also occurs between different species. Wildebeests and gazelles both eat short, tender grass. If drought kills many grass plants, gazelles and wildebeests must compete for what is left. All organisms, not just animals, compete for resources. Plants compete for water, space to grow, minerals, and sunlight. Some plants even have ways to reduce competition. They release chemicals into the soil that kill other species around them (Fath *et al.*, 2004).

Human Impact on Ecosystems

People have an enormous impact on the ecosystems we live in. Our daily activities change ecosystems in ways that make it difficult, and sometimes even impossible, for other animals and plants to survive. Landfills that we build to hold our trash change ecosystems. Each person in the United States creates about four pounds of trash every day. Together, we create about 600,000 tons per day. Some is recycled, some is burned, but most is taken to landfills. An advantage of landfills is that they reduce health hazards created by open-air dumps. However, hazardous materials, such as paint, acid from batteries, and chemicals, can leak out of landfills and harm ecosystems. People often harm the environment without even realizing it (Marques *et al.*, 2003). When fossil fuels are burned, they create air pollution. Many of our everyday activities, such as driving cars and using electricity, depend on the use of fossil fuels. Think about how many times you rode in a car or a bus this week, and how many times you used electricity. Another way people harm the environment without realizing it is by using too much water. Water is an important resource in every ecosystem. People in the United States use more water every year than people in any other country. Humans are part of the ecosystem where they live. Human activity can change the environment. Some organisms cannot survive these changes. Think of the trash you threw away so far today. In 2001 each American produced about four pounds of trash a day. Much of it ends up in landfills. The advantage of using landfills is that they reduce odor. They are safer than open dumps. But they can cause problems too. Unsafe waste, such as paint and batteries, can leak and harm ecosystems. Building landfills can cause some organisms to lose their habitats and die. Over time, landfill space gets used up, and new areas must be found for the waste. People may cause harm without even knowing it (Hall and Raffaelli, 1993). Pollution enters the air when people drive their cars. Power plants cause pollution too. Even ranching and farming can have harmful results. When livestock overgraze, plants die. The soil erodes. Fertilizers can enter the water cycle and pollute lakes and rivers.

Saving Ecosystems

During a winter storm in 1996, an oil barge ran aground off the coast of Rhode Island. About 828,000 gallons of oil spilled into the ocean. The oil spill killed more than nine million lobsters, two thousand marine birds, and about one million pounds of clams, oysters, and scallops in the ecosystem. The oil also damaged the habitat of a bird called the piping plover, which was already on the list of endangered species (Ho and Ulanowicz, 2005). Crews of workers tried to soak up the oil from the water's surface and the beaches. Ninety-seven volunteers spent hundreds of hours in an effort to seed new scallop beds. They carefully placed about eight thousand healthy scallops in the area to try to grow a new scallop population. Similar projects are underway to replace the oyster population. Unfortunately, even with the efforts of scientists and volunteers, much of the damage done to the animals and the ecosystem they live in cannot be undone. In 1989 an oil spill damaged the coast of Prince William Sound in Alaska. Millions of gallons of oil leaked from an ocean oil tanker. It polluted the land and water and put a great strain on the ecosystem. Thousands of workers helped to clean up the mess. Many washed oily animals, such as otters, with soap and water. They scrubbed oil from the rocks. But even with all the work and money spent, much of the damage could not be undone. Billions of animals died, including 22 orca whales and 250,000 sea birds. The area still has not fully recovered (Herendeen, 1981).

Preventing Problems

Crews of workers tried to soak up the oil from the water's surface and the beaches. Ninety-seven volunteers spent hundreds of hours in an effort to seed new scallop beds. They carefully placed about eight thousand healthy scallops in the area to try to grow a new scallop population. Similar projects are underway to replace the oyster population (Hannon, 1973). Unfortunately, even with the efforts of scientists and volunteers, much of the damage done to the animals and the ecosystem they live in cannot be undone, environment healthy for all its organisms. Become a recycling "watchdog" at home and at school. More than four billion individual drink boxes are thrown away each year in the United States. These can sit in a landfill for more than three hundred years before they decompose. Setting up a program to recycle just these small items is a good way to begin. It's better to prevent problems before they happen in the first place. Here are ways you can help:

☆ Understand how you affect your ecosystem.

☆ Learn how to reduce the harm you cause.

☆ Reuse, recycle, or reduce your use of natural resources.

☆ Know how ecosystems work.

☆ Get involved. Join environmental groups to help.

As an adult, you will make decisions that affect yourself and your community. And what your community does will affect other regions in your state and country. These choices may even affect the world at large. Learn to be an informed citizen now. It will make it easier for you to become a responsible adult (Gunderson *et al.,* 2000).

Freshwater Ecosystems

Some ecosystems have fresh water. Other ecosystems have salt water. In some places fresh water and salt water come together. Lakes, ponds, rivers, and streams are all freshwater ecosystems. Lakes and ponds have land all around them (Reynolds, 1984). In rivers and streams, the water moves from one place to another. The water in some lakes and rivers comes from under the ground. The water in others comes from rain or melting snow. A wetland is land that is covered by water most of the time. Trees, grasses, and plants grow in a wetland. Many animals live there too. Some wetland birds have long legs that help them walk in the water. They have beaks for catching fish to eat. Wetland frogs and toads can live in the water and on the land. The largest freshwater wetland is in Brazil. Many large rivers run through Brazil. When it rains, these rivers can overflow, flooding the surrounding land. The flooded land becomes a wetland habitat for many plants and animals. Many birds stay in the wetlands of Brazil for a short time while they are traveling to other places. Many fish live there too. The wetlands of Brazil are also home to capybaras. They have webbed feet, like a duck. Their webbed feet help them swim (Simon and Townsend, 2003).

Saltwater Ecosystems

Earth's oceans contain almost all of its salt water. They cover most of the planet. Near the land, the ocean is not very deep. Clams, crabs, and some kinds of fish live there. Far from land, the ocean water is deep. Large fish, sharks, and whales can live in deep water. The deepest parts of the oceans are dark and cold. Very few plants can grow there because there is little sunlight (Shugart, 1998). Rivers flow into oceans. The fresh water from rivers mixes with salt water from the ocean. When this happens, salt marshes are formed. A salt marsh is a type of wetland. Most of the salt marsh is covered with water. Many grasses grow in the salt marsh. These grasses can live in water and soil that is salty. Some of the animals in the salt marsh are so small that you can't see them. Many sea animals start their life in salt marshes before moving out to the ocean (Engelhardt and Ritchie, 2002).

Conclusions

Ecosystems are conceptual and functional units of study that entail the ecological community together with its abiotic environment. Implicit in the concept of any system, such as an ecosystem, is that of a system boundary which demarcates objects and processes occurring within the system from those occurring outside the system. Furthermore, as open systems, energy–matter fluxes occur across the boundary; these in turn provide the ecosystem with an available source of energy input such as solar radiation and a sink for waste heat. All ecosystems are open systems embedded in an environment from which they receive energy–matter input and discharge energy–matter output. The earth is a non-isolated system. There is almost no exchange of matter with the outer space. To be able to utilize the matter many times during the evolution or from one year and decade to the next, cycling is necessary. Cycling implies that the ecosystem components are linked in an interacting network. The flow of energy from the sun to the ecosystems is also limited. It is important that an ecosystem captures as much sunlight as possible to cover its energy needs. Therefore, ecosystems, with increased biomass, can increase net primary productivity. The development of the life forms that we know from the earth has been possible because the earth has the elements that are needed to build the biochemical compounds that explain the life processes. As ecosystems are valuable commodity, it is important to protect them with its entire species and prevent them from the harmful impacts of humans.

References

1. Berryman, A.A., 1993. Food web connectance and feedback dominance, or does everything really depend on everything else? *Oikos*, 68: 183–185.

2. DeAngelis, D.L., 1980. Energy flow, nutrient cycling and ecosystem resilience. *Ecology*, 61: 764–771.

3. Dierssen, K., 2000. Ecosystems as states of ecological successions. In: *Handbook of Ecosystem Theories and Management*, (Eds.) S.E. Jørgensen and F. Müller. Boca Raton, FL, pp. 427–446.

4. Downing, A.L. and Leibold, M.A., 2002. Ecosystem consequences of species richness and composition in pond food webs. *Nature*, 416: 837–840.

5. Dunne, J.A., Williams, R.J. and Martinez, N.D., 2002. Food-web structure and network theory: The role of connectance and size. *Proc. Natl. Acad. Sci., USA*, 99: 12917–12922.

6. Engelhardt, K.A.M. and Ritchie, M.E., 2002. The effect of aquatic plant species richness on wetland ecosystem processes. *Ecology*, 83: 2911–2924.

7. Fath, B.D., Jørgensen, S.E., Patten, B.C. and Straš kaba, M., 2004. Ecosystem growth and development. *BioSystem*, 77: 213–228.

8. Gunderson, L.H., Holling, C.S. and Peterson, G., 2000. Resilience in ecological systems. In: *Handbook of Ecosystem Theories and Management*, (Eds.) S.E. Jørgensen and F. Müller. Boca Raton, FL, pp. 385–394.

9. Hall, S.J. and Raffaelli, D.G., 1993. Food webs: theory and reality. *Advances in Ecological Research*, 24: 187–239.

10. Hannon, B., 1973. The structure of ecosystems. *J. Theor. Biol.*, 41: 535–546.

11. Herendeen, R.A., 1981. Energy intensity in ecological and economic systems. *J. Theor. Biol.*, 91: 607–620.

12. Ho, M.W. and Ulanowicz, R., 2005. Sustainable systems as organisms. *BioSystems*, 82: 39–51.

13. Jonsson, M. and Malmqvist, B., 2000. Ecosystem process rate increases with animal species richness: evidence from leaf-eating, aquatic insects. *Oikos,* 89: 519–523.

14. Jørgensen, S.E., Patten, B.C. and Straš kraba, M., 2000. Ecosystems emerging for growth. *Ecol. Model.,* 126: 249–284.

15. Knops, J.M.H., Bradley, K.L. and Wedin, D.A., 2002. Mechanisms of plant species impacts on ecosystem nitrogen cycling. *Ecology Letters,* 5: 454–466.

16. Krause, A.E., Frank, K.A., Mason, D.M., Ulanowicz, R.E. and Taylor, W.W., 2002. Compartments revealed in food-web structure. *Nature,* 426: 282–285.

17. Levine, S.H., 1980. Several measures of trophic structure applicable to complex food webs. *J. Theor. Biol.,* 83: 195–207.

18. Marques, J.C., Nielsen, S.N., Pardal, M.A. and Jørgensen, S.E., 2003. Impact of eutrophication and river management within a framework of ecosystem theories. *Ecol. Model.,* 166: 147–168.

19. Polis, G.A. and Strong, D., 1996. Food web complexity and community dynamics. *American Naturalist,* 147: 813–846.

20. Post, D.M., Pace, M.L. and Hairston, N.G., Jr., 2000. Ecosystem size determines food-chain length in lakes. *Nature,* 405: 1047–1049.

21. Pullin, A.S., 2002. *Conservation Biology.* Cambridge University Press, Cambridge.

22. Reynolds, C.S., 1984. *The Ecology of Freshwater Phytoplankton*. Cambridge University Press, Cambridge, MA, 384 pp.

23. Shugart, H.H., 1998. *Terrestrial Ecosystems in Changing Environments.* Cambridge University Press, New York, NY, 534 pp.

24. Simon, K.S. and Townsend, C.R., 2003. The impacts of freshwater invaders at different levels of ecological organisation, with emphasis on ecosystem consequences. *Freshwater Biology,* 48: 982–994.

25. Ulanowicz, R.E. and Kay, J.J., 1991. A package for the analysis of ecosystem flow networks. *Environ Software,* 6: 131–142.

Biotechnology: An Overview (2015)
Editors: Rajan Kumar Gupta, Nasim Akhtar and Deepak Vyas
Published by: DAYA PUBLISHING HOUSE, NEW DELHI

Pages 31–51

Chapter 4

Microbial Diversity: A Review of Different Approaches for its Study

Ranjana Bhatia[1]*, Gaurav Kakkar[1],
Varun Bansal[1] and Neeru Narula[2]

[1]*Biotechnology Branch, University Institute of Engineering and Technology,*
Panjab University, Chandigarh – 160 014
[2]*Department of Microbiology, CCS Haryana Agricultural University,*
Hisar – 125 004, Haryana

ABSTRACT

Micro-organisms play many essential roles both on land and in water. However, there is an apparent gap in knowledge regarding the diversity of bacteria and other prokaryotic organisms. Too small to be seen no longer means too small to be studied or valued. Microbial diversity encompasses the spectrum of variability among all types of micro-organisms (bacteria, fungi, viruses and many more). Microbial diversity studies are important not only for basic research but also to understand the link between diversity and community structure and function, which influence plant growth and soil properties It is important to conserve and characterize the variability of various agriculturally important micro-organisms for their optimal usage in future agriculture. Microbial diversity has been long studied using traditional as well as molecular methods. The choice of method depends upon the type of information required for the specific study. However, it is experienced that molecular techniques coupled with traditional methods can greatly facilitate diversity studies and helps us to get the broadest picture possible.

Keywords: Agriculturally important micro-organisms, Microbial diversity, PCR, Polymorphism, Soil, 16S rRNA gene.

---

* *Corresponding Author:* E-mail: ranjanabhatia20@yahoo.com.

Introduction

Micro-organisms are essential for the earth to function. They play many roles both on land and in water. Because micro-organisms are small, they are least known, and this gap in knowledge is particularly apparent for bacteria and other prokaryotic organisms. Studying microbial diversity, therefore, deserves greater attention. Too small to be seen no longer means too small to be studied or valued.

Microbial diversity encompasses the spectrum of variability among all types of micro-organisms (bacteria, fungi, viruses and many more). Studying microbial diversity is important for various reasons: micro-organisms are important sources of knowledge about the strategies and limits of life, micro-organisms are of critical importance to the sustainability of life on our planet, the untapped diversity of micro-organisms is a resource for new genes and organisms of value to biotechnology, diversity patterns of micro-organisms can be used for monitoring and predicting environmental change, micro-organisms play a role in conservation and restoration biology of higher organisms, and microbial communities are excellent models for understanding biological interactions and evolutionary history.

Plant microflora is of great importance as it has both beneficial and detrimental effects on the plant rhizosphere. Therefore, it is important to conserve and characterize the variability of various agriculturally important micro-organisms for their optimal usage in future agriculture. Diversity can be defined as the presence of different types of micro-organisms in a given habitat. In microbial terms, diversity of a community is expressed as the total number of species (species richness) and relative abundance of these species (species evenness). Microbial diversity has been studied for several years by defining several criteria such as physiology, phylogeny, metabolism and genomics. The choice of method depends upon the type of information required for the specific study.

Each method that describes diversity has its advantages and disadvantages. Traditional techniques of diversity studies are selective but do not represent the complete extent of microbial communities. Molecular methods, on the other hand, encounter problems at the stage of DNA/RNA extraction, which includes reliable and reproducible lysis of bacterial cells, extraction of intact nucleic acids and removal of substances such as humic acids, bacterial exo-polysaccharides and proteins which may inhibit polymerase activity during PCR and DNA digestion with restriction enzymes. However molecular techniques coupled with traditional methods can greatly facilitate diversity studies.

Traditional Methods for Microbial Diversity Analyses

Plate Counts

To gain a comprehensive understanding of microbial physiology and to access microbial metabolic pathways, cultivation of micro-organisms is required. Micro-organisms can be isolated from soil using plate count and enrichment culture techniques. In plate count/viable count technique, microbes are isolated by culturing the soil dilutions on different selective growth media whereas in enrichment culture technique, population density of microbes existing in low proportion is favoured prior to isolation on growth medium (Figure 4.1). Plate counts of culturable species do not provide community structure information found in other methods (Kennedy and Smith, 1995). So they have been relied on less in recent researches. However, they still provide a preliminary measure of diversity on subsets of population (Kennedy and Smith 1995). They offer a quick, inexpensive, reliable method of obtaining data. Plate counts also enable isolation and identification of species able to perform desired functions such as biological control (Kennedy *et al.*, 1991) or bioremediation of polluted areas (Kuritz and Wolk, 1995). Isolated species are also needed in libraries for DNA measures of phenotype.

However, this approach has become less popular not only because a limited number of micro-organisms can be cultured but also because the procedures are laborious and there occurs variation in microbial growth rate.

Most Probable Number (MPN)

A variation of viable count procedure is most probable number (MPN) estimations of microbial populations. The MPN procedure is based on the determination of dilution of the soil sample beyond which any microbial population cannot be detected. It is sometimes also called the method of ultimate or extinction dilution or end point dilution (Alexander, 1982).

This technique involves the determination of micro-organisms present or absent in several individual aliquots of each of the several consecutive dilutions of soil. Micro-organisms whose population is to be determined must be able to bring about some characteristic and readily recognizable transformation in the medium when inoculated or the micro-organisms undergone multiplication must be easily recognizable on the substrate (Figure 4.2). Population densities are determined using MPN tables relating to the statistical probability of the presence of the organisms of interest to the number of positive samples for each dilution (Alexander, 1982). MPN has been used in enumerating rhizobial population in tropical soils and in the rhizosphere of woody legumes (Odee *et al.*, 1995). Further, this method has been used to estimate the population of associated diazotrophs in the rhizosphere. This procedure, however, retains the problems of selectivity of media and incubation time.

Substrate Utilization

Substrate utilization pattern or the BIOLOG metabolic assay can provide "fingerprints" of community structure. The method was originally developed for identification of microbial isolates and has been adapted to investigate the functional diversity of soil microbial communities (Garland and Mills 1991; Flies bach and Mäden, 1996; Di Giovanni *et al.*, 1999). It is therefore also called community level physiological profiling (CLPP) in this context. CLPP's have been used to provide useful information for the assessment of soil microbial community diversity (O'Connel and Garland, 2002). These measures may also provide indications of metabolic potential (Garland and Mills, 1991; Haack *et al.*, 1995; Garland, 1996) and nutritional strategies within feeding guilds (Zak *et al.*, 1994).

BIOLOG is an easy to use yet advanced tool for identifying and characterizing micro-organisms. By using its patented technology (Microlog™, Microbial

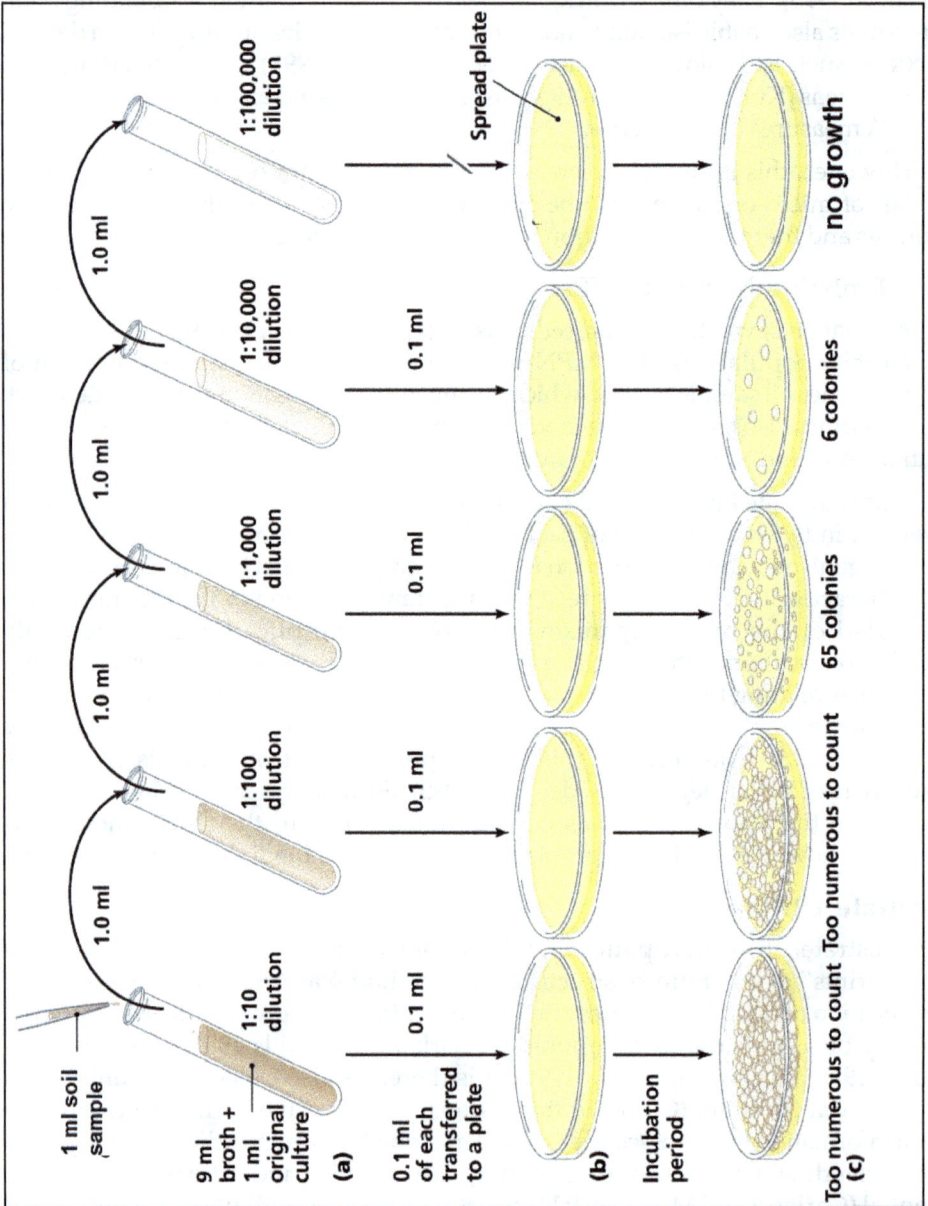

Figure 4.1. Plate count technique to assess microbial population in soil.

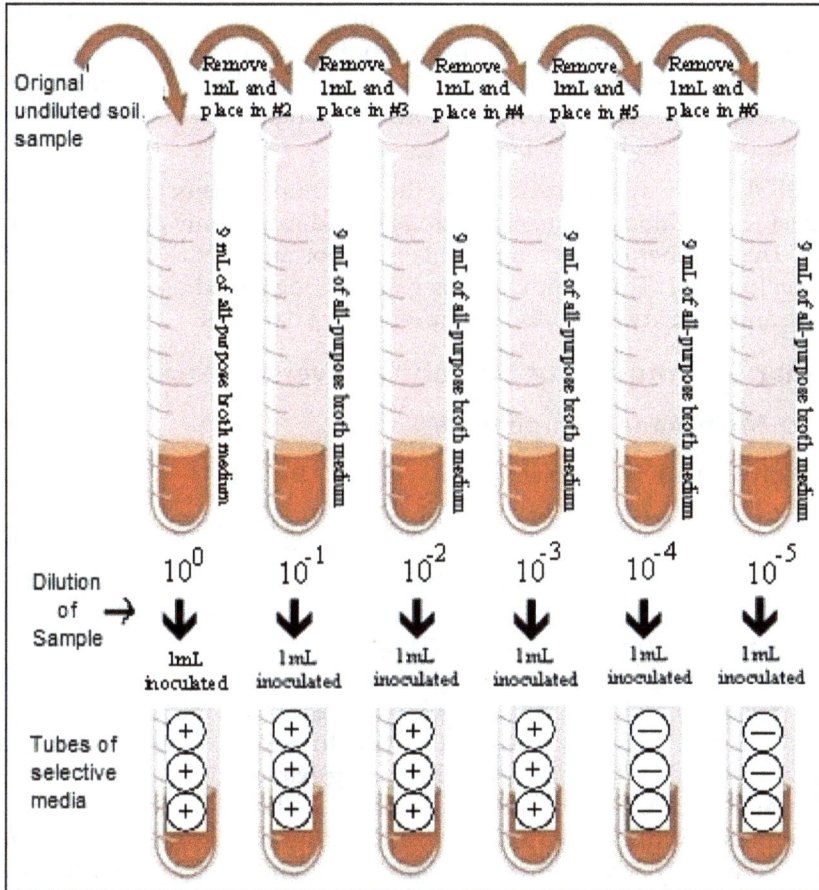

Figure 4.2. Most probable number (MPN) analysis of microbial population in a soil sample.

Identification System, Biolog Inc., California, USA) with 95 carbon source utilization tests in a microtitre plate format, the BIOLOG microbial identification system can recognize over 4×10^{28} possible metabolic patterns. BIOLOG's technology uses each microbe's ability to use particular carbon sources to produce a unique pattern or "fingerprint" for the microbe (Garland and Mills, 1991). As a micro-organism begins to use the carbon sources present in the wells of the microplate, it respires. This respiration process reduces the tetrazolium violet (redox dye) present in the wells and those wells turn purple. The end result is a pattern of purple wells which is readable either visually or by a Plate Reader and then evaluated by multivariate statistical analyses to classify the samples. Colour development in each well reflects the activity and density of each bacterial isolates and also the ability of the isolate to respond to a particular substrate. Thus, by developing a simple tool that allows 95 simultaneous carbon source utilization tests, BIOLOG enables an efficient, easy to use, powerful and reliable microbial identification system.

Substrate utilization pattern reflects a community's qualitative and quantitative composition and is able to characterize differences among habitats and between different samples within the same habitat. Shifts within substrate utilization patterns may reflect shifts in community composition and change in environmental conditions and seasons (Bossio and Scow, 1995)

The BIOLOG assay is dependant on the inoculum density and growth of cells under specific conditions in the microtitre plate indicating only potential functional diversity. The number of inoculated cells is important for the rate and degree of colour development. This technique has gained widespread use, primarily due to ease of use and capacity to produce comprehensive data sets.

Molecular Techniques for Microbial Diversity Analyses

Based on Membrane Components

Phospholipid Fatty Acid Analysis (PLFA)

The PLFA (phospholipid fatty acid analysis) profiles of soil samples offer sensitive, reproducible measurements for characterizing the dominant portion of soil microbial communities without cultivating the organisms (Zelles, 1999). Phospholipids are stable components found only in the membranes of living micro-organisms and are not storage products. They are polar lipids and some of them are specific for different groups of micro-organisms. Therefore, individual PLFA can be related to specific microbial populations. For this assay, soil microbial phospholipid fatty acids are extracted from soil with solvent mixtures such as chloroform, methanol and phosphate buffer (White *et al.*, 1979). Following extraction, the chloroform fraction is concentrated and fatty acids are separated on a silicic acid column (King *et al.*, 1977). The polar lipid fraction is collected, dried under nitrogen, saponified and methylated. The fatty acid methyl esters are then identified *e.g.* with MIDI method (MIDI, 1995). The MIDI system identifies and quantifies fatty acids of 9 to 20 carbon chain length (Figure 4.3). However, this method is time consuming. PLFA analysis has been used for characterizing soil and rhizosphere microbial communities (Soderberg *et al.*, 2002), in separating microbial communities in bulk soil and rhizosphere soil (Thirup *et al.*, 2003) and in determination of effect exerted by application of N-fertilizers on microbial communities (Clegg *et al.*, 2003).

Based on Nucleic Acid Analysis

Techniques based on nucleic acid sequences can help in a more targeted and detailed analysis of composition and diversity of micro-organisms.

Nucleic Acid Re-Association and Hybridization

DNA re-association is a measure of genetic complexity of the microbial community and has been widely used to estimate diversity (Torsvik *et al.*, 1996). It involves extraction and purification of DNA from the environmental samples followed by its denaturation and re-annealing. The rate of hybridization or re-association is a function of the complexity of DNA which can indicate the sequence diversity and size of bacteria. As the complexity or diversity of DNA sequence

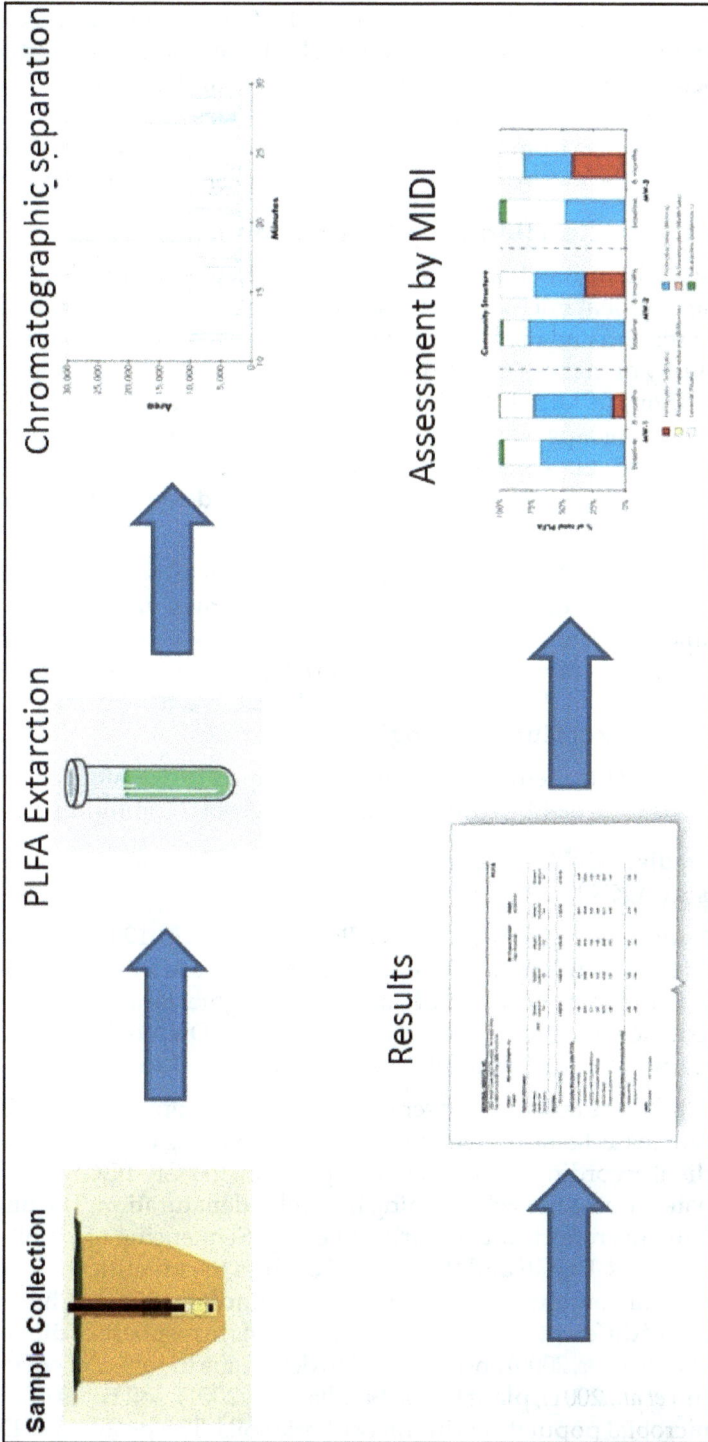

Figure 4.3. Phospholipid fatty acid analysis (PLFA) for characterizing microbial population.

increases, the rate at which DNA re-associates decreases (Theron and Cloete, 2000). Under specific conditions, the time needed for half of the DNA to re-associate can be used as a diversity index as it takes into account both the amount and distribution of DNA re-association (Torsvik *et al.*, 1998). Griffiths and Glover (1996) applied hybridization technique that provided similarity indices and measures of relative diversity of two samples from whole soil community DNA. Extracted bacteria and whole community DNA was observed to have 75 per cent similarity to each other.

Polymerase Chain Reaction (PCR) Based Methods

Polymerase chain reaction based methods have enormously affected our understanding of global microbial diversity because they have contributed to both the fast differentiation and identification of cultivated micro-organisms and access to the vast majority of micro-organisms which are yet uncultured. PCR is a technique used for amplifying a target DNA to study prokaryotic diversity and predict phylogenetic relationships. DNA is extracted from the environmental samples or bacteria and purified, targeted region of the DNA is amplified using universal or specific primers and the resulting products are separated and analyzed in different ways.

16S rRNA gene (16S rDNA) is the most common target DNA to study the genetic diversity of bacteria. It occurs in all bacteria and shows variations in the base composition among species. It consists of variable and conserved regions for targeting a majority of the members from different bacterial groups. In addition to 16S rDNA, 16S-23S intergenic region (IGS), 18S rDNA, *nifH*, *nodC* etc. have been used to study the microbial diversity of nitrogen fixing bacteria.

Several comparable methods have been developed to examine the diversity of these sequences in total DNA extracted from soil microbial communities.

Denaturing Gradient Gel Electrophoresis (DGGE) and Temperature Gradient Gel Electrophoresis (TGGE)

DGGE (Muyzer *et al.*, 1993) and TGGE (Smalla *et al.*, 1998) are two similar methods for studying microbial diversity. They are based on variation in base composition and secondary structure of same length fragments of 16S rDNA molecule. Initial step in both techniques is the amplification of 16S rDNA of known size by PCR with primers principally targeting all eubacteria or selected sub-groups.

Following PCR, the products are separated by gel electrophoresis. In PCR-DGGE, the gel itself contains a chemical-denaturing gradient, making the fragments denature along the gradient according to their base composition. In PCR-TGGE, a temperature gradient is created across the gel, resulting in similar denaturation. The number and position of fragments reflects the bacterial diversity. Sequencing of visible bands on the gel following PCR-DGGE and PCR-TGGE enables the identification of different species in microbial communities (Riemann and Windling, 2001). It has been used for identification of different microbial communities in rhizosphere of different plants (Normander and Prosser, 2000) and also used to determine the effect of environmental conditions (Smit *et al.*, 2001), plant types (Smalla *et al.*, 2001), soil types (Gelsomino *et al.*, 1999) on microbial populations in rhizospheric soils. The problem in DGGE and

TGGE is the occurrence of more number of bands in gel electrophoresis. Soil communities may easily contain several hundred bacterial strains, while the resolution of more than 20-50 bands in a gel is difficult (Heuer and Smalla, 1997).

Single Strand Conformation Polymorphism (SSCP)

It is a technique in which PCR products of 16S rRNA genes are denatured into single strands and directly electrophoresed on a non-denaturing gel. Separation is based on differences in the folded conformations of single stranded DNA, which influences the electrophoretic mobility (Lee *et al.*, 1996). They demonstrated that a bacterial strain generates a characteristic band pattern from which relative diversity in bacterial communities can be measured. This method was sensitive enough to detect bacterial population that made up even less than 1.5 per cent of a bacterial community. But a potential problem is the reannealing of the DNA during electrophoresis. Schwieger and Tebbe (1998) used one phosphorylated primer in PCR and the phosphorylated strand was later digested by a lambda exonuclease, hence removing one of the two DNA strands before electrophoresis. Scheinert *et al.* (1996) used one biotinylated primer to perform magnetic separation of one single strand after denaturation to avoid the reannealing of DNA strands during electrophoresis. To assess the identity of the predominant bacterial populations in the community, bands can either be excised and sequenced, or SSCP-patterns can be hybridized with taxon specific probes. SSCP has been used to characterize the diversity and succession of microbial communities during composting of agricultural substrates (Peters *et al.*, 2000) and also to detect differences in bacterial communities in rhizospheres of different genetically modified plants (Schmalenberger and Tebbe, 2003).

Ribosomal Intergenic Spacer Analysis (RISA)/Automated Ribosomal Intergenic Spacer Analysis (ARISA)

Heterogeneity in the length and sequence of intergenic spacer region (IGS) between the 16S-23S rRNA is useful for differentiating between bacterial strains and closely related species by the RISA (ribosomal intergenic spacer analysis) and ARISA (automated ribosomal intergenic spacer analysis) techniques. The IGS region is amplified, denatured and separated on a polyacrylamide gel under denaturing conditions (Fisher and Triplett, 1999). Sequence polymorphism is detected by silver staining in RISA, while in ARISA the forward primer is fluorescently labelled and automatically detected. RISA has been used to compare microbial diversity in soil and rhizosphere (Borneman and Triplett, 1997) and contaminated soil (Ranjard *et al.*, 2000).

Highly Repetitive Sequence or Microsatellite Region Characterization

Prokaryotic and eukaryotic organisms contain highly repetitive, short DNA sequence (1-10 bp long) throughout their genomes. These are called Microsatellites or Short Tandom Repeat Polymorphism (STRP) or Simple Tandom Repeats (STR) or Simple Sequence Repeat (SSR). These may be used as a diagnostic tool that allow differentiation upto the species or strain level . Different bacterial genomes may have different repetitive patterns in their sequence. This provides a specific genomic fingerprint of chromosome structure to each strain. Hence differences in strains can be thus identified using these repetitive sequences.

Restriction Fragment Length Polymorphism (RFLP)

Restriction fragment length polymorphism (RFLP) analysis is another tool to characterize microbial communities.RFLP is based on the PCR amplification of target region (16S, IGS, *nif*H, *nod*A, *nod*C etc.) from the sample DNA followed by digestion of the amplified product with different restriction enzymes. Different species have different sites for cleavage by different restriction enzymes. The number of these specific sites also varies among species and strains. This variation might have occurred due to evolutionary differences among the population. Hence, studying this polymorphism enables us to differentiate between communities present in a particular sample.The fragments thus obtained are of varying lengths and are detected by their separation on agarose gels or non-denaturing acrylamide gels. The banding patterns in the gel reflect the population of all restriction fragments of at least the major members of the community (Massol-Deya *et al.*, 1995). However, the technique has little utility in quantitating diversity in complex communities.

Terminal Restriction Fragment Length Polymorphism (tRFLP)

The initial description of terminal restriction fragment length polymorphism (tRFLP) was given by Liu *et al.* (1997) and is based on the PCR amplification of 16S rDNA with specific primers. The primers are labelled with a fluorescent tag at the terminus resulting in labelled PCR-products. The products are cut with several restriction enzymes, one at a time. The digested PCR products are subsequently loaded on ABI sequencer. Since the PCR products are labelled at the terminus, only the terminal fragments of a restriction digest are detected by the sequencer. The output of the machine includes peak (fragment) height, area and fragment size in graphical and tabulated form. The rate limiting step in the analysis is the extraction and purification of community DNA of uniform quality. As the ABI sequencer can discriminate between different fluorescent tags in a single gel lane, a gel can be double or triple loaded by employing differently tagged primers in the PCR amplification step. This improves the cost effectiveness of the technique. The method, however, needs delicate and expensive instrument (Tiedje *et al.*, 1999). t-RFLP has been used for defining microbial communities in soil (Dunbar *et al.*, 2000), aquatic systems (Kent *et al.*, 2003) and termit gut (Liu *et al.*, 1997; Ohkuma *et al.*, 1999).

Marker Genes Commonly Used for Diversity Analyses of Diazotrophic Bacteria

Small Subunit rRNA/16S rRNA Gene

Until 1980, the determination of microbial community structure and the identification of micro-organisms depended on culture based techniques. These were both time consuming and cumbersome with application in only a limited part of the microbial community. But with the development of selective amplification of DNA based PCR reactions, the method of phylogenetic analysis revolutionized. 16S rRNA gene remains most useful for estimating evolutionary relationships among bacteria (Woese, 1987). This gene is a very suitable molecular marker. The gene is present in all prokaryotes and shows functional consistency; 16s rRNA is part of ribosomes that are required by all organisms to synthesize proteins. The gene is ~1.5kb long which is sufficient to be used as a document of evolutionary history as the horizontal transfer of rRNA genes is limited. Due to the functional necessity of rRNA, their primary and secondary structure is constrained. 16s rRNA consists of several sequence domains that have evolved at different rates; some domains have remained almost universally conserved and are interspaced by more variable regions, specific for phylum up to subspecies level. This permits unambiguous alignment of homologous positions in a sequence and the identification of near-universally conserved and taxon specific 'signature' sequence motifs.

A large number of 16S sequences of different organisms are stored in databases. RFLP analysis of amplified 16S rDNA, termed amplified ribosomal DNA restriction analysis (ARDRA), has proved to be very effective for discriminating among rhizobacterial species (Scortichini *et al.*, 2002). Scientists have described certain set of oligonucleotide primers capable of initiating enzymatic amplification of nearly full length 16S rDNA on a phylogenetically and taxonomically wide range of bacteria (Weisburg *et al.*1991). Its application was further authenticated by Laguerre *et al.* (1994) who characterized different reference strains of *Rhizobium* by PCR-RFLP of 16S rDNA and reported that results of classification based PCR-RFLP were consistent with the results obtained from taxonomic classification of rhizobia based on DNA-DNA homology. Thus PCR-RFLP method provides a rapid tool for identification of root nodule isolates.

Different scientists all around the world have successfully used ARDRA as an effective tool to study diversity patterns at different levels. Ross *et al.* (2000) studied the genetic diversity of closely related pseudomonads isolated from the wheat field soils in South Australia by ARDRA, BIOLOG and GC-FAME. BIOLOG and GC-FAME clustering showed a >70 per cent match to ARDRA profiles. Schwieger and Tebbe (2000) studied the effect of field inoculation with *S. meliloti* L33 on the composition of bacterial communities in the rhizospheres of a target plant (*M. sativa*) and a non-target weed (*Chenopodium album*). ARDRA revealed plant specific fragment size frequencies. Impact of two herbicides (propanil and prometryne) on soil micro-organisms was studied by DGGE and ARDRA and it was found that propanil did not affect soil bacteria significantly (Crecchio *et al.*, 2001). Despite a lower utilization of prometryne, a more diverse banding was shown than with propanil. Smit *et al.* (2001) showed even distribution of *Acidobacterium*, *Proteobacterium*, *Nitrospira*, *Cyanobacterium* and green sulphur bacteria based on the analysis of 16S rDNA sequences.

The genetic diversity of fast growing rhizobia nodulating soybean was assessed by REP, RAPD, ERIC and ARDRA (Saldana *et al.*, 2003). ERIC, REP and RAPD markers revealed a considerable genetic diversity among fast-growing rhizobia. However, ARDRA showed different genotypes among fast-growing rhizobia. Rosenblueth and Martinez-Romero (2004) analyzed the genetic diversity of *R. etli* strains nodulating maize by ARDRA and found some of these strains to be more competitive as maize root colonizers rather than *R. etli* strains from the rhizosphere or from bean nodules.

*nif*H

The nif gene is the gene responsible for the coding of proteins related and associated with the fixation of atmospheric nitrogen into a form of nitrogen available to plants. These genes are found in nitrogen fixing bacteria and cyanobacteria.The detection of such a marker gene that is unique and is required for nitrogen fixation may provide a way to analyze the nitrogen fixing potential of an ecosystem.

Micro-organisms catalyze biological nitrogen fixation with an enzyme complex, nitrogenase, which has been highly conserved through evolution (Howard and Rees, 1996). The high degree of protein sequence similarity of nitrogenases among micro-organisms suggests an early origin or lateral gene transfer among prokaryotic lineages (Postgate and Eady, 1988).

Nitrogenases are enzymes that carry out nitrogen fixation in nitrogen fixing organisms. Dinitrogen is quite inert because of the strength of its N-N triple bond. To break one nitrogen atom away from another requires breaking all three of these chemical bonds. It is the only known family of enzymes that accomplish this process.This enzyme is composed of two multisubunit metallo-proteins. Component I or the dinitrogenase contains the active site for N_2 reduction, has a molecular weight of about 250 KDa and is composed of two heterodimers, encoded by the *nifD* and *nifK* genes. Component II or the dinitrogenase reductase (about 70 KDa) couples ATP hydrolysis to interprotein electron transfer and is composed of two identical subunits encoded by the *nif*H gene. Fe-S centres are present in both component I and II proteins and are co-ordinated between the subunits. 'Conventional' nitrogenases contain Mo (molybdenum) in the Fe-S centre bridging the subunits. 'Alternative' nitrogenases replace Mo with V (vanadium) (*vnfH*) and 'second alternative' nitrogenases replace Mo with Fe (iron) (*anfH*). The enzymes have somewhat different reaction kinetics and specificities (Burgess and Lowe, 1996; Eady, 1996). The nitrogenase reaction is energetically expensive (16 ATP and eight electrons per molecule reduced) and the enzyme *in vitro* is sensitive to inactivation by oxygen.

Thus *nif*H gene is integral for the synthesis and subsequently the functioning of nitrogenase enzyme.The functional *nif*H gene has highly conserved regions as well as great divergence in other regions and can therefore be used to evaluate the phylogenetic relationships among diazotrophs. Since the *nif*H gene occurs only in nitrogen fixing micro-organisms, it has been used to monitor the presence of these diazotrophs using different fingerprinting techniques (DGGE, TGGE, RFLP, sequencing) *e.g.* in pure cultures (Franke *et al.*, 1998), in soil (Widmer *et al.*, 1999) as well as in plants (Ueda *et al.*, 1995; Lovell *et al.*, 2000). It has also been detected in arbuscular mycorrhizal fungus spores (Minerdi *et al.*, 2001), marine environments (Zehr *et al.*, 1998) and termite guts (Ohkuma *et al.*, 1999).

Degenerate oligonucleotide primers can be used to amplify, clone and sequence a segment of the *nif*H gene from a natural assemblage. Zehr and McReynolds (1989) used same technique on *Trichodesmium thiebautii* and concluded it to be a cyanobacterial *nif*H gene as it was most closely related to *Anabaena* spp. Diazotrophic diversity in rice roots was investigated by Ueda *et al.* (1995). They characterized phylogenetically 23 *nif*H gene sequences by PCR amplification of mixed organisms' DNA extracted directly from rice roots.

Poly *et al.* (2001) reported soil management to be the major parameter influencing differences in the *nif*H gene pool structure among soils by controlling inorganic nitrogen content. By studying the diverse assemblage of *nif*H genes, Mehta *et al.* (2003) provided the first genetic evidence of potential nitrogen fixers in hydrothermal vent environments of Juan de Fuca Ridge. Yeager *et al.* (2004) reported that ecological factors such as soil instability and water stress may constrain the growth of nitrogen-fixing micro-organisms in mature and poorly developed soil crusts from Colorado Plateau and Chihuahuan Desert. Bürgmann *et al.* (2005) studied the activity and composition of asymbiotic soil diazotrophs following additions of artificial root exudates and single carbon sources using universal and group specific *nif*H-PCR and DGGE analysis. Their results indicate that the active nitrogen-fixing population represents only a fraction of the total diaztroph diversity. The active species identified belonged to the genus *Azotobacter*. Only sugar-containing substrates were able to induce nitrogen fixation.

Factors Influencing Soil Microbial Diversity

Soil is irreplaceable and its microbial community dynamic. The abundance and activities of soil micro-organisms are influenced by various environmental and edaphic factors. Soil type and texture, nutrient status, soil pH, soil moisture, site temperature, latitude, organic carbon content and C:N ratio, agricultural management practices and the species and age of the plant cover are some of the factors affecting microbial populations. Many of these factors interact with each other and have both direct and indirect effects on the soil microbial community. It is difficult to predict which environmental determinants are most important in determining the composition of a soil microbial community. An understanding of the role of each factor that governs microbial communities is equally important as the extent of their influence. Some of these are discussed below:-

Soil Type

Microbial diversity is best exemplified by soil diversity. Differences in soil type and its properties such as particle size distribution, aggregate stability, pH, organic matter content, water holding capacity, influence the community structure of micro-organisms. Yohalem and Lorbeer (1994) observed a strong influence of soil type on the diversity of *Burkholderia* strains. Strains isolated from the same field were found to be similar but those isolated from different sites with similar cropping history were not. Latour *et al.* (1996) also observed that the effect of soil type was more pronounced than plant type on the diversity of *Pseudomonas* spp. Chiarini *et al.* (1998) and Dalmastri *et al.* (1999) studied the influence of soil type, cultivar, root location and growth stages of maize on microbial communities in the rhizosphere and found the soil to be the dominating factor affecting genetic diversity.

Sessitsch *et al.* (2001) reported that particle size had highest impact on microbial diversity than factors like pH, the type and amount of organic input.

Soil Management Practices

Common management practices that may impact microbial diversity include tillage (Kennedy and Smith, 1995), fertilizer and pesticide regimes (Wander *et al.*, 1995) and cropping rotations (Zelles *et al.*, 1992). Crop management practices may cause community shifts by increasing resource heterogeneity and microbial diversity. Palmer and Young (2000) studied the effect of soil management practices on rhizobial diversity and observed a higher diversity of *R. leguminosarum* in arable soils than in grassland soil.

The effect of N-fertilization and soil drainage was studied on the microbial community in a grassland (Clegg *et al.*, 2003). While N-fertilizer application affected the bacterial and actinomycetes population, soil drainage affected actinomycetes and *Pseudomonas* population. Inputs of manure and cover crops can have large impacts on the size and activity of soil microbial communities. Studies are limited due to an inability to control many of the factors influencing the microbial community and evaluation must rely on alternative metagenomic approaches which can be cost prohibitive (Shi *et al.*, 2009).

Other investigations also indicated that the type and amount of available organic substrates strongly influence the abundance of microbial groups and their functional diversity in soil (Grayston *et al.*, 2001; Smit *et al.*, 2001).

Plant Type

Soil micro-organisms play an important role in nutrient acquisition for plants and to enhance plant growth. Microbial growth in turn, is dependant on the presence of organic matter in the soil. *Pseudomonas, Flavobacterium, Alcaligenes* and *Agrobacterium* species have been shown to be particularly stimulated in the rhizosphere due to the release of exudates and lysates (Curl and Truelove, 1986). Therefore, variety of organic compounds released by plants is a major factor influencing microbial diversity in different plant rhizospheres

Rooting systems of plants are also responsible for different types of root exudates (sugars, amino acids, organic acids, etc.) and have a different microflora (Miller *et al.*, 1989). Micro-organisms utilize these sources of energy and carbon to carry out their functions. The availability of substrates may have a greater impact on diversity than the number of substrates. Korner and Laczko (1992) found less microbial diversity associated with vitamin deficiencies, possibly resulting from reduced plant secretions. DiCello *et al.* (1997) observed that population of *Burkholderia cepacia* in the rhizosphere of maize decreased with plant growth.

The degree of variation of dominant bacterial species in respect of soil type (silt, sand and loamy soil), plant type (clover, bean and alfalfa) and developmental stage of plant was assessed by Wieland *et al.* (2001). Plant species showed greatest effect and plant developmental stage the lowest effect on bacterial diversity. The effect of soil type exceeded that of plant type in the soil habitat only, as different clustering of microbial community of alfalfa was seen in loamy sand and in silt sand.

Competition

Competitive interactions also influence microbial diversity (Tiedje *et al.*, 2001). Soil structure and water regime cause spatial isolation within communities and influence interactions among micro-organisms. High spatial isolation within soils showed high microbial diversity and low spatial isolation showed low diversity of micro-organisms.

Environmental Conditions

Climate, temperature, water content and nutrient availability also affect microbial community structure and diversity. Studies carried out by Smit *et al.* (2001) show that the population of fast-growing bacteria was lowest in winter and highest in summer, and that the highest species richness was found in spring and autumn. Also, though the dominant populations are stable, less-abundant populations show distinct seasonal variations.

Microbial diversity studies are important not only for basic research but also to understand the link between diversity and community structure and function, which influence plant growth and soil properties. Although methods to study diversity are improving, yet the association between diversity and function is still not clearly understood. Therefore, the best way to study soil microbial diversity would be to use a variety of tests with different end points and degrees of resolution to get the broadest picture possible.

References

1. Alexander, M., 1982. Most probable number method for microbial populations. In: *Methods of Soil Analysis*, Part 2, (Ed.) A.L. Page. Agronomy Society of America, Madison, WI. pp. 815–820.

2. Borneman, J. and Triplett, E., 1997. Molecular microbial diversity in soils from eastern Amazonia: evidence for unusual micro-organisms and microbial shifts associated with deforestation. *Appl. Environ. Microbiol.*, 63: 2647–2653.

3. Bossio, D.A. and Scow, M., 1995. Impact of carbon and flooding on the metabolic diversity of microbial communities in soils. *Appl. Environ. Microbiol.*, 61: 4043–4050.

4. Burgess, B.K. and Lowe, D.J., 1996. Mechanism of molybdenum nitrogenase. *Chem. Rev.*, 96: 2983–3011.

5. Bürgmann, H., Meier, S., Bunge, M., Widmer, F. and Zeyer, J., 2005. Effects of model root exudates on structure and activity of a soil diazotroph community. *Environ. Microbiol.*, 7: 1711–1724.

6. Chiarini, L., Bevivino, A., Dalmastri, C., Nacamulli, C. and Tabacchioni, S., 1998. Influence of plant development, cultivar and soil type on microbial colonization of maize root. *Appl. Soil Ecol.*, 8: 11–18.

7. Clegg, C.D., Lovell, R.D.L. and Hobbus, P.J., 2003. The impact of grassland management regime on the community structure of selected bacterial groups in soil. *FEMS Microbiol. Ecol.*, 43: 263–270.

8. Crecchio, C., Curci, M., Pizzigallo, M.D.R., Ricciuti, P. and Ruggiero, P., 2001. Molecular approaches to investigate herbicide-induced bacterial community changes in soil microcosms. *Biol. Fertil. Soils*, 33: 460–466.

9. Curl, E.A. and B. Truelove, 1986. *The Rhizosphere*. Springer-Verlag, Berlin.

10. Dalmastri, C., Chiarini, L., Cantale, C., Bevivino, A. and Tabacchioni, S., 1999. Soil type and maize cultivar affect the genetic diversity of maize root-associated Burkholderia cepacia populations. *Microb. Ecol.*, 38: 273–284.

11. Di Cello, F., Bevivino, L., Chiarini, R., Fani, R. and Paffetti, D., 1997. Biodiversity of a Burkholderia cepacia population isolated from the maize rhizosphere at different plant growth stages. *Appl. Environ. Microbiol.*, 63: 4485–4493.

12. Di Giovanni, G.D., Wastrud, L.S., Seidler, R.J. and Widmer, F., 1999. Comparison of parental and transgenic alfalfa rhizosphere bacterial communities using Biolog GN metabolic fingerprinting and enterobacterial repetitive intergenic consensus sequence PCR (ERIC-PCR). *Microb. Ecol.*, 37: 129–139.

13. Dunbar, J., Ticknor, L.O. and Kuske, C.R., 2000. Assessment of microbial diversity in four southwestern United States soils by 16S rRNA gene terminal restriction fragment analysis. *Appl. Environ. Microbiol.*, 66: 2943–2950.

14. Eady, R.R., 1996. *Structure-function relationships of alternative nitrogenases. Chem. Rev.*, 96: 3013–3030.

15. Fisher, M.M. and Triplett, E.W., 1999. Automated approach for ribosomal intergenic spacer analysis of microbial diversity and its application to freshwater bacterial communities. *Appl. Environ. Microbiol.*, 65: 4630–4636.

16. Fleisbach, A. and Mäden, P., 1996. Microbial biomass and size-density fractions differ between soils of organic and conventional agricultural systems. *Soil Biol. Biochem.*, 32: 757–768.

17. Franke, I.H., Fegan, M., Hayward, A.C. and Sly, L.L., 1998. Nucleotide sequence of the nifH gene coding for nitrogen reductase in the acetic acid bacterium Acetobacter diazotrophicus. *Lett. Appl. Microbiol.*, 26: 12–16.

18. Garland, J.L., 1996. Patterns of potential C-source utilization by rhizosphere communities. *Soil Biol. Biochem.*, 28: 223–230.

19. Garland, J.L. and Mills, A.L., 1991. Classification and characterization of heterotrophic microbial communities on the basis of patterns of community level sole-carbon-source utilization. *Appl. Environ. Microbiol.*, 57: 2351–2359.

20. Gelsomino, A., Keijzer-Wolters, A., Cacco, G. and van Elsas, J.D., 1999. Assessment of bacterial community structure in soil by polymerase chain reaction and denaturing gradient gel electrophoresis. *J. Microbiol. Methods*, 38: 1–15.

21. Grayston, S.J., Griffith, G.S., Mawdsley, J.L., Campbell, C.D. and Bardgett, R.D., 2001. Accounting for variability in soil microbial communities of temperate upland grassland ecosystems. *Soil Biol. Biochem.*, 33: 533–551.

22. Griffiths, B.S. and Glover, L.A., 1996. Broad-scale approaches to the determination of soil microbial community structure: Application of the community DNA hybridization technique. *Microb. Ecol.*, 31: 269–280.

23. Haack, S.K., Garchow, H., Klug, M.J. and Forney, L.J., 1995. Analysis of factors affecting the accuracy, reproducibility and interpretation of microbial community carbon source utilization patterns. *Appl. Environ. Microbiol.*, 61: 1458–1468.

24. Heuer, H. and Smalla, K., 1997. Application of denaturing gradient gel electrophoresis and temperature gradient gel electrophoresis for studying soil microbial communities. In: *Modern Soil Microbiology*, (Eds.) J.D. van Elsas, J.T. Trevors and E.M.H. Wellington. Marcel Dekker, Inc., New York, pp. 353–373.

25. Howard, J.B., and Rees, D.C., 1996. Structural basis of biological nitrogen fixation. *Chem. Rev.*, 96: 2965–2982.

26. Kennedy, A.C. and Smith, K.L., 1995. Soil microbial diversity and the sustainability of agricultural soils. *Pl. Soil*, 170: 75–86.

27. Kennedy, A.C., Elliott, L.F., Young, F.L. and Douglas, C.L., 1991. Rhizobacteria suppressive to the weed downy brome. *Soil Sci. Soc. Am. J.*, 55: 722–727.

28. Kent, A.D., Smith, D.J., Benson, B.J. and Triplett, E.W., 2003. Web–based phylogenetic assignment tool for analysis of terminal restriction fragment length polymorphism profiles of microbial communities. *Appl. Environ. Microbiol.*, 69: 6768–6776.

29. King, J.D., White, D.C. and Taylor, C.W. 1977. Use of lipid composition and metabolism to examine structure and activity of estuarine detrial microflora. *Appl. Environ. Microbiol.*, 33: 1177–1183.

30. Korner, J. and Laczko, E., 1992. A new method for assessing soil micro-organisms diversity and evidence of vitamin deficiency in low diversity communities. *Biol. Fertil. Soils*, 13: 58–60.

31. Kuritz, T. and Wolk, C.P., 1995. Use of filamentous cyanobacteria for biodegradation of organic pollutants. *Appl. Environ. Microbiol.*, 61: 234–238.

32. Laguerre, G., Allard, M.R., Revoy, F. and Amarger, N., 1994. Rapid identification of rhizobia by restriction fragment length polymorphism analysis of PCR-amplified 16S rRNA genes. *Appl. Environ. Microbiol.*, 60: 56–63.

33. Latour, X., Corberand, T., Laguerre, G., Allard, F. and Lemanceau, P., 1996. The composition of fluorescent *Pseudomonas* population associated with roots is influenced by plant and soil type. *Appl. Environ. Microbiol.*, 62: 2449–2456.

34. Lee, D.H., Zo, Y.G. and Kim, S.J., 1996. Nonradioactive method to study genetic profiles of natural bacterial communities by single stranded conformation polymorphism. *Appl. Environ. Microbiol.*, 63: 3233–3241.

35. Liu, W.T., Marsh, T.L., Cheng, H. and Forney, L.J., 1997. Characterization of microbial diversity by determining terminal restriction fragment length polymorphisms of genes encoding 16S rRNA. *Appl. Environ. Microbiol.*, 63: 4516–4522.

36. Lovell, C.R., Piceno, Y.M., Quattro, J.M. and Bagwell, C.E., 2000. Molecular analysis of diazotroph diversity in the rhizosphere of smooth cordgrass, *Spartina alterniflora*. *Appl. Environ. Microbiol.*, 66: 3814–3822.

37. Massol-Deya, A.A., Odelson, D.A., Hickey, R.F. and Tiedje, J.M., 1995. Bacterial community fingerprinting of amplified 16S and 16S–23S ribosomal DNA gene sequences and restriction endonuclease analysis (ARDRA). In: *Molecular Microbial Ecology Manual*, (Eds.) A.D. Akkermans *et al.* Kluwer Academic Publishers, Dordrecht, The Netherlands, pp: 1–8.

38. Mehta, M.P., Butterfield, D.A. and Baross, J., 2003. Phylogenetic diversity of nitrogenase (nifH) genes in deep-sea and hydrothermal vent environments of the Juan de Fuca Ridge. *Appl. Environ. Microbiol.*, 69: 960–970.

39. MIDI., 1995. Sherlock microbial identification system operating manual: version 5. MIDI, Newark D.E.

40. Miller, H., Henken, J.G. and Van Veen, J.A., 1989. Variations and composition of bacterial populations in the rhizosphere of maize, wheat and grass cultivars. *Can. J. Microbial.*, 35: 656–660.

41. Minerdi, D., Fani, R., Gallo, R., Boarino, A. and Bonfante, P., 2001. Nitrogen fixation genes in an endosymbiotic *Burkholderia* strain. *Appl. Environ. Microbiol.*, 67: 725–732.

42. Muyzer, G.A., de Waal, E.C. and Uitterlinden, A.G., 1993. Profiling of complex microbial populations by denaturing gradient gel electrophoresis analysis of polymerase chain reaction-amplified genes coding for 16S rRNA. *Appl. Environ. Microbiol.*, 59: 695–700.

43. Normander, B. and Prosser, J.I., 2000. Bacterial origin and community composition in the barley phytosphere as a function of habitat and presowing conditions. *Appl. Environ. Microbiol.*, 66: 4372–4377.

44. O'Connell, S.P. and Garland, J.L., 2002. Dissimilar response of microbial communities in Biolog GN and GN2 plates. *Soil Biol. Biochem.*, 34: 413–416.

45. Odee, D.W., Sutherland, J.M., Kimiti, J.M. and Sprent, J.I., 1995. Natural rhizobial populations and nodulation strains of woody legumes growing in diverse Kenyan conditions. *Pl. Soil*, 173: 211–224.

46. Ohkuma, M., Noda, S. and Kudo, T., 1999. Phylogenetic diversity of nitrogen fixation genes in the symbiotic microbial community in the gut of diverse termites. *Appl. Environ. Microbiol.*, 65: 4926–4934.

47. Palmer, K.M. and Young, J.P.W., 2000. Higher diversity of *Rhizobium leguminosarum* biovar *viciae* populations in arable soils than in grass soils. *Appl. Environ. Microbiol.*, 66: 2445–2450.

48. Peters, S., Koschinsky, S., Schwieger, F. and Tebbe, C.C., 2000. Succession of microbial communities during hot composting as detected by PCR-single strand conformation polymorphism-based genetic profiles small-subunit rRNA genes. *Appl. Environ. Microbiol.*, 66: 930–936.

49. Poly, F., Ranjard, L., Nazaret, S., Gourbiere, F. and Monrozier, L.J., 2001. Comparison of nifH gene pools in soils and soil microenvironments with contrasting properties. *Appl. Environ. Microbiol.*, 67: 2255–2262.

50. Postgate, J.R. and Eady, R.R., 1988. The evolution of biological nitrogen fixation. In: *Nitrogen Fixation: Hundred Years After*, (Eds.) H. Bothe, F.J. de Bruijn and W.E. Newton. Stuttgart: Gustav Fischer, pp. 31–40.

51. Ranjard, L., Brothier, E. and Nazaret, S., 2000. Sequencing bands of ribosomal intergenic spacer analysis fingerprints for characterization and microscale distribution of soil bacterium populations responding to mercury spiking. *Appl. Environ. Microbiol.*, 66: 5334–5339.

52. Riemann, L. and Windling, A., 2001. Community dynamics of free–living and particle associated bacterial assemblages during a freshwater phytoplankton bloom. *Microb. Ecol.*, 42: 274–285.

53. Rosenblueth, M. and Martinez-Romero, E., 2004. *Rhizobium etli* maize populations and their competitiveness for root colonization. *Arch. Microbiol.*, 181: 337–344.

54. Ross, I.L., Alami, Y., Harvey, P.R., Achouak, W. and Ryder, M.H., 2000. Genetic diversity and biological control activity of novel species of closely related pseudomonads isolated from wheat field soils in South Australia. *Appl. Environ. Microbiol.*, 66: 1609–1616.

55. Saldana, G, Martinez-Alcantara, V., Vinardell, J.M., Bellogin, R., Ruiz-Sainz, J.E. and Balatti, P.A., 2003. Genetic diversity of fast-growing rhizobia that nodulate soybean (*Glycine max* L. Merr). *Arch. Microbiol.*, 180: 45–52.

56. Scheinert, P., Kruse, R., Ullmann, U., Soller, R. and Krupp, G., 1996. Molecular differentiation of bacteria by PCR amplification of the 16S–23S rRNA spacer. *J. Microbiol. Methods*, 26: 103–117.

57. Schmalenberger, A. and Tebbe, C.C., 2003. Bacterial diversity in maize rhizospheres: conclusions on the use of genetic profiles based on PCR-amplified partial small subunit rRNA genes in ecological studies. *Mol. Ecol.*, 12: 251–262.

58. Schwieger, F. and Tebbe, C.C., 1998. A new approach to utilize PCR-single strand conformation polymorphism for 16S rRNA based microbial community analysis. *Appl. Environ. Microbiol.*, 64: 4870–4876.

59. Schwieger, F. and Tebbe, C.C., 2000. Effect of field inoculation with *Sinorhizobium meliloti* L33 on the composition of bacterial communities in rhizospheres of a target plant (*Medicago sativa*) and a non-target plant (*Chenopodium album*)-linking of 16S rRNA gene-based-single-strand conformation polymorphism community profiles to the diversity of cultivated bacteria. *Appl. Environ. Micorbiol.*, 66: 3556–3565.

60. Scortichini, M., Marchesi, U., Rossi, M.P. and Diprospero, P., 2002. Bacteria associated with hazelnut (*Corylus avellana* L.) decline are of two groups: *Pseudomonas avellanae* and strains resembling *P. syringae* pv. *syringae*. *Appl. Enviorn. Microbiol.*, 68: 476–484.

61. Sessitsch, A., Weilharter, A., Gerzabek, M.H., Kirchmann, H. and Kandeler, E., 2001. Microbial population structures in soil particle size fractions of long-term fertilizer field experiments. *Appl. Environ. Microbiol.*, 67: 4215–4224.

62. Shi, Y., Tyson, G., DeLong, E., 2009. Metatranscriptomics reveals unique microbialsmall RNAs in the ocean, water column. *Nature*, 459: 266–269.

63. Smalla, K., Wachtendorf, U., Heuer, H., Liu, W.T. and Fooney, F., 1998. Analysis of BIOLOG GN substrate utilization patterns by microbial communities. *Appl. Environ. Microbiol.*, 64: 1220–1225.

64. Smalla, K., Wieland, G., Buchner, A., Zock, A. and Parzy, J., 2001. Bulk and rhizosphere soil bacterial communities studied by denaturing gradient gel electrophoresis: Plant-dependent enrichment and seasonal shifts revealed. *Appl. Environ. Microbiol.*, 67: 4742–4751.

65. Smit, E., Leeflang, P., Gommans, S., van den Broek, J., Van, M.S. and Wernars, K., 2001. Diversity and seasonal fluctuations of the dominant members of the bacterial soil community in a wheat field as determined by cultivation and molecular methods. *Appl. Environ. Microbiol.*, 67: 2284–2291.

66. Soderberg, K., Olsson, P.A. and Baath, E., 2002. Structure and activity of the bacterial community in the rhizosphere of different plant species and the effect of arbuscular mycorrhizal colonization. *FEMS Microbiol. Ecol.*, 40: 223–231.

67. Thernon, J. and Cloete, T.E., 2000. Molecular techniques for determining microbial diversity and community structure in natural environments. *Critical Rev. Microbiol.*, 26: 37–57.

68. Thirup, L., Johansen, A. and Windling, A., 2003. Microbial succession in the rhizosphere of live and decomposing barley roots as affected by the antagonistic strain *Pseudomonas fluorescens* DR54-BN14 or the fungicide imazalil. *FEMS Microbiol. Ecol.*, 43: 383–392.

69. Tiedje, J.M., Cho, J.C., Murray, A., Treves, D., Xia, B. and Zhou, J., 2001. Soil teeming with life: new frontiers for soil sciences. In: *Sustainable Management of Soil Organic Matter*, (Eds.) R.M. Rees, B.C. Ball, C.D. Campbell and C.A. Watson. CAB International, pp. 393–412.

70. Tiedje, J.M., Asuming-Brempong, S., Nusslein, K., Marsh, T.L. and Flynn, S.J., 1999. Opening the black box of soil microbial diversity. *Appl. Soil Ecol.*, 13: 109–122.

71. Torsvik, V., Daae, F.L., Sandaa, R.A. and Ovreas, L., 1998. Novel techniques for analyzing microbial diversity in natural ad perturbed environments. *J. Biotechnol.*, 64: 53–62.

72. Torsvik, V., Salte, K., Sorheim, R. and Goksoyr, J., 1996. Total bacterial diversity in soil and sediment communities: A review. *J. Indust. Microbiol.*, 17: 170–178.

73. Ueda, T., Suga, Y., Yahiro, N. and Matsuguchi, T., 1995. Remarkable N_2-fixing bacterial diversity detected in rice roots by molecular evolutionary analysis of nifH gene sequences. *J. Bacteriol.*, 177: 1414–1417.

74. Wander, M.M., Hedrick, D.S., Kaufman, D., Traina, S.J., Stinner, B.R., Kehrmeyer, S.R. and White, D.C., 1995. The functional significance of the microbial biomass in organic and conventionally managed soils. In: *The Significance and Regulation of Soil Biodiversity*, (Eds.) H.P. Collins, G.P. Robertson and M.J. Klug. Kluwer Academic Publishers, The Netherlands, pp. 87–97.

75. Weisburg, W.G., Barns, S.M., Pelletier, D.A. and Lane, D.J., 1991. 16S ribosomal DNA amplification for phylogenetic study. *J. Bacteriol.*, 173: 697–703.

76. White, D.C., Davis, W.M., Nickels, J.S., King, J.D. and Bobbie, R.J. 1979. Determination of sedimentary microbial biomass by extractable lipid phosphate. *Oecologia*, 40: 51–62.

77. Widmer, F., Shaffer, B.T., Porteous, L.A. and Seidler, R.J., 1999. Analysis of nifH gene pool complexity in soil and litter at a Douglas fir forest site in the Oregon Cascade Mountain range. *Appl. Environ. Microbiol.*, 65: 374–380.

78. Wieland, G., Neumann, R. and Backhaus, H., 2001. Variation of microbial communities in soil, rhizosphere and rhizoplane in response to crop species, soil type and crop development. *Appl. Environ. Microbiol.*, 67: 5849–5854.

79. Woese, C.R., 1987. Bacterial evolution. *Microbiol. Rev.*, 51: 221–271.

80. Yeager, C.M., Kornosky, J.L., Housman, D.C., Grote, E.E., Belnap, J. and Kuske, C.R., 2004. Diazotrophic community structure and function in two successional stages of biological soil crusts from the Colorado Plateau and Chihuahuan deserts. *Appl. Environ. Microbiol.*, 70: 973–983.

81. Yohalem, D.S. and Lorbeer, J.W., 1994. Intraspecific metabolic diversity among strains of *Burkholderia cepacia* isolated from decayed onions, soils and the clinical environment. *Antonie van Leeuwenhoek*, 65: 111–131.

82. Zak, J.C., Willig, M.R., Morrehead, D.L. and Wildman, H.G., 1994. Functional diversity of microbial communities: A quantitative approach. *Soil Biol. Biochem.*, 26: 1101–1108.

83. Zehr, J.P. and McReynolds, L.A., 1989. Use of degenerate oligonucleotides for amplification of the nifH gene from the marine cyanobacterium *Trichodesmium thiebautii*. *Appl. Environ. Microbiol.*, 55: 2522–2526.

84. Zehr, J.P., Mellon, M.T and Zani, S., 1998. New nitrogen-fixing micro-organisms detected in oligotrophic oceans by amplification of nitrogenase (nifH) genes. *Appl. Environ. Microbiol.*, 64: 3444–3450.

85. Zelles, L., 1999. Fatty acid patterns of phospholipids and lipopolysaccharides in the characterization of microbial communities in soil: A review. *Biol. Fertil. Soil*, 29: 111–129.

86. Zelles, L., Bai, Q.Y., Beck, T. and Beese, F., 1992. Signature fatty acids in phospholipids and lipopolysacccharides as indicators of microbial biomass and community structure in agricultural soils. *Soil Biol. Biochem.*, 24: 317–323.

Biotechnology: An Overview (2015)   *Pages 53–65*
Editors: Rajan Kumar Gupta, Nasim Akhtar and Deepak Vyas
Published by: DAYA PUBLISHING HOUSE, NEW DELHI

Chapter 5

Bioinformatics Resources for Plant Science

*Raghunath Satpathy**

Department of Biotechnology,
MITS Engineering College, Rayagada – 765 017

ABSTRACT

Genome based study in case of plants is an emerged discipline that analyses the genetic and molecular basis of all biological processes in plants of a particular species. This fundamental understanding allows for an efficient exploitation of plants as a resource for the development of new economically important varieties, with improved nutrient quality and environmental costs. The various important genetic traits considered in case of plants are, biotic (pathogen) and abiotic stress resistance, nutrient quality traits for plant, and reproductive traits determining yield. The molecular basis of study generates huge amount of data which are further annotated by bioinformatics analysis. Bioinformatics is a new field of science which is being utilized in every field of biotechnology. Many specific databases and tools have been developed, that contains enormous amount of molecular level data and the method of annotation. The present analysis is about some of the major bioinformatics resources (databases) used in case of plant sciences, which might be useful for an investigator who is interested in investigating about the plant genes and proteins.

Key words: Bioinformatics resources, Databases, Gene expression, Plant genomics.

Introduction

Bioinformatics is the application of computer technology to the management of biological information. The term management means to acquire, store, organize, archive, analyze and visualize the biological data at molecular level (Altman and Dugan 2003). Bioinformatics is now used by mathematicians, computer scientist,

---

* *Author:* E-mail: rnsatpathy@gmail.com

statisticians, engineers along with life scientists to develop technologies for supporting information management related to genomes with an aim to applying this information in agriculture, medicine etc (Hack and kendell 2005).Bioinformatics provides the opportunity to analyse the data in case of plant genomes for improved crop production. Currently many genome project programs are going on and many are finished (Fleishmann *et al.*, 1995).Due to large amount of data are generated from the genome sequencing projects led to an absolute requirement for computerized databases to store, organize, and index the data and for specialized tools to view and analyze the data. Various tools and databases of bioinformatics are playing significant role in providing the information about the genes thereby predict the function of different genes and factors affecting these genes (Benton 1996). Bioinformatics study in case of plant genome is a great concern, as the life on earth is based on plant as they produce the life-supporting oxygen. They are also essential for our nutrition and health, and they provide an environment for the vast biodiversity on earth. The information generated about the genes by the bioinformatics tools makes the scientists to produce enhanced species of plants which have drought, herbicide, and pesticide resistance, with improved nutrient quality in them (Lee 1998).For centuries, humans have selected plant varieties that best fit their purposes and developed many nutritional rich plants that have many advantages compared to natural, wild plants in terms of quality, quantity and farming practices (Vassilev *et al.*, 2006). The scale and high resolution power of genomics makes possible a broad and detailed genetic understanding of plant performance at multiple levels of aggregation. The complex biological processes that determine pathogen resistance and crop quality are now open for systematic functional analysis. In plant bioinformatics, these analyses are made with the help of special software on huge amounts of data in databases (Roos 2001).

Progress and interest in plant genomics have been accelerating since the time in late 2000 when the genome of *Arabidopsis thaliana* was published .Since the many plant genome sequencing projects have been undertaken that include poplar (*Populus*), grape (*Vitis*), the moss *Physcomitrella*, and several globally crucial crop plants such as corn (Maize) and rice (*Oryza*) (Weigel and Mott 2009).However determining the sequence of a genome is only the first step toward understanding genome organization. Further annotation provides insight to the gene structure, gene expression patterns, disease pathogenesis and a host of other features of both scientic and commercial interests. Computational tools of genomic annotation and comparative genomics must be applied to gain a useful understanding of any genome (Figure 5.1).

Databases and Tools Used in Plant Science

Due to the rise of genomics and bioinformatics based analysis increases the dependency of biology on the result available in the electronic form (Thudi *et al.*, 2012). Most of the useful genomic data like genetic map, physical map, gene and protein sequences are available freely on World Wide Web. The information of non sequence data like mutant phenotype, gene expression pattern among organisms, gene and protein interactions etc.The amount of information about sequence and the functions of genes are to be shared among the scientific community throughout the world. It is also useful to pull these information and discovery in a centralised form

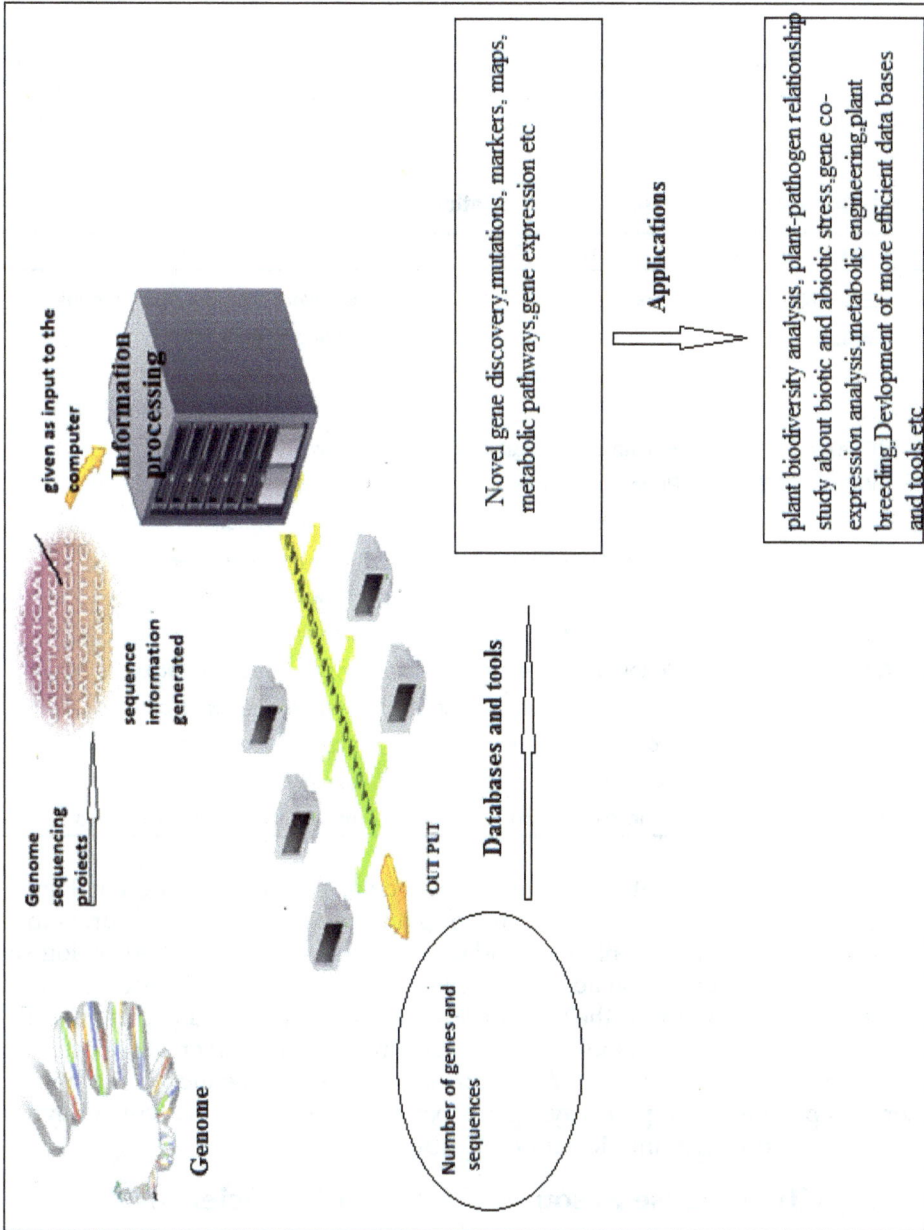

Figure 5.1. Some major applications of bioinformatics in plant science (from genome to application).

and with an universally accessible mode. This approach will lead to generate more accurate study and sharing ideas by the researcher which ultimately encourages them for their research work.

There are different types of tools and databases are used for annotation purpose in case of plant genome. These resources are based on various types of information such as published papers, gene sequences, pathways, and ontologies etc. Some databases are species independent and widely used by researchers for analysis and understanding of cellular process (Table 5.1) and some are specific to the organism or species.

Table 5.1. Some commonly used bioinformatics databases.

Sl.No.	Name of Database	Type of Data	Availability
1	GENBANK	Gene sequences	www.ncbi.nlm.nih.gov/genbank/
2	EMBL	Gene sequences	www.ebi.ac.uk/embl/
3	DDBJ	Gene sequences	www.ddbj.nig.ac.jp/
4	dbEST	Expressed sequence tags	www.ncbi.nlm.nih.gov/dbEST/
5	PRODOM	Protein domain families	www.prodom.prabi.fr/
6	SWISS PROT	Protein sequence	www.expasy.org/
7	GOLD	Genome	www.genomesonline.org/
8	KEGG	Biochemical pathways	www.genome.jp/kegg/
9	METACYC	Biochemical pathways	www.metacyc.org/
10	PDB	3D structure	www.rcsb.org/
11	CATH	Protein structure classification	www.biochem.ucl.ac.uk/bsm/cath/
12	SCOP	Protein structure classification	www.scop.mrclmb.cam.ac.uk/scop/
13	DIP	Protein-protein interaction	http://dip.doembi.ucla.edu/dip/Main.cgi
14	SMD	Gene expression	http://smd.stanford.edu/
15	ArrayExpress	Gene expression	www.ebi.ac.uk/arrayexpress/

Depending on type of data many categories of databases and tools present (Figure 5.2). In this chapter some of the selected major plant specific database resources that contains types of data, their availability and utility have been discussed. As biologists increase their dependencies on advanced resources to conduct their research, they need to be provided with all the data and analysis method by the user-friendly softwares.In this context use of bioinformatics resources provides an great opportunity to generate and integrate more accurate information regarding sequence, mutations, markers, maps, metabolic pathways, gene expression analysis etc, which having potential application in plant biotechnology (Stein 2003).

Some Specific Databse Resources Used in Plant Science

1. NIASGBdb (http://www.gene.affrc.go.jp/databases_en.php)

The National Institute of Agrobiological Sciences (NIAS) is implementing the NIAS Genebank Project for conservation and promotion of agrobiological genetic

```
┌──────────────────────────────────────────────────────────────────┐
│              Major types of Bioinformatics databases and tools     │
└──────────────────────────────────────────────────────────────────┘
        │                          │                         │
   Data base                                             Data
   Resources                  Data analysis             annotation
                                 Tools                    Tools
```

Data base Resources	Data analysis Tools	Data annotation Tools
✓ Gene and protein sequence ✓ Gene ontology ✓ Genome ✓ Proteome ✓ Phenome ✓ Metabolic Pathways ✓ Plant microbe interactions ✓ Plant molecular marker ✓ Transcription factor ✓ Stress and disease resistance gene	✓ Statistical analysis of Gene expression data ✓ Visualisation of DNA microarray data ✓ Gene pattern analysis ✓ Gene class testing ✓ Correlating expression data with visual parameters ✓ Co-expression data analysis	✓ Annotation from publication (literature based) ✓ Annotation from various data bases (knowledge based)

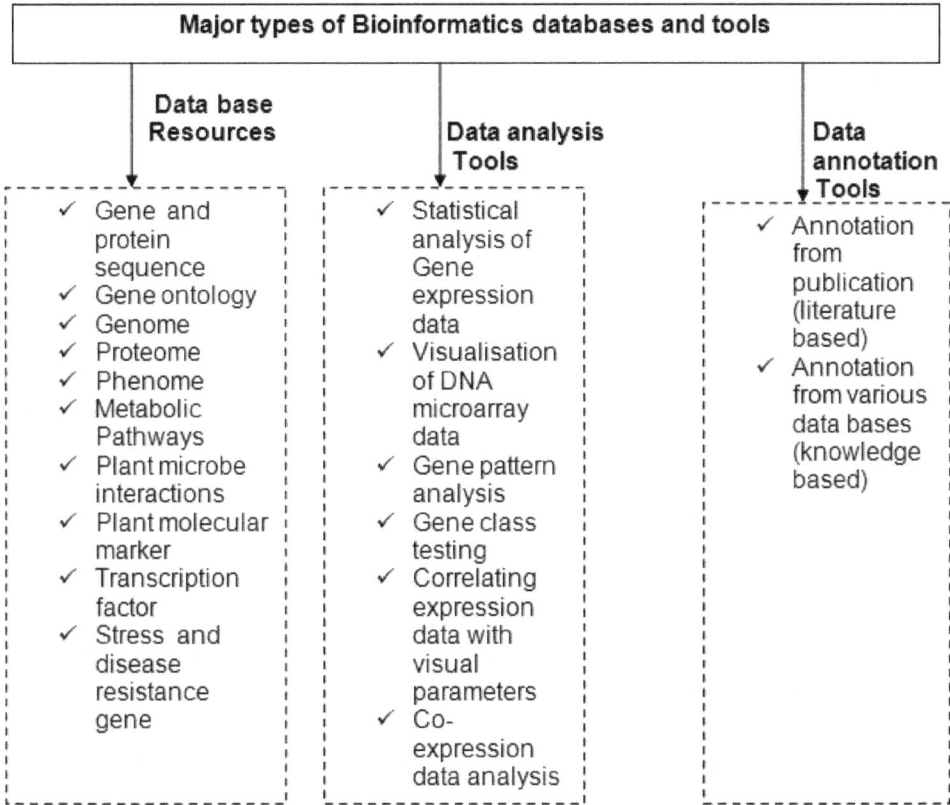

Figure 5.2. Showing different categories of databases and tools use in plant science research.

resources to contribute to the development and utilization of agriculture and agricultural products. The databases consist of a genetic resource database and a plant diseases database, linked by a web retrieval database. The NIAS Rice (*Oryza sativa*) Core Collection contains information on simple sequence repeat (SSR) polymorphisms. SSR marker information for azuki bean (*Vigna angularis*) and black gram (*V. mungo*) and DNA sequence data from some selected Japanese strains of the genus *Fusarium* are also available (Takeya *et al.*, 2010).

2. PHYTOPROT (http://genoplante-info.infobiogen.fr/phytoprot)

The database contains the protein sequences that are available from plants (including *Arabidopsis thaliana*) from SwissProt/TrEMBL have been the subject of a systematic comparison and grouped into clusters of related proteins. In addition, the domains that are common to two or more sequences within a cluster were determined and displayed *at* ProDom. The resulting graphical representations proved to be quite efficient in pinpointing those protein sequences suffering from a probable error in the annotation of their genes. The user can also launch a BLAST search of a query sequence against all the clusters (Mohseni-Zadeh *et. al* 2004).

3. PLANT-PIs (http://bighost.area.ba.cnr.it/PLANT-PIs)

PLANT-PIs is a database developed to facilitate retrieval of information on plant protease inhibitors (PIs) and related genes. For each PI, links to sequence databases are reported together with a summary of the functional properties of the molecule (and its mutants) as deduced from literature. PLANT-PIs currently having information for 351 plant PIs (De Leo *et. al* 2002).

4. PHYTOME (http://www.phytome.org)

Phytome is an online comparative genomics resource that can be applied to functional plant genomics, molecular breeding and evolutionary studies. It contains predicted protein sequences, protein family assignments, multiple sequence alignments, phylogenies and functional annotations for proteins from a large, phylogenetically diverse set of plant taxa. Phytome serves plant gene databases both by identifying the evolutionary relationships among orthologous and paralogous protein sequences from different species and by enabling cross-references between different versions of the same gene curated independently by different database groups (Stefanie *et al.*, 2006).

5. Plant snoRNA database (http://www.scri.sari.ac.uk/plant_snoRNA)

The Plant snoRNA database provides information on small nucleolar RNAs from Arabidopsis and eighteen other plant species. Information includes sequences, expression data, methylation and pseudouridylation target modification sites, initial gene organization (polycistronic, single gene and intronic) and the number of gene variants. The Arabidopsis information is divided into box C/D and box H/ACA snoRNAs, and within each of these groups, by target sites in rRNA, snRNA or unknown. Alignments of orthologous genes and gene variants from different plant species are available for many snoRNA genes (Brown *et al.*, 2003).

6. PlantMarkers (http://markers.btk.fi)

PlantMarkers is a genetic marker database that contains a comprehensive pool of predicted molecular markers Molecular markers are required in a broad spectrum of gene screening approaches, ranging from gene-mapping within traditional 'forward'-genetics approaches through QTL identification studies to genotyping and haplotyping studies. We have adopted contemporary techniques to identify putative single nucleotide polymorphism (SNP), simple sequence repeat (SSR) and conserved orthologue set markers (Stephen, Heiko and Klaus 2005).

7. PlantProm (http://www.softberry.com/)

PlantProm database is a plant promoter database, is an annotated, non-redundant collection of proximal promoter sequences for RNA polymerase II with experimentally determined transcription start site(s), TSS, from various plant species. It provides DNA sequence of the promoter regions, taxonomic/promoter type classification of promoters and Nucleotide Frequency Matrices (NFM) for promoter elements: TATA-box, CCAAT-box and TSS-motif. Analysis of TSS-motifs revealed that their composition is different in dicots and monocots, as well as for TATA and TATA-less promoters. The database serves as learning set in developing plant promoter prediction programs (Shahmuradov *et al.*, 2005).

8. The TIGR Plant Repeat Databases (www.tigr.org/tdb/e2k1/plant.repeats/index.shtml)

To better understand the nature of repetitive sequences in plants the database has been constructed contains repetitive sequences for 12 plant genera. The repetitive sequences within each database have been coded into super-classes, classes and sub-classes based on sequence and structure similarity. To further the utility for comparative studies and to provide a resource for searching for repetitive sequences in other genera within these families. These databases provide a resource for the identification, classification and analysis of repetitive sequences in plants (Shu O and Robin B 2004).

9. PlantTFDB 2.0 (http://planttfdb.cbi.pku.edu.cn)

The plant transcription factor (TF) database to version 2.0 (PlantTFDB 2.0,) contains 53319 putative TFs predicted from 49 plant species. The detailed annotation including general information, domain feature, gene ontology, expression pattern and ortholog groups, as well as cross references to various databases and literature citations for these TFs classified into 58 newly defined families with computational approach and manual inspection (Zhang *et. al* 2010).

10. PlantGDB (http://www.plantgdb.org/)

PlantGDB is a database of molecular sequence data for all plant species with significant sequencing efforts. The database organizes EST sequences into contigs that represent tentative unique genes. Contigs are annotated and, whenever possible, linked to their respective genomic DNA. Genome sequence fragments are assembled similarly. The goal of the PlantGDB web site is to establish the basis for identifying sets of genes common to all plants or specific to particular species by integrating a number of bioinformatics tools that facilitate gene prediction and cross species comparisons (Qunfeng *et al.*, 2004).

11. TropGENE-DB (http://tropgenedb.cirad.fr)

TropGENE-DB is a crop information system created to store genetic, molecular and phenotypic data of the numerous yet poorly documented tropical crop species. The most common data stored in TropGENE-DB are information on genetic resources (agro-morphological data, parentages, and allelic diversity), molecular markers, genetic maps, results of quantitative trait loci analyses, data from physical mapping, sequences, genes, as well as the corresponding references (Ruiz *et al.*, 2004).

12. PMRD: Plant microRNA database (http://bioinformatics.cau.edu.cn/PMRD)

The plant miRNA database (PMRD) integrates available plant miRNA data deposited in public databases, gleaned from the recent literature, and data generated in-house. This database contains sequence information, secondary structure, target genes, expression profiles and a genome browser. In total, there are 8433 miRNAs collected from 121 plant species in PMRD, including model plants and major crops such as *Arabidopsis*, rice, wheat, soybean, maize, sorghum, barley, etc. PMRD can be used as a useful tool for scientists in the miRNA field in order to study the function of miRNAs and their target genes, especially in model plants and major crops (Zhenhai *et al.*, 2010).

13. PathoPlant (http://www.pathoplant.de)

PathoPlant is a database on plant-pathogen interactions and signal transduction reactions. Plants react to pathogen attack by expressing specific proteins directed toward the infecting pathogens. This involves the transcriptional activation of specific gene sets. A user-friendly web interface was created for the submission of gene sets to be analyzed. This results in a table, listing the stimuli that act either inducing or repressing on the respective genes. Up to three stimuli can be combined with the option of induction factor restriction to determine similarly regulated genes (Bülow, Schindler and Hehl 2007).

14. PlantCARE (http://sphinx.rug.ac.be:8080/PlantCARE/)

PlantCARE is a database of plant *cis*-acting regulatory elements, enhancers and repressors. Regulatory elements are represented by positional matrices, consensus sequences and individual sites on particular promoter sequences. Links to the EMBL, TRANSFAC and MEDLINE databases are provided when available. Data about the transcription sites are extracted mainly from the literature, supplemented with an increasing number of *in-silico* predicted data. Apart from a general description for specific transcription factor sites, levels of confidence for the experimental evidence, functional information and the position on the promoter are given as well (Magali *et al.*, 2002).

15. AgBase (www.agbase.msstate.edu)

AgBase contains the data about structural annotation of agriculturally important genomes by experimentally confirming the *in vivo* expression of electronically predicted proteins and by proteogenomic mapping. The Gene Ontology (GO) annotations provide a two tier system of GO annotations for users. The data analysis from the database would provide an online mechanism for agricultural researchers to submit requests for GO annotations (Fiona *et al.*, 2007).

16. GreenPhylDB (http://greenphyl.cirad.fr)

GreenPhylDB is a database which provides comprehensive platform designed to facilitate comparative functional genomics in *Oryza sativa* and *Arabidopsis thaliana* genomes. The main functions of GreenPhylDB are to assign *O. sativa* and *A. thaliana* sequences to gene families using a semi-automatic clustering procedure and to create 'orthologous' groups using a phylogenomic approach. Cluster results are finally processed by phylogenetic analysis to infer orthologs and paralogs that will be particularly helpful to study genome evolution. This interface, accessible on

GreenPhylDB centralizes external references (*e.g.* InterPro, KEGG, Swiss-Prot, PIRSF, and Pubmed) (Conte *et al.*, 2008).

17. AutoSNPdb (http://autosnpdb.qfab.org.au/)

AutoSNPdb is an integrated SNP discovery pipeline, which identifies SNPs from assembled EST sequences. The results are maintained in a custom relational database along with EST source and annotation information. The current database hosts data for the important crops rice, barley and Brassica. Users may rapidly identify polymorphic sequences of interest through BLAST sequence comparison, keyword searches of annotations derived from UniRef90 and GenBank comparisons, GO annotations or in genes corresponding to syntenic regions of reference genomes. In addition, SNPs between specific varieties may be identified for targeted mapping and association studies. SNPs are viewed using a user-friendly graphical interface (Duran et al., 2009).

18. Plant Stress Gene Database (http://ccbb.jnu.ac.in/stressgenes/frontpage.html)

The plant stress gene database contains stress gene information of 259 stress-related genes of 11 species along with all the available information about the individual genes. Stress related ESTs were also found for *Phaseolus vulgaris*. Database also includes ortholog and paralog of proteins which are coded by stress related genes (Ratna, Indira, and Dhananjaya 2011).

19. Gramene (http://www. gramene.org/)

Gramene is a curated, open-source, data resource for genome analysis in the grasses. The information stored in the database is derived from public sources and includes genomes, EST sequencing, protein structure and function analysis, genetic and physical mapping, interpretation of biochemical pathways, Gene Ontologies, gene and QTL localization and descriptions of phenotypic characters and mutations. Extensive information is provided for *Oryza, Zea, Triticum, Hordeum, Avena, Setaria, Pennisetum, Secale, Sorghum, Zizania,* and *Brachypodium* (Jaiswal 2011).

20. TAIR (http://www.arabidopsis.org/)

The Arabidopsis Information Resource (TAIR) maintains a database of genetic and molecular biology data for Arabidopsis thaliana. Data available from TAIR includes the complete genome sequence along with gene structure, gene product information, metabolism, gene expression, DNA and seed stocks, genomemaps, genetic and physicalmarkers, and publication (Eva *et al.*, 2001).

21. PlnTFDB (http://plntfdb.bio.uni-potsdam.de.)

PlnTFDB integrated plant transcription factor database system provides a web interface to access large sets of transcription factors of several plant species, currently encompassing *Arabidopsis thaliana, Populus trichocarpa, Oryza sativa, Chlamydomonas reinhardtii* and *Ostreococcus tauri*. It also provides an accesspoint to its daughter databases of a species-centered representation of transcription factors(OstreoTFDB, ChlamyTFDB, ArabTFDB, PoplarTFDB and RiceTFDB). Information includingprotein sequences, coding regions, genomic sequences, expressed sequence tags (ESTs), domain architecture and scientific literature is provided for each family. Cross-species comparison and identification of regulatory modules and hence TFs is thought to become increasingly important for the rational design of new plant biomass (Diego M *et al.*, 2007).

22. PPDB : The Plant Proteome Database (http://ppdb.tc.cornell.edu)

PPDB is a Plant Proteome database for *Arabidopsis thaliana* and maize (*Zea mays*). Initially PPDB was dedicated to plant plastids, but has now expanded to the whole plant proteome. All protein-encoding gene models in the Arabidopsis, maize and rice, as assembled respectively. Thus **every predicted protein** in both species can be searched for experimental and other information (even if not experimentally identified). The PPDB stores experimental data from in-house proteome and mass spectrometry analysis, curated information about protein function, protein properties and sub-cellular localization (Sun *et al.*, 2008).

23. P3DB: A Plant Protein Phosphorylation Database (http://www.p3db.org/)

P (3) DB provides a resource of protein phosphorylation data from multiple plants. The database contains with a dataset from oilseed rape, including 14,670 non-redundant phosphorylation sites from 6382 substrate proteins, representing the largest collection of plant phosphorylation data to date. Additional protein phosphorylation data are being deposited into this database from large-scale studies of *Arabidopsis thaliana* and soybean.A BLAST search utility has been integrated along with a phosphopeptide BLAST browser to allow users to query the database for phosphopeptides similar to protein sequences of their interest. P (3) DB is a valuable resource for both plant and non-plant biologists in the field of protein phosphorylation (Jianjiong *et al.*, 2009).

24. BarleyBase (www.barleybase.org)

BarleyBase (BB) is an online database for plant microarrays with integrated tools for data visualization and statistical analysis. BB contains the raw and normalized expression data from the Affymetrix genome arrays. BB contains a broad set of query and display options at all data levels, ranging from experiments to individual hybridizations to probe sets down to individual probes. The result for gene and plant ontologies, are also an interconnecting links to physical or genetic map and other genomic data in PlantGDB, Gramene and GrainGenes. It also allows users to perform EST alignments and gene function prediction enhancing cross-species comparison (Lishuang *et al.*, 2005).

25. ATTED (http://atted.jp/)

Atted is a database that contains the known protein-protein-interaction information on co expressed gene network. Gene co-expression and protein-protein-interaction indicate other layer of regulation, and thus complementary information to understand gene function network. External links for 32 Arabidopsis databases are available. Gene co-expression provides the powerful information to identify new gene functionally related (Obayashi and Kinoshita 2010).

Future Directions

Based on the bioinformatics mode of analysis it is possible to handle very large and complex data, which can be further annotated for the applications as below:

1. Comprehensively identification of the structural and functional components encoded in the plant genomes.
2. Elucidation of organisation of genetic networks and protein pathways based on expression data.
3. Detailed understanding of diversity study among species.
4. Identification of genes contribution to disease and stress resistance.
5. Development of genome based approach for prediction of disease susceptibility and molecular taxonomy.

Here in the above sections, an attempt has been taken to highlight some of the major advance resources of bioinformatics in the plant science, in the specific areas like sequence analysis, gene expression, protein, and metabolite analyses, and ontologies etc. A number of unsolved problems exist in bioinformatics today that includes data and database integration, automated knowledge extraction. Bioinformatics currently is an essential part of plant science research and it is expected that every plant biology researcher will incorporate more bioinformatics tools and approaches in their research work to facilitate their molecular level study. In this way integration of basic research with applied research in the plant biotechnology will play an essential role in solving the required problems in our society such as developing renewable energy, reducing world hunger and poverty, and preserving the environment. Bioinformatics has the potential to provide such platform with which all of these types of integration can occur. Moreover the Bioinformatics based resources are essential for efficient research which having significant implications in plant sciences, ultimately lead to betterment of human lives.

References

1. Altman, R.B. and Dugan, J.M., 2003. Defining bioinformatics and structural bioinformatics. *Methods Biochem. Anal.*, 44: 3–14.
2. Brown, J.W. *et al.*, 2003. Plant snoRNA database. *Nucleic Acids Res.*, 3(1): 432–435.
3. Benton, D., 1996. Bioinformatics: Principles and potential of a multidisciplinary tool. *Trends Biotech*, 14(8): 261–272.
4. Conte, M.G., Gaillard, S., Lanau, N., Rouard, M. and Périn, C., 2008. GreenPhylDB: A database for plant comparative genomics. *Nucleic Acids Res.*, 36(Database issue): D991–D998.
5. Bülow, L., Schindler, M. and Hehl, R., 2007. PathoPlant: A platform for microarray expression data to analyze co-regulated genes involved in plant defense responses. *Nucleic Acids Res.*, 35(Database issue): D841–D845.

6. De Leo, F. *et al.*, 2002. PLANT–PIs: A database for plant protease inhibitors and their genes. *Nucleic Acids Research*, 30(1): 347–348.

7. Duran, C. *et al.*, 2009. AutoSNPdb: An annotated single nucleotide polymorphism database for crop plants. *Nucleic Acids Res.*, 37(Database issue): D951–D953.

8. Diego, M. *et al.*, 2007. PlnTFDB: An integrative plant transcription factor database. *BMC Bioinformatics*, 8: 42.

9. Eva, H. *et al.*, 2001. The Arabidopsis Information Resource (TAIR): A comprehensive database and web-based information retrieval, analysis, and visualization system for a model plant. *Nucleic Acids Res.*, 29(1): 102–105.

10. Fiona, M. *et al.*, 2007. AgBase: A unified resource for functional analysis in agriculture. *Nucleic Acids Res.*, 35(Database issue): D599–D603.

11. Fleishmann, R. *et. al.*, 1995. Whole genome random sequencing and assembly of *Haemohilus influenza* Rd. *Science*, 269: 496–512.

12. Hack, C. and Kendell, G., 2005. Bioinformatics: Current practices and future challenges for life science education. *Biochemistry and Molecular Biology Education*, 33: 82–85.

13. Jaiswal, P., 2011. Gramene database: A hub for comparative plant genomics. *Methods Mol Biol.*, 678: 247–275.

14. Jianjiong, G. *et al.*, 2009. P3DB: A plant protein phosphorylation database. *Nucleic Acids Res.*, 37(Database issue): D960–D962.

15. Lee, M., 1998. Genome projects and gene pools: New germplasm for plant breeding? *Proc. Natl. Acad. Sci.*, 95(5): 2001–2004.

16. Lishuang, S. *et al.*, 2005. BarleyBase: An expression profiling database for plant genomics. *Nucleic Acids Res.*, 33(Database issue): D614–D618.

17. Mohseni-Zadeh., S. *et al.*, 2004. PHYTOPROT: A database of clusters of plant proteins, A. *Nucleic Acids Res.*, 32(Database issue): D351–D353.

18. Magali, L. *et al.*, 2002. PlantCARE: A database of plant *cis*–acting regulatory elements and a portal to tools for *in silico* analysis of promoter sequences. *Nucleic Acids Res.*, 30(1): 325–327.

19. Obayashi, T. and Kinoshita, K., 2010. Coexpression landscape in ATTED–II: Usage of gene list and gene network for various types of pathways. *J. Plant Res.*, 123(3): 311–319.

20. Qunfeng, D. *et al.*, 2004. PlantGDB: Plant genome database and analysis tools. *Nucleic Acids Research*, 32(1): D354–D359.

21. Ratna, P., Indira, G. and Dhananjaya, P., 2011. Plant stress gene database: A collection of plant genes responding to stress condition. *ARPN Journal of Science and Technology*, 1(1): 28–31.

22. Roos, D.S., 2001. Bioinformatics: Trying to swim in a sea of data. *Science*, 291: 1260–1261.

23. Ruiz, M. *et al.*, 2004. TropGENE–DB: A multi-tropical crop information system. *Nucleic Acids Res.*, 32(Database issue): D364–D367.

24. Shahmuradov, I.A., Gammerman, A.J., Hancock, J.M., Bramley, P.M. and Solovyev, V.V., 2003. PlantProm: A database of plant promoter sequences. *Nucleic Acids Res.*, 31(1): 114–117.

25. Stein, L.D., 2003. Integrating biological databases. *Nat. Rev. Genet.*, 4(5): 337–345.

26. Shu, O. and Robin, B., 2004. The TIGR Plant Repeat Databases: A collective resource for the identification of repetitive sequences in plants. *Nucleic Acids Res.*, 32(Database issue): D360–D363.

27. Stefanie, H. *et al.*, 2006. Phytome: A platform for plant comparative genomics. *Nucleic Acids Res.*, 34(Database issue): D724–D730.

28. Stephen, R., Heiko, S. and Klaus, M., 2005. PlantMarkers: A database of predicted molecular markers from plants. *Nucleic Acids Res.*, 33(Database Issue): D628–D632.

29. Sun, Q. *et al.*, 2008. PPDB: The Plant Proteomics Database at Cornell. *Nucleic Acids Res.*, 37(Database issue): D969–D974.

30. Takeya, M. *et al.*, 2010. NIASGBdb: NIAS Genebank databases for genetic resources and plant disease information. *Nucleic Acids Res.*, 39(Database issue): D1108–D1113.

31. Thudi, M., Li, Y., Jackson, S.A., May, G.D. and Varshney, R.K., 2012. Current state-of-art of sequencing technologies for plant genomics research. *Brief Funct Genomics*, 11(1): 3–11.

32. Vassilev, D. *et al.*, 2006. Application of bioinformatics in fruit plant breeding. *Journal of Fruit and Ornamental Plant Research*, 14(1) : 145–162.

33. Weigel, D. and Mott, R., 2009. The 1001 genomes project for *Arabidopsis thaliana*. *Genome Biol.*, 10(5): 107.

34. Zhang, H. *et al.*, 2010. PlantTFDB 2.0: Update and improvement of the comprehensive plant transcription factor database. *Nucleic Acids Research*, 29(1): 111–113.

35. Zhenhai, Z. *et al.*, 2010. PMRD: Plant microRNA database. *Nucleic Acids Res.*, 38(Database issue): D806–D813.

Biotechnology: An Overview (2015)   Pages 67–78
Editors: Rajan Kumar Gupta, Nasim Akhtar and Deepak Vyas
Published by: DAYA PUBLISHING HOUSE, NEW DELHI

Chapter 6

Nanotechnology and its Applications

Akhilesh Kumar*

Department of Physics,
Govt. P.G. (Autonomous) College, Rishikesh – 249 201,
Dehradun, Uttarakhand

Introduction

A nanoparticle (or nanopowder or nanocluster or nanocrystal) is a small particle with at least one dimension less than 100 nm. Nanoparticles research is currently an area of intense scientific research, due to a wide variety of potential applications in biomedical, optical, and electronic fields. Nanocrystals of various metals have been shown to be 100 per cent, 200 per cent and even as much as 300 per cent harder than the same materials in bulk form [1]. Because wear resistance often is dictated by the hardness of a metal, parts made from nanocrystals might last significantly longer than conventional parts. One of the characteristic properties of all nanoparticles has been used from the outset in the manufacture of automotive catalytic converters. The surface area of the particles increases dramatically as the particle size decreases and the weight remains the same. A variety of chemical reactions take place on the surface of the catalyst, and the larger the surface area, the more active the catalyst [2, 3]. Nanoscale catalysts thus open the way for numerous process innovations to make many chemical processes more efficient and resource-saving – in other words more competitive. These are just a few of the many ways in which nanotechnology is working itself into our everyday lives. At present, there are no nanobots, no molecular-scale machines, and no assemblers - these are still in the basic research stages, and may not be seen for decades (although many would argue that a concerted effort would bring them to fruition in just a few years). Nowadays, it is established fact that when one of the dimensions of material is less than 100 nm, it's mechanical, thermal,

---

* *Author:* E-mail: singhakhilesh26@rediffmail.com

optical, magnetic and other physical properties change at some size characteristic to the material under investigation[4,5]. Incidentally, the NANO-prefix, became a host of buzzwords, is being used in Industry and Medicine field, *e.g.*, Nanocar, Nanomedicine, Nanotex, Nanocells, Nanophosphors in HDTVs etc. Scientifically, the term Nanoscience and Nanotechnology combines Chemistry (especially Catalytic Chemistry, Synthetic Chemistry, Sol-Gel Chemistry), with Material Science, Engineering and Physics.

The last two decades saw great interest in the physical, optical and transport properties of nanometer-sized semiconductor particles or quantum dots. The optical and electrical properties, crystallinity, melting point and phase transition temperature of nanocrystallites are significantly different from the corresponding bulk and depend on crystallitesize, due to quantum confinement effect. When the dimension of the semiconductor quantum dots approaches the Bohr exciton radius, there will be large changes in their properties; this effect changes the surface to volume ratio and it also shifts electronic energy levels towards higher energy, leading to an increase in the band gap. In case of semiconductor, the artificially made semiconductor structure shows a surprising variety of new interesting properties, which are completely different from the solid state bulk material after reduce its dimension. These size effects can be observed, when the average size of the crystalline grains does not exceed 100 nm and evident most vividly for grain sizes smaller than 10 nm. Polycrystalline materials with an average grain size below 100 nm are called nanocrystalline materials. These materials represent a special state of matter, which consists of matter ensembles of extremely small particles with size of few nanometers.

The properties and preparation of different nanoparticles have been studied. As the semiconductor particles exhibit size dependent properties like scaling of the energy gap and corresponding change in the optical properties, they are considered as the front runners in the technologically important materials [6]. The concept of all optical devices has attracted attention because of its potential for extreme speed and parallel processing capabilities. Basic requirements placed on material for such a device are very rapid switching speed and very high amount of photostability. Such a device involves nonlinear optical properties and gets enhanced with decreasing particle size making them important for application in area such as nonlinear optical devices.

In nanomaterials with grains sizes ranging from 100 down to 10nm, the interfaces contain 10 per cent to 50 per cent of all atoms of the nanocrystalline solid [7,8]. The grains themselves have various atomic defects *e.g.* vacancies and dislocations, whose no. and distribution differ considerably from those in big grains, which are5 to 100 micrometer (µm) in size or larger. Because of these features of the structure of the nanoparticles, the properties of such a material differ from those of ordinary bulk polycrystalline. Due to this reason at present decreasing the grains size is considered as an effective method of changing the properties of solids. In semiconductors, chemical reactivity, band gap energy, optical properties, electrical properties etc. alter considerably with variation in the size of the nanoparticles.

In field of Nanotechnology, it is well known fact that if *any one dimension of a material is less than 100 nm, it's mechanical, thermal, structural, optical, electrical, magnetic and other physical properties change at its some size characteristic to the material under investigation.* For example *bulk CdS semiconductor normal color is red, but in nano CdS semiconductor, various size nanomaterials emit various colors i.e. by changing size of materials, we can tune the energy band gap. So, it is possible to achieve variant properties through single material by altering only their size.* Scientifically, the term Nanoscience and Nanotechnology combines chemistry (especially catalytic chemistry, synthetic chemistry, Sol-Gel chemistry) with material science, engineering and physics.

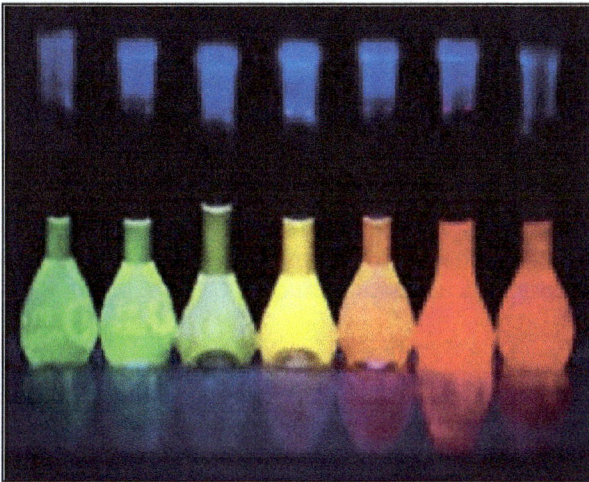

Figure 6.1. CdS nanoparticles.

10nm: Red Color; 6nm: Orange; 3nm: Yellow; 2nm: White

Color of Bulk Silver: Bright White; Color of Nano Silver: Golden *i.e.* bulk material has one color, where as nanoparticle of one material of different size have different color.

Nanotechnology Applications

With nanotechnology, a large set of materials and improved products rely on a change in the physical properties when the feature sizes are shrunk. Nanoparticles, for example, take advantage of their dramatically increased surface area to volume ratio. Their optical properties, *e.g.* fluorescence, become a function of the particle diameter. When brought into a bulk material, nanoparticles can strongly influence the mechanical properties of the material, like stiffness or elasticity. For example, traditional polymers can be reinforced by nanoparticles resulting in novel materials which can be used as light weight replacements for metals. Therefore, an increasing societal benefit of such nanoparticles can be expected. Such nanotechnologically enhanced materials will enable a weight reduction accompanied by an increase in stability and improved functionality. Practical nanotechnology is essentially the increasing ability to manipulate (with precision) matter on previously impossible scales, presenting possibilities which many could never have imagined - it therefore seems unsurprising that few areas of human technology are exempt from the benefits which nanotechnology could potentially bring.

1.Medicine

A. Nanomedicine

The biological and medical research communities have exploited the unique properties of nanomaterials for various applications (*e.g.*, contrast agents for cell imaging and therapeutics for treating cancer). Terms such as *biomedical nanotechnology*, *nanobiotechnology*, and *nanomedicine* are used to describe this hybrid field. Functionalities can be added to nanomaterials by interfacing them with biological molecules or structures. The size of nanomaterials is similar to that of most biological molecules and structures; therefore, nanomaterials can be useful for both in vivo and in vitro biomedical research and applications.

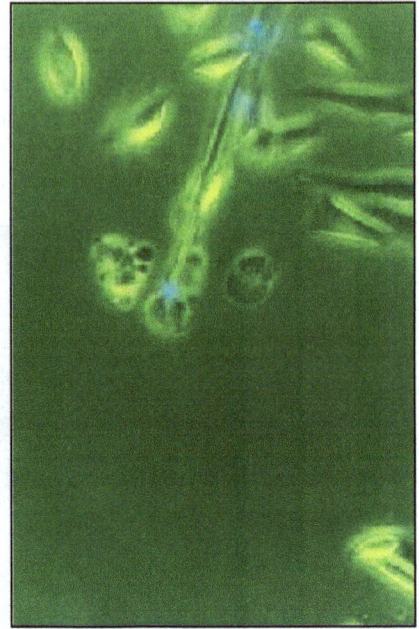

B. Diagnostics

Nanotechnology-on-a-chip is one more dimension of lab-on-a-chip technology. Magnetic nanoparticles, bound to a suitable antibody, are used to label specific molecules, structures or micro-organisms. Gold nanoparticles tagged with short segments of DNA can be used for detection of genetic sequence in a sample. Multicolor optical coding for biological assays has been achieved by embedding different-sized quantum dots into polymeric microbeads.

C. Drug Delivery

Nanotechnology has been a boon for the medical field by delivering drugs to specific cells using nanoparticles. The overall drug consumption and side-effects can be lowered significantly by depositing the active agent in the morbid region only and in no higher dose than needed. This highly selective approach reduces costs and human suffering. An example can be found in dendrimers and nanoporous materials. Another example is to use block co-polymers, which form micelles for drug encapsulation[9]. They could hold small drug molecules transporting them to the desired location. Another vision is based on small electromechanical systems; nanoelectromechanical systems are being investigated for the active release of drugs. Some potentially important applications include cancer treatment with iron nanoparticles or gold shells. A targeted or personalized medicine reduces the drug consumption and treatment expenses resulting in an overall societal benefit by reducing the costs to the public health system. Buckyballs can "interrupt" the allergy/ immune response by preventing mast cells (which cause allergic response) from releasing histamine into the blood and tissues, by binding to free radicals "dramatically better than any anti-oxidant currently available, such as vitamin E[10].

D. Tissue Engineering

Nanotechnology can help reproduce or repair damaged tissue. "Tissue engineering" makes use of artificially stimulated cell proliferation by using suitable nanomaterial-based scaffolds and growth factors. For example, bones can be regrown on carbon nanotube scaffolds. Tissue engineering might replace today's conventional treatments like organ transplants or artificial implants. Advanced forms of tissue engineering may lead to life extension.

Environment

Filtration

A strong influence of photochemistry on waste-water treatment, air purification and energy storage devices is to be expected. Mechanical or chemical methods can be used for effective filtration techniques. Nanoporous membranes are suitable for a mechanical filtration with extremely small pores smaller than 10 nm ("nanofiltration") and may be composed of nanotubes. Nanofiltration is mainly used for the removal of ions or the separation of different fluids. On a larger scale, the membrane filtration technique is named ultrafiltration, which works down to between 10 and 100 nm. One important field of application for ultrafiltration is medical purposes as can be found in renal dialysis. Magnetic nanoparticles offer an effective and reliable method to remove heavy metal contaminants from waste water by making use of magnetic separation techniques. Using nanoscale particles increases the efficiency to absorb the contaminants and is comparatively inexpensive compared to traditional precipitation and filtration methods.

Some water-treatment devices incorporating nanotechnology are already on the market, with more in development. Low-cost nanostructured separation membranes methods have been shown to be effective in producing potable water in a recent study[11].

Energy

The most advanced nanotechnology projects related to energy are: storage, conversion, manufacturing improvements by reducing materials and process rates, energy saving (by better thermal insulation for example), and enhanced renewable energy sources.

A. Reduction of Energy Consumption

A reduction of energy consumption can be reached by better insulation systems, by the use of more efficient lighting or combustion systems, and by use of lighter and stronger materials in the transportation sector. Currently used light bulbs only convert approximately 5 per cent of the electrical energy into light. Nanotechnological approaches like light-emitting diodes (LEDs) or quantum caged atoms (QCAs) could lead to a strong reduction of energy consumption for illumination.

B. Increasing the Efficiency of Energy Production

Today's best solar cells have layers of several different semiconductors stacked together to absorb light at different energies but they still only manage to use 40 per cent of the Sun's energy. Commercially available solar cells have much lower efficiencies (10-15 per cent). Nanotechnology could help increase the efficiency of light conversion by using nanostructures with a continuum of bandgaps.

The degree of efficiency of the internal combustion engine is about 30-40 per cent at present. Nanotechnology could improve combustion by designing specific catalysts with maximized surface area. In 2005, scientists at the University of Toronto developed a spray-on nanoparticle substance that, when applied to a surface, instantly transforms it into a solar collector[9].

C. Nuclear Accident Cleanup and Waste Storage

Nanomaterials deployed by swarm robotics may be helpful for decontaminating a site of a nuclear accident which poses hazards to humans because of high levels of radiation and radioactive particles. Hot nuclear compounds such as Corium or melting fuel rods may be contained in "bubbles" made from nanomaterials that are designed to isolate the harmful effects of nuclear activity occurring inside of them from the outside environment that organisms inhabit.

Information and Communication

Current high-technology production processes are based on traditional top down strategies, where nanotechnology has already been introduced silently. The critical length scale of integrated circuits is already at the nanoscale (50 nm and below) regarding the gate length of transistors in CPUs or DRAM devices.

Memory Storage

Electronic memory designs in the past have largely based on the formation of transistors. Recently, Nantero has developed a carbon nanotube based crossbar memory called Nano-RAM and Hewlett-Packard which has proposed the use of memristor material as a future replacement of Flash memory.

Novel Semiconductor Devices

A. Novel Optoelectronic Devices

In the modern communication technology traditional analog electrical devices are increasingly replaced by optical or optoelectronic devices due to their enormous bandwidth and capacity, respectively. Two promising examples are photonic crystals and quantum dots. Photonic crystals are materials with a periodic variation in the refractive index with a lattice constant that is half the wavelength of the light used. They offer a selectable band gap for the propagation of a certain wavelength, thus they resemble a semiconductor, but for light or photons instead of electrons. Quantum dots are nanoscaled objects, which can be used, among many other things, for the construction of lasers. The advantage of a quantum dot laser over the traditional semiconductor laser is that their emitted wavelength depends on the diameter of the dot. Quantum dot lasers are cheaper and offer a higher beam quality than conventional laser diodes.

B. Displays

The production of displays with low energy consumption could be accomplished using carbon nanotubes (CNT). Carbon nanotubes are electrically conductive and due to their small diameter of several nanometers, they can be used as field emitters with extremely high efficiency for field emission displays (FED). The principle of operation resembles that of the cathode ray tube, but on a much smaller length scale.

C. Quantum Computers

Entirely new approaches for computing exploit the laws of quantum mechanics for novel quantum computers, which enable the use of fast quantum algorithms. The Quantum computer has quantum bit memory space termed "Qubit" for several computations at the same time. This facility may improve the performance of the older systems.

Heavy Industry

A. Aerospace

Lighter and stronger materials will be of immense use to aircraft manufacturers, leading to increased performance. Spacecraft will also benefit, where weight is a major factor. Nanotechnology would help to reduce the size of equipment and thereby decrease fuel-consumption required to get it airborne.

B. Catalysis

Chemical catalysis benefits especially from nanoparticles, due to the extremely large surface to volume ratio. The application potential of nanoparticles in catalysis ranges from fuel cell to catalytic converters and photocatalytic devices. Catalysis is also important for the production of chemicals.

C. Construction

Nanotechnology has the potential to make construction faster, cheaper, safer, and more varied. Automation of nanotechnology construction can allow for the creation of structures from advanced homes to massive skyscrapers much more quickly and at much lower cost. In the near future Nanotechnology can be used to sense cracks in foundations of architecture and can send nanobots to repair them[9].

Nanotechnology and Constructions

Nanotechnology is one of the most active research areas that encompass a number of disciplines Such as electronics, bio-mechanics and coatings including civil engineering and construction materials[10].

The use of nanotechnology in construction involves the development of new concept and understanding of the hydration of cement particles and the use of nano-size ingredients such as alumina and silica and other nanoparticles. The manufactures also investigating the methods of manufacturing of nano-cement. If cement with nano-size particles can be manufactured and processed, it will open up a large number of opportunities in the fields of ceramics, high strength composites and electronic applications. Since at the nanoscale the properties of the material are different from that of their bulk counter parts[4,5]. When materials becomes nano-sized, the proportion of atoms on the surface increases relative to those inside and this leads to novel properties. Some applications of nanotechnology in construction are described [11].

A. Nanoparticles and Steel

Steel has been widely available material and has a major role in the construction industry. The use of nanotechnology in steel helps to improve the properties of steel. The fatigue, which led to the structural failure of steel due to cyclic loading, such as in bridges or towers.The current steel designs are based on the reduction in the allowable stress, service life or regular inspection regime. This has a significant impact on the life-cycle costs of structures and limits the effective use of resources.The Stress risers are responsible for initiating cracks from which fatigue failure results .The addition of copper nanoparticles reduces the surface un-evenness of steel which then limits the number of stress risers and hence fatigue cracking. Advancements in this technology using nanoparticles would lead to increased safety, less need for regular inspection regime and more efficient materials free from fatigue issues for construction[11].

The nano-size steel produce stronger steel cables which can be in bridge construction. Also these stronger cable material would reduce the costs and period of construction, especially in suspension bridges as the cables are run from end to end of the span. This would require high strength joints which leads to the need for high strength bolts. The capacity of high strength bolts is obtained through quenching and tempering. The microstructures of such products consist of tempered martensite. When the tensile strength of tempered martensite steel exceeds 1,200 MPa even a very small amount of hydrogen embrittles the grain boundaries and the steel material may fail during use. This phenomenon, which is known as delayed fracture, which hindered the strengthening of steel bolts and their highest strength is limited to only around 1,000 to 1,200 MPa[11].

B. Nanoparticles in Glass

Glass is also an important material in construction. Research is being carried out on the application of nanotechnology to glass. Titanium dioxide (TiO_2) nanoparticles are used to coat glazing, since it has sterilizing and anti-fouling properties. The particles catalyze powerful reactions which break down organic pollutants, volatile organic compounds and bacterial membranes. The TiO_2 is hydrophilic (attraction to water) which can attract rain drops which then wash off the dirt particles. Thus the introduction of nanotechnology in the Glass industry, incorporates the self cleaning property of glass [11].

C. Nanoparticles in Coatings

Coatings is an important area in construction, coatings are extensively use to paint the walls, doors, and windows. Coatings should provide a protective layer which is bound to the base material to produce a surface of the desired protective or functional properties. The coatings should have self healing capabilities through a process of "self-assembly." Nanotechnology is being applied to paints to obtained the coatings having self healing capabilities and corrosion protection under insulation. Since these coatings are hydrophobic and repels water from the metal pipe and can also protect metal from salt water attack[11].

D. Nanoparticles in Fire Protection and Detection

Fire resistance of steel structures is often provided by a coating produced by a spray-on-cementitiousprocess. The nano-cement has the potential to create a new paradigm in this area of application because the resulting material can be used as a tough, durable, high temperature coating. It provides a good method of increasing fire resistance and this is a cheaper option than conventional insulation.

E. Risks of Using Nanoparticles in Construction

If the nanosensors and nanomaterials becomes a everyday part of the buildings to make them intelligent, the consequences of these materials on human beings are.

1. Effect of nanoparticles on health and environment: Nanoparticles may also enter the body if building water supplies are filtered through commercially available nanofilters. Airborne and waterborne nanoparticles enter from building ventilation and wastewater systems [11].
2. Effect of nanoparticles on societal issues: As sensors become more common place, a loss of privacy may result from users interacting with increasingly intelligent building components.The technology at one side has the advantages of new building material. The otherside it has the fear of risk arises from these materials. However, the overall performance of nanomaterials to date, is that valuable opportunities to improve building performance, user health and environmental quality[11].

F. Vehicle Manufacturers

Much like aerospace, lighter and stronger materials will be useful for creating vehicles that are both faster and safer. Combustion engines will also benefit from parts that are more hard-wearing and more heat-resistant.

Consumer Goods

Nanotechnology is already impacting the field of consumer goods, providing products with novel functions ranging from easy-to-clean to scratch-resistant. Modern textiles are wrinkle-resistant and stain-repellent; in the mid-term clothes will become "smart", through embedded "wearable electronics". Already in use are different nanoparticle improved products. Especially in the field of cosmetics, such novel products have a promising potential.

A. Foods

Complex set of engineering and scientific challenges in the food and bioprocessing industry for manufacturing high quality and safe food through efficient and sustainable means can be solved through nanotechnology. Bacteria identification and food quality monitoring using biosensors; intelligent, active, and smart food packaging systems; nanoencapsulation of bioactive food compounds are few examples of emerging applications of nanotechnology for the food industry [10]. Nanotechnology can be applied in the production, processing, safety and packaging of food. A nanocomposite coating process could improve food packaging by placing anti-microbial agents directly on the surface of the coated film. Nanocomposites could increase or decrease gas permeability of different fillers as is needed for different products. They can also improve the mechanical and heat-resistance properties and lower the oxygen transmission rate. Research is being performed to apply nanotechnology to the detection of chemical and biological substances for sensanges in foods.

B. Nano-Foods

New foods are among the nanotechnology-created consumer products coming into the market at the rate of 3 to 4 per week, according to the Project on Emerging Nanotechnologies (PEN), based on an inventory it has drawn up of 609 known or claimed nano-products.

According to company information posted on PEN's Web site, the canola oil, by Shemen Industries of Israel, contains an additive called "nanodrops" designed to carry vitamins, minerals and phytochemicals through the digestive system and urea.

The shake, according to U.S. manufacturer RBC Life Sciences Inc., uses cocoa infused "NanoClusters" to enhance the taste and health benefits of cocoa without the need for extra sugar[11].

C. Household

The most prominent application of nanotechnology in the household is self-cleaning or "easy-to-clean" surfaces on ceramics or glasses. Nano ceramic particles have improved the smoothness and heat resistance of common household equipment such as the flat iron.

D. Optics

The first sunglasses using protective and anti-reflective ultrathin polymer coatings are on the market. For optics, nanotechnology also offers scratch resistant surface coatings based on nanocomposites. Nano-optics could allow for an increase in precision of pupil repair and other types of laser eye surgery.

E. Textiles

The use of engineered nanofibers already makes clothes water- and stain-repellent or wrinkle-free. Textiles with a nanotechnological finish can be washed less frequently and at lower temperatures. Nanotechnology has been used to integrate tiny carbon particles membrane and guarantee full-surface protection from electrostatic charges for the wearer. Many other applications have been developed by research institutions such as the Textiles Nanotechnology Laboratory at Cornell University, and the UK's Dstl and its spin out company P2i.

F. Cosmetics

One field of application is in sunscreens. The traditional chemical UV protection approach suffers from its poor long-term stability. A sunscreen based on mineral nanoparticles such as titanium dioxide offer several advantages. Titanium oxide nanoparticles have a comparable UV protection property as the bulk material, but lose the cosmetically undesirable whitening as the particle size is decreased.

Agriculture

Applications of nanotechnology have the potential to change the entire agriculture sector and food industry chain from production to conservation, processing, packaging, transportation, and even waste treatment. NanoScience concepts and nanotechnology applications have the potential to redesign the production cycle, restructure the processing and conservation processes and redefine the food habits of the people.

Major challenges related to agriculture like low productivity in cultivable areas, large uncultivable areas, shrinkage of cultivable lands, wastage of inputs like water, fertilizers, pesticides, wastage of products and of course Food security for growing numbers can be addressed through various applications of nanotechnology.

Sports

Nanotechnology may also play a role in sports such as soccer, football[13], and baseball[14]. Materials for new athletic shoes may be made in order to make the shoe lighter and the athlete faster. Baseball bats already on the market are made with carbon nanotubes which reinforce the resin, which is said to improve its performance by making it lighter. Other items such as sport towels, yoga mats, exercise mats are on the market and used by players in the National Football League[15], which use antimicrobial nanotechnology to prevent parasuram from illnesses caused by bacteria such as Methicillin-resistant Staphylococcus aureus (commonly known as MRSA)[12].

References

1. There is very little information on Taniguchi, except cited references to two of his works : (i) Nano technology: Integrated processing systems for Ultra-precission and Ultra-fine products, and (ii) On the basic concept of Nanotechnolgy.

2. Drexler, K.E., 1991, MIT, USA, *Ph.D. Thesis* entitled 'Nanosystems: Molecular Machinery, Manufacturing and Computation.

3. Drexler, K.E., 1992. *Nanosystems : Molecular Machinaery, Manufacturing and Computation*, John Wiley.

4. Tripathi, R., Kumar, A. and Sinha, T.P., 2009. Dielectric properties of CdS nanoparticles synthesized by soft chemical route. *Pramana Journal of Physics*, 72(6): 969–978.

5. Tripathi, Ramna, Kumar, Akhilesh, Bharti, Chandrahas and Sinha, T.P., 2010. *Journal of Current Applied Physics*, Science Direct (Elsevier) (An International Journal), 10(2): 676–681.

6. Cui, Hontao, *et al.*, 2009. Strategies of large scale synthesis of monodisperse nanoparticles. *Recent Patents on Nanotechnology*, 3: 32–41.

7. Tang, Fangquiong, *et al.*, 2006. Novel and facile method for the preparation of *Monodispersed titania* hollow spheres. *Langmuir*, 22(8): 3858–3863.

8. Dave, Shivang R., *et al.*, 2009. Overview, monodisperse magnetic nanoparticles for biodetection, imaging and drug delivery: A versatile and evolving technology, John Wiley and Sons, Inc. *WIRES NanomedNanobiotechnol.*, p. 583–609.

9. University of Waterloo, Nanotechnology in Targeted Cancer Therapy, http: // www.youtube.com/watch?v=RBjWwlnq3cA 15 January 2010.

10. Abraham, SathyaAchia (20). "Researchers Develop Buckyballs to Fight Allergy". Virginia Commonwealth University Communications and Public Relations.http: //www.news.vcu.edu/news/Researchers_Develop_ Buckyballs_to_ Fight_Allergy. Retrieved 4 November 2010.

11. Mann, Surinder (31 October 2006). "Nanotechnology and Construction". Nanoforum.org European Nanotechnology Gateway. http: // www.nanoforum.org/dateien/temp/Nanotech per cent 20and per cent 20Construction per cent 20Nanoforum per cent 20report.pdf?08112010050156. Retrieved 2 January 2012.

12. Waldner, Jean-Baptiste, 2007. *Nanocomputers and Swarm Intelligence*. London: ISTE. p. 26.

13. "Antimicrobial Nanotechnology Used by NFL Teams and Promoted to Professional Football Athletic Trainers". Azonano. 2007–06–27. http: // www.azonano.com/news.asp?newsID=4363. Retrieved 2009–11–06.

14. "Easton Integrates Nanotechnology into Baseball Bats". Nanopedia. 2006–06–05. http: //nanopedia.case.edu/NWPrint.php?page=nw.emw15.008. Retrieved 2009–11–06.

15. "Nanocomposite Cushions Make Lighter Athletic Shoes". AllBusiness. http: // www.allbusiness.com/company–activities–management/product–management–product/7794245–1.html08. Retrieved 2009–11–02.

Biotechnology: An Overview (2015)   *Pages 79–88*
Editors: Rajan Kumar Gupta, Nasim Akhtar and Deepak Vyas
Published by: DAYA PUBLISHING HOUSE, NEW DELHI

Chapter 7

Genetically Modified Foods: Harmful or Helpful?

S.K. Katiyar*

*Laboratory of Ecotechnology and Biodiversity Conservation,
Department of Botany, Government P.G. College, Lansdowne,
Pauri Garhwal – 246 139, Uttarakhand*

Genetically Modified Foods (GM foods) have made a big splash in the news lately. European environmental organizations and public interest groups have been actively protesting against GM foods for months, and recent controversial studies about the effects of genetically-modified corn pollen on monarch butterfly caterpillars[1,][2] have brought the issue of genetic engineering to the forefront of the public consciousness in the India. In response to the up swelling of public concern, the Indian Food and Drug Administration (FDA) held three open meetings in New Delhi, to solicit public opinions and begin the process of establishing a new regulatory procedure for government approval of GM foods[3].

Will GM Food Reduce Hunger in Developing Countries Like India?

If hunger could be addressed by technology, green revolution would have done it long ago. The fact is that hunger has grown in India in absolute terms - some 320 million people go to bed hungry every night. Two years back, India had a record food grain surplus of 65 million tones. If 65 million tones surplus could not feed the 320 million hungry, how will GM food remove hunger? In reality, GM food diverts precious financial resources to an irrelevant research, comes with stronger intellectual property rights, and is aimed at strengthening corporate control over agriculture.

---

* *Corresponding Author:* E-mail: drsunilkumarkatiyar@gmail.com

But What About Malnutrition? Crops like Golden Rice can Help Remove Blindness

This again is the result of misplaced thinking. There are 12 million people in India who suffer from Vitamin A deficiency. These people primarily live in food deficit areas or are marginalized. These are people who cannot buy their normal requirement of food, including rice. If they were adequately fed, there would be no malnutrition. If the poor in Kalahandi, for instance, can't buy rice that lies rotting in front of their eyes, how will they buy golden rice?

Then Why is the Indian Government Experimenting with GM Crops and Foods?

For two reasons: First, India is under tremendous pressure from the biotechnology industry to allow GM crops. These companies have the financial resources to mobilise scientific opinion as well as political support. Second, agricultural scientists are using biotechnology as a Trojan horse. With nothing to show by way of scientific breakthrough in the past three decades, GM research will ensure livelihood security for the scientists.

What GM Crops and Food Items is India Experimenting with?

Besides cotton, genetic engineering experiments are being conducted on maize, mustard, sugarcane, sorghum, pigeon pea, chickpea, rice, tomato, brinjal, potato, banana, papaya, cauliflower, oilseeds, castor, soybean and medicinal plants. Experiments are also underway on several species of fish. In fact, such is the desperation that scientists are trying to insert Bt gene into any crop they can lay their hands on, not knowing whether this is desirable or not.

What Does the Field Trial Data of GM Products, Including *Bt* Cotton, in India Reveal?

Bt cotton field trials were a sham. In three years of research trials, the experiments were not conducted as per scientific norms. And yet, the GEAC (Genetic Engineering Approval Committee, ministry of environment and forests) had approved the results. The experiment only showed that such products are not suitable for Indian conditions. If only the same attention had gone to more sustainable farming systems, India would have been able to create a unique model of agriculture where farmers are not forced to commit suicide, where the land is not polluted, and where water is not poisoned. GM crops experiments show that the country is fast moving into a hitherto unforeseen era of biological pollution, which will be more unsustainable and also destructive to human health and environment.

But India's Biological Diversity Act 2003 does Provide for an Environmental Assessment of GM Crops?

No, not at all. Genetic engineering is moving several times faster than the legal instruments. Transgenic crops and animals in essence go against the very foundation of the biological diversity that we are trying to protect.

What Role Should the GEAC Play?

GEAC should emphasise biological risk assessment. GEAC should regulate genetic techno-logy like the US Recombinant Advisory Committee (RCA) does for genetically engineered drugs. RCA makes it mandatory for companies to provide a list of negative and harmful impacts and minimises that impact before approving for commercial sale. As a result, the approval process takes 25 years. Unfortunately, GM research in India is not being made to evaluate potential harm to human health and environment. This is because the GEAC does not want the companies to spend more on research.

Does GM Technology Threaten our Genetic Resources and Traditional Knowledge?

We have already lost control over our plant, animal and microbial genetic resources. A copy of roughly 1,50,000 plant accessions that have been collected in India, are with the US department of agriculture. India has no control over these resources. At the same time, India is now busy documenting traditional knowledge, so as to help the American companies know the uses of the plant species they have got from us. Further, Trade-related Intellectual Property Rights (TRIPs) allows patents on genes and cell lines, which will block India's agricultural research leading to what I have always termed as a scientific apartheid against the developing countries.

What are Genetically-Modified Foods?

One of the first questions you might be asking about GM foods is "what is genetically modified food"? The terms genetically-modified (GM) or genetically-engineered (GE) foods and genetically-modified organisms (GMOs) refer to crop plants created for human or animal consumption using the latest molecular biology techniques. Genetically modified food techniques of modern genetics have made possible the direct manipulation of the genetic makeup of organisms. Combining genes from different organisms is known as recombinant DNA technology and the resulting organism is said to be "genetically modified," "genetically engineered," or "transgenic." Cambridge Scientific Abstracts has an excellent introduction to this topic entitled: *Genetically Modified Foods: Harmful or Helpful* ? (by Deborah B. Whitman, April 2000). Like most human planetary management issues today, such as global climate change, the genetically modified foods issue is hugely complex. Genetically modified foods have great promise and great dangers. AAEA leans in the direction of aggressive market production with needed oversight regulations in a global management context. If all goes well, one thing is certain, we will have to feed about 12 billion people every day in the next 50 years.

Good and evil are moral choices humans are free to make. As applied to technology, these moral choices present great opportunities and great dangers.

We manipulate atoms to light our buildings and to make weapons of mass destruction. Companies produce chemicals to make our lives easier, but sometimes cut corners in the management, storage and disposal to maximize profits. We utilize coal, oil and gas for our cars, businesses and utility needs, but these same natural resources pollute our air and water without adequate protections. Twenty first century choices face us in stem cell research, cloning and genetically modified foods. Proponents and opponents present their cases and policy-makers are faced with protecting the public interest. Unfortunately, human history is littered with cases of indiscretions by people with evil intentions. It is within this context that we look at the case for genetically modified and engineered organisms and foods.

We support prudent use of genetically modified foods. We believe that labels should be placed on all GM foods. We also understand the risks involved, but believe the benefits far outweigh the costs. Starvation is much more dangerous to more people than any threat presented by GM food. Droughts and famine are increasing throughout the world, particularly on the continent of Africa. Although some traditional environmental groups insist that they are simply providing facts about potential health and environmental effects of genetically modified foods, others oppose it as a Frankenstein product. Of course, none of these groups have programs to feed the world's hungry. Some USA based social justice groups, such as the Africa Faith and Justice Network are opposing USA policies that impose GM food aid on southern African countries facing severe drought and famine. In addition to concerns about health effects, they think it is a tactic to blatantly benefit agri-business, not poor and hungry people. We understand the health concerns, but see nothing wrong with agri-business profiting from such exchanges. Captialism feeds America. In fact, Americans are suffering more from overeating than lack of food. As planetary managers, we must understand that there are no benign systems that can provide for human needs and we are obligated to protect the planet to the maximum extent possible. One major advantage of GM food is that crops genetically engineered to resist weeds and bugs enable farmers to decrease pesticide and herbicide use. Of course, superweeds and bugs could also be inadvertently created. Planetary management is very complex and serious business.

Genetically modified technology will not eliminate hunger and malnutrition because dysfunctional govern-ments and economies create problems with production, access and distribution of food.Flawed policies, greed and imcompetence will always keep some people in ignorance and poverty. However, GM foods can improve survivability and increase productivity of plants in inhospital conditions. GM foods can also reduce the need to use large quantities of herbicides and pesticides. Of course, this does not stop Mendocino County, California – considered by some to be the center of America's anti-biotechnology movement– from holding a vote to prohibit GM

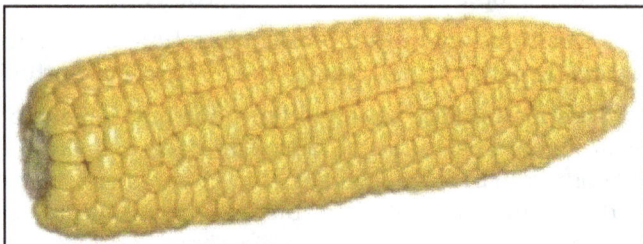

plants and animals from being raised or kept in the county. Such anti-GM entitites consider it to be the biggest uncontrolled biological experiment going on in the world today. Although proof of serious harm to humans, animals and plants has yet to be definitively proven, opponents fear that humans and the environment could be damaged through accidental cross-pollination of GM products with natural plants. This is a legitimate fear, but is not sufficient to ban the use of all GM products. Proponents point out that negative effects are nonexistent, pointing out that not a single stomach ache has been reported since the Food and Drug Administration first approved genetically engineered crops for human consumption in 1994. Great Britain's Food Standards Agency also favors the use of GM foods. Of course, most health effects of concern, including cancer and the results of long-term damage to the immune system take years to become evident. And then there would be the complex task of directly associating any damaging effects with GM products.

All types of foods and organisms have been genetically engineered: corn, cotton, tomatoes, soybeans, sugarbeets, oilseed rape, maize, salmon, pigs, cows, and the list goes on. With about 6 billion people eating everyday, we need every reasonable tool known to man to assure adequate nutrition for Earth's residents. GM foods, property utilized, can help meet these needs in a number of ways: pest resistance, herbicide tolerance, disease resistance, cold tolerance, drought tolerance and salinity toleranc, among others. Many countries are growing GM crops: U.S., Canada, China, Argentina, Australia, Bulgaria, France, Germany, Mexico, Romania, South Africa, Spain and Uruguay. Interestingly, according the USDA approximately 54 per cent of all soybeans cultivated in the U.S. in 2000 were genetically-modified. In the U.S., three government agencies have jurisdiction over GM foods: EPA evaluates GM plants for environmental safety, the USDA evaluates whether the plant is safe to grow, and the FDA evaluates whether the plant is safe to eat. Mandatory food labeling is also a complex issue. The FDA's current position on food labeling is governed by the Food, Drug and Cosmetic Act, which is only concerned with food additives, not whole foods or food products that are considered GRAS (Generally Recognized As Safe). The FDA contends that GM foods are substantially equivalent to non-GM foods, and therefore not subject to more stringent labeling. If all GM foods and food products are to be labeled, Congress must enact sweeping changes in the existing food labeling policy. The Genetically Engineered Food Right to Know Act (HR 2916) is probably a good place to start for food labeling.

Just as AAEA supports nuclear power with the belief that there should be serious oversight, we support the use of modified foods in the same way. We believe that traditional environmental groups go too far in calling for a ban on nuclear power and GM.They could still provide 95 per cent of the same constructive criticisms and oversight in these areas, but are extremist when calling for bans on useful, relatively safe products. We understand that part of this extremism partially comes as a reaction to the extremism of greedy, unscrupulous capitalists abusers. As part of a minority group with a long history of disadvantage, we do not have time for these games. However, we have serious concerns about human genetic engineering, particular cross species modifications and cloning. We fear that the Hitlerian contingent will take experiments with human DNA into an area of manufacturing humans for some ungodly reason and mad scientists will inexorably attempt to pierce the species genetic barrier and mix humans with animals FOR IMPROVEMENTS. Cinema has caught these images in *The Matrix* and *The Island of Dr. Moreau*. We would join our extremist colleagues in the traditional environmental movement in calling for a total ban on this type of unethical and immoral activity.

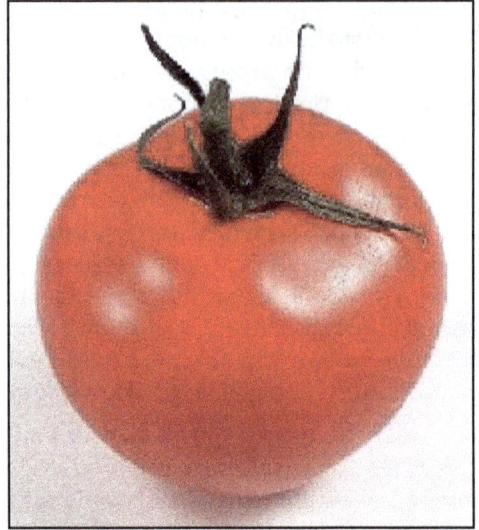

The term GM foods or GMOs (genetically-modified organisms) is most commonly used to refer to crop plants created for human or animal consumption using the latest molecular biology techniques. These plants have been modified in the laboratory to enhance desired traits such as increased resistance to herbicides or improved nutritional content. The enhancement of desired traits has traditionally been undertaken through breeding, but conventional plant breeding methods can be very time consuming and are often not very accurate. 6. A farmer grows these soybeans which then only require one application of weed-killer instead of multiple applications, reducing production cost and limiting the dangers of agricultural waste run-off[7].

- ☆ Disease resistance There are many viruses, fungi and bacteria that cause plant diseases. Plant biologists are working to create plants with genetically-engineered resistance to these diseases[8,9].

- ☆ Cold tolerance Unexpected frost can destroy sensitive seedlings. An antifreeze gene from cold water fish has been introduced into plants such as tobacco and potato. With this antifreeze gene, these plants are able to tolerate cold temperatures that normally would kill unmodified seedlings[10]. (Note: I have not been able to find any journal articles or patents that involve fish antifreeze proteins in strawberries, although I have seen such reports in newspapers. I can only conclude that nothing on this application has yet been published or patented.)

- ☆ Drought tolerance/salinity tolerance As the world population grows and more land is utilized for housing instead of food production, farmers will need to grow crops in locations previously unsuited for plant cultivation.

Creating plants that can withstand long periods of drought or high salt content in soil and groundwater will help people to grow crops in formerly inhospitable places[11, 12].

☆ Nutrition Malnutrition is common in third world countries where impoverished peoples rely on a single crop such as rice for the main staple of their diet. However, rice does not contain adequate amounts of all necessary nutrients to prevent malnutrition. If rice could be genetically engineered to contain additional vitamins and minerals, nutrient deficiencies could be alleviated. For example, blindness due to vitamin A deficiency is a common problem in third world countries. Researchers at the Swiss Federal Institute of Technology Institute for Plant Sciences have created a strain of "golden" rice containing an unusually high content of beta-carotene (vitamin A)[13]. Since this rice was funded by the Rockefeller Foundation[14], a non-profit organization, the Institute hopes to offer the golden rice seed free to any third world country that requests it. Plans were underway to develop a golden rice that also has increased iron content. However, the grant that funded the creation of these two rice strains was not renewed, perhaps because of the vigorous anti-GM food protesting in Europe, and so this nutritionally-enhanced rice may not come to market at all[15].

☆ Pharmaceuticals Medicines and vaccines often are costly to produce and sometimes require special storage conditions not readily available in third world countries. Researchers are working to develop edible vaccines in tomatoes and potatoes[16, 17]. These vaccines will be much easier to ship, store and administer than traditional injectable vaccines.

Governments around the world are hard at work to establish a regulatory process to monitor the effects of and approve new varieties of GM plants. Yet depending on the political, social and economic climate within a region or country, different governments are responding in different ways.

In Japan, the Ministry of Health and Welfare has announced that health testing of GM foods will be mandatory as of April 2001[18, 19]. Currently, testing of GM foods is voluntary. Japanese supermarkets are offering both GM foods and unmodified foods, and customers are beginning to show a strong preference for unmodified fruits and vegetables.

India's government has not yet announced a policy on GM foods because no GM crops are grown in India and no products are commercially available in supermarkets yet[20].

As an example the EPA regulatory approach, consider B.t. corn. The EPA has not established limits on residue levels in B.t corn because the B.t. in the corn is not sprayed as a chemical pesticide but is a gene that is integrated into the genetic material of the corn itself. Growers must have a license from the EPA for B.t corn, and the EPA has issued a letter for the 2000 growing season requiring farmers to plant 20 per cent unmodified corn, and up to 50 per cent unmodified corn in regions where cotton is also cultivated[21]. This planting strategy may help prevent insects from developing resistance to the *Bt.* pesticides as well as provide a refuge for non-target insects such as Monarch butterflies.

The USDA has many internal divisions that share responsibility for assessing GM foods. Among these divisions are APHIS, the Animal Health and Plant Inspection Service, which conducts field tests and issues permits to grow GM crops, the Agricultural Research Service which performs in-house GM food research, and the Cooperative State Research, Education and Extension Service which oversees the USDA risk assessment program. The USDA is concerned with potential hazards of the plant itself. Does it harbor insect pests? Is it a noxious weed? Will it cause harm to indigenous species if it escapes from farmer's fields? The USDA has the power to impose quarantines on problem regions to prevent movement of suspected plants, restrict import or export of suspected plants, and can even destroy plants cultivated in violation of USDA regulations. Many GM plants do not require USDA permits from APHIS. A GM plant does not require a permit if it meets these 6 criteria: 1) the plant is not a noxious weed; 2) the genetic material introduced into the GM plant is stably integrated into the plant's own generally recognized as safe. The FDA contends that GM foods are substantially equivalent to non-GM foods, and therefore not subject to more stringent labeling. If all GM foods and food products are to be labeled, Congress must enact sweeping changes in the existing food labeling policy.

There are many questions that must be answered if labeling of GM foods becomes mandatory. First, are consumers willing to absorb the cost of such an initiative? If the food production industry is required to label GM foods, factories will need to construct two separate processing streams and monitor the production lines accordingly. Farmers must be able to keep GM crops and non-GM crops from mixing during planting, harvesting and shipping. It is almost assured that industry will pass along these additional costs to consumers in the form of higher prices.

Secondly, what are the acceptable limits of GM contamination in non-GM products? The EC has determined that 1 per cent is an acceptable limit of cross-contamination, yet many consumer interest groups argue that only 0 per cent is acceptable. Some companies such as Gerber baby foods[42] and Frito-Lay[43] have pledged to avoid use of GM foods in any of their products. But who is going to monitor these companies for compliance and what is the penalty if they fail? Once again, the FDA does not have the resources to carry out testing to ensure compliance.

What is the level of detectability of GM food cross-contamination? Scientists agree that current technology is unable to detect minute quantities of contamination, so ensuring 0 per cent contamination using existing methodologies is not guaranteed. Yet researchers disagree on what level of contamination really is detectable, especially in highly processed food products such as vegetable oils or breakfast cereals where the vegetables used to make these products have been pooled from many different sources. A 1 per cent threshold may already be below current levels of detectability.

Finally, who is to be responsible for educating the public about GM food labels and how costly will that education be? Food labels must be designed to clearly convey accurate information about the product in simple language that everyone can understand. This may be the greatest challenge faced be a new food labeling policy: how to educate and inform the public without damaging the public trust and causing alarm or fear of GM food products.

In January 2000, an international trade agreement for labeling GM foods was established. More than 130 countries, including the US, the world's largest producer of GM foods, signed the agreement. The policy states that exporters must be required to label all GM foods and that importing countries have the right to judge for themselves the potential risks and reject GM foods, if they so choose. This new agreement may spur the U.S. government to resolve the domestic food labeling dilemma more rapidly.

Conclusion

Genetically-modified foods have the potential to solve many of the world's hunger and malnutrition problems, and to help protect and preserve the environment by increasing yield and reducing reliance upon chemical pesticides and herbicides. Yet there are many challenges ahead for governments, especially in the areas of safety testing, regulation, international policy and food labeling. Many people feel that genetic engineering is the inevitable wave of the future and that we cannot afford to ignore a technology that has such enormous potential benefits. However, we must proceed with caution to avoid causing unintended harm to human health and the environment as a result of our enthusiasm for this powerful technology.

References

1. ____1999. Transgenic pollen harms monarch larvae. *Nature*, 399(6733): 214.

2. ____2001. Assessing the impact of Cry1Ab-expressing corn pollen on monarch butterfly larvae in field studies. *Proceedings of the National Academy of Sciences*, 98, No 21, p11931–11936, Oct 2001.

3. Bioengineered Foods transcripts from the public meetings are available to download (http://www.fda.gov/oc/biotech/default.htm)

4. ____1999. The use of cytochrome P450 genes to introduce herbicide tolerance in crops: a review. *Pesticide Science*, 55(9): 867–874.

5. ____2001. Transgenic approaches to combat fusarium head blight in wheat and barley. *Crop Science*, 41(3): 628–627.

6. ____2001. Post-transcriptional gene silencing in plum pox virus resistant transgenic European plum containing the plum pox potyvirus coat protein gene. *Transgenic Research*, 10(3): 201–209.

7. ____1999. Type II fish antifreeze protein accumulation in transgenic tobacco does not confer frost resistance. *Transgenic Research*, 8(2): 105–117.

8. ____2001. Transgenic salt-tolerant tomato plants accumulate salt in foliage but not in fruit. *Nature Biotechnology*, 19(8): 765–768.

9. ____2000. Peroxidase activity of desiccation-tolerant loblolly pine somatic embryos. *In Vitro Cellular and Developmental Biology Plant*, 36(6): 488–491.

10. Genetic engineering towards carotene biosynthesis in endosperm (Swiss Federal Institute of Technology Institute for Plant Sciences

11. New rices may help address vitamin A and iron deficiency, major causes of death in the developing world (Rockefeller Foundation)

12. ____1999. Rice Biotechnology: Rockefeller to end network after 15 years of success. *Science*, 286(5444): 1468–1469.

13. ____2001. Medical molecular farming: Production of antibodies, biopharmaceuticals and edible vaccines in plants. *Trends in Plant Science*, 6(5): 219–226.

14. ____2001. Oral immunization with hepatitis B surface antigen expressed in transgenic plants. *Proceedings of the National Academy of Sciences*, USA, 98(20): 11539–11544.

15. ____1999. Japan to bring in mandatory tests for GM food. *Nature*, 402: 846.

16. ____2000. Japan steps up GMO tests. *Nature Biotechnology*, 18: 131.

17. ____1999. India intends to reap the full commercial benefits. *Nature*, 402: 342–343.

18. Letter to Bt Corn Registrants 12/20/1999 from the EPA (http://www.epa.gov/pesticides/biopesticides/otherdocs/bt_corn_ltr.htm)

19. Consumer Pressure Forces Gerber Baby Foods to Eliminate GE Corn and Soybeans from US Products (AP Online http://www.purefood.org/ge/nobabyge.cfm)

20. Frito-Lay's Halfway Measures Banning GE Corn Freak Out Their Competitors: New Seed Planted in Genetic Flap (Washington Post http://www.purefood.org/ge/fritolayhalf.cfm)

21. ____2000. Biotechnology: Both sides claim victory. *Science*, 287: 782–783.

22. ____2000. Rules agreed over GM food exports. *Nature*, 402: 473.

Biotechnology: An Overview (2015)   Pages 89–99
Editors: Rajan Kumar Gupta, Nasim Akhtar and Deepak Vyas
Published by: DAYA PUBLISHING HOUSE, NEW DELHI

Chapter 8

Role of Biofertilizers in Agriculture

U.K. Chaturvedi[1] and Iqbal Habib[2]*

[1]Department of Botany,
Bareilly College, Bareilly – 253 005, Uttar Pradesh
[2]Department of Botany
Government Degree College, Budaun – 243 601, Uttar Pradesh

ABSTRACT

Biofertilizers are a group of micro-organisms which are biologically active and which help in N_2 fixation for the benefit of plants. Organic manures are also a kind of biofertilizers which are produced by the interaction of micro-organisms in soil. Now a days the demand of biofertilizers has increased considerably, first, because their use in agriculture leads to enhancement in the crop productivity and, second, their use reduce the dependency on chemical fertilizers which are costly as well as they damage the soil texture and microflora of the soil. Different groups of Bacteria, algae, fungi, VAM fungi, *Azolla* etc. are helpful in biological N_2 fixation and their residuse increase the organic matter in the soil. Further algae and cyanobacteria have been in use for predicting pollution hazards and as indicators of new substances entering into the aquatic environment. Their trophic independence for C and N enables them a greater flexibility towards adopting to varied environmental conditions.

Keywords : Micro-organisms, Biofertilizers, Productivity, Agriculture.

Introduction

Many revolutionary changes are taking place in agriculture so that agriculturists may get a much higher yield of their crops. Scientists are busy in bringing out new food crops that can take in nitrogen directly from the air rather than to rely on the chemical fertilizers presently in use. They are also experimenting with the plant genes that may provide resistance to herbicides and can control on the spread of virus and fungal diseases. They are also experimenting with some crop plants which may grow in water and salt stress conditions. So their main moto is to grow such plants which can grow faster, are disease resistant, and are better able to withstand stresses like heat, cold, salinity and drought. Recently, they have isolated a gene for the enzyme that enhances the formation of cellulose in plants that will help in the pulp and paper industry.

Conventional farming has resulted in negative effect on the environment, availability of chemical residues in food which have led to a marked increase in various diseases, use of fertilizers and pesticides. The commercialization of farming also has a negative effect on the soil. The use of pesticides has led to an increase in chemicals in our soils, water and air. Fertilizers have a short term effect on productivity and their chemicals remain in soil and water for longer period of time, thus contaminating them.

Visualising the high cost of fertilizers and the erection of fertilizer plants, scientists and agriculturists are now trying to manipulate the soil microflora to produce more fixed nitrogen. Further, increased usage of chemical fertilizers lead to damage in soil texture and raises other environmental problems. Therefore the use of biofertilizers is both economical and environmentally friendly. As micro-organisms are well adapted to recycle the mineral supply of soil, therefore their biological activities are to be harnessed in this direction. To increase the soil fertility, organic manures along with nitrogenous fertilizers are being encouraged in several countries, including India. The micro-organisms present in organic soils are called as 'biofertilizers'. The biofertilizers are premium natural fertilizers composed of micro-organisms and ogranic ingredients special to nutrient poor soils.

Biotechnologically, biofertilizers are the most advanced type of fertilizers which support developing organic agriculture, sustainable agriculture, green agriculture and non polluted agriculture. It helps in increasing the yield, improve the quality and is responsible for healthy agricultural environment. It is 100 per cent natural. It has been widely used with excellent results in all kinds of agricultural land and in several countries (Vaishampayan, *et al.*, 2001) According to Subba Rao (1982) the appropriate term for biofertilizers should be 'microbial inoculants. Therefore, biofertilizers are defined as biologically active products or microbial inoculants of bacteria, algae and fungi, which may help in biological N_2 fixation.

Although biofertilizers do not directly increase soil fertility but they in general, accelerate the process of mineralization in the soil. The results have been summerized by Venkataraman (1970) and Subba Rao (1982). In the production of biofertilizers several activities of *Rhizobium, Azotobacter, Azospirillum,* Cyanobacteria and phosphate solubilizing bacteria are helping in decomposing the organic matter present in the soil. Biologically fixed nitrogen is such a source which can supply an adequate amount of nitrogen to plants and other nutrients to some extent. Newer techniques are being involved and are in practice at Indian Agricultural Research Institute, New Delhi and also in various agricultural institutes in India. Mass production of nitrogen fixing cyanobacteria and *Anabena azollae* at I.A.R.I., New Delhi and Central Rice Research Institute, Cuttack, respectively are in progress (R.P. Singh, 1989). As *Rhizobium* bacteria and cyanobacteria are known to fix atmospheric nitrogen and hence are widely used as biofertilizers. The agronomic potential of cyanobacteria, either free living or in symbiotic association water fern *Azolla,* has long been recognised

(Singh, 1961; Venkataraman, 1972; Whitton and Roger, 1989; Rai, 1990). Micro-organisms in organic soils not only mineralize more actively but also contribute to the build up of stable soil organic matter. Thus, nutrients are recycled faster and soil structure is improved. So, in this article, an attempt has been made to discuss some important roles of micro-organisms, particularly with reference to their use as 'biofertilizers.'

Biofertilizers are also known as 'microbial cultures' or 'microbial inoculants', they, therefore have been categorized into several groups. Following are some of their important activities:

Several free living and symbiotic bacteria fix atmospheric N_2 and most important of them are *Rhizobium, Azotobacter, Azospirillum* etc. *Rhizobium* spp. are undoubtedly the most important and best studied nitrogen fixers. They are gram negative soil bacteria. They thrive well in symbiotic relationship with several leguminous plants. Although, they occur in free living state in the soil but do not fix N_2 in this state. Several species of *Rhizobium* have been described but their main distinction is in the range of host plants which they infect to produce nodules. The yield of pulse crops can substantially be increased by rhizobial inoculation. The nitrogen fixing ability of legumes inoculated with these Rhizobia ranges from 50 kg to 150 kg. per hectare. Therefore, to increase this ability to fix N_2, newer techniques are to be employed. Leguminous crops, thus get benefit from rhizobial symbiosis. A certain amount of N_2 is left over in soil which is utilized by other crop plants which are grown in the same plot after the legume crop has been harvested. Subba Rao and Tilak (1977) studied the effect of residual nitrogen of several legumes on the yield of subsequent crops in *Rhizobium* in oculated fields than in uninoculated control.

Another group of bacteria-the *Azotobacter* and *Azospirillum* are equally important in fixing the N_2 in soil and make it available to crop plants. They also synthesize growth promoting antibiotic substances which are helpful to plants. Some reports are available from Russia on the activities of these bacteria and upto 20 per cent increase in yield were claimed by Russian workers on a wide variety of crops. Most efficient strains of *Azotobacter* fix 30 kg of N_2 from 1000 kg. of organic matter. In India, Gaur and Algawadi (1989) studied the interaction of N_2 fixing bacteria *Azotobacter chroococcum* and phosphate solubilizing bacteria (*Bacillus megatherium* and *Pseudomonas striata*) and their effect on sorghum and Rice crops in green house experiments. They found a significant increase in root nitrogenase activity, dry matter and seed yield, compared to control. When applied to fields, positive responses by field crops were observed leading to saving of 10-25kg/ha. Of N_2. Similarly, *Azospirillum* with farm yard manure (FYM) led to saving of 15-25kg. equivalent of N_2/ha in crops like sorghum and other millets. Reynders *et al.* (1982) reported the use of *Azospirillum brasilense* as biofertilizer in intensive wheat cropping and found an average green field of 9.14 per cent and 14.82 per cent respectively. Similarly, studies conducted by Vijay Kumari *et al.* (2009) on *Azospirillum* and phosphobacteria with recommended levels of N PK and FYM increased vigour and growth of brinjal plants. Field experiments conducted at I.A.R.I., New Delhi have shown that when sorghum and cotton seeds are inoculated by *A. chroococcum*, the yield was increased substantially. The yield of wheat was also increased to about 10 per cent at the same conditions. These experiments may help small and marginal farmers which can follow seed bacterization practice.

The role of 'Cyanobacteria' *i.e.* the blue green algae (BGA) in the rice fields was realised much earlier (Singh, 1961). Cyanobacteria are of great importance whenever algae and prokayotic photo autotrophs are considered. Infact, they are our 'living resources.' They are the only prokaryotes which carry out phototrophic mode of nutrition by utilizing carbon dioxide, water and sun light from the nature and capable of fixing nitrogen. Therefore, diazotrophic cyanobacteria have a great potential as biofertilizers and their use will certainly decrease the demand of chemical fertilizer (Choudhary *et. al* 2007). The agronomic potential of cyanobacteria, either free living or in symbiotic association with *Azolla*-a water fern, has long been recognised (Singh, 1961; Venkataraman 1972; Whittan and Roger, 1989; Rai, 1990).

Several species of *Nostoc, Anabaena, Aulosira, Cylindrospermum, Tolypothrix, Gloeotrichia, Plectonema* etc. are capable of N_2 fixation in soil. On the surface of moist soil, they may add a good amount of N_2 into the soil. N_2 in the atmosphere is nearly ¾ and a large amount of it can be biologically fixed. Similarly carbon is also in abundance on this earth and this too can be fixed. Here the role of cyanobacteria becomes important as they are the simplest micro-organisms on this earth which can carryout both these functions at ease. They make use of both these elements in an efficient manner in presence of sunlight, which is also in plenty in a tropical country like India. In the tropical countries where rice or paddy crop is grown, the whole N_2 requirement of the crop may be available through these nitrogen fixers. Experiments conducted by Dart and Day (1977) and Peters (1978) at the International Rice Research Institute, Philippines, showed that for several consecutive years rice crop was grown on soils unfertilized with nitrogen. The stagnant waters of paddy fields harvour a rich mass of these algae which in turn fix atmospheric N_2. Successful experiments were conducted in the state of Tamil Nadu by propagating and using cyanobacteria as a valuable and renewable source in rice cultivation on commercial scale. A number of rice fields of India and in other countries have been surveyed for the presence of cyanobacteria and several papers have been published from various parts of the world (Venkataraman, 1972; Singh *et al.*, 2000). In majority of the cases studied so far regarding the presence of cyanobacteria in the rice fields, the heterocystous forms *i.e.* nitrogen fixers are in abundance. Besides the known species of *Nostoc* and *Anabaena*, some other cyanohycean members such as *Lyngbya, Phormidium, Chroococcus, Wollea, Aphanothece, Aphanocapsa, Calothrix, Gloeocpsa* are equally important as N_2 fixers. The 'heterocystous forms' consist of heterocyts, a specialized cell that contains the nitrogen fixing mechanism (Steward, 1980). Helerocystous forms of cyanobacteria have been studied extensively for the development and function of heterocyst that contain an enzyme 'nitrogenase.' This enzyme is sensitive to oxygen and reduces atmospheric N_2 to ammonia under aerobic conditions only. Thus N_2 is reduced to NH_3 and is made available to rice plants through microbial decomposition (Roger *et al.*, 1987). The study reveals that the inoculation of cyanobacteria as a nitrogen fertilizer contributes about 25-30 kg of N_2/ha/season and increases productivity of rice upto 10-15 per cent. The results encouraged 'cyanobacterial biofertilizer programmes

(Patterson, 1996) and farmers were distributed cyanobacterial biofertilizers. Now scientists are attempting to categorise the factors that are influencing the efficiency of cyanobacterial strains as a biofertilizer in the field and to develop genetically improved resistant strains. Further, to improve the N_2 fixation efficiency of heterocystous cyanobacteria, we have to see the heterocystous frequency so that the nitrogenase enyme is increased. Heterocysts arranged in chains have been reported in *Nostoc linckia* which fixes more N_2 as compared to wild type (Singh and Tiwari, 1969). Such strains with elevated level of N_2 fixation ability can be used as an efficient biofertilizer. Some studies have revealed that herbicides used to control the weeds are damaging the photosynthetic apparatus of cyanobacteria and thus reduce the N_2 fixing activity. To overcome this problem, selection of genetically viable strains, which can efficiently fix N_2 in the presence of field dose concentrations of herbicide commonly used in rice fields, is needed. In this connection, a number of herbicide resistant mutant of heterocystous and non-heterocystous cyanobacteria have been reported by several workers. Work done by Dadhich *et al.* (1969) on the blue-green alga *Calothrix anomala* have shown an increase in yield and nitrogen uptake by certain vegetable crops like *Capsicum annuum* and *Lactuca sativa*. Earlier report by Fuller *et al.* (1960) showed favourable effect of algal application on Barley seedlings and tomato plants.

As per the reports available, as much as 15 tons of wet cyanobacteria/hactare can be produced in three weeks of time. Application of dried blue-green algal flakes at the rate of 10 kg/ha. is recommended ten days after transplantation of rice plants. Besides being a source of N_2, BGA provides other advantages such as accumulation of algal biomass as organic matter, growth promoting substances are produced which stimulate the growth of rice seedlings and in the reclaiming of saline and alkaline soils. Mass cultivation of blue-green algae, and its application in the fields is called as 'Algalization' (Venkataraman, 1970). Infact he initiated algalization technology in India and demonstrated the way how this technology could be transferred to farmer level who hold small lands. Algalization is being practised in Philipines, China, Egypt and in Russia. Four major centres in different paddy growing areas of the country were established in 1990 by the Department of Biotechnology, Govt of India, N. Delhi so as to accelerate the process and to use cyanobacteria as biofertilizers. Algalization of rice fields with cyanobacterial biofertilizers resulted into the enhanced, grain yield. Singh (1961) has reported that 'alkaline and user soils' of Uttar Pradesh could be reclaimed by using cyanobacteria so as to neutralize the pH of soil. Cyanobacteria are also known to increase soil fertility by enhancing the available N and P (Roger and Kulsasooriya, 1980). They reported the relative increase to grain yield over control by 28 per cent in pot experiments and 15 per cent yield in field experiments. Further, studies on the application of cyanobacterial biofertilizers in the fields reveal that in addition to increasing the grain and straw yield, the increase of nitrogen content of grain and straw has also been reported. The other benefits are increase in plant height, number of spikelets/panicle and amount of dry matter (Singh, 1961; Rao *et al.*, 1977).

Nitrogenage is an enzyme that is directly involved in hydrogen fixation and consequently help in the fixation of nitrogen and convert it into ammonia. The overall reaction of nitrogen fixation can be written as follows :

$$N_2 + 8H + 16ATP = 2NH_3 + H_2 + 16 \text{ ADP} + 16P$$

Further substantial quantities of aminoacids like aspartic and glutamic acids and alamine, vitamins like B_{12} are liberated by nitrogen fixing blue green algae (Gupta and Shukla, 1969). These extra metabolites benefit the crop growth thus enabling the plants to utilize more of the available nutrients. Cyanobacteria are enveloped by mucilagenous sheath which is of polysaccharide in nature and it protect them from dessication (Hill *et al.*, 1995). The polysaccharides are highly charged particles that help in the removal of toxic metals from the polluted water by forming stable gel and binding metallic ions. The viscous nature of polysaccharides enables them to make the soil more fertile and increase the water holding capacity of soil.

'*Azolla*', an aquatic small fern plant is being currently used in rice cultivation in China, Indonesia, Korea, Philippines, Japan, Bangaladesh and in India. This plant contains an endoppytic cyanobacterium *Anabaena azollae* in its leaf cavity. There are few species of *Azolla* reported from different countries of the world but in India *A. pinnata* is a commonly accuring water fern. Plants are harvested and dried to use as green manure. Reports available from IRRI (Philippines) and from CRRI (Cuttak) reveal that when *Azolla* is grown in rice fields before rice transplantation than there was considerable increase in the yield. Further, *Azolla* has great capacity to absorb heavy metals *viz.* As, Hg, Ph, Cu, Cd etc from the polluted waters discharged from various industries. *Anabaena azollae* serves as a potential biofertilizer for rice crop (Nierzwicki-Baur, 1990; Watanbe and Liu, 1992; Singh and Singh, 1997). However, the *Azolla Anabaena* symbiosis fixes more N_2 as compared to the free living cyanobacteria and it could fix as much as 3.6 kg N/ha/day. There are some other reports available as regards the beneficial aspects of Azolla on the growth and yield of rice (Singh and Singh, 1990 and 1992). Its application in paddy fields also improve the soil fertility and has a residual effects on the yield of other crops (Kannaiyan, 1990 and 1992). Moreover, it suppresses weed growth (Satapathy and Singh, 1985) and methane efflux from the flooded rice fields (Bharti *et al.*, 2000). *Azolla* has some limitations as it is labour intensive to grow and it needed adequate supply of water and optimum temperature is required for its multiplication.

Root fungus 'Mycorrhiza' shows potential to act as 'biofertilizer'. Some fungi live in symbiotic relationship with plant roots. Some higher plants including Gymnosperms and Angiosperms which suffer from nutrient scarcity especially P and N_2, develop mycorrhiza in their roots. Use of fungi as biofertilizers will also reduce the farming industry's reliance on phosphate and nitrogen fertilizers that pollute water supplies also. Since mycorrhizal fungi are more efficient in the uptake of specific nutrients and more resistant against soil borne pathogens, interest in using these fungi as 'biofertilizers' or 'bioprotectors' is increasing. So, by promoting the proliferation of mycorrhizal fungi through diminished fertilizer input, farmers would make more efficient use of the N_2 stores in the soils. Although, mycorrhiza are of several types but vesicular arbuscular mycorrhiza (VAM) is gaining importance, recently. This mycorrhiza is coiled with intracellular hyphae, vesicles and arbuscules in it. Researches are being carried out in U.S.A. and Athens for its role as biofertilizer in horticulture and forestry. In India also, F.R.I. Dehradun has established a mycorrhizal bank and inocula of VAM fungi are available as and when needed in agriculture and forestry programmes. Mycorrhizal fungi play a role in selective absorption of certain mineral elements, *viz.*, P, Zn, Cu, Ca, k, Fe, Mn, Cl, Br, and N_2 to plants. They increase resistance in plants against pathogens and pests in soil. Further, Pinus trees cannot grow in new areas or grounds unless the soil has mycorrhizal inocula.

Many cyanobacteria have also shown to mobilize the insoluble phosphate in the soil, thereby increasing its availability to the crop plants (Roger *et al.*, 1987). Therefore, phosphate solubilising bacteria (PSB) *e.g. Thiobacillus, Bacillus* spp. etc. and plant growth promoting rhizobacteria (PGPR), including *Pseudomonas fluorescens* and *P. putida* are important new biofertilizers. PS B's convert non-available inorganic phosphates into soluble organic phosphate which can be utilized by crop plants. Reports available showed that PGPRs have resulted in an increase in yields of potato to 30 per cent, radish 60-40 per cent and sugarbeet 4-8t/ha. In India these biofertilizers are yet to be utilized commercially.

Future Role

While perusing through the above account, it seems that the use of biofertilizers have several benefits for the small and marginal farmers. The low cost and easy technique will help them to use biofertilizers thus reducing the dependency on chemical or synthetic fertilizers. Use of biofertilizers is free from pollution hazards and increases soil fertility. Biofertilizers harness atmospheric N_2 with the help of specialized Micro-organisms which may be free living in soil or symbiotic with plants. 'microbial inoculants' are carrier based preparations containing beneficial micro-organisms in a viable state, intended for seed or soil application, designed to improve soil fertility and help plant growth by increasing the number of desired micro-organisms in plant rhizosphere.

Algalization in rice fields will increase paddy crop to almost double and the left over of N_2 in soil will in turn be used by the subsequent crops (Venkataraman, 1972). Cyanobacteria, besides fixing N_2 in soil also secrete some growth promoting substances such as IAA, IBA, aminoacids, proteins and vitamins etc. Also, they add a good amount of organic matter is soil. Further, cyanobacteria can thrive well under wide pH range of 6.5-8.5 and thus they can help in reclaiming the saline and user soil. Rhizobial biofertilizer can fix a very large amount of N_2, in the soil while *Azotobacter* and *Azospirillum* secrete some antibiotic substances besides N_2 fixation. The role of *Azolla* as biofertilizer are many fold. The cyanobacteria which thrive in its leaf fixes atmospheric N_2. Plant show tolerance against heavy metals and its dried masses can be used as organic manure. Therefore, it needed much importance in this direction. Although benefits of mycorrhiza as biofertilizers are not well known amongst the horticulturists and foresters, however, it is quite evident that they help in increasing the availability of certain mineral elements to the plants and also reduce plant response to soil stresses and increase resistance in plants. In general, they effect in increasing the plant growth, survival and yield. Eventually, biofertilizers provide several mineral elements and sufficient amount of organic matter in soil and thus help in increasing the fertility of soil and in crop production. Further, one should take into account the unique characteristics of cyanobacteria and to make cyanobacterial biotechnology as a significant presence in the commercial sector. The future of cyanobacterial biotechnology is promising and with emergence and acceptance of 'greener technology' and with change in environment, the photosynthetic micro-organisms will receive more and more attention. In developing countries like India, it can solve the problem of high cost of fertilizers and thus help savings in large investments.

References

1. Aizawa, K., 1982. In: *Advances in Agricultural Microbiology,* (Ed.) N.S. Subba Rao. Oxford and IBH Publishing Co., New Delhi.

2. Alexander, M., 1977. *Introduction to Soil Microbiology*, John Wiley and Sons, New York.

3. Anderson, D.A. and Sobieski, R.J., 1980. *Introduction to Microbiology.* C.V. Mosby, St. Louis.

4. Bajaj, Y.P.S., 1987. *Biotechnology in Agriculture,* (Eds.) S. Natesh, V.L. Chopra and S. Rama Chandran. Oxford and IBH Publ. Co., New Delhi.

5. Bharati, K., Mohanty, S.R., Singh, D.P., Rao, V.R. and Adhya, T.K., 2000. Influence of incorporation of dual cropping of *Azolla* on methane emission from a flooded alluvial soil planted rice in Eastern India. *Agric. Ecosyst. Environ.,* 79: 73–83.

6. Chahal, D.S., 1982. *Advances in Agricultural Microbiology,* (Ed.) N.S. Subba Rao. Oxford and IBH Co., New Delhi.

7. Chakraborty, P.K., Mandal, L.N. and Majumdar, A., 1988. Organic and chemical sources of nitrogen : Its effect on nitrogen-transformation and rice productivity under submerged conditions. *J. of Agri. Sci. U.K.,* 11(1): 91–94.

8. Choudhary, K.K., Singh and Mishra, A.K., 2007. Nitrogen fixing cyanobacteria and their potential applications. In: *Advances in Applied Phycology.* Daya Publ. House, New Delhi., 23: 40–42.

9. Cooper, R., 1959. *Soil Fertilizers,* 22 : 227–233.

10. Dadhich, K.S., Verma, A.K. and Venkataraman, G.S., 1969. The effect of *Calothrix* inoculation on vegetable crops. *Plant and Soil,* 31(2): 377–378.

11. Dart, P.J. and Day, J.M. 1977. Non-symbiotic nitrogen fixation in soil. *Soil Microbiology,* Wiley, New York, pp. 225–252.

12. Dubey, R.C., 1996. *A. Textbook of Biotechnology.* S. Chand and Co. Ltd., New Delhi, pp. 152–168.

13. Duwiedi, R.S., Dubey, R.C. and Duwivedi, S.K., 1989. In: *Plant Microbe Interactions,* (Ed.) K.S. Bilgrame. Focal Theme (Botany) ISCA Symposium. pp. 217–238. Narendra Publ. House, Delhi.

14. Edward, J.C. and Singh, R.B., 1976. *An Introduction to Microbiology.* Central Book Depot, Allahabad.

15. FAO, 1997. *Biological Farming in Europe.* REU. Technical Series 54, Rome.

16. Fuller, W.H., Cameron, R.E. and Nicholas Raica Jr., 1960. Fixation of nitrogen in desert soils by algae. *7th Interntl. Cong. Soil Sci.*, Adison, Wisco. U.S.A., 3(22): 617.

17. Gupta, A.B. and Shukla A.C., 1967. Studies on nature of algal growth promoting substance and their influence on growth, yield and protein content of rice plants. *J. Sci. Tech.*, 5: 162–163.

18. Gupta, A.B. and Shukla, A.C., 1969. Effect of algal extract of *Phormidium* sp. On growth and development of rice seedlings. *Hydrobiologia*, 34 : 77–84.

19. Gupta, P.K., 1995. *Elements of Biotechnology*. Rastogi and Co., Meerut, pp. 498–503.

20. Haselkom, R. and Bulkema, W.J., 1992. Nitrogen fixation in Cyanobacteria. In: *Biological Nitrogen Fixation*, (Eds.) G. Staccy, R.H. Burris and H. Evans, pp. 166–190. Champman and Hall, London.

21. Herrero, A., Muropastor, A.M. and Flores, E., 2001. Nitrogen control in Cyanobacteria. *J. Bacteriol.*, 183 : 411–425.

22. Hill, D.R., Keenan, I.W., Helm, R.F., Potts, M., Crowe, L.M. and Crowe, J.H., 1995. Extracullar polysaccharide of *Nostoc commune* (Cyanobacteria) inhibits fusion of membrane vesicles during dessication. *J. Appl. Phycol.*, 9: 237–248.

23. Kannaiyan, S., 1990. *Biotechnology of Biofertilizer for Rice Crop*. Tamil Nadu Agric. Univ. Tamil Nadu, pp. 1–225.

24. Kannaiyan, S., 1992. *Azolla Biofertilizer Techniques for* Rice. Tamil Nadu Agric. Univ. Mag., pp. 1–56.

25. Kumar, H.D. and L.C. Rai, 1986. *Microbes and Microbial Processes*. Affiliated East-West Press, N. Delhi, pp. 213.

26. Minathan, M.S., 1980. In: *Biotechnology in Agriculture*, (Eds.) S. Natesh, V.L. Chopra and S. Ramachandran, pp. 3–11. Oxford and IBH Publ. Co., New Delhi.

27. Nandi, S.K. and Palni, L.M.S., 1992. *Microbial Activity in Himalaya*, (Ed.) R.D. Khulbe, pp. 419–428. Shree Almora Book Depot, Almora.

28. Patterson, G.M.L., 1996. Biotechnological applications of cyanobacteria. *J. Sci. Ind. Res.*, 55: 669–684.

29. Peters, G.A., 1978. Blue Green algae and algal associations. *Biosciences*, 28: 580–583.

30. *Plant Biotechnology*, Today and Tomorrow. Special Volume of Vegetos Journal 13: 2000. Society For Plant Research, Bareilly (U.P.) India

31. Powar, C.B. and Daginawala, H.F., 1991. *General Microbiology*, Vol. II. Himalaya Pub. House, Bombay.

32. Rai, 1990. *Handbook of Symbiotic Cyanobacteria*. CRC Press, Boca Raton, FL. U.S.A.

33. Reynders, L. and Vlassak, K., 1982. *Plant and Soil*, 66(2): 217–273.

34. Roger, P.A. and Kulasooriya, S.A., 1980. *Blue Green Algae and Rice*. Int. Rice Res. Inst. Los Banos, Philippines.

35. Roger, P.A., Santiago-Ardales, S., Reddy, M.M. and Watanabe, I., 1987. The abundance of heterocystous blue-green algae in rice soils and inocula used for application in rice fields. *Biol. Fertility Soils,* 5: 98–105.

36. Satpathy, K.B. and Singh, P.K., 1985. Control of weeds by *Azolla pinnata* in rice fields. *J. Aquat Plant Manag.,* 23: 40–42.

37. Singh, A.L. and Singh, P.K., 1990. Phosphorus fertilization on growth and nitrogen fixation of *Azolla* and BGA in rice fields. *Ind. J. Plant. Physiol.,* 33: 21–26.

38. Singh, A.L. and Singh, P.K., 1987. Nitrogen fixation and balance studies of rice soils. *Biology and Fertility of Soil,* 4(1–2): 15–19.

39. Singh, P.K., 1976. Algal inoculation and its growth in water logged rice fields. *Phykos,* 15: 5–10.

40. Singh, P.K. and Bisoyi, R.N., 1989. Blue green algae in rice fields. *Phykos,* 28: 181–195.

41. Singh, P.K. and Singh, D.P., 1992. *Azolla*: A potential way towards organic farming in rice cultivation in Indian condition. In: *Proc. Natl. Seminar Organic Farming,* (Eds.) M.M. Rai and L.N. Verma. Ravi Printing Press, Jabalpur, pp. 233–240.

42. Singh, P.K. and Singh, D.P., 1997. *Azolla* and rice cultivation. In: *Biofertilizers,* (Eds.) L.L. Sowani, S.C. Bhandari, S.N. Saxena and K.K. Vyas. Scientific Publ. Jodhpur, pp. 39–45.

43. Singh, R.N., 1961. *The Role of Blue-green Algae in Nitrogen Economy of Indian Agriculture.* ICAR, New Delhi.

44. Singh, R.N., 1950. Reclamation of usar lands in India through blue green algae. *Nature,* 165: 325–326.

45. Singh, R.N. and Tiwari, D.N., 1969. Induction of Ultraviolet irradiation of mutation in the blue green alga *Nostoc linckia* (Roth.) *Nature (Lond.),* 221: 62–64.

46. Singh, R.P., 1989. *An Intro to Microbiology.* Central Book Depot, Allahabad, p. 367–374.

47. Singh, R.P. and Kawal, 1988. *An Introduction to Fungi.* Central Book Depot, Allahabad.

48. Singh, Y.V. and Mandal, B.K., 1997. Nutrition of rice (*oryza sativa*) through *Azolla,* organic materials and urea. *Indian Agronomy,* 42(4): 626–633.

49. Srinivasan, S., 1980. Blue-green algae: A renewable resource in rice cultivation, Farmers Training Centre, Aduthurai (Tamil Nadu) India

50. Steward, W.D.P., 1980. Some aspects of structure and function of nitrogen fixing Cyanobacteria. *Annual Rev. Microbiol.,* 34: 497–536.

51. Swaminathan, M.S., 1980. In: *Biotechnology in Agriculture,* (Eds.) S. Natesh, V.L. Chopra and S. Ramachandran, p. 3–11. Oxford and IBH Publ. Co., New Delhi.

52. Subba Rao, N.S., 1982. Biofertilizers. In: *Advances in Agricultural Microbiology,* (Ed.) N.S. Subba Rao. Oxford and IBH, New Delhi.

53. Subba Rao, N.S. and Tilak, K.B.R., 1977. *Souvenir Bull*. Directorate of Pulses Development, Govt. of India.

54. Thiel, T. and Pratte, B., 2001. Effect on heterocyst differentiation of nitrogen fixation in vegetative cells of the cyanobacterium *Anabaena variabilis* ATCC 27413. *J. Bacterial.*, 183: 280–286.

55. Tilak, K.V. and Rao, Subba, 1987. *Biol. Ferti. Soil.*, 4: 97–102.

56. Vaishampayan, A., Sinha, R.P., Hader, D.P., Dev, T., Gupta, A.K., Bhan, U. and Rao, A.L., 2001. *The Botanical Review*, 67(4).

57. Venkataraman, G.S., 1972. *Algal Biofertilizers and Rice Cultivation*. Today and tomorrow Printers and Publ., New Delhi.

58. Venkataraman, L.V. and Becker, E.W., 1985. *Biotechnology and Utilization of Algae: The Indian Experience*. CFTRI, Mysore.

59. Vijaykumari, B., Hiranmai Yadav and Sowmya, M., 2009. *J. Env. Sci. and Engg.*, 51(1): 13–16.

60. Vyas D. and Gupta, R.K., 2007 *Advances in Applied Phycology*. Daya Publ. House, New Delhi, p. 155–168.

61. Wantanabe, I. and Lice, C.C., 1992. Improving nitrogen fixing system and integrating them into successful rice farming. *Pl. Soil*, 141: 57–67.

62. Watanabe, I., 1987. Summary report of the *Azolla* programme of the International Network on soil fertility and fertilizer evaluation for rice. In: *Azolla Utilization*, (Ed.) I. Watanabe. International Rice Research Institute, Manila, p. 197–201.

63. Watson, C.A., 2001. Agronomic and environmental implications of organic farming systems. *Advances in Agronomy*, 70: 261–311.

64. Whitton, B.A. and Roger, P.A., 1989. Use of blue green algae and *Azolla* in rice culture. In: *Microbial Inoculation of Crop Plants*, (Eds.) R. Campbell and R.M. Macdonald, 25: 89–100.

Biotechnology: An Overview (2015)   *Pages* 101–111
Editors: Rajan Kumar Gupta, Nasim Akhtar and Deepak Vyas
Published by: DAYA PUBLISHING HOUSE, NEW DELHI

Chapter 9

Biochemical Approaches for Next Generation Biofuel Production

Akhlash Pratap Singh*

Genomics and Proteomics Lab, Department of Biochemistry,
G.G.D. S.D. College (Punjab University),
Sector 32, Chandigarh – 160 030

ABSTRACT

India is an emerging economic power in the world, but its energy demand and supply is completely mismatched. India's energy fulfillments are virtually depending upon imported fossil-based crude oils which are limited and adversely affect its economy and environmental conditions. Therefore, there is urgent need to find out cost-effective, reliable, renewable source of biofuel production. Presently, bio-ethanol is only the major biofuel used in automobiles, but, suffering some inherent problems like, low energy content, lack of distribution and storage infrastructure, produced from food crops, moreover, its incompatibility with pre-existing fuels also hinder its application. Therefore, need to explore new alternative avenues for bio-fuels production from renewable source of lignin-cellulosic material. The next generation biofuel are like, long chain alcohols, Iso-prenoids and fatty acid-based biochemical. Genetically manipulated plants with lignin modification, high content of polysaccharides will pave the way for next generation of biofuels. The new interdisciplinary approaches of system and synthetic biology have unrevealed and manipulate complex gene expression system for cell-wall and biomass production in energy crop plants. But, next generation biofuel must also possess the important desirable physico-chemical properties of transport fuel and yield are the biggest challenge before us.

Keywords: Biofuel, Lignin-cellulosic material, Synthetic biology, System biology.

---

* *Author:* E-mail: akhlash@gmail.com, akhlash@ggdsd.ac.in

Introduction

India is an emerging economic power in the world, but its energy demand and supply is completely mismatched. India's energy fulfillments are virtually depending upon imported fossil based crude oils which are limited and adversely affect its economy and environmental conditions. Therefore, there is urgent need to find out cost-effective, reliable, renewable source of biofuel production. Biofuels include bioethanol, biodiesels and biogas. In this view India and other countries have launched various programs like, National Biofuel Mission in 2003 and National Biofuel Policy in 2009 respectively. Recently, India government has set target of blending of 5 per cent ethanol with hydrocarbon based fuel by 2011–12 and on later stage this limit will raised upto 10 per cent by 2016–17 and to 20 per cent after 2017 under The Ethanol Blended Petrol Programme (Anonymous, 2012)). The US government aimed production of 36 billion gallons of biofuel production by 2022 (Shinoj, 2011). There are several reasons which have forced to India to focused on biofuel such as, huge bill of fossil fuels, environmental issue, carbon emissions, global warming, and climate changes. Advantages linked with biofuels production are renewal, efficient use of waste unfertile land, improve in the rural income, and clean environment and less reliance on fossil fuels.

Biofuels are 'green' substitute to fossil fuels generated from biomass, a renewable source come from crop materials (sugar and starch, edible and non edible vegetable oils), animal and biological waste. India has deliberately chosen biofuel production based upon molasses and seeds of jatropha and pongamia rather from food crops as in case of developed countries, because scarcity and rise in prices of food already an issue here. Today, Indian biofuel industry is facing many challenges like; inconsistent, erratic monsoon adversely affected the sugarcane and other crops farming, higher production cost, higher ethanol consumptions in industries and economic inviability. These factors have substantially spoiled the possibility to meet the envisaged ethanol mixing target in India. The economics and other issues related to biofuel in Indian and world scenario have been discussed by many authors and interested reader may consult these references (Goldemberg, 2007; Shinoj, 2011). If we see the vital statistics related to biofuel production in India, the total ethanol production was 1073 million tons, which is generally produced from molasses, and (Table 9.1) future demand is much higher than production (Shinoj, 2011) . Due to paucity of space, this chapter provides only brief account of emerging developments in area of biofuel research.

First Generation of Biofuel (2000-2010)

(a) Bioethanol

The first generation biofuels include bioethanol, biodiesel produced, generally from crop plants like, corn, sugar cane, rape seed etc. In the span of last two decades, scientific efforts have substantially increased the amount of starch and sugar in crop plants. Brazil is pioneer country in bioethanol production and obtains 40 per cent and next is the USA generates its 20 per cent bioethanol of its road transport fuel requirement from sugar cane crops (Shinoj, 2011). Continent wise ethanol production has been given in Figure 9.1.

Table 9.1. Projected ethanol requirement in India for various applications (million tons).

Year	Amount of Petrol Requirement	Fuel Ethanol Demand @ 5 per cent Blending with Petrol Fuel	Ethanol Used in other Industries	Potable Demand	Total Ethanol Requirement
2008–09	11.25	0.56	0.60	0.65	1.80
2011–12	14.37	0.72	0.65	0.71	2.08
2016–17	21.61	1.08	0.76	0.84	2.68
2020–21	29.94	1.50	0.85	0.96	3.31

Source: Modified from (P.P. Shinoj, S.S Raju and P.K. Joshi, "India's biofuels production programme: need for prioritizing the alternative options", Indian J Agri Sci,81 (5): 391–397, 2011).

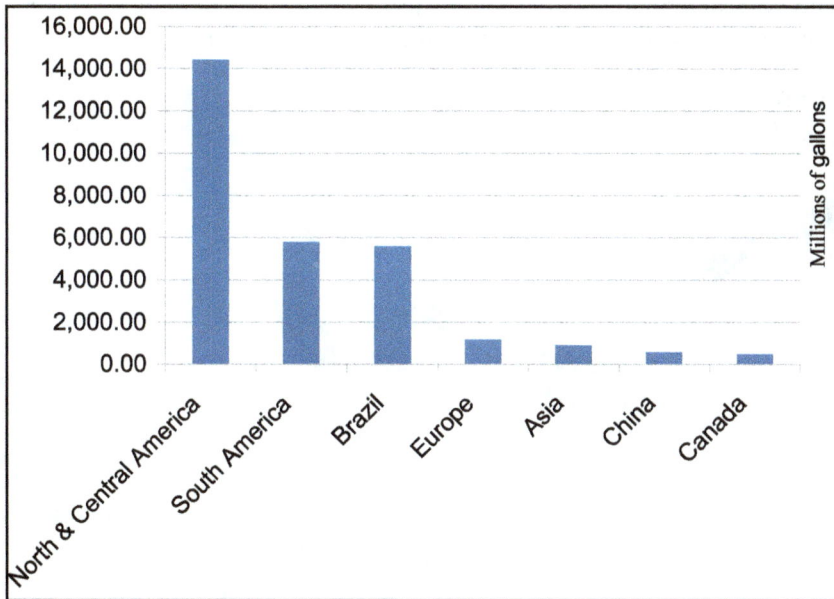

Sl.No.	Continent	Ethanol Production in Million of Gallons
1.	North & Central America	14,401.34
2.	South America	5,771.90
3.	Brazil	5,573.24
4.	Europe	1,167.64
5.	Asia	889.7
6.	China	554.76
7.	Canada	462.3

Figure 9.1: Worldwide ethanol production in 2011 (millions of gallons).
Source: http://ethanolrfa.org/pages/World-Fuel-Ethanol-Production

The general bioethanol formation methodology is that, variety of plant sugars and carbohydrates, mainly cellulose; hemicellulose can be converted into simple sugars such as glucose by costly methods of hydrolysis or various types of pretreatments. Then anaerobic fermentation of sugars is carried out by the specially selected micro-organisms *e.g.* Yeast and *E. coli*, convert sugar into ethanol (Figure 9.2). For this purpose, particularly in developed countries raw substrates are used wheat, sugar beet, corn, straw, and wood (Sticklen, 2008; Vega-Sánchez and Ronald, 2010). Many authors have reviewed the various aspects of bioethanol production process in detailed (Sticklen, 2008; Pauly and K. Keegstra, 2008; Peralta-Yahya and Keasling, 2010). Even today, Ethanol is the only major biofuel used in automobiles but, suffering many inherent drawbacks like, low energy content, lack of distribution and storage infrastructure, produced from grain starch and sugarcane sucrose, moreover, its incompatibility with pre-existing fuel hinder its application as a fuel (King *et al.*, 2009). The bioethanol production can be improved with the modern biotechnological tools and techniques including plant genetic engineering, metabolic engineering and synthetic biology (Vega-Sánchez and Ronald, 2010).

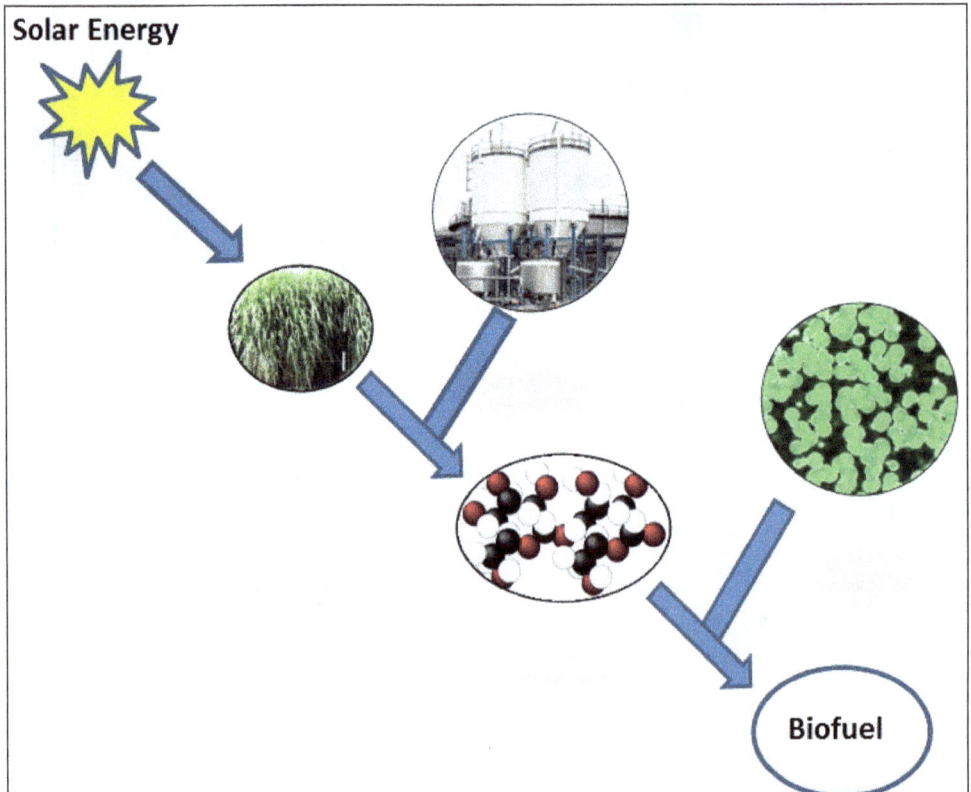

Figure 9.2. Basic outline of biofuel production at industrial scale. The Solar energy is trapped by the energy crops plants by photosynthesis process. The lingo-cellulosic material is breakdown into simple pentose and hexose sugars in large biofuel pretreatments plants. Then sugars subjected to various types of biomass deconstructing micro-organisms and sugars can convert into biofuel.

(b) Biodiesel

Biodiesel is considered better than bioethanol, because it provide 25 per cent more energy, with no corrosion problem, low inputs required, net energy balance ratio is more favorable. Biodiesel consists of the long chain alkyl (methyl, propyl or ethyl) esters of fatty acids, originated from plant, and vegetable oils and alternatively from animal fats. The major source of biodiesel is *Jatropha curcas* plant oil (King *et al.*, 2010) and vegetable oils can be used when mixed with hydrocarbon based diesel. The pure seed or vegetable oil cannot be used in direct-injection diesel engines. Presently, other new emerging sources of biodiesel include soya bean, rapeseed and palm oils. The merits of biodiesel are renewable, higher oxygenated state, no sulfur or aromatic compounds as by product, higher flashpoint, biodegradable, greater lubricity, safe handling and storage. Simultaneously, limited and costly supply of raw material is major hurdle in replacement of fossil fuel with biodiesel. Therefore, need to explore more avenues for next generation bio-energy production from lipids, lignocellulose material mainly from non-food crop plants, perennial grasses and trees.

Second Generation of Biofuel (2010-2030)

The second generation biofuels will obtain from lignin-cellulosic raw materials; nonfood crops agricultural waste material and forest residues. The three main components of lignin-cellulosic biomass are cellulose, hemicellulose and lignin acquire from corn, rice, sugarcane, fast-growing perennial grasses *e.g.* switch grass and giant miscanthus, fast-growing woody poplar and shrub willow. Lignocellulosic biofuel production process include collection of biomass, deconstruction of cell wall polymers into component sugars (by pretreatment and saccharification), and conversion of the sugars to biofuels (anaerobic fermentation). Various scientific strategies have been applied to increase and improve raw material for biofuels production.

(a) Modifications and Use of Plant Cell Wall Carbohydrates for Biofuel

They are most abundant bio-source on the earth but, meagerly exploited. Plants have been genetically manipulated as result they produce cell-wall-degrading enzymes cellulases and hemicellulases indigenously in plant itself (Sticklen, 2008). The heterologous expression of cell-wall-deconstructing proteins in form of total soluble proteins (TSP) in cytosol has been introduced in arabidopsis, tobacco, potato, maize and rice (Ziegler *et al.*, 2000). It will decrease the external use of enzymes and pretreatments that ultimately reduced the production cost.

(b) Lignin Modifications

The cell wall lignin is the major hindrance which undermines the action of enzymes cellulases and hemicellulases to lignin-cellulosic material, therefore two main strategies adopted by researchers, either modify the lignin structure or decrease its amount. The latter option is more viable for industrial point view and scientifically as well. Therefore, several efforts have been made to switch off genes with RNA interference (RNAi) technology which is responsible for lignin biosynthesis. Simultaneously, same methodology was applied to up regulated the metabolic pathway of cellulose and hemicelluloses biosynthesis to increase the amount of polysaccharides in arabidopsis (Ralph, 2006).The chemical structures of Lignin structure is also modifying to reduced the pretreatment cost during bioethanol production. Such types of efforts undertaken in case of alfalfa (*Medicago sativa*) where, enzyme 4coumarate 3hydroxylase (C3H) down regulated that led to change in lignin structure. Lignin centric research is most predominate in Europe and initiated by British Petroleum.

(c) Increase Total Biomass in Plants

According to an estimation, plants generate almost 200 million tons biomass in form of carbohydrates, oils, agricultural waste and other products that can be used as a raw material for biofuel production (http://faostat.fao.org/site/567/default.aspx) In order to increase the total biomass in poplar plant genetic manipulation were introduced to up regulated the gibberellins biosynthesis pathway which result improved growth and an increase in biomass. In rice, an enzyme, ADP-glucose pyrophosphorylase (AGP) responsible for starch biosynthesis in endosperm cause the unexpected 20 per cent increase in plant biomass. The cell wall is interwoven network of sugars and proteins, in this view, a group of enzymes has been identified which increased the accessibility of carbohydrates to cell wall degrading enzymes. Currently, the great impediment is incomplete knowledge related to cell wall structure; biosynthesis and its regulation create major difficulty which prevents to increase the cell wall biomass. Additionally cell wall composition is extremely variable across the plant kingdom (Pauly and Keegstra, 2008). The tempering with cell wall must take with great care it may cause plant more susceptible to pathogen and morphological change.

(d) Applications of TAG for Biofuel

As mentioned above oil seeds from food and nonfood crops can used for biofuel production *e.g.* jatropha and pongamia, rapeseed, sunflower and palm. Oils in form TAGs contain reduced form of energy, therefore, released much higher energy as compared to petroleum products. Traditionally, TAGs energy rich lipids can used as biofuel because of similarity with traditional diesel. But, newly discovered low-molecular-weight TAGs, such as tributyrin (4:0) up to tricaprin (10:0), may be better fuel than normal TAGs from the atomization point view. The genes responsible for improvement and enhancement of amount of TAGs are transferred to *Cuphea*, an oil crop plant. The biosynthetic pathway for TAG synthesis is highly complex, therefore metabolic engineering is difficult. There are certain areas which need to be address like, raw material for biodiesel, deep understanding related to TAGs biosynthetic pathways and their regulation, physical attributes of biodiesel, storage and oxidative stability (Durrett, 2008). The efforts are also on to improve the fatty acid compositions, viscosity of biodiesel via genetic manipulations experiments. A transgenic line of soybean is developed which contains high content of oleic acid.

(e) Terpeniods as a Biofuels

An interdisciplinary approach is needed to avail possibility that, terpeniods can use as alternative to gasoline, diesel, and jet fuel. The biochemical and structural analysis reflect that, terpeniods are more resembling to advanced generation of biofuel like short and long chain terpene (pinene, sabinene, and terpinene). Recently, isoprenoid biosynthetic pathway is studied in E. coli which indicate that, farnesene yield can increased upto 14 g/L, but overproduction of terpene toxic to E. coli itself. To overcome the problem expression of two genes yhfR and nudF increased as consequence isoprene synthase able to convert dimethylallyl pyrophosphate (DMAPP) to isoprene and relieve the prenyl diphosphate toxicity (Bohlmann and Keeling, 2008) and increased the amount of isopentenol upto 1.2 g/L.

Third Generation of Biofuel (After 2030)

Third generation of biofuels will used photoautotrophic microbes, microalgae for catching solar energy for growth and production of hydrogen and lipids. To avail this possibility, best model system is unicellular green algae, *Chlamydomonas reinhardtii* and its genome level information is also available now. The merits linked with algae based biofuel are cultivation in liquid environment, therefore, no competition with food crop for land use. The large scale biology (discussed below) approach has provided the wealth of informations related to genes, proteins and metabolites and network studies in Chlamydomonas. It will make easy to manipulate genes and enzymes involved of biosynthetic pathways in carbohydrate, lipids. The most recent approach of carbon concentration mechanism (CCM) has been experimented in Chlamydomonas. Researchers have annotated 12 isoform of enzyme carbonic anhydrases by using mass western method. Because carbonic anhydrases can bind with CO_2 very fast so utilize in CO_2-sensing pathways. More recently a genetic tool DNA Assembler applied for heterologous expression of complete biosynthetic pathways for biofuel production in chloroplast of *Chlamydomonas reinhardtii* (Noor-Mohammadi, 2012). Genome-scale level metabolic network is constructed for algal metabolism which will proved very helpful to understand light-driven metabolism and quantitative systems biology for biofuel production.

Fourth Generation Biofuel (after 2030)

Fourth generation or carbon negative biofuels will obtained from the process of sequestering (absorb higher amount of CO_2) by using high solar efficient manipulated plants ultimately, effective CO2 removal from the Earth's atmosphere. Especially design plants and micro-organism will be used for this purpose (Lu *et al.*, 2011). The interconversion of formate and potential applications for biofuels production has reported by some authors. Formate is a main metabolite in the methanogenic conversion of complex organic material and can serve as precursors of biofuels (Crable *et al.*, 2011).

Emerging Technologies for Biofuel Production

System Biology

The latest progress in the field of high-throughput experimental technologies like, genomics, proteomics and other 'omics' analysis enable us to switch from reductionist to more holistic approach *i.e.* systems biology. These technologies have made possible to investigate the behaviour of thousands of genes, proteins, metabolites and ions in a single experiment conducted under controlled condition. It has paved the way for quantitively characterization of individual components of biological systems particularly, in model organisms, thus, generation of huge amount of reliable biological data in terms of genomes sequence, mRNA transcripts, census of proteins, metabolites, ions and other biological components. Applications of system level approaches to plant biology have provided much needed help to address most complicated questions in biology like biofuel (Weckwerth, 2011). Global level of gene expression study has conducted in developing physic nut (Jatropha curcas L.) seeds, it will help to understand for genes that influence de novo lipid synthesis, accumulation and their regulatory networks in developing physic nut seeds, and other oil seeds (Jiang *et al.*, 2012).

The plants, micro-organisms and fungus play a significant role in biomass- to- bio-energy conversion processes. These biological conversions require participation of intricate networks of biomolecules, their functioning and regulatory mechanisms controlled by genome (blue print of life). Therefore, genomic information related to plant and micro-organisms will prompted the understanding of genes functions, their regulation and manipulation of metabolic pathways in variety of organisms to enhance biofuel precursors. The next generation of DNA sequencing has make task very easy and sequence of plant genomes, micro-organisms, algae and fungus can read in a span of day or week. The complete genomes have been sequenced for number of plants, fungus, micro-organisms (http://www.genomesonline.org/cgi-bin/GOLD/index.cgi) involved in biofuel production. Additionally, metagenomic studies of various microbial communities also going on for example termite, cattle hind gut community. The recent sequencing of the genomes will facilitate wide array of genomic studies and favorable genetic manipulations in energy crops and other organisms. The another important methodology of system approach is proteomics, deals with systematic analysis of all proteins at organism level including, abundance, structure, modifications, localizations and interactions of protein species and their relationships with other biomolecules(Laurence, 2011). Like other omics technologies, metabolomics is the high throughput study of metabolites in an organism and tissue. System biology knowledge can applied to applied to energy crops, photoautotrophic microbes, algae as consequence biofuel production can be increased with active contribution of genome assisted breeding and genome wide engineering for biomass recalcitrance.

Synthetic Biology

The omics technologies have enabled us to characterize the biological components (genes, proteins, metabolites, RNAs, regulators, promoters in a great detail. In view of these informations, synthetic biology can design novel functional biological organisms and system which can synthesize precursors or biofuels. For example, the tools of synthetic biology such as, switches, oscillators can design a micro-organism which produces nonnative novel metabolites *e.g.* heterologous 1-butanol production. The new interdisciplinary system and synthetic biology approaches has unrevealed complex regulatory system of transcription factors, small interfering RNAs, and regulatory proteins that play crucial role in biomass production in terms of cell wall. Therefore, we can exploit this newly acquired knowledge in construction of partial or complete artificial biological systems will be able to synthesize efficient, inexpensive biofuel(Lamsen and Atsumi, 2012).

The desirable characteristics of a cellular system design for biofuel production are efficient substrate conversion rate, consistent expression of proteins involved in sugar transport and high flux as result over expression of end products. Synthetic biology can contribute to produce such a novel synthetic system. There are two most predominant ways to engineered the organisms, first, model organisms (*e.g.* Yeast, E Coli) (Pauly and Keegstra, 2008) and second, to make use of the new organisms involved in novel functions and for this number of bacterial strains have been targeted on the basis of their biochemical and physiological properties, *Zymomonas mobilis, Pichia stipitis, Clostridium thermocellum* and *Clostridium phytofermentans*. To overcome the problems of ethanol, bacterial strains of *Clostridium acetobutylicum* have engineered for three genes *ctfA, ctfB* and *aad* to produced n- butanol moreover, improvements also introduced by using bioprospecting process (King, 2009).Additionally, multigenic or global transcriptome reprogramming enhance cellular traits in *E. coli, yeasts* and *Lactobacillus*. All types of desirable traits for biofuel production cannot be generate in a single bacteria but, modifications can bring in a better way if, metabolic engineering suitably applied with synthetic biology and artificial biology. There are number of example of these approaches, where synthetic pathways have been constructing for hydrogen production by synthetic pathways in cells. Same types of methods have applied for the production of Iso-prenoids-based fuels which is alternative to traditional method.

We suggest that available bioprospecting tactic should centered to acquire new biological functions rather than on research on new micro-organisms. Recently, the *E coli* is manipulated for the production of fatty esters (biodiesel), fatty alcohols, and waxes from the simple substrate like sugar and moreover, hemicellulase also expressed in cell to breakdown hemicellulose (http://faostat.fao.org/site/567/default.aspx)

Nanotechnology in Biofuel Production

In recent past nanotechnology has revolutionized the every field of science. The Nano approach in biofuel production includes applications of reusable nanocatalyst in bioconversions. The animal butchery waste and waste frying oils containing high level fatty acids can use in biofuel production. In two step process, high quality biodiesel can produced, in first step butchery waste change into oil, solid residue and hydrocarbon gases, then transesterified to biodiesel with NaOH based reaction (Hussain, 2011).

Conclusion

In the coming years, interdisciplinary approach is required to overcome present limitation of biofuels. The latest developments in area of functional genomics, system biology and synthetic biology with metabolic engineering will provide deep understanding and great help in design new biological system and metabolic routes in plants and micro-organisms for biofuel production. But, it has to be seen that, how plant cell walls can exploit in biofuel production and whether it possess important desirable physico-chemical properties of transport fuel like density, viscosity, heat capacity and enthalpy may influence the combustion and exhaustive emission. We hope that in near future, would be able to fulfill our demand of biofuel with desire characteristics. But this is tough challenge like as man on moon and human genome sequence projects.

References

1. Anonymous, 2012. http: //articles.timesofindia.indiatimes.com/2012–07 15/india/32684264_1_ethanol–price–ethanol–supply–crore–litres

2. Anonymous, 2006. http: //faostat.fao.org/site/567/default.aspx; parameter settings: production quantity, cotton lint, world.

3. Anonymous, 2012. http: //www.genomesonline.org/cgi–bin/GOLD/index.cgi

4. Bohlmann, J. and Keeling, I.C., 2008. Terpenoid biomaterials. *The Plant J.*, (54): 656–669.

5. Crable, B.R., Plugge, C.M., McInerney, M.J. and Stams, A.J., 2011. Formate formation and formate conversion in biological fuels production. *Enzyme Res.*, p. 532–536.

6. Durrett, T.P., Christoph, B. and Ohlrogge, J., 2008. Plant triacylglycerols as feedstocks for the production of biofuels. *The Plant Journal*, 54: 593–560.

7. Goldemberg, J., 2007. Ethanol for a sustainable energy. *Future Science*, 315: 808–810.

8. Graham, I.A., 2009. Potential of *Jatropha curcas* as a source of renewable oil and animal feed. *J. Exp. Bot.*, 60(10): 2897–2905.

9. Hussain, S.T., Ali, S.A., Bano, A. and Mahmood, T., 2011. Use of nanotechnology for the production of biofuels from butchery waste. *International Journal of the Physical Sciences*, 6(31): 7271–7279.

10. Jiang, H., Wu, P., Zhang, S., Song, C., Chen, Y., Li, M., Jia, Y., Fang, X., Chen, F. and Wu, G., 2012. Global analysis of gene expression profiles in developing physic nut (*Jatropha curcas* L.) seeds. *PLoS One*, 7(5): e36522.

11. King, A.J., He, W., Cuevas, J.A., Freudenberger, M., Ramiaramanana, D., Lamsen, E.N. and Atsumi, S., 2012. Recent progress in synthetic biology for microbial production of C3–C10 alcohols .*Front Microbiol.*, 3: 196.

12. Laurence, V., Bindschedle and Cramer, R., 2011. Quantitative plant proteomics. *Proteomics*, 11: 756–775.

13. Lü, J., Sheahan, C. and Fu, P., 2011. Metabolic engineering of algae for fourth generation biofuels production. *Energy Environ. Sci.*, (4): 2451–2466.

14. Noor-Mohammadi, S., Pourmir, A. and Johannes, T.W.A., 2012. Method to assemble and integrate biochemical pathways into the chloroplast genome of *Chlamydomonas reinhardtii*", doi: 10.1002/bit.24569.

15. Pauly, M.K. and Keegstra, K., 2008. Cell-wall carbohydrates and their modification as a resource for biofuels. *Plant J.*, 54(4): 559–568.

16. Peralta-Yahya, P.P. and Keasling, J.D., 2010. Advanced biofuel production in microbes. *Biotechnology J*, 2: 147–162.

17. Ralph, J., 2006. Effects of coumarate 3-hydroxylase down-regulation on lignin structure. *J. Biol. Chem.*, 281: 8843–8853.

18. Shinoj, P., Raju, S.S. and Joshi, P.K., 2011. India's biofuels production programme: need for prioritizing the alternative options. *Indian J. Agri. Sci.*, 81(5): 391–397.

19. Steen, E.J., Kang, Y., Bokinsky, B., Hu, Z., Schirmer, A., McClure, A., DelCardayre, S.B. and Keasling, J.D., 2010. Microbial production of fatty-acid-derived fuels and chemicals from plant biomass. *Nature*, 28 463 (7280): 559–562.

20. Sticklen, M.B., 2008. Plant genetic engineering for biofuel production: towards affordable cellulosic ethanol. *Nat. Rev. Genet.*, 9(6): 433–443.

21. Vega-Sánchez, M.E. and Ronald, P.C., 2010. Genetic and biotechnological approaches for biofuel crop improvement. *Curr. Opin. Biotechnol.*, 21(2): 218–224.

22. Weckwerth, W. Green systems biology: From single genomes, proteomes and metabolomes to ecosystems research and biotechnology. *J. Proteomics*, 75(1): 284–305.

23. Ziegler, M.T., Thomas, S.R. and Danna, K.J., 2000. Accumulation of a thermostable endo1,4dglucanase in the apoplast of *Arabidopsis thaliana* leaves. *Mol. Breeding*, 6: 37–46.

Biotechnology: An Overview (2015)
Editors: Rajan Kumar Gupta, Nasim Akhtar and Deepak Vyas
Published by: DAYA PUBLISHING HOUSE, NEW DELHI

Pages 113–127

Chapter 10

Azotobacter: A Plant Growth Promoting Rhizobacteria as Biofertilizer in Organic Farming

S.K. Sethi¹ and S.P. Adhikary²

¹P.G. Department of Biotechnology, Utkal University,
Bhubaneswar – 751 004, Orissa
²Centre for Biotechnology, Visva-Bharati,
Santiniketan – 731 235, West Bengal

ABSTRACT

Nitrogen fixation is mainly responsible for improvement of crop yield. In this regard the diazotrophs like *Rhizobium*, *Azotobacter* and *Azospirillum* are important as they enrich nitrogen nutrition in deficient soils. Of these *Azotobacter* helps plant growth promotion as well as nitrogen fixation. Thus technology has been developed for making use of *Azotobacter* biofertilizer for nitrogen and non-nitrogen fixing plants and popularized by educating about their benefits in agriculture to users for practicing integrated nitrogen management.

Introduction

In agriculture one of the limiting factors is providing plant nutrients, particularly nitrogen and phosphorous to the crops. So the improvement of the crop yield by inoculation with diazotrophs like *Azotobacter*, *Rhizobium*, *Azospirillum* has been suggested as a ecofriendly technology (Choudhary and Kenedy, 2004). These micro-organisms colonize in the rhizosphere of plants, remain in close association with roots and influence their growth. Of these *Azotobacter* is one of the most extensively studied plant growth promoting micro-organism because its inoculation benefits to a wide variety of crops. These are polymorphic, possess peritrichous flagella, produce polysachharides, sensitive to acidic pH, high salts and temperature above 35°C, and can grow on a nitrogen free medium thus utilize atmospheric nitrogen for cell protein synthesis. Cell proteins are mineralized in soil after death of *Azotobacter* and contribute nitrogen availability to the crop plants. Several types of azotobacteria have been found in the soil and the rhizosphere such as *A. chroococcum* (Beijerinck, 1901), *A. nigricans* (Krassilnikov, 1949), *A. paspali* (Döbereiner, 1966), *A. armenicus* (Thompson and Skerman, 1981), *A. salinestris* (Page and Shivprasad, 1991) and *A. vinelandi* (Lipman, 1940) of which *A. chroococcum* is most commonly found in Indian soils. Plant growth promotion by *Azotobacter* may also be attributed to other mechanisms such as ammonia excretion (Narula *et al.*, 1981). Besides nitrogen fixation, they also produce siderophores and antifungal substances (Suneja *et al.*, 1994) and plant growth regulators like hormones and vitamins (Shende *et al.*, 1977; Verma *et al.*, 2001).

Azotobacter and Nitrogen Fixation

Azotobacter belongs to the family *Azotobacteriaceae*. These are gram negative, nonsymbiotic, aerobic diazotroph. The young and rod shaped cells varies from 2.0-7.0 to 1.0-2.5μm and occasionally an adult cell may increase upto 10-12 μm, oval, spherical or rod shaped cells. *Azotobacter* can grow well on simple nitrogen free nutrient medium containing phosphate, magnesium, calcium, molybdenum, iron and carbon sources. Its catabolic versatility in utilizing several aromatic compounds like protocatechuic acid, 2-4-D, 2- chlorophenol, 4-chlorophenol, 2,4,6-chlorotriphenol, aniline, lindane, toluene, p-hydroxy benzoate, benzoate, benzene etc. has been well noted (Hardisson *et al.*, 1969; Balajee and Mahadevan, 1990, Gahlot and Narula, 1996; Moreno *et al.*, 1999, Revillas, 2000; Paul and Anupama, 2008; Thakur 2007). *Azotobacter* contributes significant amounts of fixed nitrogen in, on, or near a plant. The energy requirement for the process of nitrogen fixation is met by a very high rate of aerobic metabolism. The high oxygen demands contribute for the maintenance of minimal intracellular oxygen tension, hence the oxygen sensitive nitrogenase accomplish N_2 fixation (Robson and Postgate, 1980).

Diazotrophic bacteria in the rhizosphere of plants utilize the products of nitrogen fixation for their own growth and release little while they are alive (van Berkum and Bohlool, 1980). Hence when the bacteria die, the quantity of fixed nitrogen in the cells are released and assimilated by plants. Nitrogen fixation by heterotrophic bacteria in the rice rhizosphere develops in response to a deficiency in the availability of combined nitrogen. Thus if fixed nitrogen is not readily available, the plant will be nitrogen deficient. The highest rates of root associated nitrogenase activity were measured in nitrogen deficient plants (van Berkum and Bohlool, 1980; van Berkum and Sloger, 1981). In native bacteria the process of nitrogen fixation is inhibited by combined nitrogen in the environment. These microbes possess three genetically distinct nitrogenase complexes and the expression of these nitrogenases varies with vanadium, molybdenum and ammonium in the culture medium (Bishop *et al.*, 1980). Nitrogenase-I, is expressed only when molybdenum is present in the medium, Nitrogenase-II is expressed only when vanadium is present, while Nitrogenase-III is expressed when both molybdenum and vanadium are absent (Prema Kumar *et al.*, 1988; Chisnelle *et al.*, 1988, Bishop and Joerger, 1990; Harvey *et al.*, 1990, Falik *et al.*, 1991; Joerger *et al.*, 1991). Ammonia is the responsible for the repression of synthesis of all three nitrogenases. Energy requirement for nitrogen fixation is obtained from EMP and TCA cycle (Jackson and Dawes, 1976), and the fixed nitrogen as NH_4^+ is assimilated through GS and GOGAT pathways (Kleiner and Kleinschmidt, 1976). The futuristic projection of nitrogen fixation process in these root associated bacteria is to develop genetically altered strains so that they can fix dinitrogen gas in the presence of repressive levels of combined nitrogen and export a major portion of the nitrogenase-produced ammonia or organic nitrogen by product from their cells in to the rhizosphere and roots. Thus, plants which form associations with depressed mutant bacteria will have an additional source of combined nitrogen available for growth.

Polysaccharide or gum production is one of the characteristic features of *Azotobacter* (Moulder and Brontonegoro, 1974). Some of the species produce polysaccharides in higher quantities leading to formation of capsule around the cell. These EPS (extra cellular polysaccharides) has a composition of principally neutral sugars, *e.g.* rhamnose, mannose and galactose with trace amount of glucose (Horan *et al.*, 1983; Cote and Krull, 1988). The role of EPS has not been clearly established but it has been suggested that they play a major role in providing protection against desiccation, mechnical stress, phagocytosis and phage attack, participate in uptake of metal ions as adhesive agents, ATP sinks or to be involved in interactions between plants and bacteria (Hammad, 1998; Fyfe and Govan, 1983).

The pigments are also an important characteristic of *Azotobacter* as these are produced by all species of *Azotobacter*. *A. chroococcum* produces black, brownish-black, water-insoluble melanin like pigment in old cultures (Zinovyeva, 1962 and James, 1970). A yellow-green, fluroscent pigment is excreted by *A. vinelandii* and *A. paspali*; A red-violet or brownish-black pigment is seen in *A. nigricans*, *A. armeniacus*, cyst *Azotobacter* produce in living dormant cell with two coats and are rich in poly-β-hydroxy butyric acid. With the onset favourable condition, the cysts give rise to vegetative cells. Calcium is reported to be essential for cyst formation in *Azotobacter* (Page and Sadoff, 1975).

Use of *Azotobacter* as Bioinoculants/Biofertilizers

Azotobacter is an important PGPR (plant growth promoting rhizobacteria) Kloepper and Schroth, (1978). Its use as biofertilizer was first advocated by (Gerlach and Voel, 1902) with the purpose of supplementing soil-N with biologically fixed nitrogen. Since then, these organisms have been observed to play a multifaceted role in growth of plant. They not only fix atmospheric dinitrogen under free-living condition but also possess other plant growth promoting activities like phosphate solubilization, production of plant growth hormones like auxins, gibberellins, cytokinins, vitamins and aminoacids (Apte and Shende, 1981). *Azotobacter* has been reported to possess a very high ARA activity and the range of nitrogen fixation is estimated between 2-15mg N fixed/g of glucose consumed (Apte and Shende, 1981). *Azotobacter* sp. also possesses the ability to solubilize phosphates (Shende *et al.*, 1975) and the phosphate solubilization ability ranged from 8-16 per cent . *Azotobacter* produces IAA when tryptophan was added in the medium (Brakel and Hilger, 1965) because tryptophan is the precursor of IAA and is converted to IAA through a primary Trp-aminotransferase reaction. Thus inoculation of *Azotobacter* lead to improved seed germination rate and enhancement of vegetative growth of the inoculated plants

Table 10.1. Summary of the work on use of *Azotobacter* alone or in consortium with other growth promoting rhizobacteria as biofertilizer for different crops.

Organism	Strain	Isolated from	Growth Condition	Findings	Reference
Azotobacter and *Azospirillum*		Obtained from University of Agri. Sc., GKVK, Bangalore and Research Institute, Madurai, Tamil Nadu		Application of *Azotobacter* and *Azospirillum* biofertilizer in irrigated Mulberry under graded level of nitrogen was studied. Better response to *Azotobacter* than *Azospirillum* under low nitrogen with 150 kgN/hac was observed. Leaf nitrogen and crude protein were also significantly higher in *Azotobacter*150 kg N/hac/yr inoculation.	Das et al., 1992
Azotobacter	Ale-3	HAU, Hissar		Field trials were conducted in rabi seasons of 1987-88 and 1988-89 at research farm, HAU, Hissar. Growth and yield were significantly enhanced with the application of *Azotobacter* with and with out nitrogen. Higher plant height and yield attributes of wheat as compared to other treatments or over control was observed due to *Azotobacter*	Hooda and Dahiya, 1992
Azotobacter and PSB (*Pseudomonas striata*)		Obtained from Microbiology division, IARI	Jensens and Pikovaskyas tricalcium phosphate broth	Effect of cotton seed (Var. SRT-1) inoculation with A. chroococcum and *Pseudomonas* in combination at graded doses of nitrogen and phosphorous on the uptake of N and P, Plant height, dry matter weight and yield was studied combined of fertilizer and *Azotobacter* inoculation saved half on the N and P fertilizers.	Potdukhe et al., 1992
A .chroococcum	Bl2	Soil of West Bengal	N-free Burk's medium containing 1 per cent (w/v) glucose as the carbon source	A. chroococcum Bl2 showed higher nitrogen fixation at pH-7.0 with low concentration of potassium nitrate (25mgN/L) and ammonium sulfate (100mgN/ml) at 28°C. This strain showed tolerance against NaCl (0.12 per cent). The pesticide, Rogor inhibited growth as well as acetylene reduction at very low concentrations.	Jana and Mishra, 1994

Contd...

Table 10.1–Contd...

Organism	Strain	Isolated from	Growth Condition	Findings	Reference
A. chroococcum	MKU 201 B-8005 BKMB-1030, A-41	Tropical soil	Jensens agar medium	The temperature optima for high survival and efficiency in nitrogen fixation varied among the strains of *A. chroococcum*. It ranged between 20-35°C. Performance of temperate strains 8005 and BKMB-1030 was better at low temperature (20°C) and that of tropical strains B and MKU 109 was appreciable even at high temperature (40°C) A-41 strain exhibited tolerance over a wide range of temperatures; however, extremes of temperature reduced its growth and efficiency of nitrogen fixation.	Rajkumar and Lakshmanan, 1995
A. chroococcum	RH-30, WH-147, MAC-27, E-12	Deptt. of Plant breeding CCS HAU, Hisar	Jensen medium containing trace element, sodium glutamate as 'N' source and EDA-HCl (EDA, 0.05-1 per cent) incubated at 30°C for 48-72h	Ethylene diamine (EDA) resistant mutants (MAC-27) were found to fix nitrogen in presence of high concentration of NH_4^+ and also excreted NH_4^+. E-12 exhibited low glutamine synthetase (GS) activity was reduced NH_4^+ uptake in the mutant due to GS-induced deficiency in ammonia assimilation. Yield and dry matter in mustard and grain yield of wheat were greater with E-12 inoculation than with parent MAC-27 under green house condition.	Narula *et al.*, 1999
Azotobacter and *Rhizobium*	Azotobacter strains: W-5, CBD-15 and C-11 Rhizobium strains: BG-256 (Chick pea)	Collected from division of Microbiology, IARI, New Delhi		The effect of inoculation of chickpea seeds with three strains of *A. chroococcum* in combination with *Rhizobium* in a non-sterile soil was studied. Inoculation with either *A. chroococcum* or *Rhizobium* alone increased the nodule number, weight and yield of chick pea.	Paul and Verma, 1999

Contd...

Table 10.1–Contd...

Organism	Strain	Isolated from	Growth Condition	Findings	Reference
A. chroococcum and Trichoderma viride	W-5 and ITCC 1433, 1662, 2185, 3235 and 3255	Obtained from division of Microbiology, IARI		Strains of *T. viride* were used for solid state fermentation (SSF) of Sorghum straw after adjusting the C: N ratio to 35:1 to study the effect of the fermented residues alone and in combination with *A. chroococcum* W5 as biofertilizer for wheat. Inoculation with W5 alone increased the biomass and yield by 25 per cent over that in control. Fermented residue of *T. viride* ITCC 1433 applied in combination with *A. chroococcum* decreased the yield.	Nain et al., 2000
A. chroococcum	Mala-11 and HT54	Collected from Deptt. of Microbiology, CCS, HAU, Hisar	Jensens N-free medium at 30°C for 18 days	Plant growth regulators like gibberllin, kinetin and Indole acetic acid were produced by *Azotobacter*. Out of 20 isolates of *Azotobacter* 4 isolates produced all three PGRs, 14 produced GA$_3$ and 10 produced kinetin. All the isolates except Mala-11 amd HT-54 produced one of the three phytohormones.	Verma et al., 2001
Azotobacter sp.		Soil microbiology laboratory, G.B. pant University of Agriculture and Technology, Pantnagar	Jensens broth 28±2°C for 7 days	*Azotobacter* culture differed greatly in intrinsic resistance to streptomycin, tetracycline, trimethoprin, nalidixic acid and rifampicin. 14 cultures inhibited growth of *Fusarium oxysporum*. In modified JAM-PDA medium none of the *Azotobacter* strains inhibited the growth of *Microphomina phaseolina* and *Sclerotium rolfsii*. No relationships could be observed between the fungal inhibition and the antibiotic resistance of the diazotrophs	Agrawal and Singh, 2002

Contd...

Table 10.1–Contd...

Organism	Strain	Isolated from	Growth Condition	Findings	Reference
B. japonicum and A. chroococum				The effect of inoculation of *Bradyrhizobium japonicum* and *A. chroococum* on soybean [*Glycine max* (L) Merill var. ransom] was studied. Dual inoculation proved best to enhance the plant growth parameters. Inoculation with *Azotobacter* alone was better than uninoculated control.	Bhattarai and Prasad, 2003
A .chroococum		Collected from Deptt. of Microbiology, CCS HAU, Hisar	Grown in Jensen N-free medium for 72h at 30°C	Sixteen isolates of *A. chroococum* were studied for azide resistance. Azide sensitive mutants were developed which is widely prevalent among the isolates. Azide resistance showed no significant correlation with rate of respiration, ATP concentration, activity of cytochrome-c-oxidase and nitrogen fixation.	Vasudeva *et al.,* 2003
Azotobacter and Azospirillum				Total biomass yield was increased under all the soil amendment treatments and inoculation treatments. Highest increase in biomass yield was obtained with influenced by combination with *Azotobacter* and *Azospirillum* was observed.	Pattanayak *et al.,* 2004
A. chroococum	BG-13 and BG-33		Jensens 'N' free medium with sucrose (0.25 per cent) and with 2500 ppm 2,4 D	Four *A. chroococum* strains from soils enriched with 2, 4-D were studied for metabolism of the compound. All 4 strains degraded 2,4-D to chlorocatechol even at a concentration of 2500 ppm in presence of sucrose as 'C' source and with out any additional 'C' source in soil. Chlorocatechol formation was observed even at stationary phase of cells indicating co-metabolism of 2, 4-D. Nitrogenase activity in these strains remained unaffected upto 50ppm of 2,4-D Accumulation of chlorocatechol with less cell density indicates that some strains may not metabolize the intermediary product.	Gahlot and Narula, 2004

Contd...

Table 10.1–Contd...

Organism	Strain	Isolated from	Growth Condition	Findings	Reference
Azotobacter		Rice fields, wheat fields, vegetable gardens and grass lands	Bacteria isolated by serial dilution and plating technique in nitrogen free medium and general purpose medium (glucose-yeast extract agar)	Total Azotobacter population decreased with increasing soil moisture content. Maximum Azotobacter population was recorded in month of March. When the soil moisture content was <15 per cent during May–June it showed sharp decline. Grassland field had highest Azotobacter population compared to other fields. Similar trend was also observed in rhizospheric soils collected from vegetable garden and grasslands which were not waterlogged.	Sharma and Bhatta-charjee, 2004

(Apte and Shende, 1981). *Azotobacter* inoculation also improve plant growth indirectly through suppression of phytopathogens or reducing their deleterious effects (Pandey and Kumar, 1990) and reduced the incidence of the fungal, bacterial and viral diseases of the crops (Meshram, 1984; Pandey and Kumar, 1990). *A. chroococcum* inoculation supports maximum reduction in nematode infection upto 48 per cent followed by *Pseudomonas* upto 11 per cent and *Azospirillum* (4 per cent) (Chahal and Chahal, 1988). Further, since competition of iron is one of the well known mechanisms of biocontrol under iron limiting conditions and these bacteria produce a range of iron chelating compounds such as siderophores which have high affinity to ferric ion, the siderophores produced by *Azotobacter* bind most of the iron (Fe^{3+}) available in the rhizosphere making unavailable to pathogens present in soil. The pathogens may not have the ferrisiderophore receptor for the uptake of iron-siderophore complex, hence are not proliferating due to lack of iron in *Azotobacter* inoculated soils (Ósulivan and ÓGara, 1992).

Azotobacter Inoculation vs. Crop Yield

Improvement in crop production due to *Azotobacter* inoculation has been reported extensively (Hooda and Dahiya 1992; Paul and Verma 1999; Nain *et al.*, 2000; Bhattarai and Prasad, 2003; Pattanayak *et al.*, 2004). Yield of number of crops increased by inoculation with *Azotobacter*. (Rashid *et al.*, 1999; Sethi and Adhikary, 2009) reported that artificial inoculation of seeds of wheat crop with *A. chroococum* leads to increase of dry matter in wheat by 42 per cent over control. Similarly 10-18 per cent increase in yield of bean, corn and potato due to *Azotobacter* inoculation had been observed (Sheloumova, 1935).

Azotobacter inoculation also significantly increased the weight of plant, grain yield and nitrogen content of plant in wheat, maize and cotton crops (Apte and Shende, 1981). The foliar spray of *Azotobacter* significantly increased the grain and straw yield of rice crop (Kanniyan *et al.*, 1980). The use of *Azotobacter* inoculation has a great potential in oilseeds like Mustard (Gerlach and Vogel, 1902; Schmidt, 1960). Sunflower (Badve *et al.*, 1977) and Brassica. Vegetable crop like tomato, brinjal, cabbage, onion, potato, radish, chillies and sweet potato also responded positively to azotobacterisation (Joi and Shinde, 1976; Imam and Badaway, 1978; Khuller *et al.*, 1978 Sethi and Adhikary, 2009). Similar was also for other crops like sugarcane (Agrawal *et al.*, 1987), fruit trees (Kerni and Gupta, 1986; Pandey *et al.*, 1986), pearl millet (Wani *et al.*, 1988), Sorghum (Jadav *et al.*, 1991), Jute (Poi and Kabi, 1979), and Cotton (Apte and Shende, 1981, Paul *et al.*, 2002). Synergestic effect of co-inoculation of *Azotobacter* with *Rhizobium* in pea (Paul and Verma, 1999), chickpea (Verma *et al.*, 2000), groundnut (Rashid *et al.*, 1999) was also observed.

Mass Production Methods of *Azotobacter* Biofertilizer and Field Application Protocols

Strains of *Azotobacter* were isolated from different cultivated fields in southern region of Orissa. Samples were collected from adhering soils of uprooted plants and used for isolation of strains through serial dilution and plating techniques using *Azotobacter* isolation media containing (g/l): sucrose-20.0, K_2HPO_4-1.0, $MgSO_4.7H_2O$-

0.5, Na_2MoO_4-0.001, $FeSO_4.7H_2O$-0.01 and $CaCO_3$-2.0, pH-7.0-7.2. The cultures were incubated at 28±2°C for 4-5 days. The colonies produced in agar medium at 28°C. were white, translucent, circular shape. Basing on higher growth rate and tolerance to different environmental variables, efficient strains were selected for use as biofertilizer. For this purpose loops of the respective colonies were inoculated in sterile nitrogen free medium and grown for 5-7 days. This starter culture was inoculated into a 500 ml flask with bacterial suspension 10^5 C.F.U/ml and grown in rotary shaker at 120 rpm for 5 days at 30°C. For field experiments, 20 days old healthy seedlings was taken and roots were dipped in bacterial culture suspension for 20-30 min for proper attachment and then planted immediately. The controls were treated with normal water.

Future Prospective

Quantitative understanding of the ecological factors that control the future and performance of BNF systems in crop fields is essential for promotion and successful adoption of these technologies. More information about molecular genetic studies on *Azotobacter* inoculants will improve understanding of the root behavior towards the inoculated microbes. The interactions between host plant cells and the diazotrophs that lead to the expression of the various nitrogen fixing, growth promoting genes in *Azotobacter* is a front line area of research today. To induce most of the nif genes essential for functional nitrogen fixation or signal exchanges between plant cells and bacteroids is necessary to be understood fully. The molecular biology techniques for N-fixation are now opening the way to identifying plant molecules that play an essential role during root colonization by bacteria. Mutant strains are particularly valuable for such study, including search of signals which is essential for the root colonization and better nitrogen fixation. Research in the coming decade is to elucidate the various signals and signal transduction pathways operating during the regulation of expression of *nif* gene and some other gene cascades responsible for producing growth promoting substances and siderophore production and their host specificity. Many aspects of the molecular signaling between *Azotobacter* and cereal and non-cereal plant have been already reported (Okon, 1985; Kenedy *et al.*, 1997). The lipochitin oligosaccharides (LCOs) that have been shown as the major determinant of the host specificity of the root colonization (Denarie *et al.*, 1996). It is also needed to understand the mechanisms of N gains and losses, and to identify and refine appropriate soil and crop management practices to maximize soil N and non-leguminous and leguminous BNF in crop fields.

Conclusion

Azotobacter is a broad spectrum biofertilizer and can be used as inoculants for most of agricultural crops. Earlier, its utility as a biofertilizer was neglected due to its relatively low population in the rhizosphere of plants. But at the same time, the fact can not be overlooked is that their seed inoculation in several crops brought about phenomenal increase in the yield (Apte, 1978 and Meshram, 1981). It's well known nitrogen nutritional function is now recognized to play a multiple role in helping the plant to improve its growth potential and yield, and this has revived interest in this rhizobacterium.

References

1. Agrawal, S. and Shende, S.T., 1987. Tetrazolium reducing micro-organisms inside the root of *Brassica* sp. *Curr. Sci.*, 56: 187–188.

2. Agrawal, N. and Singh, H.P., 2002 Antibiotic resistance and inhibitory effect of *Azotobacter* on soil borne plant pathogens. *Ind. J. Microbiol.*, 42: 245–246.

3. Apte, R. and Shende, S.T., 1981a. Studies on *Azotobacter chroococcum* I. Morphological, biochemical and physiological characteristics of *A. chroococcum*. *Zbl Bakt II Abt*, 136: 548–554.

4. Apte, R. and Shende, S.T., 1981b. II. Effect of *Azotobacter chroococcum* on germination of seeds of agricultural crops. *Zbl Bakt II Abt*, 136: 555–559.

5. Apte, R. and Shende, S.T., 1981c. Establishment of *A. chroococcum* on roots of crop plants. *Zbl Bakt II*, 136: 560–562.

6. Apte, R. and Shende, S.T., 1981d. Seed bacterization with strains of *A. chroococcum* and their effect on crop yields. *Zbl Bakt II*, 136: 637–640.

7. Badve, D.A., Konde, B.K. and More, B.B., 1977. Effect of Azotobacterization in combination of different levels of nitrogen on yield of sunflower (*Helianthus annuus*) Laboratory studies. *Food Farming Agric.*, 8: 23.

8. Balajee, S. and Mahadevan, A., 1990. Utilization of chloroaromatic substances by *Azotobacter chroococcum*. *Syst. App. Microbiol.*, 13: 194–198.

9. Beijerinck, M.W., 1901. Über ologonitrophile mikroben. *Zentralbl Bakteriol Parasitenkd Infektionskr II Abt*, 7: 561–582.

10. Bishop, P.E. and Joerger, R.D., 1990. Genetics and molecular biology of alternative nitrogen fixation system. *Plant Physiol. Plant Mol. Biol.*, 41: 109–125.

11. Bishop, P.E., Jarlenski, D.M.L. and Hetheringtion, D.R., 1980. Evidence for an alternative nitrogen fixation system in *Azotobacter vinelandii*. *Proc. Natl. Acad. Sci.*, *USA*, 77: 7342–7346.

12. Bhattarai, H.D. and Prasad, B. N., 2003. Effect of dual inoculation of *Bradyrhizobium japonicum* and *Azotobacter chroococcum*. *Ind. J. Microbiol.*, 43: 139–140.

13. Brakel, J. and Hilger, F., 1965. Etude qualitative et quantitative de la synthese de substances de nature auxinique par *Azotobacter chroococcum in vitro*. *Bull. Inst. Agron. Stn. Res., Gembloux*, 33: 469–487.

14. Brown, M.E., Jackson, R.M. and Burlingham, S.K., 1968. Growth and effects of bacteria introduced into the soil. In: *Ecology of Soil Bacteria*, (Eds.) T.R.G. Cray and M. Parkinson, Liverpool, p. 531–551.

15. Chahal, P.P.K. and Chahal, V.P.S., 1988. Biological control of root-knot nematode of brinjal (*Solanum melongena* L.) with *Azotobacter chroococcum*. In: *Advances in Plant Nematology*, (Eds.) M.A. Maqbool, A.M. Golden, A. Gbaffilr, L.R. Krusberg. Proceeding of the US-Pakistan International Workshop on Plant Nematology, April 6–8, 1986, Karachi, Pakistan, pp. 257–263.

16. Chisnelle, J.R., Premkumar, R. and Bishop, P.E., 1988. Purification of a second alternative nitrogenase from a nif HDK deletion strain of *A. vinelandii. J. Bact.*, 170: 27–33.

17. Cote, G.L. and Krull, L.H., 1988. Characterization of extracellular polysaccharides from *Azotobacter chroococcum. Carbohydrate Res.*, 18: 143–152.

18. Das, P.K., Ghosh, A., Choudhury, P.C., Katiyar, R.S. and Sengupta, K., 1992. In: Response of irrigated Mulberry to Azotobacter and Azospirillum biofertilizers under graded levels of nitrogen, In: *Biofertilizer Technlogy Transfer*, (Ed.) L.V. Gangawane. Associated Publishing Company, New Delhi, pp. 71–77.

19. Denarie, J., Debelle, F. and Prome, J.C., 1996. *Rhizobium* lipo-chitooligosaccharide nodulation factors: Signalling molecules mediating recognition and morphogenesis. *Ann. Rev. of Biochemistry*, 65: 503–535.

20. Fallik, E., Chan, Y.K. and Robson, R.L., 1991. Detection of alternative nitrogenase in aerobic gram-negative nitrogen fixing bacteria. *J. Bact.*, 173: 365–371.

21. Fyfe, J.A.M. and Govan, J.R.W., 1983. Synthesis, regulation and biological function of bacterial alginate. *Prog. Ind. Microbiol.*, 18: 45–83.

22. Gahlot, R. and Naurla, N., 1996. Degradation of 2,4–dicholorophenoxy acetic acid by resistant strains of *Azotobacter chrococcum. Ind. J. Microbiol.*, 36: 141–143.

23. Gerlach, M. and Vogel, J., 1902. Nitrogen fixing bacteria. *Z. Bakt Abt*, 2: 817.

24. Hammad, A.M.M., 1998. Evaluation of alginate encapsulated *Azotobacter chroococcum* as a phage resistant and an effective inoculum. *J. Basic Microbiol.*, 1: 9–16.

25. Haardisson, C., Sala-Trepat, J.M. and Stainer, R.Y., 1969. Pathways for the oxidation of aromatic compounds by *Azotobacter. J. Gen Microbiol.*, 59: 1–11.

26. Harvey, I., Arber, J.M., Eady, R.R., Smith, B.E., Garner, C.D. and Hasnain, S.S., 1990. Iron K-edge X-ray absorption spectroscopy of the iron-vanadium cofactor of the vanadium nitrogenase from *A. chroococcum. Biochem J. (London)*, 266: 929–931.

27. Hooda, I.S. and Dahiya, D.R., 1991 Effect of biofertilizer on wheat production. In: *Biofertilizer Technlogy Transfer*, (Ed.) L.V. Gangawane. Associated Publishing Company, New Delhi, pp. 67–69.

28. Horan, N.J., Jarman, T.R. and Dawes, E.A., 1983. Studies on some enzymes of alginic acid biosynthesis in *Azotobacter vinelandii* grown in continuous culture. *J. Gen Microbiol.*, 129: 2985–2990.

29. Imam, M.K. and Badaway, F.H., 1978. Response of three potato cultivars to inoculation with *Azotobacter. Potato Res.*, 21: 1–6.

30. Jackson, F.A. and Dawes, E.A., 1976. Regulation of tricarboxylic acid and poly-ß-hydroxy butyrate metabolism in *Azotobacter beijerinckii* grown under nitrogen or oxygen metabolism. *J. Gen Microbiol.*, 97: 303–312.

31. Jadav, A.S., Shaikh, A.A. and Harinarayana, G., 1991. Response of rainfed pearlmillet (*Pennisetum glaucum*) to inoculation with nitrogen fixing bacteria. *Ind J. Agri Sci.*, 61: 268–271.

32. Jana, S.C. and Mishra, A.K., 1994. Factors affecting the growth and acetylene reduction of *Azotobacter chroococcum* BI2. *Ind. J. Microbiol.*, 34: 229–232.

33. Joerger, R.D., Elizabeth, D.W. and Bishop, P.E 1991. The gene encoding dinitrogenase reductase 2 is required for expression of the second alternative, nitrogenase from *Azotobacter vinelandii*. *J. Bact.*, 173: 4440–4446.

34. Joi, M.B. and Shinde, P.A., 1976. Response of onion crops to *Azotobacterization*. *J. Maharasthra Agric Univ.*, 1: 161.

35. Kannaiyan, S., Govindarajan, K and Lewin, H. D., 1980. Effect of foliar spray of *Azotobacter chroococum* on rice crop. *Plant Soil*, 56: 487–490.

36. Kennedy, I.R., Pereg, Gerk, L.L., Wood, C., Deaker, R., Gilchrest, K. and Katupitiya, S., 1997. Biological nitrogen fixation in non-leguminous field crops: facilitating the evolution of an effective association between *Azospirillum* and wheat. *Plant Soil*, 194: 65–79.

37. Kerni, P.N. and Gupta, A., 1986. Growth parameters affected by *Azotobacterization* of mango seedlings in comparison to different nitrogen doses. *Res. Develop. Factor*, 3: 77.

38. Khuller, S., Chahal, V.P.S. and Kaur, P.P., 1978. Effect of *Azotobacter* inoculation on chlorophyll content and other characters of carrot, radish, brinjal and chillies. *Ind. J. Microbiol.*, 18: 138.

39. Kleiner, D. and Kleinschmidt, J.A., 1976. Selective inactivation of nitrogenase in *Azotobacter vinelandii* batch cultures. *J. Bact.*, 128: 117–122.

40. Kloepper, J. W. and Schroth, M. N., 1978. Plant growth promoting rhizobacteria on radish. In: *Proc of the 4ᵗʰ Internat Conf on Plant Pathogenic Bacteria* 2, Station de Pathologie Vegetale et Phytobacteriologie, INRA, Angers, France. pp. 879–882.

41. Lipman, C.B. and Mac Lees, C., 1940. Dissociation of *Azotobacter chroococcum* (Beijerinck). *Soil Sci.*, 50: 75–82.

42. Meshram, S.U., 1984. Suppressive effect of *Azotobacter chroococcum* on *Rhizoctonia solani* infestation of potatoes. *Neth. J. Pl. Path.*, 90: 127–132.

43. Moreno, J., Vargas-Gracia, C., Lopez, M. J. and Sanchez-Serrano 1999. Growth and exopolysaccharide production by *Azotobacter vinelandii* in soil phenolic compounds. *J. Appl. Microbiol.*, 86: 439–445.

44. Narula, N., Lakshminarayana, K.L. and Tauro, P., 1981. Ammonia excretion by *Azotobacter chroococcum*. *Biotech. Bioeng.*, 23: 467–470.

45. Narula, N., Kukreja, K., Suneja, S and Lakshminarayana, K., 1999. Ammonia Excretion by Ethlene diamine Resistant (EDAᴿ) Mutants of *Azotobacter chroococcum*. *Ind. J. Microbiol.*, 39: 93–97.

46. Nain, L.R., Paul, S. and Verma, O.P., 2000. Solid state fermentation of Sorghum straw with cellulolytic *Trichoderma viride* strains and its effect on wheat in conjunction with *Azotobacter chroococcum* strain W5. *Ind. J. Microbiol.,* 40: 57–60.

47. Okon, Y.P.G., 1985. *Azospirillum* as a potential inoculant for agriculture. *TIB Technology,* 3: 223–228.

48. Page, W.J. and Sadoff, H.L., 1975. Relationship between calcium and uronic acids in the encystment of *Azotobacter vinelandii. J. Bact.* 122: 145–151.

49. Page, W.J. and Shivprasad, S., 1991. *Azotobacter salinestris* sp. nov a sodium dependent, microaerophilic and aeroadaptive nitrogen fixing bacterium. *Int. J. Syst. Bacteriol.,* 41: 369–376.

50. Pattnayak, S.K., Mohanty, R.K. and Sethi, A.K., 2004. Response of Okra to *Azotobacter* and *Azospirillum* inoculation grown in acid soil amended with lime and FYM. In: *Biotechnology in Sustainable and Organic Farming,* (Ed.) Yadav, Chaudhary and Talukdar. Shree Publishers, New Delhi, pp. 67–69.

51. Pandey, A. and Kumar, S., 1990. Inhibitory effect of *Azotobacter chroococcum* and *Azospirillum brasilense* on a range of rhizosphere fungi. *Ind. J. Expt. Biol.,* 28: 52–54.

52. Pandey, R.K., Bahl, R.K. and Rao, P.R.T., 1986. Growth stimulating effects of nitrogen fixing bacteria on oak seedling. *Indian Forester,* 112: 75.

53. Paul, S. and Verma, O.P., 1999. Influence of combined inoculation of *Azotobacter* and *Rhizobium* on the yield of chickpea (*Cicer arietinum* L.). *Ind. J. Microbiol.,* 39: 249–251.

54. Paul, S., Verma, O. P., Rathi, M. S. and Tyagi, S. P., 2002a. Effect of *Azotobacter* inoculation on seed germination and yield of onion (*Allium cepa*). *Ann. Agril. Res.,* 23: 297–299.

55. Paul, S. and Verma, O.P., 1999. Influence of combined inoculation of *Azotobacter* and *Rhizobium* on the yield of Chick pea (*Cicer arietinum* L.). *Ind. J. Microbiol.,* 39: 249–251.

56. Poi, S.C. and Kabi, M.C., 1979. Effect of *Azotobacter chroococcum* inoculation on the growth and yield of jute and wheat. *Ind. J. Agric. Sci.,* 49: 478.

57. Potdukhe, S.R., Patil, A.R., Somani, R.B. and Wangikar, P.D., 1992 Effect of *Azotobacter* and Phosphate solubilizing micro-organism on yield of cotton variety SRT–1. In: *Biofertilizer Technlogy Transfer* (Ed.) L.V. Gangawane. Associated Publishing Company, New Delhi, pp. 79–83.

58. Rajkumar, K. and Lakshmanan, M., 1995. Influence of temperature on the survival and nitrogen fixing ability of *Azotobacter chroococcum* beij. *Ind. J. Microbiol.,* 35: 25–30.

59. Revilas, B., Pozo, C., Martinez-Toledo, M.V. and Gonzalez, L.J., 2000. Production of B group by two *Azotobacter* strains with phenolic compound as sole carbon source under diazotrophic and adiazotrophic conditions. *J. Appl. Microbiol.,* 89: 486–493.

60. Schimidt, O.C., 1960. *Azotobacter* inoculation. *Soil Fertilizers*, 12: 1368.

61. Sheloumova, A.M., 1985. The use of *Azotobacter* as bacterial manure for non-leguminous plants. *Bull. State. Inst. Agril. Microbiol (USSR)*, 6: 48.

62. Sumana, D.A. and Bagyaraj, D.J., 2002. Interaction between VAM fungus and nitrogen fixing bacteria and their influence on growth and nutrition of Neem. *Ind. J. Microbiol.*, 42: 295–298.

63. Suneja, S., Narula, N., Anand, R.C. and Lakshminarayana, K., 1996. Relation of *Azotobacter chroococcum* siderophores with nitrogen fixation. *Folia Microbiol.*, 4: 154–158.

64. Thompson, J.P. and Skerman, V.B.D., 1981. Validation list No.6. *Int. J. Syst. Bacteriol.*, 31: 215–218.

65. Vani, S.P., Chandrapalaih, S., Zambre, M.A. and Lee, K.K., 1988. Association between nitrogen fixing bacteria and pearl millet. *Plant Soil*, 110: 289–302.

66. Vasundhara, G., Kurup, G.M., Jacob, V.B., Sethuraj, M. R. and Kothandaraman, R., 2002. Nitrogen Fixation by *Azotobacter chroococcum* under cadmium stress. *Ind. J. Microbiol.* 42: 15–17.

67. Vasudeva, M., Kharb, P., Vashisht, R. K., Narula, N. and Merbach, W., 2003. Azide resistance in *Azotobacter chroococcum* and its relationship with respiratory activity, ATP concentration, cytochrome C-oxidase and nitrogen fixation. *Ind. J. Microbiol.* 43: 49–52.

68. Verma, A., Kukreja, K., Pathak, D.V., Suneja, S. and Narula, N., 2001. *In vitro* production of plant growth regulators (PGRs) by *Azotobacter chroococcum*. *Ind. J. Microbiol.*, 41: 305–307.

Biotechnology: An Overview (2015)   *Pages 129–155*
Editors: Rajan Kumar Gupta, Nasim Akhtar and Deepak Vyas
Published by: DAYA PUBLISHING HOUSE, NEW DELHI

Chapter 11

Arbuscular Mycorrhizal Fungi: The Geo-Engineers

Deepak Vyas[1]*, Meenakshi Singh[2], Pradeep Kumar Singh[2], Rajan Kumar Gupta[3] and Mohd. Irfan[3]

[1]*Lab of Microbial Technology and Plant Pathology, Department of Botany, Dr. H.S. Gour University, Sagar – 470 003, M.P.*
[2]*Department of Botany, Guru Ghasi Das University, Bilashpur, Chhatisgarh*
[3]*Department of Botany, Dr. P.D.B.H. Government Post Graduate College, Kotdwar, Pauri Garhwal, Uttarakhand.*

ABSTRACT

Soil is not a pile of dirt but is treasure house of earthy material and home of many micro-organisms as it was rightly said that "we think that we are standing on the earth but reality is that we are standing on the roof of another world". Rhizospheric region of the soil is one of the important rooms where plant, soil and microbes interact. However, these interactions are of many types but among these, mutulisitic interactions are of great importance. Arbuscular mycorrhizal symbioses are one of the mutualistic relationships among the plant, soil and other microbes. In this review we have tried to summerize some of the multifacet role performed by the AMF fungi as geo-engineers for the benefit and better plant growth, soil remediation, disease control, dialogue among the co-organisms and their importance in the modern biology such as molecular biology, genetics, genomics, proteomics and fluxomics.

Keywords: AM fungi, Symbiotic bio-engineers, and Rhizosphere.

---

* *Corresponding Author:* E-mail: dvyas64@yahoo.co.in

Introduction

Arbuscular mycorrhizal (AM) symbioses are the most widespread mycorrhizal association, with a very long evolutionary history. Fossil survey and molecular data have demonstrated that the evolutionary history of AM fungi goes back at least to the Ordovician, about 460 million years ago [1]. The earliest land plant show an association with fungi that formed vesicles and arbuscules that was very similar to today's AM fungi [2-3] and Vyas *et al.*, [4-5] reported occurrence of arbuscules and vesicles in bryophytes. This type of plant-fungal association is formed with 80 per cent of angiosperms [6]. The fungi are obligate symbionts and they need a living plant root (or equivalent structure) in order to grow and reproduce. Arbuscular mycorrhizal fungal development in roots of host plants starts when fungal hyphae grow from spores or from colonized roots toward the uncolonized roots. After contact of the hyphae with the root surface, the fungus is stimulated to change in morphology from an original simple branching pattern to irregularly septate pattern with reduced interhyphal spacing [7-8]. The fungus produces swollen appressoria on the root surface and spreads between and into the root cortical cells (Figure 11.1). There are two major

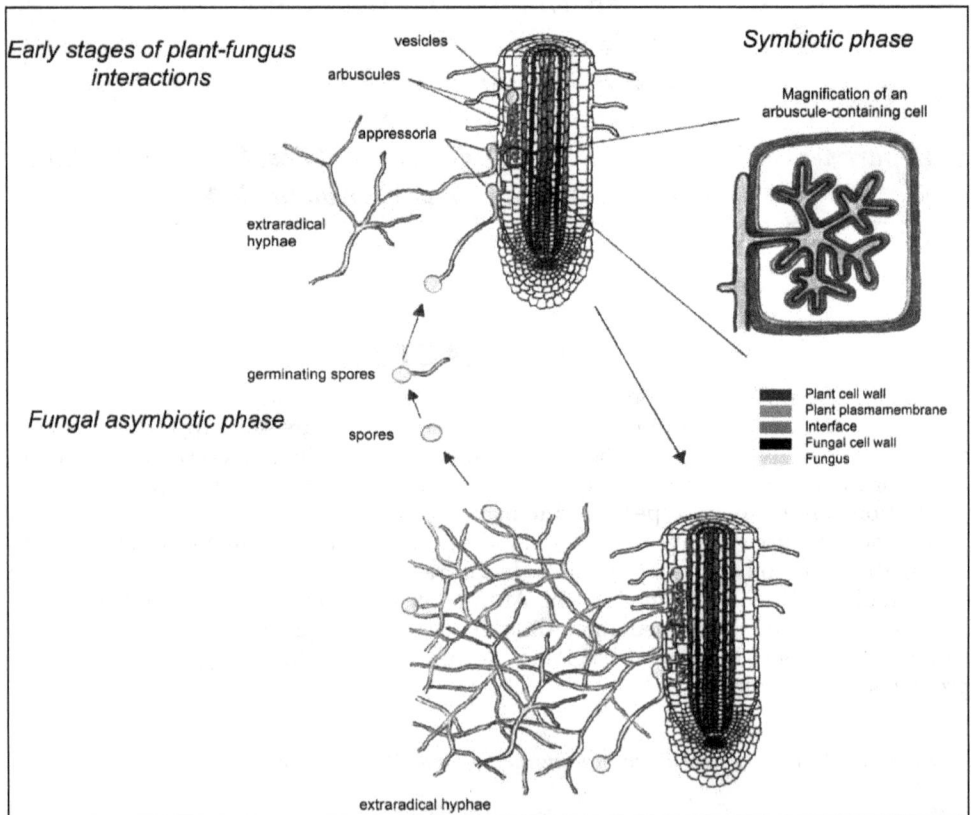

Figure 11.1. Scheme of the different stages of root colonization by an arbuscular mycorrhizal fungus[11].

morphological types of AM, *Arum* and *Paris* type, they are characterized by differences in fungal development within the roots [9-10].

In general, internal hyphae branch make four recognizable structures, intracellular hyphae that may be coiled, intercellular hyphae, arbuscules and spherical or ovoid vesicles [10]. In *Arum-type* associations, intercellular hyphae penetrate between the root cortical cells and spread rapidly and then lateral branch-hyphae penetrate the root cortical cells and branch dichotomously and produce arbuscules. In *Paris-type* associations intracellular hyphae spread from cell to cell within the root cortex and form extensive intracellular hyphal coils and arbusculate coils [9]. The types of internal structures that develop depend on the plant/fungal combination [10,12].

About 83 per cent of dicotyledonous and 79 per cent of monocotyledonous plants are associated with mycorrhizal fungi [13]. Mycorrhizal associations are found in a very wide range of habitats, including aquatic ecosystems, deserts, lowland tropical rain forests, high altitudes, high latitudes and in canopy epiphytes [14]. Bi-directional movement of nutrients characterizes the fungus-plant symbioses, where: carbon flows to the fungus and inorganic nutrients move to the plant, thereby providing a critical linkage between the plant root and soil. This association is a survival mechanism for both the fungi and plants, allowing each to survive in different environments [15]. Mycorrhizal plants, in comparison with non mycorrhizal plants, have greater nutrient uptake because they possess a network of external hyphae [16]. The hyphae are the interface between soil and plant and have a large surface area that acts as an extension of the root absorbing area [17-19]. This not only increases the volume of soil from which nutrients are absorbed, but also overcomes problem of depletion of nutrients [20-21] and water [22] depletion close to actively absorbing roots, and plays a significant role in stabilizing soil structure. Mycorrhizal fungal associations have several advantages for their hosts, including increased growth and yield and reproductive success due to enhanced nutrient acquisition [23-26]. They may also increase disease and pest resistance; improve water relations [27-29], soil structure [30-33] and tolerance of extreme pH [34-35]. This review will concentrate on structure and function of AM and their potential role in geo-engineering and commercial biotechnological product.

Arbuscular Mycorrhizal Fungi in Carbon Sequestration

Arbuscular mycorrhizal fungi (AMF) comprising fungi in the Phylum Glomeromycota from symbiosis in the majority of higher plants. AMF have been shown to have numerous effects on plant physiology and plant communities [14,21], which can lead to indirect effect on soil carbon storage. AMF are also very important in the process of soil aggregate stabilization [36]. Relatively labile carbon can be protected inside soil aggregates, which means AMF have yet another indirect influence on soil carbon storage [37].

Recently, a glycoprotein named Glomalin was discovered using specialized protocols for soil that revealed amount up to several mg proteins per gram of soil [38]. Glomalin is produced by hyphae of all members of AM genera, but not by other groups of soil fungi so for tested. Although concentration of glomalin in soil seem to be responsive to global change factors such as elevated atmospheric CO_2 [39].

ERH (Extra Radical Hyphae) and Soil Aggregation

Indeed the influences of ERH on soil aggregation might be even more important to the carbon stock than the influence of the hyphal standing and their role in soil microaggregate stabilization, the ERH appear to contribute to the certain of an aggregate hierarchy [40]. In doing so they help to create mechanism for increasing the residence time of organic debris within soil macroaggregates.

Mechanistically, the ERH contribution to soil aggregation can be viewed as 'Stioky string bag' mechanism, in which the hyphae help to etangle and enmesh soil particles to form macroaggregate structures [40-41]. The physical dimensions of ERH allow them to grow and to remify through soil pores the size of those between macroaggregates. The contribution of AMF hyphae to carbon cycling lies not only with ERH but also, with exudates from hyphae. AM fungi take up photosynthetically fixed carbon from plant roots and translocate it to their external myceliumAMF are also responsible for directing the movement of huge quantities of photosynthate to the soil [35,42]. Carbon in the root flows from plant to fungus in the form of sugars [43-44] and together with the transfer of mineral nutrients from fungus to root [45-47]. This is the nutritional mainstay of what is arguably the world's most important mutualistic symbiosis. AM fungi obtain most or all of their carbon within the host root. Here, they acquire hexose and transform it into trehalose and glycogen, typical fungal carbohydrates [43]. Triacylglyceride (TAG) is the main form of stored carbon in AM fungi [48-49] and this is mostly or exclusively made in the intraradical mycelium (IRM) [50]. Some of this storage lipid flows from the IRM to the extraradical mycelium (ERM) [50] and *in vivo* microscopic observations indicate that the rate of export is sufficient to account for the high levels of stored lipid in the ERM [50]. The glyoxylate cycle is active in the ERM [52] and this pathway appears to be important in using exported TAG to make carbohydrate in the ERM.

The structure of glomalin has not been completely defined. The molecule appears to be a complex of a repeated monomeric structures bound together by hydrophobic interactions [53] that, attaches to soil to help stabilize aggregates. The molecule contains tightly bound iron (0.04-8.8 per cent) [53] and does not contain phenolic compounds such as tannins [54]. Preliminary evidence suggests that cations are bound to glomalin in amounts that vary for different soils.

Rhizodeposition

The release of carbon compounds from living plant roots rhizodeposition into the surrounding soil is a ubiquitous phenomenon [55]. The loss of C from root epidermal and cortical cells leads to a proliferation of micro-organisms within (endorhizosphere) and on the surface (rhizoplane) and outside the root ectorhizosphere. Carbon release also result in the rhizosphere having different chemical physical and biological characteristics to the bulk soil [56]. Theoretically, almost any soluble component present inside the root can be lost to the rhizosphere; however current evidence suggests that exudation is dominated by low molecular weight solutes such as sugars, amino acids and organic acids that are present in the cytoplasm at high concentrations [57]. AMF colonization root exudation patterns may be expected to alter because the AM fungus is a considerable C sink [35,42]. Arbuscular mycorrhizal colonization alters the carbohydrate metabolism of the roots [35] and increases root respiration [35].

Arbuscular Mycorrhizal Fungi in Nutrient Sequestration

Enhanced uptake of P is generally regarded as the most important benefit that AMF provide to their host plant and plant P is often the main in plant fungal relationship [21,42,58]. Arbuscular mycorrhizal fungi can play a significant role in crop P nutrition, increasing total uptake and in some cases P use efficiency [59]. This may be associated with increased growth and yield [59-60]. Where colonization by AMF is disrupted, uptake of P, growth and in some cases yield can be significantly reduced [58,61-62]. In many cases, this is due to a high concentration of (phyto) available soil P [63-66]. Under such conditions, the colonization of roots by AMF is often suppressed [67-69]. Where strong AMF colonization still occurs under condition of high soil P concentration it may reduce crop growth [70-71].

AMF play significant role in the uptake of other nutrients by the host plant, Zinc (Zn) nutrition is most commonly reported as being influenced by the AM association, though uptake of copper (Cu), iron, N, K, calcium (Ca) and Mg have also been reported as being enhanced [20,72]. AMF may also enhance plant uptake of N from organic sources [74]. In many cases AMF cause a change in the absorption of soil nutrients by host simultaneously, though the effect on different nutrients is rarely the same [21,58,61-62,74-78].

The apparently contradictory evidence regarding the effect of AMF on plant nutrient absorption may be connected to the increasing realization that there is a degree of selectivity between host and the fungi and that different AMF have varying effects on different plant species, from strongly positive increases in nutrient uptake and/or growth to strongly negative effect [79-82]. There is a good evidence for a substantial capacity of external hyphae to absorb NH_4-N and deliver it to the host plant. In the hyphosphere as in the rhizosphere, uptake of NH_4-N is associated with substrate acidification.

Some previous studies on the effects of salinity on mycorrhizal plants have shown that AM roots had higher Na concentrations but also higher K concentrations and thus maintained a high K/Na compared to non-mycorrhizal plants. Scientists have found that mycorrhizal roots of salt grass plant (*Distichlis spicata*) had higher Na, K and P concentrations than non-mycorrhizal roots. In contrast, the results of other studies have shown that Na uptake decreased in AM plants compared to controls. Sodium content of shoots of mycorrhizal halophytic *Aster tripolium* was lower than non-mycorrhizal plants under salinity stress [83]. Potassium-Sodium (K/Na) ratio increased in mycorrhizal barley (moderate salinity tolerant plant) at high levels of salinity by decreasing Na concentration, rather than by increasing K concentration [78]. Potassium plays an important role in mycorrhization in wheat, legumes and rice plants [84-85].

Increased Uptake of other Mineral Nutrients

Concentrations of Ca, Mg and Zn in onion plants inoculated with *Glomus fasciculatum* increased in saline conditions and improved the nutritional status, which was at least partially responsible for increased plant growth [86]. The improved growth and nutrient acquisition (P, K, Zn, Cu and Fe) in tomato demonstrate the potential of AM fungi for protecting plants against salt stress in arid and semiarid areas [87-88]. Although effects of AM fungi in increasing some toxic elements (Na, Cl and Mn) have been reported [89-90]. AMF decreased sodium concentration in barley (*Hordeum vulgare*) when grow in saline conditions [78] and Mn can be reduced in mycorrhizal plants when it occurs at toxic levels compared to non-mycorrhizal plants [91-92].

Arbuscular Mycorrhizal Fungi in Metal Sequestration

The use of living organism for the remediation of soils contaminated with heavy metals, radionuclide or polycyclic hydrocarbon is known as bioremediation [93-96]. AMF are involved in bioremediation through phytoremediation, the technique based on the use of plants for soil remediation [97-98]. Depending on the type of pollutant different strategy for phytoremediation, such as phytostabilization, phytodegradation and phytoextraction has been used. For phytoremediation of soil polluted with heavy metals, the phytostabilization strategy involves the immobilization of heavy metals in the soil by establishing plants.

AMF can help phytoremediation activities, particularly in phytostabilization [97-102]. Among possible mechanism by which AM fungi improve the resistance of plants to heavy metal is the ability of AM fungi to sequester heavy metals through the production of chelates or by absorption, AM plants typically less translocates heavy metal to their shoots than the corresponding non AM control. Among the diverse type of mycorrhizosphere interactions known to benefit plant growth and health, those related to phytoremediation process and rhizobacteria and AM fungi interact synergistically to the benefit of phytoremediation. A key point in phytoremediation is the use of heavy metal adapted microbes, soil microbial diversity and activity both negative affected by excessive concentratic of heavy metals. Indigenous bacterial population [103] and AM fungi [104] must be adapted to metal toxicity and have evolved abilities to enable them to survive in polluted soil.

Various authors have reported isolating spores of AM fungal taxa such as *Glomus* and *Gigaspora* associated with most of the plants growing inheavy metal polluted habitats [105]. *Glomus mosseae* was isolated only and Duek *et al.,* [106] isolated *G. fasciculatum* alone from the heavy metal polluted soils. Pawlowska *et al.,* [107] surveyed a calamine spoil mound rich in Cd, Pb and Zn in Poland and recovered spores of *G. aggregatum, G. fasciculatum* and *Entrophospora* spp. from the mycorrhizospheres of the plants growing on spoil. Joner and Leyval [108] reported that extra-radical hyphae of AM fungus G. *mosseae* can transport Cd from soil to subterranean clover plants growing in compartmented pots, but that transfer from fungus to plant is restricted due to fungal immobilization. It was also reported that no restriction of fungal hyphal growth into soil with high extractable Cd levels. It also showed very little, if any, translocation of Zn absorbed by mycorrhizal maize seedlings grown in contaminated soil, to the shoots. Localization of heavy metals within the fungal mycelium and mycorrhizal roots of *Euphorbia cyparis-sias* from Zn contaminated wastes was studied and it was found that higher concentrations of Zn as crystaloids gets deposited within the fungal mycelium and cortical cells of mycorrhizal roots. Studies by various researchers [109-110] have shown that mycorrhizal fungal ecotypes from heavy metal contaminated sites seem to be more tolerant to heavy metals and have developed resistance than reference strains from uncontaminated soils. Galli *et al.,* [109] reported that although there was an increase in the contents of cystein, gamma EC and GSH in the mycorrhizal maize roots grown in quarto sand with added Cu, no differences in Cu uptake were detected between non-mycorrhizal and mycorrhizal plants. These results do not support the idea that AM fungi protect maize from Cu-toxicity. Mycorrhizae are also known to produce growth stimulating substances for plants, thus encouraging mineral nutrition and in-creased growth and biomass necessary for phytoremediation to become commercially viable strategy for decontamination of polluted soils. In addition to the damaging effects on plants, the effect of heavy metals on the soil micro-organisms and soil microbial activity also need to be considered. Various soil factors such as the clay contents and mobility of heavy metals affect plants and soil biota. As metal uptake by plant roots depends on soil and their associated symbionts, it is important to monitor metal mobility and availability to plant and its symbionts when assessing the effect of soil contamination on plant uptake and related phytotoxic effects. The prospect of symbionts existing in heavy metal contaminated soils has important implications for phytoremediation. The potentials of phytoremediation of contaminated soil can be enhanced by inoculating hyperaccumulator plants with mycorrhizal fungi most appropriate for contaminated site. It is further suggested that the potential of phytoremediation of contaminated soil can be enhanced by inoculating hyper-accumulator plants with mycorrhiza.

Arbuscular Mycorrhizal Fungi in Phytoprotection

The establishment of AM fungi in plant roots has been shown to reduce damage caused by soil-borne plant pathogens with an enhancement of plant resistance/tolerance in mycorrhizal plants. The effectiveness of AM in biocontrol is dependent on the AM fungus involved as well as the substrate and the host plant [111-114]. Several AM fungal species have been found to control soil-borne pathogens, for example under green house conditions. *G. faciculatum* and *Gigaspora margarita* are shown to reduce root rot disease caused by *Fusarium oxysporum* f. sp. *asparagi* and *Helicobasidium mompa* in *Asparagus officinalis* L. [115-116]. According Abdel-Fattah and Shabana [117] *G. clarum* is shown to reduce root necrosis due to *Rhizcotonia solani* in cow pea (*Vigna unguiculata* L.).

In pasteurized soil AM fungi have been shown to decrease the root damage caused by root rot fungus *Cylindrocladium spathiphylli* in banana, although the pathogen decreased the intensity of AM fungal root colonization [118]. Since AM are formed in the roots of plants, maximum attention has been paid towards mycorrhizae disease incidence interactions in relation to soil and root-borne pathogens (Table 11.1). AM fungi impart resistance to host plant against root pathogens. Root pathogens are major limiting factor for plant production. Most widespread root-borne pathogens belong to genera, *Phytophthora, Fusarium, Pythium* and *Rhizoctonia*. They kill roots or reduce their ability to absorb water and nutrients by penetrating root tissues and producing toxins.

Different AM fungi are known to confer variable tolerance to *Phytophthora parasitica*. Davis and Menge [119-120] used 5 different AM fungal species on three different plants. They observed that maximum resistance to the pathogens was due to AM fungi namely *viz. G. fasciculatum* and *G. constrictum* in *Citrus* sp. Arbuscular mycorrhizal peanut plants become resistant to root rot pathogen *Sclerotium rolfsii* [121]. Numbers of sclerotia produced by the pathogen were reduced on the mycorrhizal roots. Biomass production and phosphorus content were maximum in mycorrhizal plants and minimum in pathogen inoculated plants. Take all disease of wheat, caused by *Guamannomyces graminis* var. *tritici* was reduced when mycorrhizal fungus was inoculated in plants. The disease is favoured by inadequate plant nutrition leading to mineral deficiency especially P [122].

Table 11.1. Inhibitory influence of AM fungi on soil-borne fungal pathogens.

Host	Disease	Pathogen	Reference
Tobacco	Root rot	*Phytophthora cinnamom*	125
Allium cepa	Wilt	*Fusarium oxysporum* f. sp. *cepi*	126
Lycopersicon esculentum	Wilt	*F. oxysporum* f.sp. *lycopersici*	127
Lycopersicon esculentum	Wilt	*F. oxysporum* f.sp. *lycopersici*	123
Lycopersicon esculentum	Root rot	*F. oxysporum* f.sp. *redicis lycopersici*	128
Triticum aestivum	Take-all disease	*Gaeumannomyces graminis* var. *tritici*	122
Glycin max	Root rot	*Macrophomia phaseolina, Rhizoctonia solani, F. solani*	129
Brassica compestris	Root rot	*Olpidium brassicae*	130
Cassia tora	Wilt	*F. oxysporum*	131
Solanum melongena	Wilt	*Verticillium albo-atrum*	132
Cicer arietinum	Wilt	*F. oxysporum* f. sp. *ciceris*	124

AM fungi are able to increase nutritional status of the host plant thereby improving resistance against the pathogens. Recenlty, Singh *et al.,* [123] reported inoculations of AM fungi before *Fusarium oxysporum* causing wilt disease protect the tomato plants. Arbuscular mycorrhizal fungi and *Trichoderma* sp. produced more significant biocontrol effect [124].

Mycorrhiza-Induced Resistance

Different mechanisms have been shown to play a role in plant protection by AMF, namely, improved plant nutrition, damage compensation, competition for colonization sites or photosynthates, changes in the root system, changes in rhizosphere microbial populations, and activation of plant defense mechanisms. Several mechanisms can be operative simultaneously, with contributions depending on environmental conditions, timing of the interaction and partners involved [111].

There is evidence for the accumulation of defensive plant compounds related to mycorrhization, although to a much lower extent than in plant-pathogen interactions. Accumulation of reactive oxygen species, activation of phenylpropanoid metabolism, and accumulation of specific isoforms of hydrolytic enzymes such as chitinases and glucanases has been reported in mycorrhizal roots. Concerning aboveground effects, accumulation of insect antifeedant compounds and transcriptional regulation of defense-related genes [133] have been described in the shoots of mycorrhizal plants. Furthermore, the volatile blends released by AM plants can be more attractive to aphid paraisitism than those from non-mycorrhizas. Nevertheless, accumulation of PR proteins, salicyclic acid, or expression of marker genes associated with systemic acquired resistance has not been reported in systemic tissues.

AMF as Communicators

The most important AM fungal structure for plant nutrition is represented by the extra radical mycelium spreading from mycorrhizal roots into the surrounding soil which is able to uptake nutrients N, P, S, Ca, K, Fe, Cu, Zn and to transfer them to root cells [21,134]. Mycorrhizal mycelium has been investigated in different experimental studies, based on either destructive extraction from soil or root observation chambers or in *in vitro* system which yielded only qualitative data on its structure and growth [135–136].

The first visualization of intact AM mycelium extending from mycorrhizal roots into the extraradical environment was obtained by means of a bidimensional model system which utilized two cellulose ester members "sandwiched" around the roots of individual plantlets. After only 7 days growth, a fine network of extramatrical hyphae growing on the membranes was visible to the naked eye, and its length extended from 5169 to 7471 mm (hyphal length) in *Thymus vulgaris* and *Allium porum*, respectively.

The experimental system divided to visualize the mycorrhizal mycelium also evidenced that the mechanism allowing the formation of the network was self recognition and hyphal anastomosis. It is important to stress that the viability of mycorrhizal networks was 100 per cent and that all the anastomoses showed protoplasmic continuity and nuclear occurrence in hyphal bridges, confirming the occurrence of nuclear exchange also during fusion between extraradical symbiotic hyphae. AM fungi are known to infect a wide range of host species. They have a large geographical distribution, being found even in the Arctic tundras and the Antarctic region [137–138]. Unlike most ectomycorrhizal species, AM are not host specific. This enables them to form associations with a large number of plant species.

Arbuscular mycorrhizal fungi regulate plant communities by affecting competition, composition and succession [139]. Limited resources and the struggle of the plants for a share of these is the primary selection pressure operating on plant species [140]. In competition between plants, mycorrhizae in the soil favour the growth of one species and are detrimental to other competing species. Fitter [141] demonstrated this in competition between two grasses *Lolium perenne* and *Holcus lanatus*. Inoculation with mycorrhizae favoured the growth of *H. lanatus*. This was an indirect effect as infection with mycorrhizae reduced the root length of *L. perenne* by 40 per cent . Arbuscular mycorrhizal fungi may regulate competition between plants by making available to mycorrhizal plants, resources that are not available to non-mycorrhizal neighbours [145]. AMF symbiosis increases intraspecific competition [142–143]. As a result, density of individuals of a single species would be reduced thereby allowing the co-existence of individuals of different species. This would lead to an increase in species diversity.

Mycorrhizae govern species composition in communities by influencing plant fitness at the establishment phase. Arbuscular mycorrhizal fungi prevent non-mycorrhizal plants from growing in soils colonized by them. This has a selective advantage for the fungus. Maintaining a high proportion of compatible host species at the expense of non-compatible species provides the fungus with an undisturbed carbon supply [144]. Succession is a chain of predictable processes whose course is influenced by nutrient availability. Mycorrhizae, owing to their role in nutrient uptake, may play an important part in determining the rate and direction of the process [21,145]. They influence the outcome of succession by amending the composition of species or by affecting species diversity [146].

AM fungi have been reported to be active in mediating nutrient transfer among plants [145,147-151] mainly through the extensive mycelial networks, which due to the lack host specificity may link the roots of contiguous plant species [150,152]. According to studies showed a novel mechanism by which plants may become interconnected, that is hyphal fusion between extraradical hyphae originating from different individual plant root systems of different species, genera and families [153].

Arbuscular Mycorrhizal Fungi in Biotechnology

Introduction with AMF on micropropagate plantlets improves its establishment and growth in the field. Unfortunately not much work has been done in this. But the few convincing reports decisively prove that successful hardening and *ex vitro* establishments of plantlets could be achieved by inoculation with AM fungi.

Rai and Varma [154] reported that the growth promoting potential of *Piriformospora indica*, which is a newly discovered arbuscular mycorrhiza-like fungus. It is a facultative symbiont and unlike AM fungi, it can be cultured *in vitro*. *Adhatoda vasica* is a medicinal plant. Rapid proliferation of roots was recorded in *A. vasica* with root colonization of 95 per cent after 6 months. *P. indica* improved growth of *A. vasica*. This association forms a new host-symbiont combination.

One of the earliest reports on the effect of growth and leaf mineral content of two apple clones propagated *in vitro* were increased substantially. Similarly, enhanced effect on the growth of *in vitro* cultured strawberry planlets could be achieved due to the association of AMF [155]. The rooting of planlets of garlic regenerated from called was significantly enhanced due to inoculation with *Glomous mosseae*. The transplant success and growth of *Robus idaeus* and *Pistacia integerrima* were achieved with mycorrhiza inoculation.

The inoculation of horticulture crop with arbuscular and ericoid mycorrhizal fungi proved that the mycorrhization could be a feasible biotechnology to improve plant growth and health status of several horticulture crop species: *Cyclamen*, *Verbena* and *Rhododendrons* when appropriate strains of mycorrhizal fungi are inoculated into the peat based media [155].

Arbuscular Mycorrhizal Fungi in Genetics

Several constitutively expressed genes were shown to be expressed at the rhizodermis level, including root hairs and around the root tip. Moreover, a loss of function mutant defective in two Phtl transporters, normally expressed at the root periphery, in *Arabidopsis* exhibited a strong reduction of phosphate uptake by 75 per cent strongly indicating a role for Phtl transporters in the 'direct' uptake pathway [156]. In the AM symbiosis, phosphate uptake by the fungus from soil is the first step in the process of phosphate transport to the root. Two AMF phosphate transporters *GvPT* and *GiPT* from *Glomus versiforme* and *Glomus intraradices*, respectively have been reported [157–158]. At least two members of this clade (*PH084* and *GvPT'*) exhibit high affinity towards phosphate, with *Km* values of 8 µM and 18 µM, respectively [157,159]. The *GvPT* and *GiPT* genes are predominantly expressed in the extraradical fungal mycelium exposed to micromolar phosphate concentrations [158] and so their encoded proteins are likely to participate in phosphate uptake at the fungus soil interface.

Arbuscular Mycorrhizal Fungi in Genomics and Proteomics

In recent years, outstanding molecular approaches have been used for the identification of genes and functions involved in plant-microbe endosymbioses. Following the first completion of genome sequencing projects, biological research has developed high throughout genetic programmes with multiparallel analyses of

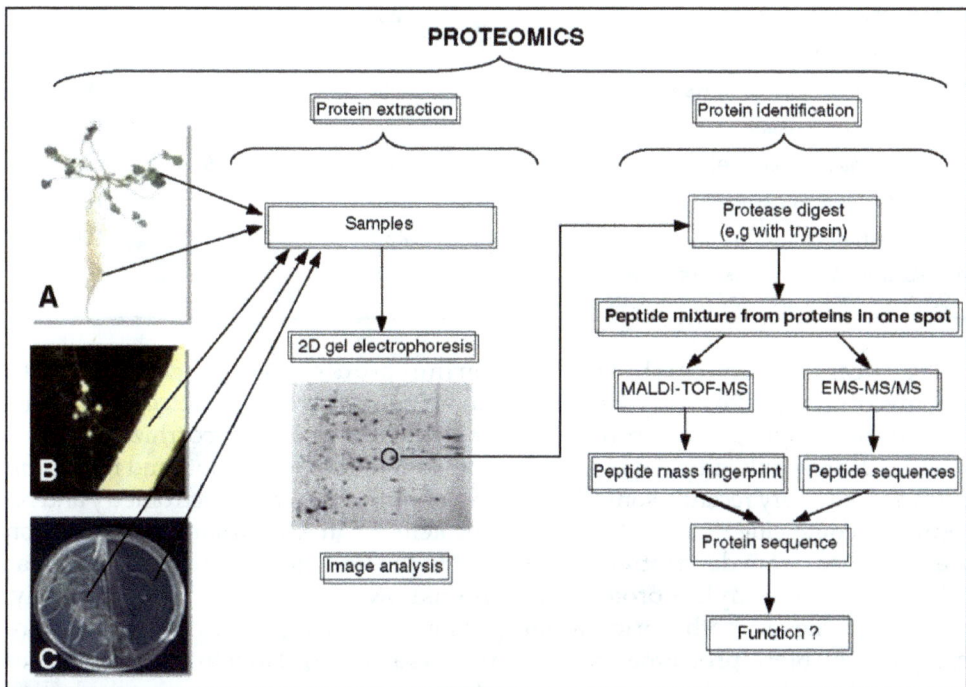

Figure 11.2. Typical flow-chart for the analysis of proteomes by mass spectrometry (Dumas-Gaudot *et al.*, 2009).

gene transcripts, proteins and metabolites, that are now tentatively transposed to the world of mycorrhizal and rhizobial symbioses [160–161].

Arbuscular mycorrhizal fungi bear large and highly repeated genomes, a feature that completely hinders the development of genomic analyses (Table 11.2). Due to their obligate symbiotic status, the amount of accessible mRNA material is very limited; however, many limitations have been overcome with the help of polymerase chain reaction (PCR)-based methods [160] and *in vitro* monoxenic cultures [161]. Several EST libraries have thus been constructed using activated spores of *Gigaspora rosea* [162–163], *G. mosseae* [164], presymbiotic mycelium of *Gigaspora margarita* [165] and extra radical hyphae of *G. intradices* [166].

Table 11.2. Proteomic studies in the field of plant-microbe symbioses.

Biological Models	Proteins Characterized by 2 DE	Identified Proteins	References
Allium cepa/Glomus mosseae	15	0	173
Nicotiana tabacum/G. mosseae/G. instraradices	34	0	174
Lycopersicon esculentum/G. intraradices	5	0	175
Pisum sativum/G. mosseae	42	0	176
L. esculentum/G. mosseae	44	0	177
L. esculentum/G. mosseae	14	0	178
L. esculentum/G. mosseae	26	1	179
Triticum aestivum/G. mosseae	1	1	180
Medicago truncatula/G. mosseae	55	8	183
P. sativum/G. mosseae	7	6	184
M. truncatula/G. mosseae	34	6	183
Ri T-DNA Daucus carota/G. intraradices	8	2	185
G. intraradices	450	6	186
Acaulospora laevis/Gigaspora rosea/ Scutellospora castanea/G. mosseae	12	0	181

Concerning mycorrhizas, pioneering studies were achieved with ectomycorrhizas in the early 1990s [167–170] allowing researchers to detect symbiosis-related (SR) polypeptides, up-regulated or newly induced in mycorrhizal roots, as well as down-regulated polypeptides, by comparison to control roots and mycelium extracts. Only very recently some SR proteins identified by mass spectrometry and N-terminal sequencing [171], as well as mycelial proteins [172]. In AMF, a similar progression was followed, from descriptive studies [173–178] to the first identification of a vacuolar H$^+$-ATPase and a Myk15 protein with an unknown function [179–180]. Additionally, protein profiles of both dormant and germinated spores of several fungi were compared [177]. Now, proteome analysis is used as a powerful tool to reveal more and more proteins involved in mycorrhiza development and functioning including proteins involved in defence response, root physiology and the respiratory pathway [182–183].

Fluxomics

The goal of fluxomics is to quantify all the metabolic fluxes in a cell, tissue or organism [187-188]. Investigation of the flow of matter through biochemical systems has always been central to the study of metabolism, and analysis of the rates of metabolic transformations. The study of enzyme kinetics is likewise a long-established aspect of understanding any biological system in detail. However, the conceptual, experimental and computational tools for quantifying the integrated functioning of metabolic networks began to become available only in recent decades, and are still very much under development. Like many other omic approaches, fluxomics has yet to attain the goal of generating comprehensive system-wide data sets. However, progress in the last 10 years has been rapid and fluxomics has grown beyond is origin in bacterial systems and has begun to make significant contributions to the study of plant systems [189-191].

The analysis of multiple flows through a network involves both direct and indirect determination of metabolic and transport fluxes. Direct determination fluxes involve individually measuring the rates of substrate uptake, product secretion, and the accumulation of storage or structural compounds (lipids, carbohydrates and proteins) of known composition. Indirect determination of fluxes is performed in two ways. In the first method, fluxes measured directly are used to deduce other net fluxes; this is done by balancing the influxes and effluxes from individual metabolite pools using the known stoichiometries of biochemical reactions (flux balancing). The second method is based on interpreting the results of labeling experiments. Labeling measurements using radioactive isotopes are made by fractionation or chromatographic separation methods followed by scintillation counting of different intermediate and product metabolites. Measurements of stable isotopic labeling usually involve ^{13}C (or less commonly ^{15}N or ^2H) and are made by nuclear magnetic resonance (NMR) spectroscopy and mass spectrometry. The interpretation of labeling and flux measurement data in terms of multiple fluxes is nontrivial and almost always involves computer-aided modeling. Models are used to estimate fluxes by finding the values that result in a best fit of computed (simulated) to experimental results.

This reliance on fitting to a metabolic model means that a fluxomic investigation requires some prior knowledge of and assumptions about the metabolic architecture of the system. This knowledge is certainly much less complete in mycorrhizas than in many bacteria or model fungi and plants. However, progress in delineating metabolic and transport networks in mycorrhizal systems has been steady in recent years and has reached a point where models can be constructed for the quantitative interpretation of labeling data in the best-studied cases [192-195]. The advent of mycorrhizal plant and fungal genome sequences will be of enormous help in building and in filling in the molecular mechanisms of such models. When transcript and proteomic data sets for mycorrhizas become much more complete than they are at present which can be expected to happen sooner rather than later they too will be important in defining and validating model networks for use in flux analysis.

Conclusion

☆ Soils are the major sink of carbon and AM fungi have shown that, they have also potential to sequester the carbon from the soil.

☆ Deposition of heavey metal in the soil is troublesome affair which resulted into the toxicity, which is not only harmful for the plants but also unsafe for organism dependent on their food from plants. Here also AMF fungi provide release from the heavy metal toxicity by sequestring heavy metals from the soil,

☆ Control of plant disease was remaining challenge for scientist and farmer both. Use of pesticide no longer remain the healthy affair, therefore, biological control was searched as a alternative, among the biocontrol agents, role of AMF have been recognized a fruitful one.

☆ As it has been established that microbes and plants have very strong communication in the form of signal transduction, here also AMF provided evidences that they are best communicators in microbial world and have developed world-wide-web (www) for their communication.

☆ Biotechnology is the talk of hour and therefore role of AMF fungi have also been explored and recognised in the field of modern biology and have been tried to elucidate their relevance in biotechnology, molecular biology, genomics, proteomics and fluxomics.

Acknowledgement

Senior author D.V. is thankful to Head, Department of Botany, Dr. H.S. Gour University, Sagar, M.S. thankfully acknowledges head *Department of Botany, Guru Ghasi Das University, Bilashpur Chhatisgarh, PKS* thankful acknowledge DST for financial assistance, RKG and Irfan Mohd. acknowledge principale Govt. P.G. College Rishikesh, for nessesory facilities.

References

1. Redecker D, Kodner R and Graham LE. Glomalean fungi from Ordovician. *Science* 2000; **289**: 1920–1921.

2. Nicoloson TH. Evolution of vesicular–arbuscular mycorrhizas In: Sanders E, Mosse B, Tinker PB. Editors. Endomycorrhizas. Academic press, London., 1975. pp. 25–34.

3. Remy W, Taylor TN, Hass H, Kerp H. Four hundred million year old vesicular arbuscular mycorrhizae. *PNAS* USA 1994; **91**: 11841–11843.

4. Vyas D, Soni A, Singh PK. Differential effects of pesticides on occurrence of Vam fungi associated with some leguminous plants. *J Basic Appl Mycol* 2007; 6: 143–150.

5. Vyas D, Dubey A, Soni A, Mishra MK, Singh PK. Arbuscular mycorrhizal fungi in early land plants. *Mycorrhiza News.*, 2007; **19**: 21–23.

6. Harley JL. Harley EL. A check list of mycorrhizae in British flora. *New Phytol* 1987; **105**: 1–102.

7. Giovannetti M, Sbrana C, Avio L, Citernesi AS, Logi C. Differential hyphal morphogenesis in arbuscular mycorrhizal fungi during pre–infection stages. *New Phytol* 1993; **125**: 587–593.

8. Harrison MJ. Development of the arbuscular mycorrhizal symbiosis. *Cur Opin Plant Biol* 1998; **1**: 360–365.

9. Gallaud I. Etudes surles mycorrhizes endotrophes. *Revue Generale de Botanique.*, 1905; **17**: 5–48,66–83,123–135; 223–239–313–325,425–433, 479–500.

10. Smith FA, Smith SE. Structural diversity in vesicular–arbuscular mycorrhizal symbiosis. *New Phytol* 1997; **151**: 469–475.

11. Balestrini R, Lanfranco L. Fungal and plant gene expression in arbuscular mycorrhizal symbiosis. *Mycorrhiza* 2006; **16**: 509–524.

12. Canvagnaro TR, Gao LL, Smith FA, Smith SE. Morphology of arbuscular mycorrhizas is influenced by fungal identity. *New Phytol* 2001; **137**: 373–388.

13. Wilcox HE. Mycorrhizae. In: Waisel Y, Eshel A, Kafkafi U. editors. The Plant Root: The Hidden Half. Marcel Dekker, New York 1991. pp. 731–765.

14. Allen MF. The Ecology of Mycorrhizae. Cambridge University Press, 1991.

15. Gupta V, Satyanarayana T, Sandeep G. General aspects of mycorrhiza. In: Mukerji KG, Chamola BP, Singh J. Mycorrhizal Biology. Kluwer Press, New York., 2000. pp. 245–266.

16. Sanders FE, Sheikh NA. The development of vesicular–arbuscular mycorrhizal in plant root systems. *Plant and Soil* 1983; **71**: 223–246.

17. Rhodes H, Gerdemann JW. Phosphate uptake zones of mycorrhizal and non-mycorrhizal onions. *New Phytol* 1975; **75**: 555–561

18. Owusu–Bennoah E, Wild A. Autoradiography of the depletion zone of phosphate around onion roots in the presence of vesicular–arbuscular mycorrhiza. *New Phytol* 1979; **82**: 133–140.

19. Li XL, George E, Marschner H. Extension of the phosphorous, depletion zone in VA–mycorrhizal white clover in calcareous soil. *Plant and Soil* 1991; **136**: 41–48.

20. Nurlaeny N, Marschner H, George E. Effects of liming and mycorrhizal colonization on soil phosphate depletion and phosphate uptake by maize (*Zea mays* L.) and soyabean (*Glycine max* L.) grown in two tropical soils. *Plant and Soil* 1996; **181**: 275–285.

21. Smith SE and Read DJ (1997). Mycorrhizal symbiosis. Academic Press, London.

22. Marulanda A, Azcon R, Ruiz–Lozano JM. Contribution of six arbuscular mycorrhizal fungal isolates to water uptake by *Lactuca sativa* plants under drought stress. *Physiologia Plantarum* 2003; **119**: 1–8.

23. Mosse B. Advances in the study of vesicular arbuscular mycorrhizal fungi. *Annu Rev Phytopathol* 1973; **11**: 171–196.

24. Diedrich C. Improved growth of *Cajanus cajan* (L.) Mill sp. In an unsterile tropical soil by three mycorrhizal fungi. *Plant and Soil* 1990; **123**: 261–266

25. Lewis JD, Koide RT. Phosphorous supply, mycorrhizal infection and plant offspring vigour. *Funct Ecol* 1990; **4**: 695–702.

26. Stanley MR, Koide RT, Shumway DL. Mycorrhizal symbiosis increases growth, reproduction and recruitment of *Abutilon theophrasti* Medicin the field. *Oecologia* 1993: **94**: 30–35.

27. Allen EB, Allen MF. Water relations of xeric grasses in the field: interaction of mycorrhizas and competition. *New Phytol* 1986; **104**: 559–571.

28. Davies FT, Jr. Potter JR, Linderman RG. Drought resistance of mycorrhizal pepper plants independent of leaf P concentration–response in gas exchange and water relations. *Physiologia Plantarum* 1993; **87**: 45–53.

29. Subramanian KS, Charest C, Dwyer LM, Hamilton RI. Effects of arbuscular mycorrhizae on leaf water potential, sugar content and phosphorous content during drought and recovery of maize. *Can J Bot* 1997; **75**: 1582–1591.

30. Tisdall JM, Oades JM. Stabilization of soil aggregates by the root system rye grass. *Aus J Soil Res* 1979; **17**: 429–441.

31. Thomas RS, Dakessian S, Ames RN, Brown MS, Bethlenfalvay GJ. Aggregation of a silty clay loam soil by mycorrhizal onion roots. *Soc Am J* 1986; **50**: 1494–1499.

32. Degans BP, Sparling GP, Abott LK. The contribution from hyphae, roots and organic carbon constituients to the aggregation of a sandy loam under long term clover based and grass pastures., *Eur J Soil Sci* 1994; **45**: 459–490.

33. Beaden BN, Peterson L. Influence of arbuscular mycorrhizal fungi on soils structure and aggregate stability of vertisol. *Plant and Soil* 2000; **218**: 173–188.

34. Sidhu OP, Behl HM. Response of three *Glomus* species on the growth of *Prosopis Juliflora* Qwartz at high pH levels. *Symbiosis* 1997; **23**: 23–24.

35. Douds DD, Gadkar V, Adholeya A. Production of VAM fungus biofertilizer. In: Mukerji KG, Chamola BP, Singh J. Editors. Mycorrhizal Biology. Academic Publisher, New York., 2000. pp., 197–215.

36. Tisdall JM, Behl HM. Organic matter and water stable aggregates in soils. *J Soil Sci* 1997; **33**: 141–163.

37. Miller RM, Jastrow JD. The role of mycorrhizal fungi in soil conservation. In: Bethlenfalvay GJ, Linderman RG. Editors. Mycorrhizae in sustainable agriculture. *Am Soc Agron* 1992; pp. 29–44.

38. Wright SF, Upadhayaya A. Extraction of an abundant and unusual protection from soil and comparison with hyphal protection of arbuscular mycorrhizal fungi. *Soil Sci* 1996; **161**: 575–586.

39. Rillig MC, Field CB, Allen MF. Soil biota responses to long term atmospheric CO_2 enrichment in two califorina annual grasslands. *Oceologia* 1999: **119**: 572–577.

40. Miller RM and Jastraw JD. Mycorrhizal fungi influence soil structure. In: Arbuscular Mycorrhizae: Physiology and Function, Kluwer Academic Publications., 2000. pp. 4–8.

41. Oades JM, Waters AG. Aggregate hierarchy in soils. *Aust J Soil Res* 1991; **29**: 815–828.

42. Graham JH. Assessing costs of arbuscular mycorrhizal symbiosis in agroecosystems. In: Podila GK, Douds DD. editors. Current advances in mycorrhizae research. Am Phytopathological Soc Press., 2000; pp 127–140.

43. Shachar–Hill Y, Pfeffer PE, Douds D, Osman SF, Doner LW, Ratcliffe RG. Partitioning of intermediate carbon metabolism in VAM colonized leek. *Plant Physiol* 1995; **108**: 7–15.

44. Solaimain MD, Satio M. Use of sugars by intraradical hyphae of arbuscular mycirrhizal fungi revealed by radiorespirometry. *New phytol* 1997; **136**: 533–538.

45. Koide RT, Schreiner RP. Regulation of the vesicular–arbuscular mycorrhizal symbiosis. *Ann Rev Plant Physiol Plant Mol Biol* 1992; **43**: 557–581.

46. George E, Marschner H, Jakobsen I. Role of arbuscular mycorrhizae fungi in uptake of phosphorus and nitrogen from soil. *Crit Rev Biotechnol* 1995; **15**: 257–270.

47. Jakobsen I. Transport of phosphorus and carbon in VA mycorrhizas. In: Varma A, Hock B. editors. Mycorrhiza: Structure, Function, Molecular Biology and Biotechnology. Springer–Verlag, Berlin., 1995. pp. 297–323.

48. Beilby JP, Kidby DK. Biochemistry of ungerminated and germinated spores of the vesicular–arbuscular mycorrhizal fungus, *Glomus caledonium*: changes in neutral and polar lipids. *J Lipid Res* 1980; **21**: 739–750.

49. Jabaji Jabaji–Hare S. Lipid and profiles of some vesicular arbuscular mycorrhizal fungi: Contribution to taxonomy. *Mycologia* 1998; **80**: 622–629.

50. Pfeffer PE, Douds DD, Becard G, Shachar–Hill Y. Carbon uptake and the metabolism and transport of lipids in and arbuscular mycorrhiza. *Plant Physiol* 1999; **120**: 587–598.

51. Bago B, Zipfel W. Williams R, Jun J. Arreola R, Lammer P, Pfeffer PE, Shachar HY. Translocation and utilization of fungal lipid in the arbuscular mycorrhizal symbiosis. *Plant Physiol*, 2002; **128**: 108–124.

52. Lammers PJ, Jun J, Abubaker J, Arreola R, Gopalan A, Bago B, Hernandez–Sebastia C, Allen JW, Douds DD, Pfeffer PE. The glyoxylate cycle in an arbuscular mycorrhizal fungus: gene expression and carbon flow. *Plant Physiol* 2001; **127**: 1287–1298.

53. Nichols K. Characterization of Glomalin–A Glycoprotein Produced by Arbuscular mycorrhizal fungi. Ph.D. Thesis, University of Maryland, College Park, Maryland., 2003.

54. Rilling MC, Wright SF, Nichols KA, Schmidt WF, Torns MS. Large contribution of arbuscular mycorrhizal fungi to soil carbon pools in tropical forest soils. *Plant Soil* 2001; **233**: 167–177.

55. Curl EA, Trueglove B. The rhizosphere, advanced series in agriculture science 15. Springer–Verlag., 1986.

56. Barber S (1975) Soil nutrient bioavailability: a mechanistic approach. John Wiley and Sons, 1975.

57. Farrar SC, Hawes M, Jones D, Lindo S. How roots control the flux of carbon to the rhizosphere. *Ecology* 2003; **84**: 827–833.

58. Thompson JP. Decline of vesicular arbuscular mycorrhizae in long fallow disorder of field crops and its expression in phosphorus deficiency of sunflower. *Aust J Agric Res* 1987; **38**: 847–86.

59. Koide RT, Goff MD, Dickie IA. Component growth efficiencies of mycorrhizal and non–mycorrhizal plant. *New Phytol* 2000; **148**: 163–168.

60. Ibijbijen J, Urguiaga S, Ismaili M, Alves BJR, Boddey MR. Effect of arbuscular mycorrhizas an uptake of nitrogen by *Brachiaria arrecta* and *Sorghum vulgaris* from soils labeled for several years with ¹⁵N. *New Phytol* 1996; 133: 487–494.

61. Thompson JP. Improving the mycorrhizal condition on the soil through cultural practices and effects on growth and P uptake in plants. In: Johansen C, Lee KK, Sahrawat KL. Editors. Phosphorous Nutrition of Grain Legumes in the Semi Arid Tropics. International Crops Research Institute for Semi Arid Tropics, Patancheru, India., 1991. pp. 117–137.

62. Thompson JP. Inoculation with vesicular–arbuscular mycorrhizal fungi from cropped soil overcomes long–fallow disorder of linseed (*Linum usitatissimum* L.) by improving P and Zn uptake. *Soil Biol Biochem* 1994; **26**: 1133–1143.

63. Bethlenfalavy GJ, Barea JM. Mycorrhizae in sustainable agricultural. J. Effects on seed yield and soil aggregation. *Am J Agric* 1994; **9**: 157–161.

64. Hetrick BAD, Wilson GWT, Todd TC. Mycorrihiza response in wheat cultivars, relationship to phosphorous, *Can J Bot* 1996; **74**: 19–25.

65. Thingstrup I, Rubaek G, Sibbesen E, Jakobsen I. Flax (*Linum uritatiusim* L.) depends on arbuscular mycorrhizal fungi for growth and P uptake at intermediate but not high soil P levels in the field. *Plant Soil* 1998; **203**: 37–46.

66. Sorensen N, Larsen J, Jakobsen I. Mycorrhiza formation and nutrient concentration in looks (*Allium porrum*) in relation to previous crop and cover crop management on high P soils. *Plant Soil* 2005; **273**: 101–114.

67. Jensen A, Jakobsen I. The occurrence of vesicular–arbuscular mycorrhiza in barley and wheat grow in some Danish soils with different fertilizer treatments. *Plant Soil* 1980; **55**: 403–414.

68. Al–Karaki GN, Clark RB. Varied rates of mycorrhizal inoculum on growth and nutrient acquisition by barley grown with drought stress. *J Plant Nutrion* 1999; **22**: 1775–1784.

69. Kahiluoto H, Vestberg M, Olesen JE, Eltun R, Gording MS, Jensen ES and Kopke U. Impact of cropping system of mycorrhizae. In: Designing and Testing Crop Rotations for Organic Farming. Proceedings from an International Workshop Danish Centre for Organic Farming (DAECOF)., 1999. pp. 305–309.

70. Gavito ME, Varela L. Response of criolla maize to single and mixed species inoculation of arbuscular mycorrhizal fungi. *Plant Soil* 1995; **176**: 101–105.

71. Kahiluoto H, Ketoja E and Vestberg M, Saarela I. Promotion of AM utilization through reduced P fertilization Z. Field studies. *Plant Soil* 2001; **231**: 65–79.

72. Clark RB, Zeto SK. Mineral acquisition by arbuscular mycorrhizal plants. *J Plant Nutri* 2000; **23**: 867–902.

73. Hodge A, Campbell CD, Fitter AH. An arbuscular mycorrhizal fungus accelerates decomposition and acquires nitrogen directly from organic material. *Nature* 2000; **413**: 297–299.

74. Lambert DH, Baker DJ, Code H. The role of mycorrhize in the interactions of phosphorous with Zn, Cu and Other elements. *Soil Sci Am J* 1979; **43**: 976–980.

75. Kothari SK, Marschner H, Romheld V. Direct and Indirect effects of VAMF and rhizospheric–micro-organism on acquisition of mineral nutrients by maize (*Zea mays* L.) in calcareous soil. *New Phytol* 1990; **116**: 637–645.

76. Wellings NP, Wearing AH, Thompson JP. Vesicular–arbusculr mycorrhizal (VAM) improve phosphorous and zinc nutrition and growth of pigeonpea in vertisol. *Aus J Agric Res* 1991; **42**: 835–845.

77. Azaizeh HA, Marschner H, Romheld V, Wittenmayer L. Effects of vesicular–arbuscular mycorrhizal fungus and other soil micro-organisms on growth, mineral nutrient acquisition and root exudation of soil grown maize plants. *Mycorrhiza* 1995; **5**: 321–327.

78. Mohammad MJ, Malkwai HI, Shibli. Effects of mycorrhizal fungi and phosphorous fertilization on growth and nutrient uptake of barley grown on soils with different levels of salts. *J Plant Nutri* 2003; **26**: 125–137.

79. Monzon A, Azcon R. Relevance of mycorrhizal fungaql origin and host plant genotype to inducing growth and nutrient uptake in *Medicago* species. *Agric Ecosyst Envion* 1996; **60**: 9–15.

80. Bever JD, Schulatz PA, Pringle A, Morton JB. Arbuscular mycorrhizal fungi: more diverse than meets the eye, and the ecological tale of why. *Bioscience*, 2001; **51**: 923–931.

81. Van der Heijden. Arbuscular mycorrhizal fungi as determinant of plant diversity, in search for underlying mechanisms and general principles. In: van der Heijden MGA, Sanders IR. Editors. Mycorrhizal Ecology. Ecological studies. 157. Springer Verlag, 2002. pp. 243–265.

82. Munkvold L, Kjoller R, Vestberg M, Rosendahl S, Jakobsen I. High functional diversity with in species of arbuscular–mycorrhizal fungi, *New Phytol* 2004; **164**: 357–364.

83. Rozema J, Arp W, Vandiggelen J, Vanesbroek M, Broekman R, Punte H. Occurrence and ecological significance of vesicular arbuscular mycorrhizae in the salt Marsh environment. *Acta Botanica Neerlandica* 1986; **35**: 457–467.

84. Dwivedi OP. Studies in soil micro-organisms with special reference to vesicular–arbuscular mycorrhizal (VAM) fungal association with wheat crop of Sagar region. Ph.D. Thesis. Dr. H.S. Gour University, Sagar (MP)., 2003.

85. Soni A .Studies on diversity of AMF with special reference to legume crops of Sagar region. Ph.D. Thesis. Dr. H.S.G. University, Sagar, Madhya Pradesh, India., 2006.

86. Ojala JC, Linderman RG. Preinoculation of lettuce and onion with VA mycorrhizal fungi reduces deleterious effect of soil salinity. *Plant and Soil* 2001; **223**: 269–281.

87. Al–Karaki GN. Gwowth of mycorrhizal tomato and mineral acquisition under salt stress, *Mycorrhiza* 2000; **10**: 51–54.

88. Al–Karaki GN, Hammad R. Mycorrhizal influence on fruit yield and mineral content of tomato grown under salt stress. *J Plant Nutrition* 2001; **24**: 1311–1323.

89. Pfeiffer CM, Bloss HE. Growth and nutrition of guayule (*Parthenium argenatum*) in a saline soil as influenced by vesicular arbuscular mycorrhiza and phosphorous fertilizer. *New Phytol* 1988; **108**: 315–321.

90. Cantrell IC, Linderman RG. Preinoculation of lettuce and onion with VA mycorrhizal fungi reduces deleterious effects of soil salinity. *Plant and Soil* 2001; **223**: 269–281.

91. Sanders IR, Fitter AH. The ecology and functioning of vesicular arbuscular mycorrhizas in coexisting grassland species I. Seasonal patterns of mycorrhizal occurrence and morphology. *New Phytol* 1992; **120**: 517–524.

92. Cardoso E, Navarro RB, Nogueira MA. Charges in manganese uptake and translocation by mycorrhizal soybean under increasing Mn doses. *Rev Bras De Ciencia Do Solo*, 2003; **27**: 415–423.

93. Kumar P, Duschenkov V, Motto H, Raskin I. Phytoextraction: the use of plants to remove heavy metals from soils. *Environ Sci Technol* 1995; **29**: 1232–1238.

94. Brooks RR, Robinson BH. The potential use of hyperaccumulators and other plants for phytomining. In: Brooks RR. Editor. Plants that hyperaccumulate heavy metals their role in phytoremediation, microbiology, archeology, mineral exploration, and phytomining. CAB International., 1998. pp. 327–356.

95. Salt DE, Smith RD, Raskin I. Phytoremediation. *Ann Rev Plant Physiol Plant Mole Biol* 1998; **49**: 643–668.

96. Baker AJM, MC Grath SP, Reeves RD, Smith JAC. Metal hyperaccumulator plants: a review of the ecology and physiology of a biological resource for phytoremediation of metal polluted soils. In: Terry N, Banuelos G, Vangronsveld J. editors. Phytoremediation of contaminated soil and water. Boca Raton, Fl, USA: CRC Press 2000. pp. 85–107.

97. Leyval C, Turnau K, Haselwandter K. Effect of heavy metal pollution on mycorrhizal colonization and function: physiological, ecological and applied aspects. *Mycorrhiza* 1997; **7**: 139–153.

98. Turnau K, Jurkiewicz A, Lingua G, Barea JM, Gianinazzi–Pearson V. Role of arbuscular mycorrhiza and associated micro-organisms in phytoremediation of heavy metal polluted sites. Trace element in the environment biochemistry, biotechnology and bioremediation. CRC Press., 2006. pp. 235–252.

99. Goncalves SC, Goncalves MT, Freitas H, Martin–S, Loucao MA. Mycorrhizae in a Portuguese serpentine community. In: Jaffre T, Reeves RD, Becquer T. editors. The ecology of ultarmafic and metalliferous areas. Proceedings of the second International Conference on Serpentine Ecology in Noumea, 1997. pp. 87–89.

100. Leyval C, Turnau K, Haselwandter K. Potential of arbuscular mycorrhizal fungi for bioremediation. In: Gianinazzi S, Schuepp H, Barea JM, Haselwandtar K. editors. Mycorrhiza technology in agriculture from genes to bioproducts. Birkhauser Verlag, 2002. pp. 175–186.

101. Orlowska E, Zubek SZ, Jurkiewic ZA, Szarek–Lukaszewska G. Influence of restoration on arbuscular mycorrhiza of *Biscutella laevigata* L. (Brassicaceae) and *Pantago lanceolata* L. (Plantaginaceae) from calamine spoil mounds. *Mycorrhiza* 2002; **2**: 153–160.

102. Regvar M, Vogel K, Irgel N, Wraber T, Hildebrandt U, Wilde P, Bothe H. Colonization of pennycresses (Th laspi spp.) of the Brassicaceae by arbuscular mycorrhizal fungi. *J Plant Physiol* 2003; **160**: 615–626.

103. Giller K. Toxicity of heavy metals to micro–organisms and microbial processes in agricultural soils: A review. *Soil Biol Biochem* 1998; **30**: 1389–1414.

104. Del Val C, Barea JM, Azcon–Aguilar C. Diversity of arbuscular mycorrhizal fungus population in heavy metal contaminated soil. *Appl Environ Microbiol* 1999; **65**: 718–723.

105. Chaudhary TM, Hill, Khan AG, Duek C. (1999). Colonization of iron and zinc–contaminated dumped filter–cake waste by microbes, plants and association mycorrhizae. In: Wong MH, Wong JWC, Baker AJM. Editors. Remediation and management of Degraded Land. CRC press, 1999. pp. 275–283.

106. Duek TA, Viser P, Ernst WHO, Schat H. Vesicular–arbuscular mycorrhizae decrease zinc toxicity to grasses growing in zinc–polluted soil. *Soil Biol Biochem* 1986; **18**: 331–333.

107. Pawloska TEJ, Blaszkowski, Riihling A. The mycorrhizal status of plants colonizing a calamine spoil mound in southern Poland. *Mycorrhiza* 1996; **6**: 499–505.

108. Joner EL, Levyal C. Uptake of ^{109}Cd by roots and hyphae of a *Glomus mosseae/Trifolium subterranean* mycorrhiza from soil amended with high and low concentrations of Cadamium. *New Phytol* 1997; **135**: 353–360.

109. Galli U, Schuepp H, Brunold C. Heavy metal binding by mycorrhizal fungi. *Physiol Plantarum* 1994; **92**: 364–368.

110. Hetrick BAD, Wilson GWT, Figge DH. The influence of mycorrhizal symbiosis and fertilizer amendment on establishment of vegetation in heavy metal mine spoil. *Environ Poll* 1994; **86**: 171–179.

111. Azcon–Aguilar C, Barea JM. Arbuscular mycorrhizas and biological control of soil borne plant pathogens: an overview of the mechanisms involved. *Mycorrhiza* 1996; **6**: 457–458.

112. Linderman RG. Role of VAM fungi in biocontrol. In: Pfleger FL, Linderman RG. Editors. Mycorrhizae and plant health. APS Press, 1994. pp. 1–25.

113. Linderman RG. Effect of mycorrhizas on plant tolerance to diseases. In : Kapulnik Y, Douds Jr DD. editors. Arbuscular mycorrhizae: Physiology and function. Kluwer Academic Publishers., 2000; pp. 345–365.

114. Elmer WH. Influence of formononetin and NaCl on mycorrhizal colonization and *Fusarium* crown and root rot of asparagus. *Plant Dis* 2002; **86**: 1318–1324.

115. Mastubara Y, Kayukawa Y, Yano M, Fukui H. Tolerance of asparagus seedlings infected with arbuscular mycorrhizal fungus to violet root rot caused by *Helicobasidium mompa. J Japn Hortic Sci* 2000; **69**: 552–556.

116. Mastubara Y, Ohba N and Fukui H. Effect of arbuscular mycorrhizal fungus infection on the incidence of *Fusarium* root rot in asparagus seedlings. *J Japn Hortic Sci* 2001; **70**: 202–206.

117. Abdel–Fattah GM, Shabana YM. Efficacy of the arbuscular mycorrhizal fungus *Glomus* in protection of cowpea plants against root rot pathogen *Rhictonia solani. JPlant Dis Protec* 2002; **109**: 207–215.

118. Declerck S, Risede JM, Rufyikiri G, Delvaux B. Effect of arbuscular mycorrhizal fungi on severity of root rot of bananas caused by *Cylindrocladium Spathiphylli. Plant Pathol* 2002; **51**: 109–115.

119. Davis RM, Menge JA. Influence of *Glomus fasciculatum* on *Phytphthora* root rot of citrus. *Phytopathol* 1980; **70**: 447–452.

120. Davis RM and Menge JA. *Phytophthora parasitica* inoculation and intensity of vesicular–arbuscular mycorrhizae in citrus. *New Phytol* 1981; **87**: 705–715.

121. Krishna KR, Bagyaraj DJ. Interaction between *Glomus fasisculatum* and *Sclerotium rolfsii* in peanut. *Can J Bot* 1983; **61**: 2349–2351.

122. Graham JH, Leonard RT, Menge JA. Membrance mediated decrease in root exudation responsible for phosphorous inhibition of vesicular–arbuscular mycorrhiza formation. *Plant Physiol* 1981; **68**: 548–552.

123. Singh Pradeep Kumar, Mishra M, Vyas D. Interaction of vesicular arbuscular mycorrhizal fungi with *Fusarium* wilt and growth of the tomato. *Indian Phytopath* 2010; **63**: 30–34.

124. Singh Pradeep Kumar. To study microbial interaction of below ground organism with special reference to *Fusarium* wilt of Chickpea. Ph.D. Thesis. Dr. H.S. Gour University, Sagar, Madhya Pradesh, India., 2008.

125. Bartschi H, Gianinazzi–Pearson V, Vegh I. Vesicular arbuscular mycorrhizae formation and root disease (*Phytophthora cinnamomi*) development in *Chamaecyparis lawso. Phytopathol Z* 1981; **102**: 213–218.

126. Dehne HW. Interaction between vesicular–arbuscular mycorrhizal fungi and plant pathogens. *Phytopathol* 1982; **72**: 1115–1119.

127. Dehne HW, Schonbeck F. Untersuchungen zun Einfluss der Endotrophen Mycorrhiza auf pflanzen krankheiten.II phenolstaff–wechsel und Linifizierung (The influence of endtrophic mycorrhiza on plant diseases. II. Phenol metabolism and lignifications. *Phytopathol Z* 1979; **95**: 210–216.

128. Caron M, Fortin JA, Richard C. Effect of phosphorous concentration and *Glomus intraradices* on *Fusarium* crown and root rot of tomatoes. *Phytopathol* 1986; **76**: 942–946.

129. Zambolim L, Schenck NC. Reduction of the effects of pathogenics root– infecting fungi on soyabean by the mycorrhizal fungus *Glomus mosseae*. *Phytopathol* 1983; **73**: 1402–1405.

130. Schoenbeck F. Endomycorrhiza in relation to plant disease. In: Schippers B, Gams. Editors. Soil–borne Plant Pathogens. Academic Press, London., 1979; pp. 271–292.

131. Chakravarty P, Mishra RR. Influence of endotrophic mycrrhizae on the Fusarial wilt of *Casia tora* L. *J Phytopathol* 1986; **115**: 130–133.

132. Melo IS, Costa CP, Silviera APD. Effect of vesicular arbuscular mycorrhizae on aubergine wilt caused by *Verticillium alboatrum* Reinke and Berth. *Summa phytopathologia* 1985; **11**: 173: 179.

133. Pozo MJ, Azcon–Aguilar C.Unraveling mycorrhiza–induced resistance. *Curr Opin Plant Biol* 2007; **10**: 393–398.

134. Cox G, Moran KJ, Sanders F, Nockolads C, Tinker PB. Translocation and transfer of nutrients in vesicular–arbuscular nutrients in vesicular arbuscular mycorrhizas; III. Polyphosphate granules and phosphorous translocation. *New Phytol* 1980; **84**: 649–659.

135. Jacobsen DJ, Beurkensk, klomparens KL. Microscopic and Ultrastructural examination of vegetative incompatibility in partial diploids heterozygous at heterozygous loci in *Neurospora crasa*. *Fung Genet Biol* 1998; **23**: 45–46.

136. Jones MD, Durall DM, Tinker PB. Comparison of arbuscular and ectomycorrhizal *Eucalyptus coccifera*: Growth response, phosphorous uptake efficiency and external hyphal production. *New Phytol* 1998; **140**: 125–134.

137. De mars BG, Boerner REJ. Arbuscular mycorrhizal development in three crucifers. *Mycorrhiza* 1995; **5**: 405–408.

138. Gardes M, Dahlberg A. Mycorrhizal diversity in arctic and alpine tundra: an open question. *New Phytol* 1996; **133**: 147–157.

139. Allen EB, Allen MF. The competition between plants of different successional stages: mycorrhizal as regulators. *Can J Bot* 1984; **62**: 2625–2692.

140. John TV, Coleman DC. The role of mycorrhizae in plant ecology. *Can J Bot* 1983; **61**: 1005–1014.

141. Fitter AH. Influence of mycorrhizal infection on competition for phosphorus and potassium by two grasses. *New Phytol* 1977; **79**: 119–125.

142. Facelli E, Facelli JM, Smith SE, Mclaughlin MJ. Interaotive effects of arbuscular mycorrhizal symbiosis intraspecific competition and resource availability on Trifolium subterranean cv. Mt. Barker. *New Phytol* 1999; **141**: 535–547.

143. Facelli E, Smith SE, Facelli JM, Christophersen HM, Smith FA. Underground friends or enemies: model plants help to unravel direct and indirect effects of arbuscular mycorrhizal fungi on plant competition. *New Phytol* 2010; **185**: 1050–1061.

144. Francis R, Read DJ. The contributions of mycorrhizal fungi to the determination of plant community structure. *Plant and Soil* 1994; **159**: 11–25.

145. Francis R, Read DJ. Direct transfer of carbon between plants connected by vesicular–arbuscular mycorrhizal mycelium *Nature* 1984; **307**: 53–56

146. Gange AC, Brown VK, Farmer LM. A test of mycorrhizal benefit in an early successional plant coumnity. *New Phytol* 1990; **115**: 85–91.

147. Chiariello N, Hickman JC, Mooney HA. Endomycorrhizal role for interspecific transfer of phosphorous in a community of annual plants, *Science* 1982; **217**: 941–943.

148. Grime JP, Mannkey JML, Hillersh, Read DJ. Floristic diversity in a model system using experimental microcosms. *Nature* 1987; **328**: 420–422.

149. Watkins NK, Fitter AH, Graves JD, Robinson D. Carbon transfer between C_3 and C_4 plants linked by a common mycorrhizal network quantified using stable carbon isotopes. *Soil Biol Biochem* 1996; **28**: 471–477.

150. Graham JH. Effect of citrus root exudates on germination of chlamydospores of the vesicular–arbuscular mycorrhizal fungus *Glomus epigacum*. *Mycologia* 1982; **74**: 831–835.

151. Lerat S, Gauci R, Catford JG, Vierheilig H, Pichey, Lapointel (2002). C–14 transfer between the spring ephemeral *Erythronium americanum* and sugar *mapl* saplings via arbuscular mycorrhizal fungi in natural sands. *Oecologia* 2005; **132**: 181–187.

152. Van der Heijden MGA, Moutoglis P, Streitwolf, Klironomos JN, Ursic M, Engel R, Boller T, Wiemken A. Mycorrhizal fungal diversity determines plant biodiversity, ecosystem variability and productivity. *Nature* 1998; **396**: 69–72.

153. Giovannetti M, Sbrana C, Avio L. Patterns of below ground plant interconnection established by means of arbuscular mycrrhizal networks. *New Phytol* 2004; **164**: 175–181.

154. Rai M, Varma A. Arbuscular mycorrhiza–like biotechnological potential of *Piriformospora indica*, which promotes the growth of *Adhatoda vasica* Nees. *Elect J Biotech* 2005; **8**: 107–112.

155. Vosatka M, Jansa J, Regvar M, Sramek F, Malcova R. Inoculation with mycorrhizal fungi–a feasible biotechnology for horticulture. *Phyton* 1999; **39**: 219–224.

156. Shin H, Shin HS, Dewbre GR, Harrison MJ. Phosphate transport in *Arabidopsis*: Pht1;1 and Pht1;4 play a major role in phosphate acquisition from both low- and high–phosphate environments. *The Plant J* 2004; **39**: 629–642.

157. Harrison MJ, Van Burren ML. A phosphate transporter from the mycorrhizal fungus *Glomus vesiforme*. *Nature* 1995; **378**: 626–629.

158. Maldnado–Mendoja IE. A phosphate transporter gene from the extraradical mycelium of an arbuscular mycorrhizal fungus *Glomus intraradices* is regulated in response to phosphate in the environment. *Mol Plant Microbe Interact* 2001; **14**: 1140–1148.

159. Bun–Ya M. The pho84 gene of *Saccharomyces cerevisiae* encodes an inorganic phosphate transporter. *Mol Cell Biol* 1991; **11**: 3229–3238.

160. Franken P, Requena N. Analysis of gene expression in arbuscular mycorrhizas: New approaches and challenges. *New Phytol* 2001; **150**: 517–523.

161. Colebatch G, Trevaskis B, Udvardi M. Functional genomics: tool of the trade. *New Phytol* 2002; **153**: 27–36.

162. St–Arnaud M, Hamel C, Vimard B, Caron M, Fortin A. Enhanced hyphal growth and spore production of the arbuscular mycorrhizal fungus *Glomus intraradices* an *in vitro* system in the absence of host roots. *Mycol Res* 1996; **100**: 328–332.

163. Stommel M, Mann P, Franken P. Construction and analysis of an EST library using RNA from dormant spores of the arbuscular mycorrhizal fungus *Gigaspora rosea*. *Mycorrhiza* 2001; **10**: 281–285

164. Tamasloukht M, Sejalon–Delmas N, Kluever A, Jauneau A, Roux C, Becard G, Franken P. Root factors induce mitochondrial related gene expression and fungal respiration during the devlopment switch from asymbiosis to presymbiosis in the arbuscular mycorrhizal fungus *Gigaspora rosea*. *Plant Physiol* 2003; **131**: 1468–1478.

165. Requena N, Mann P, Hampp R, Franken P. Early development regulated genes in the arbuscular mycorrhizal fungus *Glomus mosseae*: Identification of Gm GINI, a C–terminus of metazoan hedgehog proteins. *Plant Soil* 2002; **244**: 129–139.

166. Lanfranco L, Novero M, Bonfante P. The mycorrhizal fungus *Gigaspora margarita* possesses a Cu Zn superoxide dismutase that is up–regulated during symbiosis with legume hosts. *Plant Physiol* 2005; **137**: 1319–1330.

167. Swaki H, Satio M. Expressed genes in the extraradical hyphae of an arbuscular mycorrhizal fungus, *Glomus intraradices* in the symbiotic phase. *FEMS Microbiol Lett* 2001; **195**: 109–113.

168. Hilbert JL, Cista G, Martin F. Ectomycorrhizin synthesis and polypeptide changes during the early stage of *Eucalyptus* mycorrhiza development. *Plant Physiol* 1991; **97**: 977–984.

169. Simoneau P, Viemont JD, Moreau JC, Strulu DJ. Symbiosis related polypeptides associated with the early stages of ectomycorrhiza organogenesis in birch (*Betula pendula* Roth.) *New Phytol* 1993; **124**: 495–504.

170. Burgess T, Laurent P, Dell B, Malajczuk N, Martin F. (1995) Effect of fungal isolate aggressivity on the biosynthesis of symbiosis related polypeptides in differentiating eucalypt mycorrhizas. *Can J Bot 1995;* **195**: 408–417.

171. Burgess T, Dell B. Changes in protein biosynthesis during the differentiation of *Pisolithus–Eucalyptus grandis* ectomycorrhiza. *Planta* 1996; **74**: 553–560.

172. Tarkka MT, Nyman TA, Kallkinen N, Raudaskoski M. Scots pine expresses short–root specific peroxidases during development. *Eur J Biochem* 2000; **267**: 1–8.

173. Vallorani L, Bernardini F, Sacconic C, Pierleoni R, Pieretti B, Piccoli G, Buffalini M, Stocchi V. Identification of Tuber *Borchii vittad* mycelium proteins separated by two dimensional polyacrylamide gel electrophoresis using amino acid analysis and sequence tagging. *Electrophoresis* 2000; **21**: 3710–3716.

174. Garcia–Garrido JM, Toro N, Ocampo JA. Presence of specific polypeptides in onion roots colonized by *Glomus mosseae. Mycorrhiza* 1993; **2**: 175–177.

175. Dumas–Gaudot E, Guillaume P, Tahiri–Alaoui A, Gianinazzi–Pearson V, Gianinazzi S. Changes in polypeptide patterns in tobacco root colonized by two *Glomus* species. *Mycorrhiza* 1994; **4**: 215–221.

176. Simoneau P, Louisy–Lois N, Plenchettec, Strullu DG. Accumulation of new polypeptides in Ri–tDNA–transformed roots of tomato (*Lycopersicon esculentum*) during the development of vesicular– aruscular mycorrhizae. *Appl Environ Microbiol* 1994; **60**: 1810–1813.

177. Samara A, Dumas–Gaudot E, Gianinazzi–Pearson V, Gianinazzi S. Soluble proteins and polypeptide profiles of spores of arbuscular mycorrhizal fungi. Interspecific variability and effect of host (myc+) and non– host (myc–) *Pisum sativum* root exudates. *Agronomie* 1996; **16**: 709–719.

178. Benabdellah K, Azcon–Aguilar C, Ferrol N. Soluble and membrane symbiosis related polypeptides associated with the development of arbuscular–mycorrhizas in tomato (*Lycopersicon esculentum*) *New Phytol* 1998; **140**: 15–143.

179. Dassi B, Samara A, Dumas–Gaudat E, Gianinazzi S. Different polypepetide profiles from tomato polypeptide profiles from tomato roots following interaction with arbuscular mycorrhizal (*Glomus mosseae*) or pathogenic (*Phytophthora parasitica)* fungi. *Symbiosis* 1999; **26**: 65–77.

180. Benabdellah K, Azcon–Aguilar C, Ferrol N. Alterations in the plasma membrane polypeptide pattern of tomato roots (*Lycopersicon esculentum*) during the development of arbuscular mycorrhiza. *J Exp Bot* 2000; **51**: 747–754.

181. Fester TMK, Strack D. A mycorrhizae–responsive protein in wheat roots. *Mycorrhiza* 2003; **12**: 219–222.

182. Dumas–Gaudat E, Bestel–Corre G, Gianinazzi S. Proteomics, a powerful approach towards under standing functional plant root interactions with arbuscular mycorrhizal fungi: In: Pan–Dilai SG. Editor. Recent research developments in plant biology. Research Signopost, Trivandrum, India., 2001. pp. 95–104

183. Bestel Corre G, Dumas–Gaudot E, Gianinazzi S. Proteomics as a tool to monitor plant microbe endosymbioses in the rhizosphere., 2002; 14: 1–10.

184. Repetto O, Bestel–Corre G, Dumas–Gaudot E, Berta G, Gianinazzi–Pearson V, Gianinazzi S. Targeted proteomics to identify cadmium–induced protein modifications in *Glomus mosseae*–inoculated pea roots. *New Phytol* 2003; **157**: 555–567.

185. Bestel–Corre G, Dumas–Gaudot E, Poinsot V, Dieu M, Dierick JF, van Tuinen D, Remacle J, Gianinazzi–Pearson V, Gianinazzi S Proteome analysis and identification of symbiosis related proteins from Medicago truncatula Gaertn. by two dimensional electrophoresis and mass spectrometry. *Electrophoresis* 2002; **23**: 122–137.

186. Dumas–Gaudot E, Bestel Corre G, Valot B, Lenogue S, Amiour N, St–Arnaud M, Fontain B, Dieu M, Raes M, Gianiazzi S. Use of *in vitro* grown *Glomus intraradices* to sensor the effect of agricultural amendment. In: Proceeding of the IV International Congress on Mycorrhiza, Montreal., 2003. pp. 107–121.

187. Sauer, U, Lasko DR, Flaux J, Hochuli M, Glaser R, Szyperski T, Wuthrich K, Bailey JE. Metabolic flux ratio analysis of genetic and environmental modulations of *Escherichia coli* central carbon metabolism. *J Bacteriol* 1999; **181**: 6679–6688.

188. Sauer U. High–throughput phenomics: experimental methods for mapping fluxomes. *Curr Opin Biotech* 2004; **15**: 58–63.

189. Kruger NJ, Ratcliffe RG, Roscher A. Quantitative approaches for analyzing fluxes through plant metabolic networks using NMR and stable isotope labeling. *Phytochem Rev* 2003; **2**: 17–30.

190. Schwender J, Goffman F, Ohlrogge JB, Shachar–Hill Y. Rubisco without the Calvin cycle improves the carbon efficiency of developing green seeds. *Nature* 2004a; **432**: 779–782.

191. Raticliffe RG, Shachar–Hill Y. Measuring multiple fluxes through plant metabolic networks. *Plant J* 2006; **45**: 495–511.

192. Chalot M, Brun A. Physiology of organic nitrogen acquisition by ectomycorrhizal fungi and ectomycorrhizas. *FEMS Microbiol Rev* 1998; **22**: 21–44.

193. Bago B, Pfeffer PE, Shachar HY. Carbon metabolism and transport in arbuscular mycorrhizas. *Plant Physiol* 2000; **128**: 949–957.

194. Bucking H, Sachar–Hill Y. Phosphate uptake, transport and transfer by the arbuscular mycorrhizal fungus *Glomus* in protection of cowpea plants against root rot pathogen *Rhictonia solani. J Plant Dis Protect* 2005; **109**: 207–215.

195. Govindarajulu M, Pfeffer P, Jin HR, Abubaker J, Douds DD, Allen JW, Bucking H, Lammers PJ, Shachar–Hill Y. Nitrogen transfer in the arbuscular mycorrhizal symbiosis. *Nature* 2005; 435: 819–823.

Biotechnology: An Overview (2015)
Editors: Rajan Kumar Gupta, Nasim Akhtar and Deepak Vyas
Published by: DAYA PUBLISHING HOUSE, NEW DELHI

Pages 157–177

Chapter 12

Phytoremediation: An Ecofriendly Approach for Environmental Cleanup

Indu Bajwa¹ and Anjali Pandey²

¹Ex-Visiting Faculty, CSJM University, Kanpur, Uttar Pradesh
²Department of Chemistry, IIT, Kanpur – 208 011, Uttar Pradesh

ABSTRACT

Phytoremediation is the use of plants and their associated microbes to sequester, remove, or degrade inorganic and organic contaminants in soils, sediments, surface water and ground water. It actually refers to a diverse collection of plant- based technologies that use either naturally occurring or genetically engineered plants for cleaning contaminated environment. It is an emerging technology which offers a potentially cost-effective and environmentally sound alternative to the environmentally destructive physical methods which are currently practiced for the clean up of contaminated soil and water resources.

This chapter provides a broad and introductory description of various phytoremediation technologies, paying special attention to use of transgenic plants for remediation of organics and heavy metals contaminated sites. This covers the basic physiological processes occurring in the rhizosphere as well as in the plant system. Further more, these basic processes are related to the phytoremediation mechanism, which form the basis for the various applications used in the field today. Site-specific consideration and practical field implementation techniques are also provided for different applications.

*Keywords: **Hyperaccumulators, Phytoremediation, Rhizosphere, Transgenic Plants, Xenobiotics.***

Introduction

Phytoremediation is the name given to set of technologies that use plants to remediate contaminated environment. 'Phyto' means pertaining to plants, while 'remediation' means correcting an error or a fault.

Some plants can absorb and store toxins; other plants can break down nasty chemicals and turn them into less harmful substances. Plants grown for phytoremediation can also prevent toxins from moving from a contaminated site into other areas (USEPA 2000).

Plants act as solar – driven pumping and filtering systems as they take up contaminants (mainly water soluble) through their roots and transport/translocate them through various plant tissues where they can be metabolized, sequestered or volatilized (Conningham *et al.*, 1996; Greenberg *et al.*, 2006; Abhilash, 2007; Doty *et al.*, 2007; Van Aken, 2008).

Further, plants may survive higher concentrations of hazardous wastes than many micro- organisms used for bioremediation. Phytoremediation increases the amount of organic carbon in the soil, which can stimulate microbial activity and augment the rhizospheric degradation of pollutants. This also increases soil stability. The development of phytoremediation technologies for the plant based cleanup of contaminated soils is therefore of significant interest (Hooker and Skeen, 1999; Dietz and Schnoor, 2001; Eapen and D' souza, 2005; Kaushik, 2011).

Plant-based environmental remediation, or phytoremediation, has been widely pursued in recent years as a favorable clean-up technology and is an area of intensive scientific investigation. Phytoremediation has been considered as a cost effective method for the decontamination of soil and water resources (Salt *et al.*, 1998; Macek *et al.*, 2000; Meagher, 2000; Eapen and D' Souza' 2005; Cherian and Oliveira, 2005;Mello-Farias and Chaves, 2008). This is an eco friendly approach for remediation of contaminated soil and water.

The technology has successfully been used on many contaminated sites(Gunther *et al.*, 1996; Liste and Alexzander, 2000; Mattina *et al.*, 2000; White, 2000, 2001, 2002; Li *et al.*, 2002; Maila and Cloete, 2002; Yoon *et al.*, 2002; Singh and Jain, 2003; Sunderberg *et al.*, 2003; Gao and Zhu, 2004; Ma *et al.*, 2004; Mattina *et al.*, 2004; Suresh *et al.*, 2005; Mills *et al.*, 2006; Parrish *et al.*, 2006; Aslund *et al.*, 2007; Cook *et al.*, 2010).

Specially selected or engineered Living green plants are used for in situ risk reduction and/or removal of contaminants from contaminated soil, water, sediments, and air. Risk reduction can be through a process of removal, degradation of, or containment of a contaminant or a combination of any of these factors. Phytoremediation is an energy efficient, aesthetically pleasing method of remediating sites with low to moderate levels of contamination and it can be used in conjunction with other more traditional remedial methods as a finishing step to the remedial process.

For the vast majority of field applications, vegetative 'phyto-crops' are selected specifically for their capacity for site decontamination and not for additional concurrent or post-remediation utility. Phytoremediation provides the ancillary benefits of concurrent site stabilization through erosion mitigation and hydraulic control of solubilized contaminants

While phytotechnologies generally are applied in situ, ex situ applications (*e.g.* hydrophonic systems) are also possible. Phytoremediation can be used to remediate various contaminants including metals, pesticides, solvents, explosives, petroleum hydrocarbons, polycyclic aromatic hydrocarbons, and landfill leachates (USEPA 2006; ITRC 2009; Wu *et al.*, 2011).

Types of Phytoremediation

Phytoremediation consists of different plant-based technologies each having a different mechanism of action for the remediation of organics/heavy metals polluted soil, sediment, or water, *viz.* Phytoextraction/phytoaccumulation, Rhizofiltration, Phytostabilization, Phytovolatilization, Phytodegradation and Rhizodegradation etc (Table 12.1 and Figure 12.1).

Table 12.1. Typical plants used in various phytoremediaion applications.

Type of Phytoremediation	Contaminant Treated	Plant (s)
Phytoextraction	Cd, Cr, Pb, Ni, As, Zn, Radionuclides,	Indian mustard, Cabbage *Thlaspi caerulescens*, Willow,
	BTEX*, pentachlorophenol, short chained aliphatic compounds	Hemp, Sunflower Alfalfa, Poplar, Junifer
Rhizofilteration	Heavy metals, Radionucleides and Hydrophobic organics	Sunflowers, Indian mustard, aquatic plants (Cattail, Arrowroot, Hydrilla)
Phytostabilisation	Heavy metals	Hybrid Poplar, Grasses, *Agrostis tenuis, Festuca robra L.*
	Phenols and chlorinated solvents	Salix, Poplar, Grasses
Phytovolatilisation	Chlorinated solvents, Organic VOCs, BTEX	Poplar (Populus sp)
	Hg, Se. As	Cattail (*Typha latifolia)*, Chinese Brake Fern
Phytodegradation	Nitrobenzene, nitroethane, nitrotoluene, atrazine, chlorinatedsolvents (chloroform, carbon tetrachloride, etc)	Duckweed, Poplar, Willow, Cotton Wood, Grasses (Rye, Bermuda, Sorghum, Reed), Legumes (Clover, Alfalfa)
Rhizodegradation	Polyaromatic hydrocarbons, BTEX and other petroleum hydrocarbons, PCB and other organic compounds	Hybrid Poplar, Perennial Rye grass, Grasses, Willows and Legumes

Phytoextraction/Phytoaccumlation

This involve the use of pollutant- accumulating plants to remove metals from soil and translocating them to the harvestable shoots where they accumulate. In practice, metal accumulating plants are seeded or transplanted into metal-polluted soil and are cultivated using established agricultural practices. The roots of established plants absorb metal elements from the soil and translocate them to the aboveground shoots where they accumulate. As different plant have different abilities to uptake and withstand high levels of pollutants, many different plants may be used. This is of particular importance on sites that have been polluted with more than one type of metal contaminant.

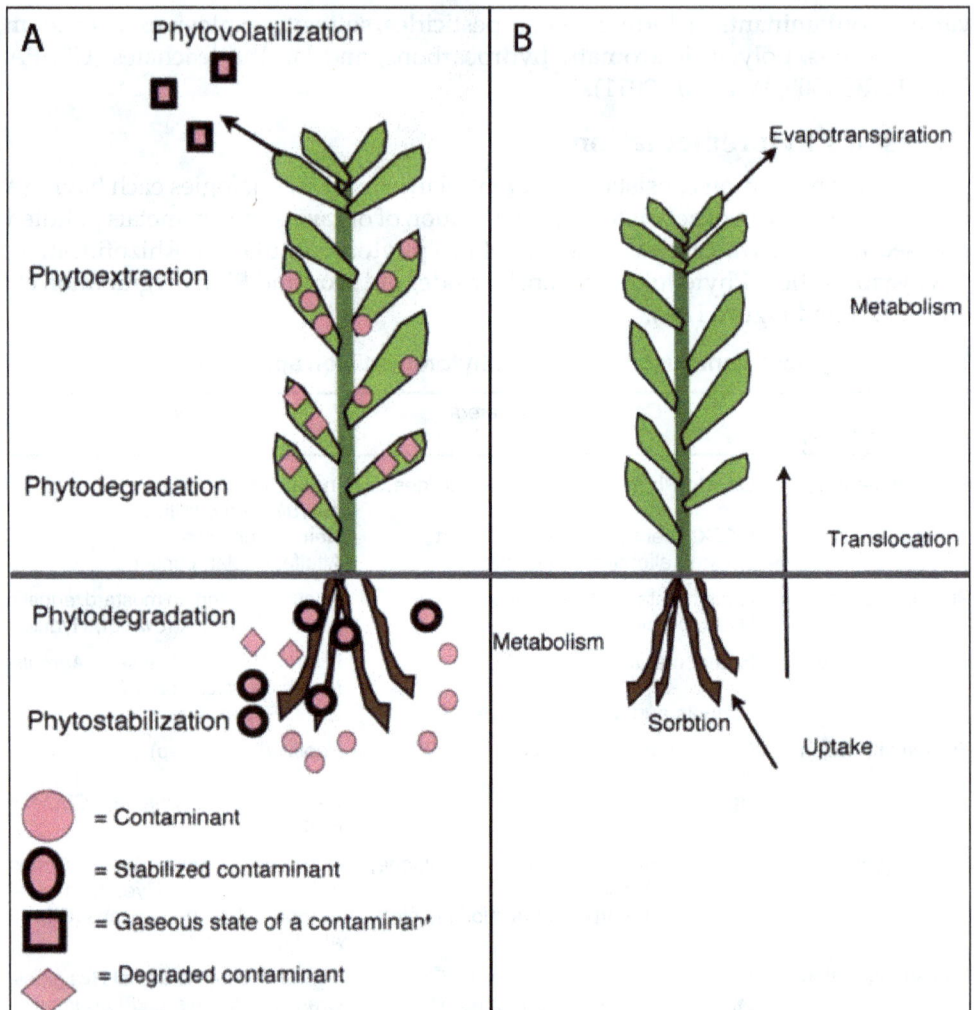

Figure 12.1. (A) Schematic model of different types of phytoremediation; (B) Physiological processes that take place in plants during phytoremediation (adopted from Greipsson, 2011).

Hyperaccumulator plant species (species which absorb higher amounts of pollutants than most other species) are used on many sites due to their tolerance of relatively extreme levels of pollution. If metal availability in the soil is not adequate for sufficient plant uptake, chelates or acidifying agents may be used to liberate them into the soil solution (Huang and Cunningham, 1996; Huang *et al.*, 1997; Lasat *et al.*, 1998; Abbaspour *et al.*, 2012).

After sufficient plant growth and metal accumulation, the above-ground portions of the plant are harvested and removed, resulting the permanent removal of metals from the site. As with soil excavation, the disposal of contaminated material is a major concern. Testing of plant tissue, leaves, roots, etc., will determine if the plant tissue is a hazardous waste. Landfilling, incineration, and composting are options to dispose of or recycle the metal, depending upon the result of toxicity testing and the cost. Some researchers suggest that the incineration of harvested plant tissue dramatically reduces the volume of the material requiring disposal (Kumar *et al.*, 1995a). In some cases valuable metals can be extracted from the metal-rich ash (phytomining) and serve as a source of revenue, thereby offsetting the expense of remediation (Comis, 1996; Robinson *et al.*, 1997; Martinez *et al.*, 2006).

This is a long term remediation process, as it may require several cycles of cropping to reduce metal concentration to acceptable levels. The time required for remediation depends on several factors:

☆ The type and extent of metal contamination

☆ Size and depth of the polluted area

☆ Type and number of plants being used

☆ The length of the growing season

☆ Type of soil and conditions present

☆ The efficiency of metal removal by plants

These factors vary from site to site. Normally it takes from 1 to 20 years to clean up a site with phytoremediation (Kumar *et al.*, 1995a; Blaylock and Huang, 2000). Metal compounds that have been successfully phytoextracted include zinc, copper, and nickel, but there is promising research being completed on lead and chromium absorbing plants.

This technology is suitable for the remediation of large areas of land that are contaminated at shallow depths with low to moderate levels of metal contaminants (Kumar *et al.*, 1995a; Blaylock and Huang, 2000), because plant growth is not sustained in heavily polluted soils. Soil metals should also be bioavailable, or subject to absorption by plant roots. The land should be relatively free of obstacles, such as fallen trees or boulders, and have an acceptable topography to allow for normal cultivation practices, which employ the use of agricultural equipment.

It is the combination of high metal accumulation and high biomass production that results in the most metal removal. Ebbs *et al.* (1997) reported that *B. juncea*, is more effective at Zn removal from soil than *T. caerulescens*, a known hyperaccumulator of Zn due primarily to the fact that *B. juncea* produces ten times more biomass than *T. caerulescens*. Plants being considered for phytoextraction must be tolerant to the targeted metal, or metals, and be efficient at translocating them from roots to the harvestable above-ground portions of the plant (Blaylock and Huang, 2000). Other desirable plant characteristics include the ability to tolerate difficult soil conditions (*i.e.*, soil pH, salinity, soil structure, water content), the production of a dense root system, ease of care and establishment, and few disease and insect problems. . Hyperaccumulators are good candidates in phytoremediation, particularly for the removal of heavy metals (Kumar *et al.*, 1995a; Cunningham and Ow, 1996; Prasad and Freitas, 2003).

11 genera and 87 species of *Brassicaceae* including *Alyssum* species, *Thlaspi* species and *Brassica juncea*, *Violaceae* such as *Viola calaminaria*, *Leguminosae* such as *Astragalus racemosus* are known to take up high concentrations of heavy metals and radionucleides (Reeves and Baker, 2000; Negri and Hinchman, 2000, Eapen and D'Souza, 2005).Brassicaceae also known to have the largest number of nickel (7 genera and 72 species) and zinc hyperacccumulators (3 genera and 20 species).

The phytoextraction of As contaminated soil by Chinese Brake Fern (*Pteris vittata* L.) has been reported to be beneficial not only for decreasing the As content of rice field but also increasing rice yield (Mandal *et al.*, 2012).

However, the remediation potential of many of these plants is limited because of their slow growth and low biomass (Chaney *et al.*, 2000; Lasat, 2002; Mc Grath *et al.*, 2002; Eapen and D'Souza, 2005). Phytoremediation efficiency of plants can be substantially improved using genetic engineering technologies (Cherian and Oliveira, 2005; Mello-Farias and Chaves, 2008).

Rhizofiltration

This involves the use of plant roots to absorb and adsorb pollutants, mainly metals, from water and aqueous waste streams. Metal pollutants in industrial-process water and in groundwater are most commonly removed by precipitation or flocculation, followed by sedimentation and disposal of the resulting sludge (Ensley, 2000). Rhizofiltration is a promising alternative to this conventional clean-up method designed for the removal of metals in aquatic environments (Prasad and Freitas, 2003).

This is a phytoremediative technique, similar to phytoextraction. In this process plants used for rhizofiltration are not planted directly in situ but are acclimated to the pollutants first. Plants are hydroponically grown in clean water rather than soil, until a large root system has developed. Once a large root system is in place the water supply is substituted for a polluted water supply to acclimatize the plant. After the plants become acclimatized they are transplanted into metal- polluted waters where plants absorb and concentrate the metals in their roots and shoots (Dushenkov *et al.*, 1995; Salt *et al.*, 1995; Flathman and Lanza, 1998; Zhu *et al.*, 1999a).

Root exudates and changes in rhizosphere pH may also cause metals to precipitate onto root surfaces. As they become saturated with the metal contaminants, roots or whole plants are harvested and disposed off safely (Flathman and Lanza, 1998; Zhu *et al.*, 1999a). Repeated treatments of the site can reduce pollution to suitable levels.

According to Dushenkov and Kapulnik (2000) plants, capable of accumulating and tolerating significant amounts of the target metals and produce significant amounts of root biomass or root surface area in conjunction with easy handling, low maintenance cost, and a minimum of secondary waste requiring disposal are ideal for rhizofiltration. Although several aquatic species like water hyacinth (*Eichhornia crassipes*) (Zhu *et al.*, 1999a), pennywort (*Hydrocotyle umbellata* L. and duckweed (*Lemna minor* L.) are effective in removing heavy metals from water, but to some extent only due to their small and slow growing roots (Dushenkov *et al.*, 1995). Further, the high water content of aquatic plants complicates their drying, composting, or incineration. Terrestrial plants are thought to be more suitable for rhizofiltration due to their longer and often fibrous root systems with large surface areas for metal sorption. The roots of Indian mustard (*Brassica juncea* Czern.) are very effective in the removal of Cd, Cr, Cu, Ni, Pb, and Zn (Dushenkov *et al.*, 1995), and sunflower(*Helianthus annuus* L.) in removal of Pb (Dushenkov *et al.*, 1995), U (Dushenkov *et al.*, 1997a), 137Cs, and 90Sr (Dushenkov *et al.*, 1997b) from hydroponic solutions. Rhizofiltration is a cost-competitive technology in the treatment of surface water or groundwater containing low, but significant concentrations of heavy metals such as Cr, Pb, and Zn (Kumar *et al.*, 1995b; Ensley, 2000). Further advantages of this technology are its applicability to many problem metals, ability to treat high volumes, reduced volume of secondary waste, possibility of recycling, and regulatory and public acceptance (Dushenkov *et al.*, 1995; Kumar *et al.*, 1995b). Perhaps the greatest benefit of this remediation method is the positive public perception because use of plants at a contaminated site conveys the idea of cleanliness.

Phytostabilisation

Phytostabilisation, also known as phytorestoration, is a plant-based remediation technique used to immobilize soil and water contaminants and thus prevent further dispersal of pollutants in the environment. Contaminants are absorbed and accumulated by roots, adsorbed onto the roots, or precipitated in the rhizosphere.This reduces or even prevents the mobility of the contaminants preventing migration into the groundwater or air, and also reduces the bioavailability of the contaminant. This also prevents exposure pathways via wind and water erosion, thus preventing spread through the food chain. . Unlike other phytoremediative techniques, the goal of phytostabilization is not to remove metal contaminants from a site, but rather to stabilize them and reduce the risk to human health and the environment (Flathman and Lanza, 1998; Berti and Cunningham, 2000; Schnoor, 2000).

Plants chosen for phytostabilization should be poor translocators of metal contaminants to aboveground plant tissues that could be consumed by humans or animals. The lack of appreciable metals in shoot tissue also eliminates the necessity of treating harvested shoot residue as hazardous waste (Flathman and Lanza, 1998). Further, the selected plants should be easy to establish and care for, grow quickly, have dense canopies and root systems, and be tolerant of metal contaminants and other site conditions which may limit plant growth.

Phytovolatilisation

Phytovolatilization is the process where plants uptake contaminaints (metals as well as organic compounds) which are water soluble and release them into the atmosphere as they transpire the water(Raskin and Ensley, 2000).The contaminant may become modified along the way, as the water travels along the plant's vascular system from the roots to the leaves, whereby the contaminants evaporate or *volatilize* into the air surrounding the plant. Transformation or degradation of the contaminant within the plant can create a less toxic product that is transpired; however, degradation of some contaminants, like trichloroethene (TCE), may produce even more toxic products (*e.g.*, vinyl chloride). Once in the atmosphere, these products may be degraded more effectively by sunlight (photodegradation) than they would be by the plant (phytodegradation), but the potential advantages and disadvantages of phytovolatilization must be assessed on a site-specific basis (USEPA 2006).

This technique is a promising tool for the remediation of Se and Hg contaminated soils (Heaton *et al.*, 1998). The release of volatile Se compounds from higher plants (especially members of the Brassicaceae) has been reported by Prasad and Freitas (2003).

Furthermore, sites that utilize this technology may not require much management after the original planting. This remediation method has the added benefits of minimal site disturbance, less erosion, and no need to dispose of contaminated plant material (Heaton *et al.*, 1998; Rugh *et al.*, 2000).

Although plants may already carry out this process at modest levels, genetic engineering will likely be necessary to volatilize metals at a sufficiently rapid rate to clean polluted soil. Though plants can move mercury from soil and even volatilize a portion of it (Moreno *et al.*, 2008), non-engineered plants are unlikely to volatilize mercury at rates that are useful for cleaning mercury contaminated soils.

Plants can concentrate selenium in aboveground tissues at different efficiencies depending on species (Banuelos *et al.*, 1996). However selenium volatilization from contaminated soils is enhanced significantly by the presence of plants on the site (Terry *et al.*, 2000).The predominant volatilized form is DMSe (dimethylselenide) of which a significant portion of the volatilized total comes either directly from plants or from soil microbes that are supported by plants.

Genetic engineering can be used to accelerate the rate-limiting steps in the biochemical pathway of selenium assimilation and volatilization, leading to a plant that is more efficient at phytovolatilization (de Souza *et al.*, 2000).

Phytodegradation (Phytotransformation)

Phytodegradation is the uptake, metabolizing, and degradation of contaminants within the plant, or the degradation of contaminants in the soil, sediments, sludge, ground water, or surface water by enzymes produced and released by the plant. The process is not dependent on micro-organisms associated with the rhizosphere. Phytodegradation is also known as phytotransformation, and is a contaminant destruction process. For phytodegradation to occur within the plant, the plant must be able to take up the compound. Plants contain enzymes that catalyze and accelerate chemical reactions. *Ex planta* metabolic processes hydrolyze organic compounds into smaller units that can be absorbed by the plant. Some contaminants can be absorbed by the plant and are then broken down by plant enzymes. These smaller pollutant molecules may then be used as metabolites by the plant as it grows, thus becoming incorporated into the plant tissues. Plant enzymes have been identified that break down ammunition wastes, chlorinated solvents such as TCE (Trichloroethane), and others which degrade organic herbicides (Newman *et al.*, 1997; ITRC, 2009).

Rhizodegradation

Rhizodegradation (also called enhanced rhizosphere biodegradation, phytostimulation, and plant assisted bioremediation) is the breakdown of organic contaminants in the soil by soil dwelling microbes which is enhanced by the rhizosphere's presence. Certain soil dwelling microbes digest organic pollutants such as fuels and solvents, producing harmless products through a process known as *Bioremediation*. Plant root exudates such as sugars, alcohols, and organic acids act as carbohydrate sources for the soil microflora and enhance microbial growth and activity. Some of these compounds may also act as chemotactic signals for certain microbes. The plant roots also loosen the soil and transport water to the rhizosphere thus additionally enhancing microbial activity.

Although plants show some ability to reduce the hazards of organic pollutants (Cunningham *et al.*, 1995; Gordon *et al.*, 1997; Carman *et al.*, 1998), the greatest progress in phytoremediation has been made with metals (Salt *et al.*, 1995; Watanabe, 1997; Blaylock and Huang, 2000). An ideal plant for environmental cleanup should have a high biomass production, combined with superior capacity for pollutant tolerance, accumulation, and degradation, depending on the type of pollutant and the phytoremediation technology of choice. Specifically, two subsets of phytoremediation are nearing commercialization. First is phytoextraction, in which high biomass metal-accumulating plants and appropriate soil amendments are used to transport and concentrate metals from the soil into the harvestable part of roots and aboveground shoots, which are harvested with conventional agricultural methods . The other is rhizofiltration, in which plant roots grown in water absorb, concentrate, and precipitate toxic metals and organics from polluted effluents (Cunningham *et al.*, 1995; Salt *et al.*, 1998). Plants can accelerate bioremediation in surface soils by stimulating the growth and metabolism of soil micro-organisms through the release of nutrients and the transport of oxygen to their roots (Schnoor *et al.*, 1995). The zone of soil closely associated with the plant root, the rhizosphere, has much higher numbers of metabolically active micro-organisms than the surrounding bulk soil. Studies have demonstrated greater than 100-fold increase in microbial counts. Plant roots release compounds including simple sugars, amino acids, enzymes, aliphatics, and aromatics that encourage growth of specific microbial communities. The interactions between plants and microbes in the rhizosphere are complex and in some cases have evolved to the mutual benefit of both organisms (Hedge and Fletcher, 1996). This mutual relationship is responsible for the accelerated degradation of soil contaminants in the presence of plants (Leigh *et al.*, 2006; Mackova *et al.*, 2007). In addition to this rhizosphere effect, plants themselves are able to passively take up a wide range of organic wastes from soil or water through their roots.

Advantages of Phytoremediation

Phytoremediation has made tremendous gains in market acceptance in recent years. In addition to its favorable economics, according to various authors (Salt *et al.*, 1995, Schnoor *et al.*, 1995) the main advantages of phytoremediation in comparison with classical remediation methods are as follows:

☆ It is potentially the least harmful method because it uses naturally occurring organisms and preserves the environment as the plants can be easily monitored.

☆ The possibility of the recovery and re-use of valuable metals (by companies specializing in "phyto mining").

☆ It is far less disruptive to the environment.

☆ There is no need for disposal sites.

☆ It has a high probability of public acceptance.

☆ It avoids excavation and heavy traffic.

☆ It has potential versatility to treat a diverse range of hazardous materials.

Considering these factors and the much lower cost expected for phytoremediation, it appears that it may be used in much larger scale clean-up operations than is possible by other methods. The process is relatively inexpensive, because it uses the same equipment and supplies that are generally used in agriculture.

Disadvantages of Phytoremediation

Like other methods of environmental remediation, phytoremediation has its disadvantages (Salt *et al.*, 1995; Schnoor *et al.*, 1995):

☆ Phytoremediation is limited to the surface area and depth occupied by the roots.

☆ It may take longer than other technologies.

☆ The solubility of some contaminants may be increased, resulting in greater environmental damage and/or pollutant migration therefore it is not possible to completely prevent the leaching of contaminants into the groundwater.

☆ As the survival of the plants is affected by extremes of environmental toxicity and the general condition of the soil, the use of phytoremediation is limited by the climatic and geological conditions of the site to be cleaned, temperature, altitude and soil type etc.

☆ Contaminants collected in leaves can be released again to the environment during litter fall.

☆ Bio-accumulation of contaminants, especially metals, into the plants which then pass into the food chain, from primary level consumers upwards or requires the safe disposal of the affected plant material.

Advances in Phytoremediation

Breeding programs and genetic engineering are powerful methods for enhancing natural phytoremediation capabilities, or for introducing new capabilities into plants. With increased understanding of the enzymatic processes involved in plant tolerance and metabolism of xenobiotic chemicals, there is new potential for engineering plants with increased phytoremediation capabilities(Raskin, 1996; Hooker, 1999; Panz and Miksch, 2012). Genes for phytoremediation may originate from a micro-organism or may be transferred from one plant to another variety better adapted to the environmental conditions at the cleanup site.

For inorganics, transgenic plants have been developed with enhanced accumulation (particularly in the harvestable shoot) and/or volatilization, as well as higher tolerance. For organics, different transgenics have been developed with the capacity to degrade and/or volatilize different pollutants, as well as with enhanced tolerance. These transgenics have mainly been tested in lab and greenhouse studies, but some have been tested in the field, with promising results. Thus, biotechnology has proven to be a promising strategy to enhance plants' ability for phytoremediation, with the big advantages of being fast and being able to introduce novel properties from other organisms that could never be introduced via classical breeding. The next challenge will be to facilitate the use of such transgenics, which requires overcoming regulatory hurdles and gaining public acceptance.

Genetic engineering of plants for synthesis of metal chelators will improve the capability of plant for metal uptake (Karenlampi *et al.*, 2000; Pilon-Smits and Pilon, 2002; Clemens *et al.*, 2002; Eapen and D'Souza, 2005).

Genetic and molecular techniques have been applied to identify a range of gene families that are likely to be involved in transition metal transport. Transferring the genes responsible for the hyper-accumulating phenotype to higher shootbiomass-producing plants has been suggested as a potential avenue for enhancing phytoremediation as a viable commercial technology (Pilon-Smits and Pilon, 2002; Yang *et al.*, 2005a, 2005b; Martinez *et al.*, 2006;, Mello-Farias and Chaves, 2008).

Some genes have been isolated and introduced into plants with increased heavy metal (Cd) resistance and uptake, like *AtNramps* (Thomine *et al.*, 2000), *AtPcrs* (Song *et al.*, 2004), and *CAD1* (Ha *et al.*, 1999) from *Arabidopsis thaliana*, library enriched in Cd-induced cDNAs from *Datura innoxia* (Louie *et al.*, 2003), *gshI, gshII* (Zhu *et al.*, 1999b) and PCS cDNA clone (Heiss *et al.*, 2003) from *Brassica juncea*. There are some examples of transgenic plants for metal tolerance/phytoremediation, as tobacco with accumulation of Cd, Ca and Mn transformed with gene *CAX-2* (vacuolar transporters) from *A. thaliana* (Hirschi *et al.*, 2000); *A. thaliana* tolerant to Al, Cu, and Na with gene *Glutathione-S-transferase* from tobacco (Ezaki *et al.*, 2000); tobacco with Ni tolerance and Pb accumulation with gene *Nt CBP4* from tobacco (Arazi *et al.*, 1999); tobacco (Goto *et al.*, 1998) and rice (Goto *et al.*, 1998; 1999) with increased iron accumulation with gene *Ferretin* from soybean; *A. thaliana* and tobacco resistant to Hg with gene *merA* from bacteria (Rugh *et al.*, 2000; Bizily *et al.*, 2000; Eapen and D'Souza, 2005); Indian mustard tolerant to Se transformed with a bacterial glutathione reductase in the cytoplasm and also in the chloroplast (D'Souza *et al.*, 2000); transgenic *A. thaliana* plants expressing SRSIp/ArsC and ACT 2p/dakua-ECS together showed high tolerance to As, these plants accumulated 4- to 17-fold greater fresh shoot weight and accumulated 2- to 3-fold more arsenic per gram of tissue than wild plants or transgenic plants expressing dakua-ECS or ArsC alone (Dhankher *et al.*, 2002; Mello-Farias and Chaves, 2008).

New metabolic pathways can be introduced into plants for hyperaccumulation or phytovolatilization as in case of *MerA* and *MerB* genes which were introduced into plants which resulted in plants being several fold tolerant to Hg and volatilized elemental mercury (Heaton *et al.*, 1998; Rugh *et al.*, 1998 Bizily *et al.*, 2000; Dhankher *et al.*, 2002;Eapen and D'Souza, 2005).

The first attempt to develop engineered plants for phytoremediation of organic pollutants in tobacco plants was reported by French *et al.* (1999) and Doty *et al.* (2000). These plants have been developed to contain either transgen responsible for the metabolization of xenobiotics or increased resistance to pollutants. Tobacco plants containing a human P450 2E1 were able to transform up to 640 times the amount of TCE compared with control plants (Doty *et al.*, 2000). They also showed increased uptake and metabolism of ethylene dibromide, another halogenated hydrocarbon commonly found in groundwater. Higher tolerance to the explosives glycerol trinitrate and 2,4,6-trinitrotoluene was achieved by transgenic tobacco plants expressing a microbial pentaerythritol tetranitrate reductase (French *et al.*, 1999). Francova *et al.* (2003, 2004) reported transgenic plants having capability of PCBs degradation.

One of the promising developments in transgenic technology is the insertion of multiple genes from microbes, human and animals into the plants for complete degradation of chlorinated solvents, xenobiotics, herbicides, explosives and PCBs (Schwitzguebel and Vanek, 2003; Mackoya *et al.*, 2006; Meagher, 2000; Bruce, 2007; Eapen *et al.*, 2007; Doty, 2008; Macek *et al.*, 2008 and Abhilash *et al.*, 2009).

Conclusions

Phytoremediation is one of the methods which can be used to remedy soil problems in order to achieve sustainable soil. This remediation technology is competitive, and may be superior to existing conventional technologies at sites where phytoremediation is applicable.

It has many different forms which will suit different kind of soil problems.

Phytoremediative technologies which are soil-focused are suitable for large areas that have been contaminated with low to moderate levels of contaminants.

Phytoremediation should be viewed as a long-term remediation solution.The time needed for successful phytoremediation is an important factor that needs to be taken into consideration for practical applications.

The low cost of phytoremediation is the main advantage of phytoremediation, however many of the pro's and cons of phytoremediation applications depend greatly on the location of the polluted site, the contaminants in question, and the application of phytoremediation.

Genetic modification has offered a new hope for phytoremediation as GM approaches are being used for engineering plants with increased phytoremediaion capabilities. This knowledge will allow phytoremediation to be applied more widely and effectively.

References

1. Abbaspour, A., Arocena, J.M. and Kalbasi, M., 2012. Uptake of phosphorus and lead by *Brassica juncea* and *Medicago satavia* from chloropyromorphite. *Inter. J. Phytorem.*, 14(6): 531–542.

2. Abhilash, P.C., 2007. Phytoremediation: an innovative technique for ecosystem clean up. *Our Earth* 4: 7–12.

3. Abhilash, P.C., Jamil, S. and Singh, N., 2009. Transgenic plants for enhanced biodegradation and phytoremediation of organic xenobiotics. *Biotechnol. Adv.*, 27: 474–488.

4. Arazi, T., Sunkar, R., Kaplan, B. and Fromm, H.A., 1999. Tobacco plasma membrane calmodulin binding transporter confers Ni+ tolerance and Pb2+ hypersensitivity in transgenic plants. *Plant Journal*, 20: 171–182.

5. Aslund, M.L.W., Zeeb, B.A., Rutter, A. and Reimer, K.J., 2007. *In situ* phytoextraction of polychlorinated biphenyl–(PCB) contaminated soil. *Sci. Total Environ.*, 374: 1–12.

6. Banuelos, G.S., Zayed, A., Terry. N., Wu, L., Akohoue, S. and Zambrzuski, S., 1996. Accumulation of selenium by different plant species grown under increasing sodium and calcium chloride salinity. *Plant Soil*, 183: 49–59.

7. Berti, W.R. and Cunningham, S.D., 2000. Phytostabilization of metals. In: *Phytoremediation of Toxic Netals: UsingPlants to Clean-up the Environment*, (Eds.) I. Raskin and B.D. Ensley. John Wiley and Sons, Inc., New York, p. 71–88.

8. Bizily, S., Rugh, C.L., Summer, A.O. and Meagher, R.B., 1999. Phytoremediation ofmethylmercury pollution: merB expression in Arabidopsis thaliana confers resistance to organomercurials. *Proc. Natl. Acad. Sci., USA*, 96: 6808–6813.

9. Bizily, S.P., Rugh, C.L. and Meagher, R.B., 2000. Phytodetoxification of hazardous organomercurials by genetically engineered plants. *Nature Biotechnology*, 18: 213–217.

10. Blaylock, M.J. and Huang, J.W., 2000. Phytoextraction of metals. In: *Phytoremediation of Toxic Netals: UsingPlants to Clean-up the Environment*, (Eds.) I. Raskin and B.D. Ensley. John Wiley and Sons, Inc., New York, p. 53–70.

11. Bruce, N. (2007) Biodegradation and phytoremediation of explosives. In: Book of Abstracts, *4th Symp. on Biosorption and Bioremediation*, (Eds.) M. Mackova *et al.*, August 26–30; Prague, 77, VSCHT Prague.

12. Carman, E., Crossman, T. and Gatliff, E., 1998. Phytoremediation of no. 2 fuel–oil contaminated soil. *J. Soil Contam.*, 7: 455–466.

13. Chaney, R.L., Li, Y.M., Brown, S.L., Homer, F.A., Malik, M., Angle, J.S., Baker, A.J.M., Reeves, R.D. and Chin, M., 2000. Improving metal hyperaccumulator wild plants to develop phytoextraction systems: Approaches and progress. In: *Phytoremediation of Contaminated Soil and Water*, (Eds.) N. Terry and G. Banuelos, p. 129–158. Lewis Publishers, Boca.

14. Cherian, S. and Oliveira, M.M., 2005. Transgenic plants in phytoremediation: recent advances and new possibilities. *Environ. Sci. Technol.*, 39: 9377–9390.

15. Clemens, S., Palmgren, M. and Krämer, U., 2002. A long way ahead: Understanding and engineering plant metal accumulation. *Trends Plant Sci.*, 7: 309–315.

16. Comis, D., 1996. Green remediation: Using plants to clean the soil. *Journal of Soil and Water Conservation*, 51(3): 184–187.

17. Cook, R.L., Landmeyer, J.E., Atkinson, B., Messier, J.P. and Nichols, E.G., 2010. Field note: Successful establishment of a phytoremediation system at a petroleum hydrocarbon contaminated shallow aquifer: Trends, trials, and tribulations. *International Journal of Phytoremediation*, 12(7): 716–732.

18. Cunningham, S.D. and Ow, D.W., 1996. Promises and prospects of phytoremediation. *Plant Physiology*, 110: 715–719.

19. Cunningham, S.D., Berti, W.R. and Huang, J.W., 1995. Phytoremediation of contaminated soils. *Biotechnology*, 13: 393–397.

20. Cunningham, S.D., Anderson, T.A., Schwab, A.P. and Hsu, F.C., 1996. Phytoremediation of soil contaminated with organic pollutants. *Adv. Agron.*, 56: 55–114.

21. de Souza, M.P., Pilon-Smits, E.A.H. and Terry, N., 2000. Phytoextraction of metals. In: *Phytoremediation of Toxic Metals*, (Eds.) I. Raskin and B.D. Ensley. John Wiley and Sons, Inc. New York, p. 171–190.

22. Dhankherz, O.P., Li, Y., Rosen, B.P., Shi, J., Salt, D., Senecoff, J.F., Sashti, N.A. and Meagher, R.B., 2002. Engineering tolerance and hyperaccumulation of arsenic in plants by combining arsenate reductase and γ-glutamylcysteine synthetase expression. *Nature Biotechnology*, 20: 1140–1145.

23. Dietz, A. and Schnoor, J.L., 2001. Advances in phytoremediation. *Environ. Health Perspect.*, 109: 163–168.

24. Doty, S.L., 2008. Enhancing phytoremediation through the use of transgenic plants and entophytes. *New Phytol.*, 179: 318–333.

25. Doty, S.L., Shang, Q.T., Wilson, A.M., Moore, A.L., Newman, L.A., Strand, S.E. and Gordon, M.P., 2000. Enhanced metabolism of halogenated hydrocarbons in transgenic plants contain mammalian P450 2E1. *Proc. Nat. Acad. Sci., USA*, 97: 6287–6291.

26. Dushenkov, S. and Kapulnik, Y., 2000. Phytofilitration of metals. In: *Phytoremediation of Toxic Netals: UsingPlants to Clean-up the Environment*, (Eds.) I. Raskin and B.D. Ensley. John Wiley and Sons, Inc., New York, p. 89–106.

27. Dushenkov, V., Kumar, P.B.A.N., Motto, H. and Raskin, I., 1995. Rhizofiltration: The use of plants to remove heavy metals from aqueous streams. *Environ. Sci. Technol.*, 29: 1239–1245.

28. Dushenkov, S., Vasudev, D., Kapulnik, Y., Gleba, D., Fleisher, D., Ting, K.C. and Ensley, B., 1997a. Removal of uranium from water using terrestrial plants. *Environ Sci. Technol.*, 31(12): 3468–3474.

29. Dushenkov, S., Vasudev, D., Kapulnik, Y., Gleba, D., Fleisher, D., Ting, K.C. and Ensley, B., 1997b. Phytoremediation: A novel approach to an old problem. In: *Global Environmental Biotechnology*, (Ed.) D.L. Wise. Elsevier Science B.V., Amsterdam, p. 563–572.

30. Eapen, S. and D'Souza, S.F., 2005. Prospects of genetic engineering of plants for phytoremediation of toxic metals. *Biotechnol. Adv.*, 23: 97–114.

31. Eapen, S., Singh, S. and D'Souza, S.F., 2007. Advances in development of transgenic plants for remediation of xenobiotic pollutants. *Biotechnol. Adv.*, 25: 442–451.

32. Ebbs, S.D., Lasat, M.M., Brandy, D.J., Cornish, J., Gordon, R. and Kochian, L.V., 1997. Heavy metals in the environment: Phytoextraction of cadmium and zinc from a contaminated soil. *J. Environ. Qual.*, 26: 1424–1430.

33. Ensley, B.D., 2000. Rational for use of phytoremediation. In: *Phytoremediation of Toxic Netals: UsingPlants to Clean-up the Environment*, (Eds.) I. Raskin and B.D. Ensley. John Wiley and Sons, Inc., New York, p. 3–12.

34. Ezaki, B., Gardner, R.C., Ezaki, Y. and Matsumoto, H., 2000. Expression of aluminium induced genes in transgenic *Arabidopsis plants* can ameliorate aluminium stress and/or oxidative stress. *Plant Physiology*, 122: 657–665.

35. Flathman, P.E. and Lanza, G.R., 1998. Phytoremediation: Current views on an emerging green technology. *J Soil Contamination*, 7(4) : 415–432.

36. Francová, K., Macková, M., Macek, T. and Sylvestre, M., 2004. Ability of bacterial biphenyl dioxygenases from *Burkholderia* sp. LB400 and *Comamonas testosteroni* B–356 to catalyse oxygenation of ortho-hydroxybiphenyls formed from PCBs by plants. *Environ. Pollut.*, 127: 41–48

37. Francova, K., Sura, M., Macek, T., Szekeres, M., Bancos, S. and Demnerova, K., 2003. Preparation of plants containing bacterial enzyme for degradation of poly chlorinated biphenyls. *Fresenius Environ. Bull.*, 12: 309–313.

38. French, C.E., Rosser, S.J., Davies, G.J., Nicklin, S. and Bruce, N.C., 1999. Biodegradation of explosives by transgenic plants expressing pentaerythritol tetranitrate reductase. *Nat. Biotechnol.*, 17: 491–494.

39. Gao, Y. and Zhu, L., 2004. Plant uptake, accumulation and translocation of phenanthrene and pyrene in soils. *Chemosphere*, 55: 1169–1178.

40. Gordon, M., Choe, N., Duffy, J., Ekuan, G., Heilman, P., Muiznieks, I., Newman, L., Ruszaj, M., Shurtleff, B.B., Strand, S. and Wilmoth, J., 1997. Phytoremediation of trichloroethylene with hybrid poplars. In: *Phytoremediation of Soil and Water Contaminants*, (Eds.) E.L. Kruger, T.A. Anderson and J.R. Coats. American Chemical Society, Washington, DC, p. 177–185.

41. Goto, F., Yoshihara, T. and Saiki, H., 1998. Iron accumulation in tobacco plants expressing soybean ferritin gene. *Transgenic Research*, 7: 173–180.

42. Goto, F., Yoshihara, T., Shigemoto, N., Toki, S. and Takaiwa, F., 1999. Iron accumulation in rice seed by soybean ferritin gene. *Nature Biotech.*, 17: 282–286.

43. Greenberg, B.M., Hunag, X.D., Gurska, Y., Gerhardt, K.E., Wang, W. and Lampi, M.A, *et al.*, 2006. Successful field tests of a multi-process phytoremediation system for decontamination of persistent petroleum and organic contaminants. *Proceedings of the Twenty-Ninth Artic and Marine Oil Spill Program (AMOP, Technical Seminar Vol. 1. Environment Canada, p. 389–400

44. Greipsson, S., 2011. Phyoremediation. *Nature Education Knowledge,* 3(10): 7.

45. Gunther, T., Dornberger, U. and Fritsche, W., 1996. Effects of rye grass on biodegradation of hydrocarbons in soil. *Chemosphere,* 33: 203–215.

46. Ha, S.B., Smith, A.P., Howden, R., Dietrich, W.M., Bugg, S., O'Connell, M.J., Goldsbrough, P.B. and Cobbett, C.S., 1999. Phytochelatin synthase genes from Arabidopsis and the yeast *Schizosaccharomyces pombe. Plant Cell,* 11: 1153–1163.

47. Heaton, A.C.P., Rugh, C.L, Wang, N.J. and Meagher, R.B., 1998. Phytoremediation of mercury- and methylmercury-polluted soils using genetically engineered plants. *J. Soil Contam.,* 7: 497–509.

48. Hedge, R.S. and Fletcher, J.S., 1996. Influence of plant growth stage and season on the release of root phenolics by mulberry as related to development of phytoremediation technology. *Chemosphere,* 32: 2471–2479.

49. Heiss, S., Wachter, A., Bogs, J., Cobbett, C. and Rausch, T., 2003. Phytochelatin synthase (PCS) protein is induced in *Brassica juncea* leaves after prolonged Cd exposure. *J. Exp. Bot.,* 54: 1833–1839.

50. Hirschi, K.D., Korenkov, V.D., Wilganowski, N.L. and Wagner, G.J., 2000. Expression of Arabidopsis CAX2 in tobacco altered metal accumulation and increased manganese tolerance. *Plant Physiology,* 124: 125–133.

51. Hooker, B.S. and Skeen, R.S., 1999. Transgenic phytoremediation blasts onto the scene. *Nat. Biotechnol.,* 17: 428.

52. Huang, J.W. and Cunningham, S.D., 1996. Lead phytoextraction: Species variation in lead uptake and translocation. *New Phytologists,* 134: 75–84.

53. Huang, J.W., Chen, J., Berti, W.R. and Cunningham, S.D., 1997. Phytoremediation of lead contaminated soil: Role of synthetic chelates in lead phytoextraction. *Environ. Sci. Technol.,* 31: 800–805.

54. ITRC (Interstate Technology and Regulatory Council, 2009). *Phytotechnologies for site cleanup.*

55. Karenlampi, S., Schat, H., Vangronsveld, J., Verkleij, J.A.C., Van Der Lelie C., Mergeay, M. and Tervahauta, A.I., 2000. Genetic engineering in the improvement of plants for phytoremediation of metal polluted soils. *Environmental Pollution,* 107(2): 225–231.

56. Kaushik, S., 2011. Bioremediation: Use of green plants to remove pollutants. www.biotecharticle.com.

57. Kumar, P.B.A.N., Dushenkov, V., Motto, H., and Raskin, I., 1995a. Phytoextraction: The use of plants to remove heavy metals from soils. *Environ. Sci. Technol.,* 29: 1232–1238.

58. Kumar, P.B.A.N., Motto, H. and Raskin, I., 1995b. Rhizofiltration: The use of plants to remove heavy metals from aqueous streams. *Environ. Sci. Technol.,* 29 (5): 1239–1245.

59. Lasat, M.M., Fuhrmann, M., Ebbs, S.D., Cornish, J.E. and Kochian, L.V, 1998. Phytoremediation of a radiocesium contaminated soil: Evaluation of cesium-137 bioaccumulation in the shoots of three plant species. *J. Environ. Qual.*, 27: 165–168.

60. Lasat, M.M., 2002. Phytoextraction of toxic metals: A review of biological mechanisms. *Journal of Environmental Quality*, 131: 109–120.

61. Lee, K.Y., Strand, S.E. and Doty, S.L., 2012. Phytoremediation of chloropyrifos by *Populus* and *Salix*. *Int. J. Phyrem.*, 14(1): 48–61.

62. Leigh, M.B., Prouzova, P., Macková, M., Macek, T., Nagle, D.P. and Fletcher, J.S., 2006. Polychlorinated biphenyl (PCB)–degrading bacteria associated with trees in a PCB–contaminated site. *Appl. Environ. Microbiol.*, 72: 2331–2342.

63. Li, H., Sheng, G., Sheng, W. and Xu, O., 2002. Uptake of trifluralin and lindane from water by ryegrass. *Chemosphere*, 48: 335–341.

64. Liste, H.H. and Alexander, M., 2000. Plant promoted pyrene degradation in soil. *Chemosphere*, 40: 7–10.

65. Louie, M., Kondor, N. and Witt, J.G., 2003. Gene expression in cadmium-tolerant *Datura innoxia*: Detection and characterization of cDNAs induced in response to Cd^{2+}. *Plant Mol. Biol.*, 52: 81–89.

66. Ma, X., Andrew, A.R., Burken, J.G. and Albers, S., 2004. Phytoremediation of MTBE with hybrid poplar trees. *Int. J. Phytoremed.*, 4: 157–167.

67. Macek, T., Mackova, M. and Kas, J., 2000. Exploitation of plants for the removal of organics in environmental remediation. *Biotechnol. Adv.*, 18: 23–34.

68. Macek, T., Kotrba, P., Svatos, A., Novakova, M., Demnerova, K. and Mackova, M., 2008. Novel roles for genetically modified plants in environmental protection. *Trends Biotechnol.*, 26: 146–152.

69. Mackova, M., Barriault, D., Francova, K., Sylvestre, M., Möder, M., Vrchotova, B., Lovecka, P., Najmanová, J., Demnerova, K., Novakova, M., Rezek, J. and Macek, T., 2006. Phytoremediation of polychlorinated biphenyls. In: *Phytoremediation and Rhizoremediation: Theoretical Background*, (Eds.) M. Mackova, D. Dowling and T. Macek. Series: Focus on Biotechnology. Springer.

70. Macková, M., Vrchotová, B., Francová, K., Sylvestre, M., Tomaniová, M., Lovecká, P., Demnerová, K. and Macek, T., 2007. Biotransformation of PCBs by plants and bacteria-consequences of plant-microbe interactions. *Eur. J. Soil Biol.*, 43: 233–241

71. Maila, M.P. and Cloete, T.E., 2002. Germination of *Lepidium sativum* as a method to evaluate polycyclic aromatic hydrocarbons (PAHs) removal from contaminated soil. *Int. Biodeterior. Biodegrad.*, 50: 107–113.

72. Mandal, A., Purakayastha, T.J., Patra, A.K. and Sanyal, S.K., 2012. Phytoremediation of arsenic contaminated soil by *Pteris bittata* L. II. Effect on arsenic uptake and rice yield. *International Journal of Phytoremediation*, 14(6): 621–628.

73. Martínez, M., Bernal, P., Almela, C., Vélez, D., García-Agustín, P. and Serrano, R., *et al.*, 2006. An engineered plant that accumulates higher levels of heavy metals than *Thlaspi caerulescens*, with yields of 100 times more biomass in mine soils. *Chemosphere*, 64: 478–485.

74. Mattina, M.I., Iannucci-Berger, W. and Dykas, L., 2000. Chlordane uptake and its translocation in food crops. *J. Agric. Food Chem.*, 48: 1909–1915.

75. Mattina, M.I., Eitzer, B.D., Iannucci-Berger, W., Lee, W.Y. and White, J.C., 2004. Plant uptake and translocation of highly weathered, soil-bound technical chlordane residues: data from field and rhizotron studies. *Environ. Toxicol. Chem.*, 23: 2756–2762.

76. McGrath, S.P., Zhao, F.J. and Lombi, E., 2002. Phytoremediation of metals, metalloids and radionuclides. *Advances in Agronomy*, 75: 1–56.

77. Meagher, R.B., 2000. Phytoremediation of toxic elemental and organic pollutants. *Curr. Opin. Plant Biol.*, 3: 153–162.

78. Mello-Farias, P.C. and Chaves, A.L.S., 2008. Biochemical and molecular aspects of toxic metals phytoremediation using transgenic plants. In: *Transgenic Approach in Plant Biochemistry and Physiology*, (Eds.) M.E. Tiznado-Hernandez, R. Troncoso-Rojas and M.A. Rivera-Domínguez, p. 253–266, Research Signpost, Kerala, India.

79. Mills, T., Arnold, B., Sivakumaran, S., Northcott, G., Vogeler, I. and Robinson, B., *et al.*, 2006. Phytoremediation and long-term site management of soil contaminated with pentachlorophenol (PCP) and heavy metals. *J. Environ. Manag.*, 79(3): 232–241.

80. Mohammadi, M., Chalavi, V., NovakovaSura, M., Laliberte, J. F. and Sylvestre, M., 2007. Expression of bacterial biphenylchlorophenyl dioxygenase genes in tobacco plants. *Biotechnol. Bioeng.*, 97: 496–505.

81. Moreno, F.N., Anderson, C.W., Stewart, N., Robert, B. and Robinson, B.H., 2008. Phytofilteration of mercury-contaminated water: Volatilisation and plant-accumulation aspects. *Environmental and Experimental Botany*, 62(1): 78–85.

82. Negri, C.M. and Hinchman, R.R., 2000. The use of plants for the treatment of radionuclides. In: *Phytoremediation of Toxic Netals: UsingPlants to Clean-up the Environment*, (Eds.) I. Raskin and B.D. Ensley. John Wiley and Sons, Inc., New York, Chapter 8.

83. Newman, L.A., Strand, S.E., Domroes, D., Duffy, J., Ekuan, G., Karscig, G., Muiznieks, I.A., Ruszaj, M., Heilman, P. and Gordon, M.P., 1997. Removal of Trichloreth-ylene from a simulated aquifer using poplar. *Fourth International in situ and onsite Bioremediation Symposium*, April 28 – May 1, New Orleans, LA. 3: 321.

84. Parrish, Z.D., White, J.C., Isleyen, M., Gent, M.P.N., Iannucci-Berger, W. and Eitzer, B.D. *et al.*, 2006. Accumulation of weathered polycyclic aromatic hydrocarbons (PAHs) by plant and earthworm species. *Chemosphere*, 64: 609–618.

85. Panz, K. and Miksch, K., 2012. Phytoremediation of explosives (TNT, RDX, HMX) by wild type and transgenic plants. *J. Environ. Mang.*, 113: 85–92.

86. Pilon-Smith, E. and Pilon, M., 2002. Phytoremediation of metals using transgenic plants. *Crit. Rev. Plant Sci.*, 21: 439–456.

87. Prasad, Freitas, 2003. Metal hyperaccumulation in plants: Biodiversity prospecting for phytoremediation technology. *Electronic Journal of Biotechnology*, 6: 285–321.

88. Raskin, I., 1996. Plant genetic engineering may help with environmental cleanup. *Proc. Natl. Acad. Sci., U.S.A.*, 93: 3164–3166.

89. Raskin I. and Ensley, B.D. (Eds.) 2000. *Phytoremediation of Toxic Netals: UsingPlants to Clean-up the Environment*. John Wiley and Sons, Inc., New York, p. 352.

90. Reeves, R. and Baker, A., 2000. Metal accumulating plants. In: *Phytoremediation of Toxic Netals: UsingPlants to Clean-up the Environment*, (Eds.) I. Raskin and B.D. Ensley. John Wiley and Sons, Inc., New York, p. 193–229.

91. Robinson, B.H., Brooks, R.R., Howes, A.W., Kirkman, J.H. and Gregg, P.E.H., 1997. The potential of the high biomass nickel hyperaccumulator *Berkheya coddii* for phytoremediation and phytomining. *J. Geochem. Explor.*, 60: 115–126.

92. Rugh, C.L., Bizily, S.P. and Meagher, R.B., 2000. Phytoremediation of environmental mercury pollution. In: *Phytoremediation of Toxic Netals: UsingPlants to Clean-up the Environment*, (Eds.) I. Raskin and B.D. Ensley. John Wiley and Sons, Inc., New York, p. 151–171.

93. Salt, D.E., Blaylock, M., Kumar, N.P.B.A., Dushenkov, V., Ensey, D., Chet, I. and Raskin, I., 1995. Phytoremediation: A novel strategy for the removal of toxic metals from the environment using plants. *Biotechnology*, 13: 468–474.

94. Salt, D.E., Smith, R.D. and Raskin, I., 1998. Phytoremediation. *Ann. Rev. Plant Physiol. Plant Mol. Biol.*, 49: 643–668.

95. Singh, O.V. and Jain, R.K., 2003. Phytoremediation of toxic aromatic pollutants from soil. *Appl. Microbiol. Biotechnol.*, 63: 128–135.

96. Schnoor, J.L., Light, L.A., Mccutcheon, S.C., Wolfe, N.L. and Carrira, L.H., 1995. Phytoremediation of organic and nutrient contaminants. *Environ Sci. Technol.*, 29: 318A–323A.

97. Schnoor, J.L., 2000. Phytostabilization of metals using hybrid poplar trees. In: *Phytoremediation of Toxic Netals: UsingPlants to Clean-up the Environment*, (Eds.) I. Raskin and B.D. Ensley. John Wiley and Sons, Inc., New York, p. 133–150.

98. Schwitzguebel, J.P. and Vanek, T., 2003. Advances for xenobiotic chemicals. *Phytoremediation: Transformation and Control of Contaminant*, (Eds.) S.C. McCutcheon and J.L. Schnoor. John Wiley and Sons Inc, New Jersey, USA, p. 123–158.

99. Song, W.Y., Martinoia, E., Lee, J., Kim, D., Kim, D.Y., Vogt, E., Shim, D., Choi, K.S., Hwang, I. and Lee, Y., 2004. A novel family of cys-rich membrane proteins mediates cadmium resistance in Arabidopsis. *Plant Physiology*, 135: 1027–1039.

100. Sunderberg, S.E., Ellington, J.J., Evans, J.J., Keys, D.A. and Fisher, J.W., 2003. Accumulation of perchlorate in tobacco plants: Development of a plant kinetic model. *J. Environ. Monit.*, 5: 505–512.

101. Suresh, B., Sherkhane, P.D., Kale, S., Eapen, S. and Ravishankar, G.A., 2005. Uptake and degradation of DDT by hairy root cultures *Cichorium intybus and Brassica juncea*. *Chemosphere*, 61: 1288–1292.

102. Tangahu, B.V., Abdullah, S.R.S., Basri, H., Idris, M., Auar, N. and Mukhisin, M., 2011. A Review on heavy metals (As, Pb, Hg,) uptake by plants through phytoremediation. *Int. J. Chem. Engg.*, 1–31

103. Terry, N., Zayed, A.M., de Souza, M.P. and Tarun, A.S., 2000. Slenium in higher plants. *Ann. Rev. Plant Physiol. Plant Mol. Biol.*, 51: 401–432.

104. Thomine, S., Wang, R., Ward, J.M., Crawford, N.M. and Schroeder, J.I., 2000. Cadmium and iron transport by members of a plant metal transporter family in Arabidopsis with homology to Nramp genes. *Proc. Nat. Acad. Sci., USA*, 97: 4991–4996.

105. U.S. Environmental Protection Agency (USEPA), 2000. *Introduction to Phytoremediation*, EPA 600–R–99–107. February.

106. U.S. Environmental Protection Agency (USEPA), 2006. *In situ* treatment technologies for contaminated soil: Engineering forum issue paper. *EPA 542–F–06–013*.

107. Watanabe, M.E., 1997. Phytoremediation on the brink of commercialisation. *Env. Sci. Technol.*, 31: 182A–186A.

108. Van Aken, B., 2008. Transgenic plants for phytoremediation: Helping nature to clean up environmental pollution. *Trends in Biotech.*, 26: 225–227.

109. White, J.C., 2000. Phytoremediation of weathered p, p'-DDE residues in soil. *Int. J. Phytoremed.*, 2: 133–144.

110. White, J.C., 2001. Plant-facilitated mobilization of weathered 2,2-bis(p-chlorophenyl)-1, 1-dichloroethylene (p, p'-DDE) from an agricultural soil. *Environ. Toxicol. Chem.*, 20: 2047–2052.

111. White, J.C., 2002. Differential bioavailability of field weathered p, p'–DDE to plants of the *Cucurbita peppo* and *Cucumis genera*. *Chemosphere*, 49: 143–152.

112. Wu, L., Li, Z., Han, C., Liu, L., Teng, Y., Sun, X., Pan, C., Huang, Y., Luo, Y. and Christie, P., 2012. Phytoremediation of soil contaminated with cadmium, copper and polychlorinated biphenyls, 14(6): 570–584.

113. Yang, X., Jin, X.F, Feng, Y. and Islam, E., 2005a. Molecular mechanisms and genetic bases of heavy metal tolerance/hyperaccumulation in plants. *J. Integrative Plant Biol.*, 47: 1025–1035.

114. Yang, X., Feng, Y., He, Z. and Stoffella, P., 2005b. Molecular mechanisms of heavy metal hyperaccumulation and phytoremediation. *Journal of Trace Elements in Medicine and Biology*, 18: 339–353.

115. Yoon, J.M., Oh, B.T., Just, C.L. and Schnoor, J.L., 2002. Uptake and leaching of octahydro-1,3,5,7,-tetranitro-1,3,5,7-tetrazocine by hybrid poplar trees. *Environ. Sci. Technol.*, 36: 4649–4655.

116. Zhu, Y.L, Zayed, A.M., Quian, J.H., De Souza, M. and Terry, N., 1999a. Phytoaccumulation of trace elements by wetland plants: II. Water hyacinth. *J. Environ. Qual.*, 28: 339–344.

117. Zhu, Y.L, Pilon-Smits, E.A.H., Tarun, A.S., Weber, S.U., Jouanin, L. and Terry, N., 1999b. Cadmium tolerance and accumulation in Indian mustard is enhanced by overexpressing γ-glutamylcysteine synthetase. *Plant Physiology*, 121: 1169–1177.

Biotechnology: An Overview (2015)   *Pages 179–184*
Editors: Rajan Kumar Gupta, Nasim Akhtar and Deepak Vyas
Published by: DAYA PUBLISHING HOUSE, NEW DELHI

Chapter 13

Biotechnological Production of Promising Anticancerous Drug Taxol

Nivedita Srivastava

Navjeevanam Kayakalp and Medical Research Centre,
Shiva Enclave, Rishikesh, Uttarakhand

ABSTRACT

Taxol is a new anti-cancer drug that is a natural product derived from the bark of the pacific yew tree. The drug promotes polymerization and stabilization of tubulin to microtubules and interferes with the mitotic spindle. Clinical trials indicate that Taxol is effective in the treatment of patient with ovarian cancer, breast cancer, malignant melanoma and probably other solid tumors

Keywords: Taxol, Anticancerous, Taxus yew.

Introduction

Taxol is an anticancer drug that is naturally occuring deterpenoid belonging to taxone group of compounds present in genus *Taxus* under family Taxaceae. This genus is also known as yew. Yews are known for their toxicity since old days and all parts are poisinous except the fleshy aril of scarlet red fruit, which is non-poisonous. The genus Taxus consist of various speices, four of which are medicinal importance. They are *Taxus cuspidata* (Japanese yew), *Taxus baccata* (European Yew), *Taxus brevifolia* (Pacific Yew) and *Taxus canadensis* (American Yew), other useful speices are *T. yunnanensis, T. wallichiana, T. chiniensis, T. media* and *T. mairei.*

Yew is a slow growing evergreen gymnospermous tree. The chief constituent Taxol is present in all parts of plants especially in leaves, root and bark of the plant. The content of Taxol and Taxanes in root cuttings, leaves and bark vary in proportion. Leaves contain a high concentration. It is reported that extraction facility should be located close to the cultivation area of plants to prevent deterioration. The drying in shade give the highest yield of Taxol. The average annual content of Taxol from shoot of Taxus baccata found to be 0.0075 per cent maximum level recorded in April and minimum in February.

Chemical Constituents

Taxanes are most important group of chemical constituents, till now 60 different taxane compound are isolated, all of which are diterpenoids in structure. Three most important members are Taxol, Cephalomannine and 10 deacety 1 baccatin. In all speices with little variations. Taxol occurs from 0.007 per cent to 0.01 per cent . The isolations of Taxol needs at least 60 years old 3-4 trees to get 1gm of the Taxol. The derivative of Taxol called Taxotere has been reported to gave better bioavailabity and pharmological properties.

Taxane derivatives with various carbon skeltons legnans, flavonoids, steriods and sugar derivatives have been obtained from different *Toxus* species. New photo reactive analogons of Taxol bearing (3H2) - 3 (4-benzol) phenyl propionl group and radiolabelling unit at 7, 10 position showed excellent results on photo affinity labelling of tubulin and P.Glycoprotein. Taxus speices contains a number of pharmaceutically important constituents. Some important constituents are listed in Table 13.1.

Pharmacological Activities

Taxol is an alkoloidal ester promotes polymerization and stablization of tululin of microtubles and interfere with the mitotic spindle. Clinical trials indicate that taxol is effective in the treatment of patient with refractory ovarian cancer, breast cancer, malignent melonama and probably other solid tumors. Toxicites includes anphyloctoid reactions, leukopenia, peripheral neuropathy and oropharangeal micositis. Increased supplies of drug and required to support further phase II and III testing.

Taxol also inhibits cell migration thus preventing spread of metastatic cancer cells. Taxol and Taxotere are reported to be similar in action against cancer cells. Binding side chain of Taxol with polymerized tuble was presented by kingston. Two flavonoids (+) catechin and (-) epicatechin seem to possesses the chetin synthoose II inhebitary activity Taxus spine D isolated from *T. cuspidata* inhibited Ca induced depolymerization of micro tubules. Its against gram (-) bacteria and some fungi. The effiacy of Docetaxel and Taxol to mobilize circulating progenitor cells in cancer patients has been evaluated high doses of chemotherapy. Incubation of J.T. Cells in Taxol produces 7 epitoxol with low anti tumour activity.

Production by Total and Semi-Synthesis

The total syntheses have been carried out of date. The Hollon group and Nicolaou out to group published their approaches in 1994 and very recently. Danish efsky and co-workers reported their route to taxol. Holton used (-) borneol as starting material which he converted to an unsaturated ketone over 13 steps. This ketone was converted to alcohol and than fragmentation to yield taxol. The method employed by Danishefsky stated with a ketone, which are elaborated to a complex enol triflate bearing an olefin on c-ring for synthesis of taxol.

Table 13.1. Chemical constituents of *Taxus* species.

Sl.No.	Name of Species	Phytoconstituent
1.	*Taxus cuspidata*	Taxol and cephalomannine, Taxinine NN7 and 3, II Cyclo-taximine NN-2, 5α, 13α diacetoxy-taxa- 4(20) 11 diene, 9α, 10β - diol, 7β, 13α diacetoxy 5α, Cinnamoyloxy - 2 (3 to 20) - abeo - taxa - 4(20), 11-diene, 2α 10β diol and 2α, 10β, 13α tri-acetoxy - taxa - 4(20) II diene 5α , 7α, 9α triol, Taxinine A 11, 12- epoxide and 3 penicillides.
2.	*Taxus yunnanensis*	Taxin B and 2- deacetyl Taxin B, Taxuyunnanine K-O (1-5), 13α -acetoxy, 5α, Cinnamoyloxy-11 (15 to 1) abeotaxa - 4(20) 11 diene, 9α, 10β, 15 triol, 20 acetoxy, 2α benzyloxy, 4α, 5α, 7β, 13α - hydroxy - 11 (15 to 1), 11 (10 to 9) - bisabeotax - 11-eve - 10, 15 lactone 2α, 10β - diacetoxy - 5α - cinnanoyloxy - 9α hydroxy, 11-cyclotoxy 4(20)- ene, 13 one, 10β acetoxy 2α, 5α, 9α trihydroxy-3, 11-cyclotax 4(20) - ene β - one, 1β, xylosyl, Taxol D, Taxuyutin G
3.	*Taxus canadensis*	5-epi-cinnamoyl canadense, 2, 9, 10, 13-tetra acetoxy, 0-cinnamoyloxy-taxa-4(5), 11(12) diene, 2-acetyl-epi-taxol, 9-deacetyltaxinine, 7-acetyl, 10-deacetyl Taxol, 2-acetyl, 7-epi-cephalomannine and 10-deacetyl glycolylbaccatin VI
4.	*Taxus wallichiana*	Taxusin, a C-14 oxygenated taxoid, a dibenzoylated taxoid and 7-xylosyl, 10 deacetyl Taxol, Dihydrotaxol and two dibenzoylated rearranged taxanes, (-) secoisolariciresinol, taxiresinol, catechin and isotaxiresinol.
5.	*Taxus brevifolia*	Compounds having phenyl-isoserine unit and oxetane ring at α and β position.
6.	*Taxus chinensis*	19-acetoxy taxagifine, Rutaecarpine, Kaempferol-4, methyl ether and sitostery 1-3-O- glucoside, 2α, 9α, 10β, 13α - tetracetoxy, 5α, 9 (3- methyl amino - 3') - propionyl - oxy- taxa 4 (20) 11 - diene, 7β, 9α, 10β, 13α - tetracetoxy, 5α, 9 (3' - methyl amino-3') phenyl propionyl-oxy-taxa 4(20) 11-diene
7.	*Taxus baccata*	2' deacetoxy austopicatine and Coniferaldehyde
8.	*Taxus media*	Taxinine M, 10-deacetyl Paclitaxel, 10-deacetyl Paclitaxel 7-xyloside, epigenin and p-hydroxybenzaldehyde
9.	*Taxus mairel*	Taxumairol N and O namely, 7β, 9α, 10β, 13, tetracetoxy- 2α, 4α, 5α, 20 - tetrahydroxytax -11- ene and 7β, 9α, 10β,13α-tetracetoxy-1β,2α,4α,5α, 20- pentahydroxytax- 11 - ene and Taxumairol M

Taxol was synthesised by inactive taxanes found in abudance in *Taxus* species syntherized taxol. Baccatin III obtained from 10-DAB found in leaves and stems of Taxus plant used as starting material. The side chain was prepared by using alpha-methyl 1 - benzylamine. Taxol was prepared by their union. The semi synthesis of Taxol isolated from *T. brevifolia* bark from Baccatin III via dioxo-oxathiazolidine intermediate is also reported.

Production by Biotechnological Pathway (Tissue Culture)

Gew et. al reported that the suspension culture of Taxus yunnanensis in B5 medium was done, the growth cycle was of 30 days. The Taxol content was 1mg/1 and the content was greatly increased by increasing the sucrose concentration to 40g/1 and also by adding coconut milk, (CM) caseinhydrolysate (CA) and Lalbumin Hydrolysate. Only CM and CA promoted cell growth of cultured cells.

The suspension culture of T. media was established in modified DCR medium. The paclitaxel content in this was 1.04 µg/g DW. After addition of phenylalanme 0.5M in the medium. The highest Taxol accumulation was observed in bioreactor after 11 days of culture. The enzyme geranyl geranyl diphosphate (GGDP) synthase is extracted by inclusion of non-ionic detergent and its activity parallels Taxene production. The enzyme activity reaches its maximum at day 3 of culture growth.

The production of Baccatin III (9.43 ± 3.8, 5.997 ± 0.5) and Deacetyl baccatin (8.86 ± 4.20, 4.77 ± 9.22) over the control cultures (0.99 ± 0.10 DAB and 1.21 ± 0.11 BA) of Taxus wallichiana is enhanced by addition of a Arachidonic acid (1mg/1) and Methayl Jasmonate (100µg) on day O. Incubation of Euculptus citridora cell culture with paclitaxel (10 µg/lt) was carried in a refrigerator shaker incubator at 120 rpm and 250C for 48 hrs. These culture led to biotransformation of paclitaxel into two known compounds (Baccatin III and DAB III) and an unknown compound.

Conclusion

This article has attempted to bring together all the information about the various species of Taxus. It includes the isolated chemical constituents, the unique pharmocological actions. Breast Cancer and ovarian cancer are the most common in women. Taxol is a strong antitumor agent, approved by FDA for the chemotherapy of these severe cancer deseases. It also shows effectiveness in advanced cases of head and neck cancer, small cell lung cancer, esophogal adenocarcinoma and prostrate cancer.

Despite in promise as a care for cancer the isolation of taxol from the bark of pacific yew tree is time consuming expensive and ineffective. Removing the bark from yew tree for isolation purpose not only destroys the trees, which are not found in abundance but also yeilds quantities of pure compound although a species of fungus has been also identified as a source of taxol, continuous low yield of taxol extracted from natural agents poses a barrier to taxol potential to server as an anticancerous drug. The use of needles from pacific yew trees for the artificial synthesis of taxol has been successful accomplished but the complex biosynthetic pathway is still undergoing studies and improvement.

Researches from Tuft university, School of Engineering and MIT have reported a new way to biosynthesize, important precursors to the potent anti cancer compound. Taxol in an engineered from strain of *E. coli* bacteria. The findings are significant steps on the way achieving cost effective large scale production of Taxol on the effect to design new Taxol like pharmaceuticals.

Acknowledgement

The author wish to thank Dr. D.K. Srivastava, an eminent ayurvedic physician, Rishikesh and Dr. R.K. Gupta, Associate Professor, Department of Botany, Government Post Graduate College, Rishikesh for providing some important inputs for this communications and facilities.

References

1. Chopra, R.N., Chopra, I.C. and Verma, B.S., 1974. *Supplement to Glossary of Indian Medicinal Plant*. Publications and Information Directorate, New Delhi, p. 95.

2. El-Sohly, H.N., Croom, E.M (Jr.), Kopycki, W.J., Joshi, A.S., Moraes, R.M. and McCherney, J.D., 2000. *Journal of Herbs Spices and Medicinal Plant*, 54(1): 14–17.

3. Kikuchi, Y, Kawamura, F., Ohira, T. and Yatagai, M., 2000. *Natural Medicines*, 54(1): 14–17.

4. Griffin, J. and Hook, I., 1996. *Planta Medica*, 62(4): 370–372.

5. Parmar, V.S., Jha, A., Bisht, K.S., Taneja, P., Singh, S.K., Kumar, A., Jain, R. and Olsen, C.E., 1999. *Phytochemistry*, 50(8): 1267–1304.

6. Ojima, I., Bounaud, Y. and Ahern, G., 1999. *Bio-organic and Medical Chemistry Letters*, 9(8): 1184–1194.

7. Tripathi, K.D., 1999. *Essentials of Medical Pharmacology*, 4th Edition. Jaypee Brothers, New Delhi, pp. 832.

8. Barar, F.S.K., 2000. *Essentials of Pharmacotherapeutics*, 3rd Edition. S. Chand and Company Ltd., New Delhi, pp. 484.

9. Erdenmoglu, N. and Sener, B., 200. *Journal of Faculty of Pharmacy of Ankara University*, 29(1): 77–90.

10. Kingston, D.G.I., 2000. *Journal of Natural Products*, 63(5): 726–734.

11. Kim, S.U., Hwang, E.I., Nam, J.Y., Son, K.H., Bok, S.H., Kim, H.E. and Kwon, B.M., 1999. *Planta Medica*, 65(1): 97–98.

12. Kobayashi, J., Hosoyama, H., Shigemori, H., Koiso, Y. and Iwasaki, S. 1995. *Experientia*, 51(6): 592–595..

13. Erdenmoglu, N. and Sener, B., 2001. *Fitoterapia*, 72(1): 59–61.

14. Biamonte, R., Murabita, F., Pucei, G., Rovito, A., Fillippelli, G., Conforti, S., Liguori, V. and Palazzo, S., 1997. *Tumori*, 83(4): 47.

15. Hamada, H., Sanada, K., Ikeda, S., Oda, T., Williams, H.J., Moyna, G. and Ian Scott, A., 1996. *Natural Product Letters*, 8(2): 113–117.

16. Jung, H., Sok, D.F., Kim, Y., Min, B.S., Lee, J. and Bac, K., 2000. *Planta Medica*, 66(1): 74–76.

17. Favaretto, A., Paccagnella, A., Oniga, F., Ossana, L., Kussis, H. and Forentino, M., 1997. *Tumori*, 83(4): 20.

18. Valenza, R., Gebbia, V., Agostara, B., Gebbia, N., Borsellino, N. and Testa, A., 1997. *Tumori*, 83(4): 46.

19. Pinguet, F., Culine, S. and Sauvaire, Y., 2001. *Planta Medica*, 67(1): 79–81.

20. Donati, S., Pazzagli, I. and Conte, P.F., 1997. *Tumori*, 83(4): 46.

21. Bergnolo, P., Donadro, M., Conrandone, A., Bertetto, A., Boglione, A., Oliva, C. and Bumma, 1997. *Tumori*, 83(4): 106.

22. Goodman and Gilman, 1996. *The Pharmacological Basis of Therapeutics*, 9th Edition. Mc Graw Hill Publisher, New York, pp. 1260.

23. Yue, Q., Fang, Q.C. and Liang, X.T., 1996. *Acta Pharmaceutica Sinica*, 31(12): 911–917.

24. Baloglu, E. and Kingston, D.G.I., 1999. *Journal of Natural Products*, 62(7): 1068–1071.

25. Laskaris, G., De Jang, C.F., Jaziri, M., Van Der Heijden, R., Theodoridis, G. and Veropoorte, R., 1999. *Phytochemistry*, 50(6): 939–946.

26. *Ibid.*

27. Prasad Babu Chittimala, Mamtha Routhu, Kokate Chandrakant and Veershan Ciddi, 2001. *Indian Drugs*, 38(10): 502.

28. *Ibid.*

29. Veershan, C., Rao, M.A., Chittimala, P.B., Mamatha, R. and Kokate, C.K., 2000. *Indian Drugs*, 37(2): 86–89.

30. *Ibid.*

Biotechnology: An Overview (2015)   Pages 185–202
Editors: Rajan Kumar Gupta, Nasim Akhtar and Deepak Vyas
Published by: DAYA PUBLISHING HOUSE, NEW DELHI

Chapter 14

Application of Genetic Engineering in Fungal Biotechnology

Koushalya Dangwal*

Department of Biotechnology, Modern Institute of Technology (MIT), Dhalwala, Rishikesh – 249 201, Uttarakhand

ABSTRACT

This chapter discusses the various gene cloning strategies employed in fungi and illustrated specific applications of gene cloning with selected examples in several important areas of fungal biotechnology. Fungi as model organisms for cloning genes has proven to be an important experimental tool that has provided valuable information for understanding basic biochemical and cellular processes of the eukaryotic cell. In the near future the number of fungi with fully sequenced genomes is likely to increase. The information from these genomes will show the structure of the genetic material, but important questions such as the identification of functional units, *i.e.*, genes and the assignment of biological and/or biochemical functions to these genes, still remain to be solved. Some of the gene cloning techniques discussed in the chapter will certainly play an important role in the functional characterization of the open reading frames (ORFs) identified in model organisms.

Furthermore, gene cloning techniques remain valuable experimental tools in studying the vast majority of fungi that are not subjects of genome sequencing programs. Many of these organisms contain genes that control cellular processes affecting traits of significant values in terms of industrial, agricultural, or medical biotechnology. Gene cloning methods will continue to be exploited in the identification of the key genetic components of processes of potential biotechnological significance.

Keywords: Biotechnology, Cloning strategy, Fungi, Genetic engineering.

* *Corresponding Author:* E-mail: kdangwal1@yahoo.co.in

Introduction

Genetic engineering offers cloning, characterization and modification of genes, regulatory regions, or any other kind of DNA sequences. Model organisms for such studies include bacteria, fungi, plants and animal cells. Amongst these, fungi are especially suitable as model organisms for a number of reasons. Fungi have relatively small and compactly organized genomes. Some fungi, like most species of yeasts are unicellular organisms, whereas others are filamentous. The spores and vegetative cells can also be genetically manipulated for clonal propagation. Many fungi can be maintained indefinitely as haploid, mitotic lineages. These lineages can also be broken by initiating mating and sexual recombination. Almost all fungi can easily be manipulated by using molecular genetic techniques, including DNA-mediated transformation and site-specific mutagenesis.

In addition, fungi could serve as excellent experimental organisms in biotechnology owing to their outstanding metabolic versatility. Most filamentous ascomycetes, members of the genera *Aspergillus*, *Neurospora*, or *Podospora*, can utilize a wide range of compounds as sole carbon or nitrogen sources and they are capable of degrading a number of highly polymerized compounds, like cellulose, pectin, or starch. Moreover, their intermediary metabolism shows remarkable similarities with that of the higher eukaryotes. In some cases, the metabolic pathways are controlled by functionally interchangeable genes in fungi and higher eukaryotes. Finally, many fungi have been awarded the generally regarded as safe (GRAS) status, therefore they are regarded as favorable hosts for producing compounds, including novel type or recombinant molecules for human utilization.

Both basic molecular cell biology research and modern production biotechnology require the use of cloned genes. This chapter describes the basic strategies of gene cloning with selected examples of utilization of cloned fungal genes in agricultural, industrial, and medical biotechnology.

Cloning Strategies

The fungal genome size range between 12 to 42 Mb corresponding to 6000 to 9000 genes that control overall cellular, physiological, and morphological processes in fungi (Kupfer *et al.*, 1997). Understanding of fungal biology ultimately requires the functional characterization of the genes which involves cloning and further identification of the same.

Various approaches are available for the cloning of genes in fungi. The choice depends on several aspects, including the experimental purpose and the applicability of techniques to the target organism. The experimental strategies are based on the following methods; (a) selection for the phenotype characteristic to the biological function of the gene, (b) detection of the biochemical activity of the gene product, (c) the structure of the gene, (d) transcriptional regulation of the gene, and (e) chromosomal position of the gene.

Cloning Based on Biological Function

Two alternative gene cloning strategies based on the biological functions are available: complementation of mutants and generation of tagged insertional mutants.

Complementation of Mutants

Complementation requires the availability of appropriate stable mutants and an efficient gene transfer system. Stable mutants may not be always available in case of specific target organisms; therefore complementation experiments are frequently performed in heterologous hosts. *Saccharomyces cerevisiae* is the most frequently used host organism for this purpose. In complementation studies mutant strains are transformed with a gene library (*c*DNA or a genomic fragment of the gene) from the target organism, and complementing phenotypes indicative of the restoration of gene function are selected. Transformation vectors purified from these strains contain the gene(s) of interest. If the vector contains large genomic fragment carrying gene of interest then, identification of an open reading frame (ORF) responsible for the appropriate phenotypic expression requires further subcloning and complementation experiments. This strategy is most appropriate for the isolation of a single gene responsible for a given biological function, and an example is the cloning of the uac1 gene (encoding an adenylate cyclase) from *Ustilago maydis* by complementing a mutant yeast strain capable of budding growth (Barrett *et al.*, 1993). However, this approach is also suitable for the cloning of a series of genes acting downstream of the inactivated gene in a biochemical pathway.

Insertional Mutagenesis

Insertional mutagenesis, also known as gene tagging, is an elegant strategy to generate loss-of-function mutants which further allow identification of genes responsible for specific biological functions. It can therefore be applied as an alternative gene cloning strategy in fungi. In general, a foreign sequence is introduced into the target organism and, as a consequence of random chromosomal integrations of foreign DNA sequences, host gene(s) may become inactivated. The collection of transformants is subsequently screened for a specific mutant phenotype, usually for the loss of an ability of the target organism, *e.g.*, impairment in pathogenicity. Regions, flanking the insertion sites are then identified by PCR-based techniques or plasmid rescue (Walser *et al.*, 2001). This approach requires efficient transformation systems for inserting sequences into the target organism.

Restriction Enzyme Mediated Integration (REMI)

It is the most widely used method for insertional mutagenesis. In REMI, fungal cells are transformed with a vector DNA (plasmid) in the presence of a restriction endonuclease. The endonuclease facilitates integration of the vector DNA into the chromosome. The precise mechanism and critical parameters of the procedure are not fully understood, and the results obtained thus far are in part contradictory (Walser *et al.*, 2001), but in a number of experiments, the addition of restriction enzymes increased the transformation frequency and decreased the number of integrated copies of foreign DNA per transformant. This approach has mainly been used to study pathogenicity determinants in plant pathogens (Bolker *et al.*, 1995; Sweigard *et al.*, 1998).

Signature Tagged Mutagenesis (STM)

It is a method originally developed to identify genes essential for in vivo colonization of bacterial pathogens in animal tissues (Hensel *et al.*, 1995). Separate insertional mutant pools are generated with tagged sequences (usually with transposons) and then a group of mutants tagged with distinct sequences are used to co-inoculate the host. In parallel, the same subpool of mutants is subjected to in vitro culturing. After an appropriate incubation time, the tagged sequences are quantitatively determined in both the host tissues by hybridization to oligonucleotide arrays or PCR assay. Mutant strains that fail to proliferate in vivo but are capable of reproduction in vitro contain an inactivated gene essential for the colonization of specific niches. This method has been earlier used to identify a gene (pabaA) encoding para-aminobenzoic acid synthetase in an opportunistic fungal pathogen *Aspergillus fumigates* causing pulmonary aspergillosis (Brown *et al.*, 2000). An increasing number of genes are expected to be identified and cloned from pathogenic fungi using this approach in the near future.

Agrobacterium tumefaciens Mediated Transformation (ATMT)

Agrobacterium tumefaciens, the tumor-inducing bacterium is widely used for transformation and gene tagging of plants owing to its ability to transfer distinct parts of its Ti-plasmid DNA into the plant cells and integrate the plasmid DNA into plant chromosomes in a random fashion (Koncz *et al.*, 1992). Transformation of yeast and a wide range of filamentous fungal species by *A. tumefaciens* have been reported in many studies (deGroot *et al.*, 1998a, b; Piers *et al.*, 1996). Mullins *et al.* (2001) determined the critical parameters affecting transformation efficiency and the copy number of T-DNA inserted into the genome of the wilt-causing fungus, *Fusarium oxysporum* using novel vectors.

Cloning Based on Biochemical Function

Detection of the biochemical activity of a protein may also be employed to clone its encoding gene. This could be achieved by purifying the protein of interest from complex protein mixtures. Instead, various heterologous expression systems could also be used that have been developed to detect either enzymatic activities or specific DNA-binding abilities of the gene products.

Protein Purification

In this approach, a crude protein extract is prepared from disrupted cells or from the extracellular environment of the target organism. The extract is then subjected to a number of purification steps needed to obtain homogeneously purified protein. A partial amino acid sequence of the purified protein is then determined and used to design degenerate oligonucleotide primers. Afterward, a portion of the gene is amplified by PCR to generate a gene specific probe. Alternatively, antibodies raised against the purified gene products could be used to screen expression libraries.

Heterologous Expression

An efficient yeast expression system using *S. cerevisiae* as heterologous host has been developed to clone fungal enzyme genes (Dalboge and Heldt-Hansen, 1994). A cDNA library is constructed using *Escherichia coli*–yeast shuttle vector containing an inducible yeast promoter favoring enzyme production. Plasmid DNA isolated from pools of transformant *E. coli*, is then used to transform yeast cells. Transformant yeast colonies are then replica-plated onto the medium, containing substrates, which allow the detection of the enzyme of interest. This methodology requires appropriate post translational processing into functional fungal enzyme and its secretion in to sufficient quantity suitable for plate assay. The advantages of this method over traditional ones are that it requires no preliminary information on the properties of the protein, needs no laborious protein purification procedures, and allows the cloning of genes whose products are synthesized in limited amounts by the source organism. Moreover, several different enzyme genes can be simultaneously cloned by replica-plating the yeast library onto media containing different substrates. Hundreds of enzyme genes including several enzyme families from a wide range of filamentous fungi have already been cloned using this strategy (Dalboge and Lange, 1998).

Another cloning strategy based on yeast expression systems was developed to clone genes on the basis of specific DNA binding and transcription activating properties of their products (Saloheimo *et al.*, 2000). The method seems to be of general applicability, however, an important requirement is that the yeast strain used as host should not express transcription factors that activate the promoter of interest. In this technique the target promoter is fused with a selectable yeast reporter gene (encoding the auxotrophic marker X), this plasmid is then introduced into the yeast host to give a reporter strain. A constitutively expressed cDNA library is prepared from the fungus in a yeast–*E. coli* shuttle vector containing another auxotrophic marker (Y). Subpools of the library are then used to transform the reporter strain and colonies that can grow on a medium lacking amendments for both the X and Y markers are isolated. Plasmids from these colonies that promote yeast growth in a medium lacking X, only in the presence of the reporter plasmid, contain transcription factors positively acting on promoter sequences introduced into the reporter strain. This method has been used to clone two transcriptional activator genes, ace1 and ace2 of the cellobiohydrolase gene and cbh1 of *Trichoderma reesei* (Aro *et al.*, 2001; Saloheimo *et al.*, 2000).

Cloning based on Sequence Similarity

This methodology involves gene cloning on the basis of their sequence similarities. This strategy is quite suitable especially when the target organism is difficult to culture or an efficient gene transfer system is lacking. Previously cloned gene as a heterologous probe can also be used in library screening to isolate its homologous sequence. However, the success of the method can be severely limited due to insufficient homologies or the limited stretch of the homologous regions. Nevertheless, there are many examples in the literature reporting the successful application of heterologous gene probes to isolate their homologues from distantly related fungal species. When sequence similarities between genes (or gene products) are confined to short regions, *e.g.*, regions corresponding to specific functional domains, such as catalytic or substrate binding domains of enzyme families, alignment of these sequences give information that can be used to design degenerate oligonucleotide primers. Using these primers, portions of the genes of the targeted family can be amplified and then used as probes in library screening. With the growing accumulation of sequence data in gene banks, this approach is likely to gain broader application.

Cloning based on the Transcriptional Regulation of the Gene

A number of approaches have been developed for cloning genes based on their regulation. Essentially, all these techniques involve the comparison of the transcribed (or translated) sequences present during different physiological, morphological, or developmental stages of an organism. Transcripts that appear exclusively or more abundantly in a particular stage are assumed to originate from genes whose activity contributes to bring about or maintain the condition studied. In terms of experimental methodology, three main approaches, namely (a) hybridization-based techniques, (b) PCR-based techniques and (c) protein-based techniques are used to detect differences in transcript abundance. However, a large variety of modifications of the basic protocols have recently been developed. The border line between the methods listed above is not always sharp, and, for example, some methods combine hybridization and PCR technology.

Hybridization-Based Techniques

Differential hybridization involves the preparation of a cDNA library from mRNAs obtained from cells grown under the conditions of interest. This is followed by a differential screening of replicas of this library by hybridization to cDNA probes obtained from different conditions that are subjects of comparison. Clones that give unique signals, or significantly more intense signals, when hybridized to a specific probe, are then selected and analyzed further.

A more sophisticated version of the hybridization-based approach is subtractive hybridization, which involves the selective enrichment of specific sequences present in one sample but absent from another one. The population of cDNA molecules in the sample originating from the condition of interest (tester) is mixed with excessive amounts of nucleic acids from the other sample (driver), and these are then hybridized to each other. This is followed by physical separation of the single-stranded molecules. The single stranded material contains sequences from genes with stage-specific expression properties different from the double-stranded nucleic acids that contain transcripts present in both samples (Duguid and Dinauer, 1990; Hedrick *et al.*, 1984). In an advanced version of this method, physical separation is substituted by adaptor ligation and selective PCR amplification of transcripts originating only from the desired sample (Diatchenko *et al.*, 1996).

PCR-Based Differentiation

Differential display of mRNA is a powerful method to identify and clone differentially transcribed genes (Liang and Pardee, 1992). In general, mRNA populations of samples obtained from different conditions are reverse transcribed, and then PCR-amplification is performed with an arbitrary primer and an 'anchor' primer, homologous to the polyadenylated 3' end of mRNA species but containing extra bases at the 3' end allowing arbitrary selection of the template molecules.

Amplification then generates specific fragments characteristic of the samples obtained from different conditions. These fragments can be separated by electrophoresis. Differentially expressed fragments, assumed to be portions of genes with stage-specific expression are then cloned and used as probes to screen gene libraries. A major advantage of this strategy over subtractive hybridization is that several differentially treated samples can be analyzed simultaneously.

Protein-Based Differentiation

Comparison of the abundance of gene products rather than transcripts in samples obtained from different conditions is another important approach for identifying and cloning differentially regulated genes. Furthermore, in specific cases this approach can offer a more relevant solution in circumstances when (a) protein abundance does not always correspond to $mRNA$ abundance, (b) the samples to be compared may not always contain $mRNA$ (*e.g.*, when studying secreted gene products), and (c) the regulation of genes may occur at the post-translational level (proteolysis, glycosylation, etc.) (Pandey and Mann, 2000). In general, protein extracts from different conditions are separated by two-dimensional gel electrophoresis. This involves a first dimension isoelectric focusing of the proteins in a gel strip followed by separation in a second dimension according to the size of the polypeptides. Separated and fixed proteins are then transferred to a solid surface and stained resulting in condition specific patterns. These patterns can then be analyzed quantitatively with appropriate image analysis software. Molecules that show significantly different staining intensities or only occur in a specific condition denote polypeptides translated from differentially expressed messages or proteins subject to differential post-translational modification. These can be further analyzed by excision and the determination of partial amino acid sequence. The information obtained can be used to design oligonucleotide primers for the amplification of sequences that can be used as a gene-specific probe for screening libraries. A good example of making use of this strategy in gene cloning from fungi was the isolation from *Penicillium decumbens* of the epoA gene encoding an epoxidase enzyme which plays a key role in the biosynthesis of fosfomycin, an antibiotic compound of commercial importance (Watanabe *et al.*, 1999).

Cloning by Chromosomal Position

A number of cloning methodologies including map-based cloning and synteny based cloning strategies exploit information of the chromosomal position of a gene.

Map-Based Cloning

Map-based cloning (or positional cloning) involves the construction of a genetic map of the target organism and determine the position of the target gene on the map. Molecular markers that co-segregate with highest frequency with the trait can then be used as probes to initiate chromosome walking in order to clone the genomic region containing the gene of interest. This approach is only applicable to fungal species that are amenable to classical genetic studies. The construction of a complete genetic map involves the assessment of the relative distances between phenotypic and molecular markers. This is achieved by calculating the frequency of recombination between the respective markers. Different types of molecular markers are used comprehensively to generate highly saturated maps. Restriction length polymorphisms (RFLPs) are detected by hybridizing randomly cloned DNA fragments to restriction digested DNA samples of fungal specimens.

Randomly amplified polymorphic DNA fragments (RAPDs) are products of arbitrarily primed polymerase chain reactions. Amplified fragment length polymorphisms (AFLPs) are generated by ligating adaptors to restriction digested DNA fragments and selectively amplifying a subset of the fragments (Vos *et al.*, 1995). The generation of a fungal genetic map highly saturated with molecular markers increases chances of finding two closely linked markers on each side of the target gene. Such markers can then be used as probes in chromosome walking experiments to obtain overlapping clones containing the gene of interest. Once these clones are isolated the gene responsible for the trait can be identified by subcloning followed by sequencing and/or mutant complementation.

Synteny

The co-linearity of genes, *i.e.*, the conservation of gene orders among fungal genomes, can also be used to clone a gene. This strategy uses a conserved gene or a molecular marker found to be closely linked to the gene of interest in a model organism with a highly saturated genetic map. This gene or marker is then used as a probe in chromosome walking experiments on a library from the target organism.

The success of this approach depends on the similarity of the genomes of the model and the target fungi. To date there has only been limited information published on the extent of synteny between fungal genomes. The existence of syntenic regions has been observed between *Magnaporthe grisea* and *Neurospora crassa* but no synteny was detected between *M. grisea* and the yeasts *S. cerevisiae* and *Candida albicans* (Hamer *et al.*, 2001). As fungal genome projects proceed and, as a consequence, comparative genomics development, the applicability and usefulness of this approach is likely to increase.

Applications of Gene Cloning in Fungal Biotechnology

Cloning of fungal genes has made a major contribution to a better understanding of the biochemical and cellular mechanisms that occur in the eukaryotic cell, it has helped to reveal the recognition process between cells of different origins and opened up new possibilities for improving productivity of fungi and plants. Some particular examples of these are highlighted below.

Towards the Understanding of the Reproduction Strategies in Fungi

A better knowledge on the reproduction strategies of fungi is needed in order to monitor changes in fungal populations or select the most appropriate disease control measures. In many fungi, the mating system is heterothallic, based on the interaction of two compatible strains of opposite mating type. Some species, however, are homothallic, and are capable of sexual reproduction within a single line. A third group of fungi, comprising thousands of morphological species, are considered as asexual taxa, that have either lost the ability to reproduce sexually or have a cryptic sexual cycle elicited by unusual environmental conditions or other unknown stimuli.

Mating type is determined by MAT genes that encode transcriptional regulators, confer mating type identity, and control the development of fruiting bodies and meiospores. MAT genes have been cloned by using diverse strategies, including genomic subtraction, complementation of MAT deletion mutants, and PCR-based techniques (Turgeon, 1998). The latter proved to be especially useful for cloning mating type genes in a wide range of ascomycetes, as the MAT idiomorphs of these fungi contain two highly conserved sequences, called the a-domain and HMG-box sequences bound in MAT-1 and MAT-2 strains, respectively. These are easily amplified by using either group specific or degenerate primers (Arie *et al.*, 1997). Cloning of MAT genes has provided a rapid and robust diagnostic procedure for mating type identification in the *Gibberella fujikuroi* species complex, including the determination of biological species and asexual lineages (Steenkamp *et al.*, 2000).

Recognition Between Plants and Pathogenic Fungi

Cloned fungal genes helped to understand the recognition specificity between plants and pathogenic fungi and reveal the molecular background of the gene-for-gene relationship, a theory first proposed more than 40 years ago (Flor, 1955). The tomato–*Cladosporium* interaction has been widely used as a model system in these investigations. *Cladosporium fulvum* (now correctly named *Mycovellosiella fulva*), a biotrophic fungal pathogen causes leaf mold on tomato. Plant breeders have introduced several resistance genes (called Cf genes) against this pathogen from wild *Lycopersicon* species into commercial tomato cultivars. As a result, many races have emerged in *C. fulvum* that are able to overcome the Cf genes built in tomato in various combinations. A typical gene-for-gene relationship has been seen in the tomato–*C. fulvum* system, in which each plant resistance gene (Cf) has its avirulence gene (Avr) counterpart in the fungal population. Interaction between tomato cultivars and *C. fulvum* races that contain properly matching Cf and Avr genes, respectively, results in recognition, leading to a hypersensitive defense reaction (HR) by the plant. If partners containing non matching Cf and Avr genes meet, recognition fails to occur and the fungus successfully invades the plant tissues. Experiments on the tomato–*Cladosporium* system resulted in the cloning of the first fungal avirulence gene Avr9 gene which was capable of inducing HR in tomato cultivars carrying Cf9 resistance gene (Van Kan *et al.*, 1991).

Fungal genes could also be used to improve plant resistance against bacteria and fungi either by enhancing or supplementing the existing plant defense system. The plant– pathogen interaction triggers a cascade of defense reactions such as oxidative burst, *i.e.*, a rapid and transient production of active oxygen species (superoxide anionradical, hydroxyl radical, and hydrogen peroxide), followed by cell-wall strengthening, phytoalexin synthesis, and accumulation of pathogenesis-related proteins. Enhanced H_2O_2-generation has been shown to be one of the means by which plant resistance could be improved by using a gene of fungal origin (Wu *et al.*, 1995). A glucose oxidase gene, GO, has been cloned from *Aspergillus niger* for transforming potato. Consequently, leaves and tubers of the transgenic plants showed accumulation of high levels of GO. Since, GO catalyzes the generation of H_2O_2 from glucose, transgenic tubers were found to harbor strong H_2O_2-mediated resistance against the soft rot pathogen, *Erwinia carotovora*.

Production of Valuable Pharmaceuticals

This yeast can be grown rapidly and to a high cell density, can secrete heterologous proteins into the extracellular broth, and knowledge of its genetics is more advanced than that of any other eukaryote (Romanos *et al.*, 1992). Fungi are known to produce a wide range of antibiotics β-Lactams antibiotics (penicillins and cephalosporins), are the most important class of antibiotics secreted by fungi. These compounds are synthesized by condensation of L-α-aminoadipic acid, L-cystein, and L-valine to form (L-α-aminoadipyl)-L-cysteinyl-D-valine (ACV), which is further converted to isopenicillin N (IPN). Subsequent steps include acetylation or epimerization yielding penicillin G (produced by *Penicillium chrysogenum*) and penicillin N (produced mainly by *Acremonium chrysogenum*), respectively. Penicillin N is then converted to cephalosporin by a bifunctional expandase-hydroxylase and an acetyltransferase. Almost all genes involved in β-lactam biosynthesis have been cloned (Alvarez *et al.*, 1993; Dý´ez *et al.*, 1990; Gutie´rrez *et al.*, 1992; Samson *et al.*, 1985; 1987). The genes pcbAB and pcbC, which control ACV and IPN synthesis are common in all β-lactam producing fungi; penDE coding for an IPN acetyltransferase, which is the first specific enzyme of the penicillin G biosynthetic pathway is found exclusively in *P. chrysogenum*, whereas *A. chrysogenum* carries a cefEF–cefG cluster, that comprise of genes encoding for the later steps of cephalosporin synthesis. Cloned β-lactam biosynthesis genes have been utilized to engineer overproducing strains by overexpressing either the cefG gene in *A. chrysogenum* (Gutie´rrez *et al.*, 1997) or the acvA gene (a homologue of pcbAB) in *A. nidulans* (Kennedy and Turner, 1996). An additional gene, named cahB, and that not linked to the cef cluster, has also been found to affect cephalosporin C production in *A. chrysogenum* (Velasco *et al.*, 2001). In addition to antibiotic genes, the mammalian genes encoding valuable pharmaceuticals have also -been cloned and expressed in *S. cerevisiae*, including human interferon, human epidermal growth factor, and human hemoglobin (reviewed by Adrio and Demain, 2003). The most commercially important yeast recombinant process has been the production of genes encoding surface antigens of the hepatitis B virus (Adrio and Demain, 2003). Besides *S. cerevisiae*, *Pichia pastoris* is the second most extensively used expression systems for the production of valuable heterologous pharmecutical (Adrio and Demain, 2003).

Genes of Potential Uses in Yeast Improvement

The recombinant DNA technology has proved to be most suitable for producing improved strains for brewing, wine-making, or baking industries. A number of genes, that have been cloned recently for strain improvement. The major targets for manipulation in *S. cerevisiae* strains are improved flocculation, elimination of undesired by-products, extension of substrate utilization, increased ethanol and organic acid tolerance, as well as enhanced tolerance against osmotic stress. Genes controlling these traits are therefore of commercial interest.

Flocculation promotes rapid aggregation of yeast cells that can easily be removed by sedimentation which is an important step for acquiring clear, well-flavored beers. Several flocculation specific genes have been isolated including FLO1, to improve the timing of flocculation (Kobayashi *et al.*, 1998). The FLO1 promoter was substituted by promoter of the heat-shock protein gene, HSP30 (Verstrepen *et al.*, 2000), which is induced by sugar exhaustion and high ethanol concentration that occurs at the end of the brewing process. Engineered strains carrying the HSP30-FLO1 fusion construct resulted in optimal flocculation allowing for an appropriate timely sedimentation under laboratory conditions.

Glycerol and isoamyl acetate are important flavoring components of beverages. GPD1, encoding glycerol-3-phosphate dehydrogenase has been successfully cloned to enhance glycerol production of wine yeast strains (Michnick *et al.*, 1997). The AFT1 gene, encoding an alcohol transferase which catalyzes the formation of ester bonds between acetyl CoA and alcohols, has been overexpressed resulting in substantial increase in isoamyl-acetate production in both sake and wine yeast strains (Fuji *et al.*, 1994; Lilly *et al.*, 2000).

S. cerevisiae has been employed for commercial production for ethanol by fermenting sugar components of lignocelluloses hydrolysates. However, the presence of phenolic compounds in the hydrolysates inhibits the fermentation process and yeast growth. Therefore, an enzymatic detoxification method, based on a laccase extract prepared from the lignin degrading basidiomycete *Trametes versicolor* has been developed for removing phenolic compounds prior to fermentation. The use of genetically modified yeast strains with phenol degrading capability provides a promising single step approach to achieve this. Two laccase genes (lac1, lac2) cloned from *T. versicolor* have been used to transform two *S. cerevisiae* strains (Cassland and Jo"nsson, 1999). Stable transformants capable of secreting heterologous laccase were obtained only from one that had been transformed by lac2, indicating that both the recipient strain and the laccase gene should be carefully chosen for the strain-improvement program.

Yeast has also been genetically modified for the increased production of acetate esters Isoamyl acetate (banana-like aroma), which is an important determinant for beer, sake and young wines. This ester is produced by yeast from isoamyl alcohol, which itself is a by-product of leucine synthesis. Overexpression of one of the genes responsible for leucine synthesis (*LEU4* encoding α-isopropylmalate synthase) in a sake yeast strain resulted in a very slight increase in isoamyl alcohol concentrations and the corresponding ester (Hirata *et al.*, 1992). Another strategy developed in a brewing strain was to increase the formation of acetate esters by overexpression of the *ATF1* gene, which encodes the alcohol acetyltransferase catalyzing the formation of esters from acetyl CoA and the relevant alcohols. This led to a 27-fold increase in the production of isoamyl-acetate during fermentation on a laboratory scale (Fuji *et al.*, 1994). The strategy has been also successfully applied to wine yeast strains (Lilly *et al.*, 2000). However, a major disadvantage of this approach is that it generates a simultaneous increase in the formation of esters, *e.g.* ethyl acetate, which can be undesirable in wine above a critical amount. Another approach developed on sake yeast strains was to disrupt the *EST2* gene, which codes for the major esterase hydrolyzing isoamyl acetate (Fukuda *et al.*, 1998). However, this only resulted in a limited increase (approximately 2-fold) in isoamyl acetate production

Cloned Fungal Sequences and Improved Heterologous Protein Production

Enzymes are progressively being employed in the paper, textile, fodder, or detergent industries. The majority of these enzyme preparations are generated as heterologous proteins secreted by fungi, including *Aspergillus, Fusarium, Kluyveromyces,* and *Trichoderma* (Domínguez, 1998; Adrio and Demain, 2003; Patrick *et al.*, 2005; Ooyen *et al.*, 2006), Even though fungi employed for heterologous enzyme production secrete remarkably large amounts of proteins into the growth medium, this secretion ability still needs improvement. The major limitation of heterologous protein production by fungi is improper post-translational modification process of these proteins. Therefore, genetic regulation of protein folding, glycosylation, and the whole secretory machinery are of major thrust areas for improvement of heterologous enzyme production in fungi.

Chaperones and foldases are enzymes that assist in protein secretion and have been collectively termed helper proteins. A number of genes encoding helper proteins have been cloned and characterized. For example, overexpression of KAR2, a protein-binding chaperone gene, resulted in a 20-fold higher bovine prochymosin production in yeast (Harmsen *et al.*, 1996). Similarly, increased PDGF (human platelet-derived growth factor) production was obtained in *S. cerevisiae* by overexpressing the foldase encoding the PDI gene (Robinson *et al.*, 1994)

Fungal Genes for Bio-Remediation Processes

Industrialization has resulted in increased accumulation of dangerous pollutants into the environment such as heavy metals, including arsenic, cadmium, copper, iron, lead, and mercury. Plants can hyper accumulate these heavy metals from polluted soils and hence augmenting phytoremediation. The efficiency of phytoremediation could be further enhanced by producing genetically engineered plants, capable of increased metal uptake that could accumulate exceptionally high levels of heavy metals in their tissues or may transform and volatilize the pollutants (Kra"mer and Chardonnens, 2001). Heavy metal tolerance in oil seed rape (*Brassica oleracea*) has been improved by transformation, with the methallothionein gene (CUP1), from *S. cerevisiae*. The *B. oleracea* plants harboring the CUP1 gene accumulated 10–70 per cent higher concentrations of cadmium in their leaves than the wild-type plants when both were grown in a medium containing 25 mM cadmium (Hasegawa *et al.*, 1997). Similarly, iron tolerance in tobacco has been considerably improved by introducing two ferric reductase encoding genes (FRE1, FRE2) from yeast. The transformants showed 50 per cent higher iron accumulation in the leaves, than that in control plants (Samuelson *et al.*, 1998).

Use of Cloned Genes in Mushroom Improvement

Molecular breeding strategies utilizing cloned genes provide potential for the production of transgenic mushroom varieties with improved traits, including productivity, disease resistance, or marketability. A number of genes, encoding various lignocellulose degrading enzymes including two laccase genes (LCC1 and LCC2) as well as a cellulose (CEL1), a cellobiohydrolase (CEL2), an endoxylanase (XLNA), and a mannosidase (CEL4) gene have been cloned from *Agaricus bisporus*, to impart broad substrate utilization ability for improved mushroom production (Chow *et al.*, 1994; deGroot *et al.*, 1998a, b; Raguz *et al.*, 1992; Yagu¨e *et al.*, 1997).

Another target of mushroom breeding is to improve the marketability of the product. Polyphenoloxidases, most notably tyrosinases are responsible for this brown discoloration in mushroom leading to their reduced marketability. These enzymes catalyze the conversion of tyrosine to o-diphenol-3,4-dihydroxyphenylalanine and then to dopaquinone, which is further converted to melanin and melanochrome, two major pigments found at high concentrations in browned fruiting bodies. The antisense of two tyrosinase encoding genes, AbTYR1 and AbTYR2 has been cloned from *A. bisporus* to obtain stable transformants with silenced activity of the native tyrosinase genes (Stoop and Mooibroek, 1999). Therefore, the antisense strategy could be one of the most rational ways of delaying discoloration.

Fungal Genes with Human Relations

Many fungal genes have been shown to possess homology with their human counterparts such as homogentisate dioxygenase (hereditary deficiency causes Alkaptonuria), RAS2 (affects cellular ageing), LAG1 and LAG2 (LAG ¼ longevity assurance gene affecting ageing). Therefore, inherited human disorders linked to these genes could be investigated in the fungal model (Ferna´ndez-Can on and Pen alva, 1995; Rodriguez *et al.*, 2000; Pen alva, 2001; Sinclair, *et al.*, 1998). Investigations of organization, function, and regulation of certain fungal genes may further allow a better insight into the human inherited disorders and ageing process of cells and organisms, and may therefore be helpful in understanding the genetic control of human disorders.

References

1. Adrio, J.L. and Demain, A.L., 2003. Fungal biotechnology. *Int. Microbiol.* 6: 191–199.

2. Alvarez, E., Meesschaert, B., Montenegro, E., Gutie´rrez, S., Dý´ez, B., Barredo, J.L. and Martý´n, J.F., 1993. The isopenicillin N acyltransferase of *Penicillium chrysogenum* has isopenicillin N amidohydrolase, 6-aminopenicillanic acid acyltransferase and penicillin amidase activities, all of which are encoded by the single penDE gene. *Eur. J. Biochem.*, 215: 323–332.

3. Arie, T., Christiansen, S.K., Yoder, O.C. and Turgeon, B.G., 1997. Efficient cloning of ascomycete mating type genes by PCR amplification of the conserved MAT HMG box. *Fungal Genet. Biol.*, 21: 118–130.

4. Aro, N., Saloheimo, A., Ilmen, M. and Penttila, M., 2001. ACEII, a novel transcriptional activator involved in regulation of cellulase and xylanase genes of *Trichoderma reesei*. *J. Biol. Chem.*, 276: 24309–24314.

5. Barrett, K.J., Gold, S.E. and Kronstad, J.W., 1993. Identification and complementation of a mutation to constitutive filamentous growth in *Ustilago maydis*. *Mol. Plant Microbe. Interact.*, 6: 274–283.

6. Bolker, M., Bohnert, H.U., Braun, K.H., Gorl, J. and Kahmann, R., 1995. Tagging pathogenicity genes in *Ustilago maydis* by restriction enzyme-mediated integration (REMI). *Mol. Gen. Genet.*, 248: 547–552.

7. Brown, J.S., Aufauvre-Brown, A., Brown, J., Jennings, J.M., Arst, H. Jr. and Holden, D.W., 2000. Signature-tagged and directed mutagenesis identify PABA synthetase as essential for *Aspergillus fumigates* pathogenicity. *Mol. Microbiol.*, 36: 1371–1380.

8. Cassland, P. and Jonsson, L.J., 1999. Characterization of a gene encoding *Trametes versicolor* laccase A and improved heterologous expression in *Saccharomyces cerevisiae* by decreased cultivation temperature. *Appl. Microbiol. Biotechnol.*, 52: 393–400.

9. Chow, C., Yague, E., Raguz, S., Wood, D.A. and Thurston, C.F., 1994. The CEL3 gene of *Agaricus bisporus* codes for a modular cellulose and is transcriptionally regulated by the carbon source. *Appl. Environ. Microbiol.*, 60: 2779–2785.

10. Dalboge, H. and Heldt-Hansen, H.P., 1994. A novel method for efficient expression cloning of fungal enzyme genes. *Mol. Gen. Genet.*, 243: 253–260.

11. Dalboge, H. and Lange, L., 1998. Using molecular techniques to identify new microbial biocatalysts. *Trends Biotechnol.*, 16: 265–272.

12. deGroot, M.J., Bundock, P., Hooykaas, P.J. and Beijersbergen, A.G., 1998a. *Agrobacterium tumefaciens*–mediated transformation of filamentous fungi. *Nat. Biotechnol.*, 16: 839–842.

13. deGroot, P.W.J., Basten, D.E.J.W., Sonnenberg, A.S.M., van Griensven, L.J.L.D., Visser, J. and Shaap, P.J., 1998b. An endo-1,4-bxylanase-encoding gene from *Agaricus bisporus* is regulated by compost specific factors. *J. Mol. Biol.*, 277: 273–284.

14. Dyez, B., Gutierrez, S., Barredo, J.L., Van Solingen, P., van der Voort, L.H.M. and Martýn, J.F., 1990. The cluster of penicillin biosynthetic genes. Identification and characterization of the pcbAB gene encoding the α-aminoadipyl-cysteinyl-valine synthetase and linkage to the pcbC and penDE genes. *J. Biol. Chem.*, 265: 16358–16365.

15. Diatchenko, L., Lau, Y.F., Campbell, A.P., Chenchik, A., Moqadam, F., Huang, B., Lukyanov, S., Lukyanov, K., Gurskaya, N., Sverdlov, E.D. and Siebert, P.D., 1996. Suppression subtractive hybridization: A method for generating differentially regulated or tissue-specific cDNA probes and libraries. *Proc. Natl. Acad. Sci.*, 93: 6025–6030.

16. Domínguez, A., Ferminán, E., Sánchez, M., González, F.J., Pérez-Campo, F.M., Garcíal, S., Herrero, A.B., Vicente, A.S., Cabello, J., Prado, M., Iglesias, F.J., Choupina, A., Burguillo, F.J., Fernández-Lago, L. and López, M.C., 1998. Non-conventional yeasts as hosts for heterologous protein production. *International Microbiol.*, 1: 131–142.

17. Duguid, J.R. and Dinauer, M.C., 1990. Library subtraction of *in vitro* cDNA libraries to identify differentially expressed genes in scrapie infection. *Nucleic Acids Res.*, 18: 2789–2792.

18. Fernandez-Canon, J.M. and Penalva, M.A., 1995. Molecular characterization of a gene encoding a homogentisate dioxygenase from *Aspergillus nidulans* and identification of its human and plant homologues. *J. Biol. Chem.*, 270: 21199–21205.

19. Flor, H.H., 1955. Host-parasite interaction in flax rust: Its genetics and other implications. *Phytopathology*, 45: 680–685.

20. Fuji, T., Nagzawa, N., Iwamatsu, A., Bogaki, T., Tamai, Y. and Hamachi, M., 1994. Molecular cloning, sequence analysis and expression of the yeast alcohol acetyltransferase gene. *Appl. Environ. Microbiol.*, 60: 2789–2792.

21. Fukuda, K., Yamamoto, N., Kiyokawa, Y., Yanagiuchi, T., Wakai, Y., Kitamoto, K., Inoue, Y. and Kimura, A., 1998. Brewing properties of sake yeast whose *EST2* gene encoding isoamyl acetate-hydrolysing esterase was disrupted. *J. Ferm. Bioeng.* 85: 101–106.

22. Gutierrez, S., Velasco, J., Fernandez, F.J. and, Martýn, J.F., 1992. The cefG gene of *Cephalosporium acremonium* is linked to the cefEF gene and encodes a deacetylcephalosporin C acetyltransferase closely related to homoserine O-acetyltransferase. *J. Bacteriol.*, 173: 3056–3064.

23. Gutierrez, S., Velasco, J., Marcos, A.T., Fernandez, F.J., Fierro, F., Barredo, J.L., Dýez, B., and Martýn, J.F., 1997. Expression of the cefG gene is limiting for cephalosporin biosynthesis in *Acremonium chrysogenum*. *Appl. Microbiol. Biotechnol.*, 48: 606–614.

24. Hamer, L., Pan, H., Adachi, K., Orbach, M.J., Page, A., Ramamurthy, L. and Woessner, J.P., 2001. Regions of microsynteny in *Magnaporthe grisea* and *Neurospora crassa*. *Fungal Genet. Biol.*, 33: 137–143.

25. Harmsen, M.M., Bruyne, M.I., Raue, H.A. and Maat, J., 1996. Overexpression of binding protein and disruption of the PMR1 gene synergistically stimulate secretion of bovine prochymosin but not plant thaumatin in yeast. *Appl. Microbiol. Biotechnol.*, 46: 365–370.

26. Hasegawa, I., Terada, E., Sunairi, M., Wakita, H., Shinmachi, F., Noguchi, A., Nakajima, M. and Yakayi, J., 1997. Genetic improvement of heavy metal tolerance by transfer of the yeast metallothionein gene (CUP1). *Plant Soil*, 196: 277–281.

27. Hedrick, S.M., Cohen, D.I., Nielsen, E.A. and Davis, M.M., 1984. Isolation of cDNA clones encoding T cell-specific membrane-associated proteins. *Nature*, 308: 149–153.

28. Hensel, M., Shea, J.E., Gleeson, C., Jones, M.D., Dalton, E. and Holden, D.W., 1995. Simultaneous identification of bacterial virulence genes by negative selection. *Science*, 269: 400–403.

29. Hirata, D., Aoki, S., Watanabe, K., Tsukioka, M. and Suzuki, T., 1992. Stable overproduction of isoamyl alcohol by *Saccharomyces cerevisiae* with chromosome-integrated multicopy *LEU4* genes. *Biosci. Biotechnol. Biochem.*, 56: 1682–1683.

30. Kennedy, J. and Turner, G., 1996. d-(L-a-Aminoadipyl)-L-cysteinyl-D-valine synthetase is a rate limiting enzyme for penicillin production in *Aspergillus nidulans*. *Mol. Gen. Genet.*, 253: 189–197.

31. Kobayashi, O., Haayashi, N., Kuroki, R. and Sone, H., 1998. Region of FLO1 proteins responsible for sugar recognition. *J. Bacteriol.*, 180: 6503–6510.

32. Koncz, C., Nemeth, K., Redei, G.P. and Schell, J., 1992. T-DNA insertional mutagenesis in *Arabidopsis*. *Plant. Mol. Biol.*, 20: 963–976.

33. Kramer, U. and Chardonnens, A.N., 2001. The use of transgenic plants in the bioremediation of soils contaminated with trace elements. *Appl. Microbiol. Biotechnol*, 55: 661–672.

34. Kupfer, D.M., Reece, C.A., Clifton, S.W., Roe, B.A. and Prade, R.A., 1997. Multicellular ascomycetous fungal genomes contain more than 8000 genes. *Fungal Genet. Biol*, 21: 364–732.

35. Liang, P. and Pardee, A.B, 1992. Differential display of eukaryotic messenger RNA by means of the polymerase chain reaction. *Science* 257: 967–971.

36. Lilly, M., Lambrechts, M.G. and Pretorius, I.S., 2000. Effect of increased yeast alcohol acetyltransferase activity on flavor profiles of wine and distillates. *Appl. Environ. Microbiol*, 66: 744–753.

37. Michnick, S., Dequin, S., Roustan, J.L., Remize, F. and Barre, P., 1997. Modulation of glycerol and ethanol yields during alcoholic fermentation in *Saccharomyces cerevisiae* strains overexpressed or disrupted for GPD1 encoding glycerol-3-phosphate dehydrogenase. *Yeast*, 13: 783–793.

38. Mullins, E.D., Chen, X., Romaine, P., Raina, R., Geiser, D.M. and Kang, S., 2001. *Agrobacterium*-mediated transformation of *Fusarium oxysporum*: An efficient tool for insertional mutagenesis and gene transfer. *Phytopathology*, 91: 173–180.

39. Ooyen, A.J.J., Dekker, P., Huang, M., Olsthoorn M.M.A., Jacobs D.I., Colussi, P.A. and Taron, C.H., 2006. Heterologous protein production in the yeast *Kluyveromyces lactis*. *FEMS Yeast Res.*, 6: 381–392.

40. Pandey, A. and Mann, M., 2000. Proteomics to study genes and genomes. *Nature*, 405: 837–846.

41. Patrick, S.M., Fazenda, M.L., McNeil, B. and Harvey, L.M., 2005. Heterologous protein production using the *Pichia pastoris* expression system. *Yeast*, 22: 249–270.

42. Penalva, M.A., 2001. A fungal perspective on human inborn errors of metabolism: alkaptonuria and beyond. *Fungal Genet. Biol.*, 34: 1–10.

43. Piers, K.L., Heath, J.D., Liang, X., Stephens, K.M. and Nester, E.W., 1996. *Agrobacterium tumefaciens*–mediated transformation of yeast. *Proc. Natl. Acad. Sci.*, 93: 1613–1618.

44. Raguz, S., Yague, E., Wood, D.A. and Thurston, C.F., 1992. Isolation and characterization of a cellulose-growth-specific gene from *Agaricus bisporus. Gene,* 119: 183–190.

45. Robinson, A.S., Hines, V. and Wittrup, K.D., 1994. Protein disulfide isomerase overexpression increases secretion of foreign proteins in *Saccharomyces cerevisiae. Biotechnology,* 12: 381–384.

46. Rodriguez, J.M., Timm, D.E., Titus, G.P., Beltran-Valero de Bernabe., D., Criado, O., Mueller, H.A., Rodriguez de Cordoba, S. and Penalva, M.A., 2000. Structural and functional analysis of mutations in alkaptonuria. *Hum. Mol. Genet.,* 9: 2341–2350.

47. Romanos, M.A., Scorer, C.A. and Clare, J.J., 1992. Foreign gene expression in yeast: A review. *Yeast,* 8: 423–488.

48. Saloheimo, A., Aro, N., Ilmen, M. and Penttila, M., 2000. Isolation of the ace1 gene encoding a Cys(2)-His(2) transcription factor involved in regulation of activity of the cellulase promoter cbh1 of *Trichoderma reesei. J. Biol. Chem.,* 275: 5817–5825.

49. Samson, S.M., Belagaje, R., Blankeship, D.T., Chapman, J.L., Perry, D., Skatrud, P.L., Vankfrank, R.M., Abraham, E.P., Baldwin, J.E., Queener, S.W. and Ingolia, T.D., 1985. Isolation, sequence determination and expression in *Escherichia coli* of the isopenicillin N synthase gene from *Cephalosporium acremonium. Nature,* 318: 191–194.

50. Samson, S.M., Dotzlaf, J.E., Slisz, M.L., Becker, G.W., Van Frank, R.M., Veal, L.E., Yeh, W.K., Miller, J.R., Queener, S.W. and Ingolia, T.D., 1987. Cloning and expression of the fungal expandase/hydrolase gene involved in cephalosporin biosynthesis. *Biotechnology,* 5: 1207–1216.

51. Samuelson, A.I., Martin, R.C., Mok, D.W.S. and Machteld, C.M., 1998. Expression of the yeast FRE genes in transgenic tobacco. *Plant Physiol.,* 118: 51–58.

52. Sinclair, D., Mills, K. and Guarente, L., 1997. Accelerated aging and nucleolar fragmentation in yeast sgs1 mutants. *Science,* 277: 1313–1316.

53. Sinclair, D., Mills, K. and Guarente, L., 1998. Aging in *Saccharomyces cerevisiae. Ann. Rev. Microbiol.,* 52: 533–560.

54. Steenkamp, E.T., Wingfield, B.D., Coutinho, T.A., Zeller Steenkamp, E.T., Wingfield, B.D., Coutinho, T.A., Zeller, K.A., Wingfield, M.J., Marasas, W.F.O. and Leslie, J.F., 2000. PCR-based identification of MAT-1 and MAT-2 in the *Gibberella fujikuroi* species complex. *Appl. Environ. Microbiol.,* 66: 4378–4382.

55. Stoop, J.M.H. and Mooibroek, H., 1999. Advances in genetic analysis and biotechnology of the cultivated button mushroom, *Agaricus bisporus. Appl. Microbiol. Biotechnol.,* 52: 474–483.

56. Sweigard, J.A., Carroll, A.M., Farrall, L., Chumley, F.G. and Valent, B., 1998. *Magnaporthe grisea* pathogenicity genes obtained through insertional mutagenesis. *Mol. Plant Microbe. Interact.,* 11: 404–412.

57. Turgeon, B.G., 1998. Application of mating type gene terminology to problems in fungal biotechnology. *Ann. Rev. Phytopathol.*, 36: 115–137.

58. Van der Ackerveken, G.F.J.M., Van Kan, J.A.L. and De Wit, P.J.G.M., 1992. Molecular analysis of the avirulence gene avr9 of the fungal tomato pathogen *Cladopsorium fulvum* fully supports the gene-for-gene hypothesis. *Plant J.*, 2: 359–366.

59. Van Kan, J.A.L., Van der Ackerveken, G.F.J.M. and De Wit, P.J.G.M., 1991. Cloning and characterization of cDNA or avirulence gene avr9 of the fungal tomato pathogen *Cladosporium fulvum*, causal agent of tomato leaf mold. *Mol. Plant Microbe Interact.*, 4: 52–59.

60. Velasco, J., Gutierrez, S., Casquueiro, J., Fierro, F., Campoy, S. and Martýn, J.F., 2001. Cloning and characterization of the gene cahB encoding a cephalosporin C acetylhydrolase from *Acremonium chrysogenum*. *Appl. Microbiol. Biotechnol.*, 57: 350–356.

61. Verstrepen, K., Bauer, F.F., Michiels, G., Derdelinckx, F., Delvaux, F. and Pretorius, I.S., 2000. Controlled expression of FLO1 in *Saccharomyces cerevisiae*. *Eur. Brew. Conv. Monogr.*, 28: 30–42.

62. Vos, P., Hogers, R., Bleeker, M., Reijans, M., van de Lee, T., Hornes, M., Frijters, A., Pot, J., Peleman, J. and Kuiper, M., 1995. AFLP: A new technique for DNA fingerprinting. *Nucleic Acids Res.*, 23: 4407–4414.

63. Walser, P.J., Hollenstein, M., Klaus, M.J. and Kues, U., 2001. Genetic analysis of basidiomycete fungi. In: *Molecular and Cellular Biology of Filamentous Fungi*, (Ed.) N.J. Talbot. Oxford University Press, Oxford, pp. 59–90.

64. Watanabe, M., Sumida, N., Murakami, S., Anzai, H., Thompson, C.J., Tateno, Y. and Murakami, T., 1999. A phosphonate-induced gene which promotes *Penicillium*-mediated bioconversion of cis-propenylphosphonic acid to fosfomycin. *Appl. Environ. Microbiol.*, 65: 1036–1044.

65. Wu, G., Shrott, B.J., Lawrence, E.B., Levine, E.B., Fitzsimmons, K.C. and Shah, D.M., 1995. Disease resistance conferred by expression of a gene encoding H_2O_2-generating glucose oxidase in transgenic potato plants. *Plant Cell*, 7: 1357–1368.

66. Yague, E., Mahk-Zunic, M., Morgan, L., Wood, D.A. and Thurston, C.F., 1997. Expression of CEL2 and CEL4, two proteins from *Agaricus bisporus* codes with similarity to fungal cellobiohydrolase I and β-mannanase, respectively, is regulated by the carbon source. *Microbiology*, 143: 239–244.

Biotechnology: An Overview (2015)   Pages 203–222
Editors: Rajan Kumar Gupta, Nasim Akhtar and Deepak Vyas
Published by: DAYA PUBLISHING HOUSE, NEW DELHI

Chapter 15

An Overview on the Transformation and Gene Manipulation in Filamentous Fungi

Madhuri Kaushish Lily*

Department of Biotechnology, Modern Institute of Technology (MIT), Dhalwala, Rishikesh – 249 201, Uttarakhand

ABSTRACT

Noteworthy development has been made in transforming a vast variety of different filamentous fungi. This accomplishment has principally been due to the development of novel techniques such as the particle bombardment and *Agrobacterium*-mediated transformation, which has either greatly improved low transformation frequencies or made previous "non-transformable" fungi now accessible to gene transfer. However, the fate of the transforming DNA in the respective organisms is still not fully understood and recombination and re-arrangements are often observed. Therefore, the applications of such transformation systems are still limited and the interpretations of the results observed are complicated. Although a reasonable number of diverse markers are employed for transformation, however, with a few exceptions, no tools such as targeted integration systems, artificial chromosomes, or centromere vectors are yet available.

Keywords: Filamentous fungi, Gene, Genetic manipulation, Marker, Transformation.

Introduction

The development of suitable gene transfer systems is a major prerequisite for molecular genetic and biochemical investigations in various organisms, including the filamentous fungi. Molecular transformation involves two main components: (a) an appropriate vector containing a selectable marker and (b) a suitable transformation procedure for the introduction of the resultant vector into the respective fungal system. Thereafter, a sustainable replication of the transforming DNA has to be accomplished

---

* *Author:* E-mail: mklily18@gmail.com

either by integration of the vector into the genome or by applying an autonomously replicating vector system.

Over the year, substantial progress has been made in developing transformation techniques for filamentous fungi. Most of the research has focused on unicellular yeast *Saccharomyces cerevisiae* and the two main model organisms for filamentous fungi, *Neurospora crassa* and *Aspergillus nidulans*. However, there are many reports on the successful transformation of a number of other filamentous fungi and expression systems for specific groups of fungi (Ballance, 1991; Bodie *et al.*, 1994; Fincham, 1989; Frommer and Ninnemann, 1995; Goosen *et al.*, 1992; Mach *et al.*, 1994; May, 1992; Radzio and Kuck, 1997; Riach and Kinghorn, 1996; Ruiz-Diez, 2002; Timberlake and Marshall, 1989; Turner, 1994; van den Hondel and Punt, 1991; Verdoes *et al.*, 1995). The advances in the transformation systems for the two filamentous model organisms (*N. crassa* and *A. nidulans*) has offered a foundation for transferring such techniques to less investigated but economically and industrially important fungal species (Timberlake and Marshall, 1989). Mishra and Tatum (1973) reported the first fungal transformation describing the conversion of auxotrophic *N. crassa* mutants (inositol requiring) to prototrophy at very low rates by exposing them to total wild type DNA and calcium chloride. Later, Hinnen *et al.* (1978) achieved considerably improved competence of DNA uptake by removing cell walls, thereby generating protoplasts, and they became the first researchers to transform a *S. cerevisiae* leu2- mutant. The protoplast technique, which was already proven to be suitable for yeast transformation, was applied to *N. crassa* (Case, 1986; Case *et al.*, 1979) and *A. nidulans* (Tilburne *et al.*, 1983). These major breakthroughs in fungal transformation were later exploited with a variety of other fungal species.

Transformation Methods for Filamentous Fungi

A number of techniques including protoplasting, chemical treatments (calcium-chloride/PEG method), application of electric pulses, or physical damage have been employed for introduction of DNA into the nuclei of filamentous fungi. However, protoplasts are most often used for such insertions of exogenous DNA. In addition to the generally used calcium-chloride/PEG method, a liposome- (Radford *et al.*, 1981) and an electroporation-based method for protoplasts have also been described (Goldman *et al.*, 1990; Thomas and Kenerley, 1989). Some of the other methods that use intact cells for transformation, include exposure of cells to lithium acetate and transforming DNA (Dhawale *et al.*, 1984), effectual mixing of cells in the presence of glass beads and DNA (Costanzo and Fox, 1988), particle bombardment (Klein *et al.*, 1987), and *Agrobacterium tumefaciens*-mediated transformation (Bevan, 1984).

Preparation of Protoplasts and Regeneration

Protoplasts are obtained by incubating mycelia or spores of filamentous fungi with various cell-wall degrading enzymes under osmotically stabilized conditions (Peberdy, 1985). The wall structure of the particular fungal species determines the efficiency of the cell wall degradation. Enzyme cocktails are commercially available for protoplasting including enzymes isolated from snail stomach, (*e.g.*, Helicase and Gluculase), and an enzyme concentrate from *Arthrobacter luteus* (Zymolase 100T), although the most frequently (Novozyme 234) enzyme cocktail is derived from *Trichoderma viride* (Riach and Kinghorn, 1996). All such preparations contain a complex mixture of hydrolytic enzymes, consisting mainly of 1,3 glucanases and chitinases.

During their preparation, protoplasts have to be osmotically stabilized during by sodium chloride, magnesium sulphate, mannitol, or most frequently by sucrose or sorbitol. Concentrations between 0.8 and 1.2M function as good osmotic stabilizers (May, 1992) and lead to regeneration rates of up to 90 per cent.

Polyethylene Glycol-Mediated Transformation

The polyethylene glycol (PEG)/calcium chloride method is the most frequently used transformation method for filamentous fungi such as *A. nidulans*. In this method, protoplasts at a concentration of 5×10^7–5×10^8 are employed and treated with a mixture of PEG, calcium chloride, and transforming DNA (Ballance *et al.*, 1983; Tilburne *et al.*, 1983). The DNA molecules appear to be internalized during the PEG induced protoplast fusion, whereas no transformation occurs when PEG is omitted (Timberlake and Marshall, 1989). The PEG is typically used at a molecular weight of 4000; and other alkali cations such as lithium chloride instead of calcium chloride can be used in some methods (Ito *et al.*, 1983).

Liposome-Mediated Transformation

Liposome-mediated transformation has been reported for the first time for *N. crassa* (Radford *et al.*, 1981). Owing to difficulties in the reproducibility of the preparation of liposomes and the positive results obtained from other procedures this method has not been applied further.

Electroporation

Electroporation involves the electrical permeabilisation of biomembranes. This is a simple, efficient physical transformation method for prokaryotic as well as eukaryotic cells (Fo¨rster and Neumann, 1989; Potter *et al.*, 1984; Riggs and Bates, 1986). This technique has been used in filamentous fungi, with protoplasts, conidia, germinated conidia, and hyphal fragments that were exposed to high voltage-short pulses. This method has been described for several species including *Colletotrichum, Neurospora, Aspergillus, Beauveria, Penicillium, Leptosphaeria, Fusarium, Ustilago,* and *Trichoderma* (Bakkeren and Kronstad, 1993; Chakraborty *et al.*, 1991; Goldman *et al.*, 1990; Redman and Rodriguez, 1994; Ward *et al.*, 1989). The main advantages of electroporation include simplicity of preparing sensitive cells and a extensively documented increase in transformation reproducibility.

Though protoplast-dependent transformation methods have proved successful, they do carry some major drawbacks: (a) the variability of the efficiency of protoplast formation and regeneration for different strains, (b) for several fungal strains, protoplasts have to be prepared freshly for each transformation experiment, and (c) protoplasts often contain more than one nucleus, and this can result in a long and time-consuming purification of transformants to obtain homokaryons.

Lithium Acetate-Based Transformation

Lithium acetate based transformation method offers an efficient alternative to protoplast based transformation. This method was initially described for *S. cerevisiae* (Ito *et al.*, 1983), where the exposure of fungal cells to 0.1M lithium acetate and DNA gave successful results. This method has been successfully applied to *N. crassa* (Dhawale *et al.*, 1984) and *Coprinus cinereus* (Binninger *et al.*, 1987). This is carried out by treating germinating spores with 0.1M lithium acetate resulting in an alkaline metal-based increase in permeability for the transformation. However, this technique is of limited use in the filamentous fungi transformation due to very low transformation frequencies in comparison to those obtained from protoplast-based methods (Dhawale *et al.*, 1984).

Biolistic Transformation

Costanzo and Fox 1988 reported for the first time the transformation brought about by physical damage by vigorously mixing cells with transforming DNA and glass beads (Costanzo and Fox, 1988). Later, similar methods of bombarding cells with particles functioning as carriers for transforming DNA have been applied to several fungal species including *S. cerevisiae, A. nidulans, Paecilomyces fumosoroseus, Trichoderma harzianum, T. reesei, Gliocladium virens, Mucor circinelloides, N. crassa,* and *Magnaporthe grisea* (Aramayo *et al.*, 1989; Barreto *et al.*, 1997; Gomes-Barcellos *et al.*, 1998; Hazell *et al.*, 2000; Herzog *et al.*, 1996; Klein *et al.*, 1992; Lorito *et al.*, 1993; Riach and Kinghorn, 1996). In case of filamentous fungi, particles are frequently delivered to conidia placed on agar plates.

Klein *et al.* (1987) described particle bombardment for the first time. This technique employs high-velocity microprojectiles for delivering nucleic acids into intact living cells and tissues. It involves coating the DNA on small diameter microbeads (0.1–30 mm) of tungsten, gold or glass. Lambda phage or dried *Escherichia coli* cells either coated with or containing the respective transforming DNA have also been used successfully. The bombardment is generally carried out in a vacuum chamber to minimize the effect of air on the velocity of microprojectiles. This method was initially applied to plants (Klein *et al.*, 1988a, b, c; Wang *et al.*, 1988) but later it has found its application in the transformation of other cellular systems including fungi. This technique offers several advantages: 1) increased transformation frequency and transformant stability. 2) simplicity of the method permits more number of transforming experiments to be undertaken in a short time, since both steps (plating of conidia and coating of microprojectiles with DNA) are uncomplicated and easy to standardize 3) this technique is the only one described so far for transformation of mitochondria and chloroplasts (Fox *et al.*, 1988; Johnston *et al.*, 1988; Ye *et al.*, 1990).

Agrobacterium tumefaciens-Mediated Transformation

This technique was originally developed for plants, and it involves the utilization of *A. tumefaciens* to deliver plasmid DNA into fungal cells (Bundock and Hooykaas, 1996; Bundock *et al.*, 1995; de Groot *et al.*, 1998). *Agrobacterium*, Gram negative soil bacteria possess large plasmids (Ti-plasmids) and cause gall tumours at wound sites in infected dicots monocots (Frommer and Ninnemann, 1995; Hooykaas and

Beijersbergen, 1994; Linnemannstons *et al.*, 1999; Van Veen *et al.*, 1988; Zambryski, 1992). The formation of galls is due to the transfer of the T-DNA (part of the Ti-plasmid flanked by 24bp imperfect direct repeats) into the plant nuclear genome. The transfer of T-DNA is dependent on the induction of virulence (vir) genes that are also located on the Ti-plasmid. This whole process is activated in response to induction by the products secreted from the wounded plant cells, such as acetosyringone (AS), (Winans, 1992). An efficient plant transformation based on a Ti-plasmid system involves two plasmids (binary vector) functioning simultaneously for transferring DNA to the respective host. One plasmid (Bin19) contains the T-DNA, the left and the right border regions of the Ti-plasmid, and a kanamycin resistance gene. The other plasmid bears all of the vir genes needed for T-DNA transfer. After transforming both vectors into *A. tumefaciens*, T-DNA is subsequently transferred to the plant by the help of the vir gene products from the Ti-plasmid (Bevan, 1984).

A similar system has been applied to filamentous fungi using a hygromycin B-resistance expression cassette as T-DNA and a selectable marker to detect transformation events. Induction of the concerted transfer action of the binary vector system was achieved by inducing the vir genes by supplementing the media with AS (de Groot *et al.*, 1998). This technique has resulted in successful transformation fungal protoplasts as well as conidia and hyphal tissue of *A. niger, A. awamori, Colletotrichum gloeosporioides, Fusarium venenatum, T. reesei, N. crassa*, and the mushroom *Agaricus bisporus* at varying transformation frequencies. This method offers advantages 1) ease of handling, 2) possibility to transfer high molecular weight DNA into fungal chromosomes since *A. tumefaciens* can insert at least 150kb of foreign DNA to plant cells.

Transforming DNA

Transformation Vectors

Numerous diverse types of vectors, such as basic plasmids, plasmids bearing cos sequences (cosmids), yeast artificial chromosome (YAC) vectors and bacterial artificial chromosome (BAC) vectors have been employed for transferring up to several 100 kb of foreign DNA into a selected fungus (Burke *et al.*, 1987; Diaz-Perez *et al.*, 1996; Gold *et al.*, 2001).

Autonomously Replicating Vectors (ARV)

Vectors for fungal transformation are generally divided into two categories: (a) autonomously replicating vectors (ARV) and (b) integration vectors replicating within the fungal genome. In ARV, introduced sequences are recognized by the fungus and are subsequently replicated without integration into the genome. Besides the autonomously replicating sequence (ARS) from *S. cerevisiae*, a number of such elements have been described that impart autonomous replication to transformation vectors used for filamentous fungi, *e.g.*, ANS1 and AMA1 of *Aspergillus* (Aleksenko and Clutterbuck, 1995; Aleksenko *et al.*, 1996; Gems *et al.*, 1991), and telomere sequences of *Fusarium, Colletotrichum*, and *Cryptococcus*; (Edman, 1992; Garcia-Pedrajas and Roncero, 1996; Kistler and Benny, 1992; Redman and Rodriguez, 1994).

Vector Composition

In order to mediate higher integration frequencies, both functional and nonfunctional sequences are included into a vector system. In addition to the use of homologous genomic DNA coding for selectable markers or regulatory sequences (*e.g.*, promoters and terminators), the addition of non-selectable homologous sequences has been revealed to increase transformation efficiency in some filamentous fungi (Herrera-Estrella *et al.*, 1990).

Selectable Markers

The selection of transformants within a background of non-transformed cells is permitted by the gain of dominant selectable phenotypes (Timberlake and Marshall, 1989). Three types of selectable markers are used for this purpose: (a) genes coding for suppressor tRNA, (b) auxotrophic markers, and (c) dominant selection markers. A special type of selectable marker is the suppression of an auxotrophic mutation by tRNA suppressor genes. This was mutation by tRNA suppressor genes. Brygoo and Debuchy (1985) demonstrated such a mutation in a *Podospora anserina* strain carrying the nonsense mutation leu12, and that was suppressed to leucine prototrophy by the tRNA suppressor genes su4-1 and su8-1 (Brygoo and Debuchy, 1985).

The first transformations were carried out by converting auxotrophic mutants to prototrophy (Fincham, 1989). These transformations require the availability of mutants. Common mutants are those deficient in the metabolism of amino acids (leucine, arginine, and tryptophan), uracil, nitrogen, or in the catabolism of quinic acid. Vectors used for complementation contain the respective genes leu, arg, and trp, or ura and pyr, or niaD, or qa2 (Alani *et al.*, 1987; Baek and Kenerley, 1998; Ballance *et al.*, 1983; Banks and Taylor, 1988; Berges and Barreau, 1991; Case *et al.*, 1979; Gruber *et al.*, 1990; Penttila *et al.*, 1987; Smith *et al.*, 1991; Unkles *et al.*, 1989; Van Hartingsveldt *et al.*, 1987; Wada *et al.*, 1996; Weidner *et al.*, 1998; Yelton *et al.*, 1984).

Genes conferring resistance or a new metabolic activity have been widely used as dominant selection markers with several different species of filamentous fungi. The hph (hygromycin B phosphotransferase encoding) gene from *E. coli* is used as a marker in transformation by detoxifying hygromycin B, an amino glycoside antibiotic from *Streptomyces hygrosporicus* (Mach *et al.*, 1994; Punt *et al.*, 1987). G 418 resistance using a bacterial G 418 R can also be conferred to several fungi (Bull and Wootton, 1984; Jimenez and Davies, 1980; Penalva *et al.*, 1985; Revuelta and Jayaram, 1986). In most cases, these resistance genes have been fused to a variety of fungal promoter and terminator sequences to ensure expression. A dominant heterologous expression marker useful with several filamentous fungi is the *A. nidulans* amdS gene encoding acetamidase that allows growth on acetamide or acrylamide as sole carbon and nitrogen source. Reports of successfully transforming various fungal species have been published, *e.g.*, *A. niger* (Kelly and Hynes, 1985), *Penicillium chrysogenum* (Beri and Turner, 1987), *T. reesei* (Penttila *et al.*, 1987), *Cochliobolus heterostrophus* (Turgeon et al., 1987), *Glomerella cingulata* (Rodriguez and Yoder, 1987), and *T. harzianum* (Pe'er *et al.*, 1991). Resistance to phosphinothricin (PPT)–a potent inhibitor of glutamine synthetase (through bar/pat gene) has been developed as a marker for transforming *Cercospora kikuchii* (Upchurch *et al.*, 1994), Phanerochaete *chrysosporium* (Ma *et al.*,

2003), *Fusarium vaenenatum* (Yoder and Lehmbeck, 2004), *Aspergillus nidulans* (Nayak *et al.*, 2006), *A. niger* (Ahuja and Punekar, 2008). Recently, genetic transformation of *Beauveria bassiana*, a filamentous ascomycete to sulfonylurea resistance using the *Magnaporthe grisea* acetolactate synthase gene (sur), sulfonyl resistance marker has been successfully achieved (Zang *et al.*, 2010). Molecular analysis of the *B. bassiana* transformants revealed that the gene cassette was successfully integrated in the fungal genome.

Markers for Two-Way Selection

Two-way selection systems are a helpful tool to achieve appropriate mutants in addition to the reuse of the same selection system in several subsequent steps of genetic manipulation. Representative examples are uridine-negative mutants selected via loss of orotidine-50-monophosphate carboxylase activity (encoded by ura3 or pyr4 and required for uridine biosynthesis) which confers resistance to the inhibitory analogon 5-fluoro-orotic-acid (Alani *et al.*, 1987; Berges and Barreau, 1991; Gruber *et al.*, 1990; Smith *et al.*, 1991). The nitrate reductase gene (*e.g.*, niaD of *A. nidulans*) is also a striking selection marker for developing a gene transfer system in genetically poorly characterized species. In this case, loss of niaD function can be selected for by the gain of resistance to chlorate (Unkles *et al.*, 1989).

Cotransformation

Cotransformation is a technique in which recipient cells are treated with two different kinds of DNA. This offers an alternative to overcome the problem that in most cases transforming genes cannot be directly selected for. To achieve this, transformation is undertaken together with a more readily selectable marker, as it is known that the probability of taking up both transforming DNAs is quite high (Fincham, 1989). Although cotransformation frequencies range from 10 to 100 per cent for different fungi, the majority of fungi show rates higher than 50 per cent (Cooley and Caten, 1989; Harkki *et al.*, 1991; Kubicek-Pranz *et al.*, 1991; Nicolaisen and Geisen, 1996; Penttila *et al.*, 1987; Punt *et al.*, 1987). Cotransformation in *Phytophtora infestans* occurred in only 10 per cent of transformants when circular DNA was used but reached nearly 100 per cent when linearized DNA was used (Judelson, 1993).

Purification and Characterization of Transformants

Integration of DNA into the Genome

With the exception of the few examples of autonomously replicating systems, all commonly used vectors integrate into the genome of the respective transformed filamentous fungus. Upon chromosomal DNA analysis of the transformants, in broadly, integration events can be divided into three types (a) vector integration by homologous recombination, (b) ectopic integration by nonhomologous recombination, and (c) gene replacement.

The incidence of any particular integration type is greatly variable and depends upon the species to be transformed, the transformation method, and most likely on the conformation of the transforming DNA whether circular or linear. Homologous integration events are greatly variable for different species, *e.g.*, *A. nidulans* about 80 per cent (Yelton *et al.*, 1984), *N. crassa* 1–5 per cent (Case, 1986), or *Hypocrea jecorina* about 2 per cent (Mach *et al.*, 1995; Seiboth *et al.*, 1992).

In ectopic integration, small stretches of target-vector DNA identity provides integration sites for initiating recombination event. The occurrence of such sites depends upon the genome sizes and results in random integration (Timberlake, 1991). However, ectopic plasmid integration has clear site preferences into the genome of *A. nidulans* (Diallinas and Scazzocchio, 1989).

In gene replacement, the target gene seems to be replaced by the introduced copy without the additional integration into the genome of any part of the vector sequence. Gene replacement has been shown to be preferred on employing linear molecules in many cases (Aramayo *et al.*, 1989; Boylan *et al.*, 1987; Fowler and Brown, 1992; Mach *et al.*, 1995; Seiboth *et al.*, 1992).

Purification of Mitotically Stable Transformants

Since some filamentous fungi carry multinucleated cells therefore, protoplasts derived from such cells carry more than one nucleus. Furthermore, several fungal species have multinucleated conidia. Transformation of such material will consequently lead to heterokaryons since all nuclei of one cell compartment will not take up the DNA. Therefore, in order to obtain homokaryotic recombinants, a number of steps of genetic purification needs to be followed after transformation. In fungal species with uninucleated conidia, (*e.g. Aspergillus* spp.), purification can be obtained by plating conidia and isolating single colonies. In polynucleated systems, purification can be achieved by at least three plating rounds to give a high probability of a stochastic loss of one or other nucleus (Fincham, 1989). If selective conidiation steps are interchanged with non selective conidiation steps, non stable transformants will be lost, resulting in homokaryotic mitotically stable strains.

Application in Molecular Biology and Biotechnology

Transformation systems are indispensable tools for research based on molecular biology and for strain improvement for biotechnology. Some of the applications are:

Cloning by Complementation

Cloning by complementation requires respective mutants and the construction of genomic or cDNA libraries. Furthermore, the success of such strategies depends upon the transformation frequency and ease of selection of the complemented phenotype. This strategy has found its relevance in the isolation of genes involved in cell metabolism, *viz.* genes involved in amino acids and pyrimidines biosynthesis, in nitrogen assimilation, and fungicide resistance. Complementation techniques have been developed for many genes (*e.g.*, non-nutritional), for which no direct selection is possible or it is too time consuming and labor intensive. Furthermore, cloning of regulatory genes involved in gene expression is frequently vulnerable by cross feeding in traditional selection methods. A cloning system has been developed in *A. niger* based on the bi-directional selection marker pyrA, encoding orotidine-50-phosphate decarboxylase which catalyses a step in pyrimidine nucleotide synthesis (van Peij *et al.*, 1998). The pyrA is kept under the control of an in *cis* acting element of a regulated xylanolytic gene. Introduction of the selection cassette into *A. niger* permitted a pathway-specific pyrA expression and bi-directional selection, *i.e.*, uridine prototrophy for a PYR 1 phenotype or 5-fluoro-orotic acid resistance for a PYR2 phenotype. Furthermore, this strategy could be employed for the selection of mutants for positively and negatively pathway specific in trans acting factors.

Restriction Enzyme Mediated Integration (REMI)

The most common substitute to complementation cloning is gene disruption. The first attempts were carried out with occasional ectopic integrations of homologous vector systems, leading to gene disruptions in the mutants generated (Diallinas and Scazzocchio, 1989). Transposons have also been exploited to generate insertion mutations with immense success in several bacterial systems however these have seldom been used in eukaryotic systems. Transposons are not found in all fungi and particular laboratory and industrial strains often lack such elements. The so-called REMI mutagenesis was first developed in *S. cerevisiae*, an organism in which homologous integration is extremely frequent (Schiestl and Petes, 1991). Transforming *S. cerevisiae* with non-homologous DNA linearized by digestion with a restriction enzyme and then including this enzyme into the transformation mixture encouraged the integration of the transforming DNA into the respective restriction sites in the genome. The key mechanism taking place during REMI is considered to be a form of non-homologous end-joining (Thode *et al.*, 1990).

REMI has first been used in *Dictyostelium discoideum* (Kuspa and Loomis, 1992), afterwards, it has also been applied to a variety of Asco- and Basidiomycetes, including *C. heterostrophus* (Lu *et al.*, 1994); *Ustilago maydis* (Bolker *et al.*, 1995); *Magnaporthe grisea* (Liu *et al.*, 1998; Sweigard *et al.*, 1998); *Coprinus cinereus* (Granado *et al.*, 1997); *Candida albicans* (Brown *et al.*, 1996; Riggle *et al.*, 1997); *A. fumigatus* (Brown *et al.*, 1998); *A. nidulans* (Sanchez *et al.*, 1998); *Lentinus edodes* (Sato *et al.*, 1998); *A. niger* (Shuster and Bindel Connelley, 1999); *Alternaria alternata* (Tanaka *et al.*, 1999); *Giberella fujikuroi* (Linnemannstons *et al.*, 1999); *A. oryzae* (Yaver *et al.*, 2000); *C. graminicola* (Thon *et al.*, 2000); *Hansenula polymorpha* (van Dijk *et al.*, 2001) and *F. oxysporum* (Namiki *et al.*, 2001).

REMI offers numerous advantages as a method for genetic analysis: (a) physically tagged random insertion mutations, (b) enhancement of transformation frequency by several folds, and (c) establishment of single, stable, and un-rearranged genomic insertions. REMI has also been employed in RFLP mapping (Kuspa and Loomis, 1994; van Dijk *et al.*, 2001), promoter trapping (Chang *et al.*, 1995), promoter tagged REMI (Shuster and Bindel Connelley, 1999), and dominant genetics (Brown *et al.*, 1996). However, REMI transformation protocols are developed empirically and the form of transforming DNA (linear or circular) and the restriction enzyme employed varies from species to species. Stimulation of transformation (from non to 20-fold) as well as type of integration also varies extensively between different organisms.

Use of Reporter Genes for Promoter Analysis

Reporter genes have widely been used to characterize regulatory DNA sequences from bacteria to higher eukaryotes. The chief advantage of such systems is the simple identification of recombinant strains bearing such reporter systems due to detection of the activity gained from the reporter gene insertion. The most prominent reporter genes, the β-galactosidase (lacZ), the β-glucuronidase (GUS), the β-lactamase (bla), the glucose oxidase (goxA) and the green fluorescent protein (gfp). These reporter genes could be employed in systems where the respective transformants produce pigments that permit the detection and quantification of the inserted genes. In comparison to other systems that involve cultivation media supplemented with substrates for the detection of the expressed enzymes, no such substrates are necessary for the autofluorescing protein gfp and transformants could be visualized by epifluorescence (Prasher *et al.*, 1992). Gfp has been used extensively in a number of studies including the investigation of nuclear traffic in fungal hyphae in *A. nidulans* (Suelmann *et al.*, 1997), identification of a motor protein in *U. maydis* (Lehmler *et al.*, 1997), regulation of biocontrol specific genes in *T. atroviridae* during host interaction (Zeilinger *et al.*, 1999), compartmentation of the glucose repressor Cre1 in *Sclerotinia sclerotiorum* (Vautard *et al.*, 1999), protein secretion in *A. niger* (Gordon *et al.*, 2000), subcellular localization of the purine transporter in *A. nidulans* (Valdez-Taubas *et al.*, 2000), identification of components involved in polarized growth in *Ashbya gossypii* (Alberti- Segui *et al.*, 2001), and subcellular localization of homocitrate synthase in *P. chrysogenum* (Banuelos *et al.*, 2002) and protein distribution in *Pyrenophora triticirepentis and Verticillium dahliae* (Andrie *et al.*, 2005). The broad range of applications of GFP is due chiefly due to the fact that this system is non-toxic and non-invasive to the cell and can be studied in vivo. In the prospect, the use of fluorescent proteins would possibly rise, principally since a variety of recombinant versions that emit light at other wavelengths (*e.g.*, blue, red, yellow) are currently commercially available.

References

1. Ahuja, M. and Punekar, N.S., 2008. Phosphinothricin resistance in *Aspergillus niger* and its utility as a selectable transformation marker. *Fungal Genetics and Biology*, 45: 1103–1110.

2. Alani, E., Cao, L. and Kleckner, N., 1987. A method for gene disruption that allows repeated use of URA3 selection in the construction of multiply disrupted yeast strains. *Genetics* 116: 541–545.

3. Alberti-Segui, C., Dietrich, F., Altmann-Johl, R., Hoepfner, D., Philippsen, P., 2001. Cytoplasmic dynein is required to oppose the force that moves nuclei towards the hyphal tip in the filamentous ascomycete *Ashbya gossypii. J. Cell Sci.*, 114: 975–986.

4. Aleksenko, A.Y. and Clutterbuck, A.J., 1995. Recombinational stability of replicating plasmids in *Aspergillus nidulans* during transformation, vegetative growth and sexual reproduction. *Curr. Genet.*, 28: 87–93.

5. Aleksenko, A., Gems, D. and Clutterbuck, J., 1996. Multiple copies of MATE elements support autonomous plasmid replication in *Aspergillus nidulans. Mol. Microbiol.*, 20: 427–434.

6. Andrie, R.M., Martinez, J.P. and Ciuffetti, L.M., 2005. Development of ToxA and ToxB promoter-driven fluorescent protein expression vectors for use in filamentous ascomycetes. *Mycologia*, 97(5): 1152–1161.

7. Aramayo, R., Adams, T.H. and Timberlake, W.E., 1989. A large cluster of highly expressed genes is dispensable for growth and development in *Aspergillus nidulans*. *Genetics*, 122: 65–71

8. Baek, J.M. and Kenerley, C.M., 1998. The arg2 gene of *Trichoderma virens*: cloning and development of a homologous transformation system. *Fungal Genet. Biol.*, 23: 34–44.

9. Bakkeren, G. and Kronstad JW., 1993. Conservation of the b mating type gene complex among bipolar and tetrapolar smut fungi. *Plant Cell*, 5: 123–136.

10. Ballance, D.J., 1991. Transformation systems for filamentous fungi. In: *Molecular Industrial Mycology*, (Ed.) S.A. Leong. Marcel Dekker, New York, pp. 1–29.

11. Ballance, D.J., Buxton, F.P. and Turner, G., 1983. Transformation of *Aspergillus nidulans* by the orotidine-50-phosphate decarboxylase gene of *Neurospora crassa*. *Biochem. Biophys. Res. Commun.*, 112: 284–289.

12. Banks, G.R. and Taylor, S.Y., 1988. Cloning of the PYR3 gene of *Ustilago maydis* and its use in DNA transformation. *Mol. Cell Biol.*, 8: 5417–5424.

13. Banuelos, O., Casqueiro, J., Steidl, S., Gutierrez, S., Brakhage, A. and Martin, J.F., 2002. Subcellular localization of the homocitrate synthase in *Penicillium chrysogenum*. *Mol. Genet. Genomics*, 266: 711–719.

14. Barreto, C.C., Alves, L.C., Aragao, F.J., Rech, E., Schrank, A., Vainstein, M.H., 1997. High frequency gene transfer by microprojectile bombardment of intact conidia fromthe entomopathogenic fungus *Paecilomyces fumosoroseus*. *FEMS Microbiol. Lett.*, 156: 95–99.

15. Berges, T. and Barreau, C., 1991. Isolation of uridine auxotrophs from *Trichoderma reesei* and efficient transformation with the cloned ura3 and ura5 genes. *Curr. Genet.*, 19: 359–365.

16. Beri, R.K. and Turner, G., 1987. Transformation of *Penicillium chrysogenum* using the *Aspergillus nidulans* amdS gene as a dominant selective marker. *Curr. Genet.*, 11: 639–641.

17. Bevan, M., 1984. Binary *Agrobacterium* vectors for plant transformation. *Nucleic Acids Res.*, 12: 8711–8721.

18. Binninger, D.M., Skrzynia, C., Pukkila, P.J and Casselton, LA., 1987. DNA–mediated transformation of the basidiomycete *Coprinus cinereus*. *Embo. J.*, 6: 835–840.

19. Bodie, E.A., Bower, B., Berka, R.M. and Dunn-Coleman, N.S., 1994. *Apergillus: 50 Years*, Vol. 29. Elsevier Science, New York.

20. Bolker, M., Bohnert, H.U., Braun, K.H., Gorl, J. and Kahmann, R., 1995. Tagging pathogenicity genes in *Ustilago maydis* by restriction enzyme-mediated integration (REMI). *Mol. Gen. Genet.*, 248: 547–552.

21. Boylan, M.T., Mirabito, P.M., Willett, C.E., Zimmerman, C.R. and Timberlake, W.E., 1987. Isolation and physical characterization of three essential conidiation genes from *Aspergillus nidulans*. *Mol. Cell. Biol.*, 7: 3113–3118.

22. Brown, D.H., Slobodkin, I.V. and Kumamoto, C.A.., 1996. Stable transformation and regulated expression of an inducible reporter construct in *Candida albicans* using restriction enzyme-mediated integration. *Mol. Gen. Genet.*, 251: 75–80.

23. Brown, J.S., Aufauvre-Brown, A. and Holden, D.W., 1998. Insertional mutagenesis of *Aspergillus fumigatus*. Mol. Gen. Genet., 259: 327–335.

24. Brygoo, Y. and Debuchy, R., 1985. Transformation by integration in *Podospora anseria*: Methodology and phenomenology. *Mol. Gen. Genet.*, 200: 128–131.

25. Bull, J.H. and Wootton, J.C., 1984. Heavily methylated amplified DNA in transformants of *Neurospora crassa. Nature,* 310: 701–704.

26. Bundock, P. and Hooykaas, P.J., 1996. Integration of *Agrobacterium tumefaciens* T–DNA in the *Saccharomyces cerevisiae* genome by illegitimate recombination. *Proc. Natl. Acad. Sci., USA,* 93: 15272–15275.

27. Bundock, P., den Dulk-Ras, A., Beijersbergen, A. and Hooykaas, P.J., 1995. Trans-kingdom T-DNA transfer from *Agrobacterium tume*faciens to *Saccharomyces cerevisiae. Embo. J.,* 14: 3206–3214.

28. Burke, D.T., Carle, G.F. and Olson, M.V., 1987. Cloning of large segments of exogenous DNA into yeast by means of artificial chromosome vectors. *Science,* 236: 806–812.

29. Case, M.E., 1986. Genetical and molecular analyses of qa-2 transformants in *Neurospora crassa. Genetics,* 113: 569–587.

30. Case, M.E., Schweizer, M., Kushner, S.R. and Giles, N.H., 1979. Efficient transformation of *Neurospora crassa* by utilizing hybrid plasmid DNA. *Proc. Natl. Acad. Sci., USA,* 76: 5259–5263.

31. Chakraborty, B.N., Patterson, N.A. and Kapoor, M., 1991. An electroporation-based system for high-efficiency transformation of germinated conidia of filamentous fungi. *Can. J. Microbiol.,* 37: 858–863.

32. Chang, W.T., Gross, J.D. and Newell, P.C., 1995. Trapping developmental promoters in *Dictyostelium. Plasmid,* 34: 175–183.

33. Cooley, R.N. and Caten, C.E., 1989. Cloning and characterization of the b–tubulin gene and determination of benomyl resistance in *Septoria nodorum*. In: *EMBO Alko Workshop on Molecular Biology of Filamentous Fungi*, Vol. 6, (Eds.) H. Nevalainen and M. Penttila. Helsinki: Foundation for Biotechnical and Industrial Fermentation Research, pp. 207–216.

34. Costanzo, M.C. and Fox, T.D., 1988. Transformation of yeast by agitation with glass beads. *Genetics,* 120: 667–670.

35. de Groot, M.J., Bundock, P., Hooykaas, P.J. and Beijersbergen, A.G., 1998. *Agrobacterium tumefaciens*-mediated transformation of filamentous fungi. *Nat. Biotechnol.,* 16: 839–842.

36. Dhawale, S.S., Paitta, J.V. and Marzluf, G.A.., 1984. A new rapid and efficient transformation procedure for *Neurospora. Curr. Genet.,* p. 77–79.

37. Diallinas, J. and Scazzocchio, C., 1989. A gene coding for uric acid–xantin permease of *Aspergillus nidulans*: Inactivational cloning, characterization and sequence of a cis-acting mutation. *Genetics*, 122: 341–350.

38. Diaz-Perez, S.V., Crouch, V.W. and Orbach, M.J., 1996. Construction and characterization of a *Magnaporthe grisea* bacterial artificial chromosome library. *Fungal Genet. Biol.*, 20: 280–288.

39. Edman, J.C., 1992. Isolation of telomere-like sequences from *Cryptococcus neoformans* and their use in high-efficiency transformation. *Mol. Cell. Biol.*, 12: 2777–2783.

40. Fincham, J.R., 1989. Transformation in fungi. *Microbiol. Rev.*, 53: 148–170.

41. Forster, W. and Neumann, E., 1989. Gene transfer by electroporation. A practical guide. In: *Electroporation and Electrofusion in Cell Biology*, (Eds.) E. Neumann, A.E. Sowers and C.A. Jordan. Plenum Press, New York, pp. 299–318.

42. Fowler, T. and Brown, R.D., 1992. The bgl1 gene encoding extracellular β-glucosidase from *Trichoderma reesei* is required for rapid induction of the cellulase complex. *Mol. Microbiol.*, 6: 3225–3235.

43. Fox, T.D., Sanford, J.C. and McMullin, T.W., 1988. Plasmids can stably transform yeast mitochondria lacking endogenous mtDNA. *Proc. Natl. Acad. Sci., USA*, 85: 7288–7292.

44. Frommer, W.B. and Ninnemann, O., 1995. Heterologous expression of genes in bacterial, fungal, animal, and plant cells. *Ann. Rev. Plant Physiol. Plant Mol. Biol.*, 46: 419–444.

45. Garcia-Pedrajas, M.D. and Roncero, M.I., 1996. A homologous and self-replicating system for efficient transformation of *Fusarium oxysporum*. *Curr. Genet.*, 29: 191–198.

46. Gems, D., Johnstone, I.L. and Clutterbuck, A.J., 1991. An autonomously replicating plasmid transforms *Aspergillus nidulans* at high frequency. *Gene*, 98: 61–67.

47. Gold, S.E., Duick, J.W., Redman, R.S. and Rodriguez, R.J., 2001. Molecular transformation, gene cloning and gene expression systems for filamentous fungi. In: *Applied Mycology and Biotechnology*, Vol. 1, (Eds.) G.G. Khachatourians and D.K. Arora. Elsevier Science, Amsterdam, pp. 199–238.

48. Goldman, G.H., Van Montagu, M. and Herrera-Estrella, A., 1990. Transformation of *Trichoderma harzianum* by high-voltage electric pulse. *Curr. Genet.*, 17: 169–174.

49. Gomes-Barcellos, F., Pelegrinelli-Fungaro, M.H., Furlaneto, M.C., Lejeune, B., Pizzirani-Leiner, A.A. and Azevedo, J.L., 1998. Genetic analysis of *Aspergillus nidulans* unstable transformants obtained by biolistic process. *Can. J. Microbiol.*, 44: 1137–1141.

50. Goosen, T., Bos, C.J. and van den Broek, H.W., 1992. Transformation and gene manipulation in filamentous fungi: an overview. In: *Handbook of Applied Mycology, Fungal Biotechnology*, D.K. Arora, K.G. Mukerji and R.P. Elander, Marcel Dekker, New York, pp. 157–168.

51. Gordon, C.L., Khalaj, V., Ram, A.F., Archer, D.B., Brookman, J.L., Trinci, A.P., Jeenes, D.J., Doonan, J.H., Wells, B., Punt, P.J., van den Hondel, C.A. and Robson, G.D., 2000. Glucoamylase: Green fluorescent protein fusions to monitor protein secretion in *Aspergillus niger*. *Microbiology*, 146: 415–426.

52. Granado, J.D., Kertesz-Chaloupkova, K., Aebi, M. and Kues, U., 1997. Restriction enzyme-mediated DNA integration *in Coprinus cinereus*. *Mol. Gen. Genet.*, 256: 28–36.

53. Gruber, F., Visser, J., Kubicek, C.P. and de Graaff, L.H., 1990. The development of a heterologous transformation system for the cellulolytic fungus *Trichoderma reesei* based on a pyrG negative mutant strain. *Curr. Genet.*, 18: 71–76.

54. Harkki, A., Mantyla, A., Penttila, M., Muttilainen, S., Buhler, R., Suominen, P., Knowles, J. and Nevalainen, H., 1991. Genetic engineering of *Trichoderma* to produce strains with novel cellulase profiles. *Enzyme Microb. Technol.*, 13: 227–233.

55. Hazell, B.W., Teo, V.S., Bradner, J.R., Bergquist, P.L. and Nevalainen, H., 2000. Rapid transformation of high cellulase-producing mutant strains of *Trichoderma reesei* by microprojectile bombardment. *Lett. Appl. Microbiol.*, 30: 282–286.

56. Herrera-Estrella, A., Goldman, G.H. and Van Montagu, M., 1990. High efficiency transformation system for the biocontrol agents, *Trichoderma* spp. *Mol. Microbiol.*, 4: 839–843.

57. Herzog, R.W., Daniell, H., Singh, N.K. and Lemke, P.A., 1996. A comparative study on the transformation of *Aspergillus nidulans* by microprojectile bombardment of conidia and a more conventional procedure using protoplasts treated with polyethyleneglycol. *Appl. Microbiol. Biotechnol.*, 45: 333–337.

58. Hinnen, A., Hicks, J.B. and Fink, G.R., 1978. Transformation of yeast. *Proc. Natl. Acad. Sci., USA*, 75: 1929–1933.

59. Hooykaas, P.J.J. and Beijersbergen, A.G.M., 1994. The virulent system of *Agrobacterium* tumefaciens. *Ann. Rev. Phytopathol.*, 32: 157–179.

60. Ito, H., Fukuda, Y., Murata, K. and Kimura, A., 1983. Transformation of intact yeast cells treated with alkali cations. *J. Bacteriol.*, 153: 163–168.

61. Jimenez, A. and Davies, J., 1980. Expression of a transposable antibiotic resistance element in *Saccharomyces*. *Nature*, 287: 869–871.

62. Johnston, S.A., Anziano, P.Q., Shark, K., Sanford, J.C. and Butow, R.A., 1988. Mitochondrial transformation in yeast by bombardment with microprojectiles. *Science*, 240: 1538–1541.

63. Judelson, H.S., 1993. Intermolecular ligation mediates efficient cotransformation in *Phytophthora infestans*. *Mol. Gen. Genet.*, 239: 241–250.

64. Kelly, J.M. and Hynes, M.J., 1985. Transformation of *Aspergillus niger* by the amdS gene of *Aspergillus nidulans*. *Embo. J.*, 4: 475–479.

65. Kistler, H.C. and Benny, U., 1992. Autonomously replicating plasmids and chromosome rearrangement during transformation of *Nectria haematococca*. *Gene*, 117: 81–89.

66. Klein, T.M., Wolf, E.D., Wu, R. and Sanford, J.C., 1987. High-velocity microprojectiles for delivering nucleic acids into living cells. *Nature*, 327: 70–73.

67. Klein, T.M., Fromm, M.E., Weissinger, A., Thomes, T., Shaaf, S., Sletten, M. and Sanford, J.C., 1988a. Transfer of foreign genes into intact mice cells with high-velocity projectiles. *Proc. Natl. Acad. Sci., USA*, 85: 4305–4309.

68. Klein, T.M., Gratziel, T., Fromm, M.E. and Sanford, J.C., 1988b. Factors influencing gene delivery into Zea mays cells by high-velocity microprojectiles. *Biotechnology*, 6: 559–563.

69. Klein, T.M., Harper, E.C., Svab, Z., Sanford, J.C., Fromm, M.E. and Maligan, P., 1988c. Stable genetic transformation of intact *Nicotiana* cells by the particle bombardment process. *Proc. Natl. Acad. Sci., USA*, 85: 8502–8505.

70. Klein, T.M., Arentzen, R., Lewis, P.A. and Fitzpatrick-McElligott, S., 1992. Transformation of microbes, plants and animals by particle bombardment. *Biotechnology* (NY), 10: 286–291.

71. Kubicek-Pranz, E.M., Gruber, F. and Kubicek, C.P., 1991. Transformation of *Trichoderma reesei* with the cellobiohydrolase II gene as a means for obtaining strains with increased cellulose production and specific activity. *J. Biotechnol.*, 20: 83–94.

72. Kuspa, A. and Loomis, WF., 1992. Tagging developmental genes in *Dictyostelium* by restriction enzyme-mediated integration of plasmid DNA. *Proc. Natl. Acad. Sci., USA*, 89: 8803–8807.

73. Kuspa, A. and Loomis, W.F., 1994. REMI-RFLP mapping in the *Dictyostelium* genome. *Genetics*, 138: 665–674.

74. Lehmler, C., Steinberg, G., Snetselaar, K.M., Schliwa, M., Kahmann, R. and Bolker, M., 1997. Identification of a motor protein required for filamentous growth in *Ustilago maydis*. *Embo. J.*, 16: 3464–3473.

75. Linnemannstons, P., Voss, T., Hedden, P., Gaskin, P. and Tudzynski, B., 1999. Deletions in the gibberellin biosynthesis gene cluster of *Gibberella fujikuroi* by restriction enzyme-mediated integration and conventional transformation-mediated mutagenesis. *Appl. Environ. Microbiol.*, 65: 2558–2564.

76. Liu, S., Wei, R., Arie, T. and Yamaguchi, I., 1998. REMI mutagenesis and identification of pathogenic mutants in blast fungus (*Magnaporthe grisea*). *Chin. J. Biotechnol.*, 14: 133–139.

77. Lorito, M., Hayes, C.K., Di Pietro, A. and Harman, G.E., 1993. Biolistic transformation of *Trichoderma harzianum* and *Gliocladium virens* using plasmid and genomic DNA. *Curr. Genet.*, 24: 349–356.

78. Lu, S., Lyngholm, L., Yang, G., Bronson, C., Yoder, O.C. and Turgeon, B.G., 1994. Tagged mutations at the Tox1 locus of *Cochliobolus heterostrophus* by restriction enzyme-mediated integration. *Proc. Natl. Acad. Sci., USA*, 91: 12649–12653.

79. Ma, B., Mayfield, M.B. and Gold, M.H., 2003. Homologous expression of *Phanerochaete chrysosporium* manganese peroxidase, using bialaphos resistance as a dominant selectable marker. *Curr. Genet.*, 43: 407–414.

80. Mach, R.L., Schindler, M. and Kubicek, C.P., 1994. Transformation of *Trichoderma reesei* based on hygromycin B resistance using homologous expression signals. *Curr. Genet.*, 25: 567–570.

81. Mach, R.L., Seiboth, B., Myasnikov, A., Gonzalez, R., Strauss, J., Harkki, A.M. and Kubicek, C.P., 1995. The bgl1 gene of *Trichoderma reesei* QM 9414 encodes an extracellular, cellulose-inducible β-glucosidase involved in cellulase induction by sophorose. *Mol. Microbiol.*, 16: 687–697.

82. May, G.S., 1992. Fungal technology. In: *Applied Molecular Genetics of Filamentous Fungi*, (Eds.) J.R. Kinghorn and G. Turner. Blackie Academic and Professional, Glasgow, pp. 1–27.

83. Mishra, N.C. and Tatum, E.L., 1973. Non-mendelian inheritance of DNA-induced inositol independence in *Neurospora*. *Proc. Natl. Acad. Sci., USA*, 70(12): 3875–3879.

84. Namiki, F., Matsunaga, M., Okuda, M., Inoue, I., Nishi, K., Fujita, Y. and Tsuge, T., 2001. Mutation of an arginine biosynthesis gene causes reduced pathogenicity in *Fusarium oxysporum* sp. *melonis*. *Mol. Plant Microbe Interact.*, 14: 580–584.

85. Nayak, T., Szewczyk, E., Oakley, C.E., Osmani, A., Ukil, L., Murray, S.L., Hynes, M.J. and Osmani, S.A. and Oakley, B.R., 2006. A versatile and efficient gene-targeting system for *Aspergillus nidulans*. *Genetics*, 172: 1557–1566.

86. Nicolaisen, M. and Geisen, R., 1996. Transformation of *Penicillium freii* and a rapid PCR screening procedure for cotransformation events. *Microbiol. Res.*, 151: 281–284.

87. Peer, S., Barak, Z., Yarden, O. and Chet, I., 1991. Stability of *Trichoderma harzianum* amdS transformants in soil and rhizosphere. *Soil Biol. Biochem.*, 23: 1043–1046.

88. Peberdy, J.F., 1985. Mycolytic enzymes. In: *Fungal Protoplasts: Aplications in Biochemistry and Genetics*, (Eds.) J.F. Peberdy and L. Ferenczy. Marcel Dekker, New York, pp. 31–44.

89. Peberdy, J.F., 1989. Fungi without coats: Protopalsts as tools for mycological research. *Mycol. Res.*, 93: 1–20.

90. Penalva, M.A., Tourino, A., Patino, C., Sanchez, F., Fernamdez Sousa, J.M. and Rubio, V., 1985. Studies on transformation of *Cephalosporium acremonium*. In: *Molecular Genetics of Filamentous Fungi*, (Ed.) R. Alan. Liss, New York, pp. 59–68.

91. Penttila, M., Nevalainen, H., Ratto, M., Salminen, E. and Knowles, J., 1987. A versatile transformation system for the cellulolytic filamentous fungus *Trichoderma reesei*. *Gene*, 61: 155–164.

92. Potter, H., Weir, L. and Leder, P., 1984. Enhancer-dependent expression of human kappa immunoglobulin genes introduced into mouse pre-B lymphocytes by electroporation. *Proc. Natl. Acad. Sci., USA*, 81: 7161–7165.

93. Prasher, D.C., Eckenrode, V.K., Ward, W.W., Prendergast, F.G. and Cormier, M.J., 1992. Primary structure of the *Aequorea victoria* green-fluorescent protein. *Gene*, 111: 229–233.

94. Punt, P.J., Oliver, R.P., Dingemanse, M.A., Pouwels, P.H. and van den Hondel, C.A., 1987. Transformation of *Aspergillus* based on the hygromycin B resistance marker from *Escherichia coli*. *Gene*, 56: 117–124.

95. Radford, A., Pope, S., Sazci, A., Fraser, M.J. and Parish, J.H., 1981. Liposome-mediated genetic transformation of *Neurospora crassa*. *Mol. Gen. Genet.*, 184: 567–569.

96. Radzio, R. and Kuck, U., 1997. Efficient synthesis of the bloodcoagulation inhibitor hirudin in the filamentous fungus *Acremonium chrysogenum*. *Appl. Microbiol. Biotechnol.*, 48: 58–65.

97. Redman, R.S. and Rodriguez, R.J., 1994. Factors affecting the efficient transformation of *Colletotrichum* species. *Exp. Mycol.*, 18: 230–246.

98. Revuelta, J.L. and Jayaram, M., 1986. Transformation of *Phycomyces blakesleeanus* to G–418 resistance by an autonomously replicating plasmid. *Proc. Natl. Acad. Sci., USA*, 83: 7344–7347.

99. Riach, M.B.R. and Kinghorn, J.R., 1996. Genetic transformation and vector developments in filamentous fungi. In: *Fungal Genetics: Principles and Practice*, (Ed.) C.J. Bos. Marcel Dekker, New York, pp. 209–233.

100. Riggle, P.J., Slobodkin, I.V, Brown, D.H., Jr., Hanson, M.P., Volkert, T.L. and Kumamoto, C.A., 1997. Two transcripts, differing at their 3' ends, are produced from the *Candida albicans* SEC14 gene. *Microbiology*, 143: 3527–3535.

101. Riggs, C.D. and Bates, G.W., 1986. Stable transformation of tobacco by electroporation: Evidence for plasmid concatenation. *Proc. Natl. Acad. Sci., USA*, 83: 5602–5606.

102. Rodriguez, R.J. and Yoder, O.C., 1987. Selectable genes for transformation of the fungal plant pathogen *Glomerella cingulata* sp. phaseoli (*Colletotrichum lindemuthianum*). *Gene*, 54: 73–81.

103. Ruiz-Diez, 2002. Strategies for the transformation of filamentous fungi. *J. Appl. Microbiol.*, 92: 189–195.

104. Sanchez, O., Navarro, R.E. and Aguirre, J., 1998. Increased transformation frequency and tagging of developmental genes in *Aspergillus nidulans* by restriction enzyme-mediated integration (REMI). *Mol. Gen. Genet.*, 258: 89–94.

105. Sato, T., Yaegashi, K., Ishii, S., Hirano, T., Kajiwara, S., Shishido, K. and Enei, H., 1998. Transformation of the edible basidiomycete *Lentinus edodes* by restriction enzyme–mediated integration of plasmid DNA. *Biosci. Biotechnol. Biochem.*, 62: 2346–2350.

106. Schiestl, R.H. and Petes, T.D., 1991. Integration of DNA fragments by illegitimate recombination in *Saccharomyces cerevisiae*. *Proc. Natl. Acad. Sci., USA*, 88: 7585–7589.

107. Seiboth, B., Messner, R., Gruber, F. and Kubicek, C.P., 1992. Disruption of the *Trichoderma reesei* cbh2 gene encoding for cellobiohydrolase II leads to a delay in triggering of cellulase formation by cellulose. *J. Gen. Microbiol.*, 136: 1259–1264.

108. Shuster, J.R. and Bindel Connelley, M., 1999. Promoter-tagged restriction enzyme-mediated insertion (PT–REMI). mutagenesis in *Aspergillus niger*. *Mol. Gen. Genet.*, 262: 27–34.

109. Smith, J.L., Bayliss, F.T. and Ward, M., 1991. Sequence of the cloned pyr4 gene of *Trichoderma reesei* and its use as a homologous selectable marker for transformation. *Curr. Genet.*, 19: 27–33.

110. Suelmann, R., Sievers, N., Fischer, R., 1997. Nuclear traffic in fungal hyphae: *In vivo* study of nuclear migration and positioning in *Aspergillus nidulans*. *Mol. Microbiol.*, 25: 757–769.

111. Sweigard, J.A., Carroll, A.M., Farrall, L., Chumley, F.G. and Valent, B., 1998. *Magnaporthe grisea* pathogenicity genes obtained through insertional mutagenesis. *Mol. Plant Microbe Interact.*, 11: 404–412.

112. Tanaka, A., Shiotani, H., Yamamoto, M. and Tsuge, T., 1999. Insertional mutagenesis and cloning of the genes required for biosynthesis of the host-specific AK: Toxin in the Japanese pear pathotype of *Alternaria alternata*. *Mol. Plant Microbe Interact.*, 12: 691–702.

113. Thode, S., Schafer, A., Pfeiffer, P., Vielmetter, W., 1990. A novel pathway of DNA end-to-end joining. *Cell*, 60: 921–928.

114. Thomas, M.D. and Kenerley, C.M.., 1989. Transformation of the mycoparasite *Gliocladium*. *Curr. Genet.*, 15: 415–420.

115. Thon, M.R., Nuckles, E.M. and Vaillancourt, L.J., 2000. Restriction enzyme-mediated integration used to produce pathogenicity mutants of *Colletotrichum graminicola*. *Mol. Plant Microbe Interact.*, 13: 1356–1365.

116. Tilburne, J., Scazzocchio, C., Taylor, G.G., Zabicky-Zissman, J.H., Lockington, R.A. and Davies, R.W., 1983. Transformation by integration in *Aspergillus nidulans*. *Gene*, 26: 505–521.

117. Timberlake, W.E., 1991. Cloning and analysis of fungal genes. In: *More Gene Manipulations in Fungi*, (Eds.) J.W. Bennett and L.L. Lasure. Academic Press, San Diego, pp. 51–85.

118. Timberlake, W.E. and Marshall, M.A., 1989. Genetic engineering of filamentous fungi. *Science*, 244: 1313–1317.

119. Turgeon, B.G., Garber, R.C. and Yoder, O.C., 1987. Development of a fungal transformation system based on selection of sequences with promoter activity. *Mol. Cell Biol.*, 7: 3297–3305.

120. Turner, G., 1994. Vectors for genetic manipulation. *Prog. Ind. Microbiol.*, 29: 641–665.

121. Unkles, S.E., Campbell, E.I., Carrez, D., Grieve, C., Contreras, R., Fiers, W., Van den Hondel, C.A. and Kinghorn, J.R., 1989. Transformation of *Aspergillus niger* with the homologous nitrate reductase gene. *Gene*, 78: 157–166.

122. Upchurch, R.G., Meade, M.J., Hightower, R.C., Thomas, R.S. and Callahan, T.M., 1994. Transformation of the Fungal Soybean Pathogen *Cercospora kikuchii* with the Selectable Marker bar. *Appl. Environ. Microbiol.*, 60: 4592–4595.

123. Valdez-Taubas, J., Diallinas, G., Scazzocchio, C. and Rosa, A.L., 2000. Protein expression and subcellular localization of the general purine transporter UapC from *Aspergillus nidulans. Fungal Genet. Biol.*, 30: 105–113.

124. van den Hondel, C.A.M.J.J. and Punt, .PJ., 1991. Gene transfer systems and vector development for filamentous fungi. In: *Applied Molecular Genetics of Fungi*, (Eds.) J.F. Peberdy, C.E. Caten, J.E. Ogden, J.W. Bennett. Cambridge University Press, Cambridge, pp 1–28.

125. van Dijk, R., Faber, K.N., Hammond, A.T., Glick, B.S., Veenhuis, M. and Kiel, J.A., 2001. Tagging Hansenula polymorpha genes by random integration of linear DNA fragments (RALF). *Mol. Genet. Genomics*, 266: 646–656.

126. van Hartingsveldt, W., Mattern, I.E., van Zeijl, C.M., Pouwels, P.H. and van den Hondel, C.A., 1987. Development of a homologous transformation system for *Aspergillus niger* based on the pyrG gene. *Mol. Gen. Genet.*, 206: 71–75.

127. van Peij, N.N., Visser, J. and de Graaff, L.H., 1998. Isolation and analysis of xlnR, encoding a transcriptional activator co-ordinating xylanolytic expression in *Aspergillus niger. Mol. Microbiol.*, 27: 131–142.

128. van Veen, R.J.M., Hooykaas, P.J. and Schilperoort, R.A., 1988. Mechanisms of tumorigenesis by *Agrobacterium tumefaciens*. In: *Physiology and Biochemistry of Plant-Microbe Interactions*, (Eds.) N.T. Keen, T. Kosuge and L.L. Walling. Rockville, MD: American Society of Plant Physiologists, pp. 19–30.

129. Vautard, G., Cotton, P. and Fevre, M., 1999. The glucose repressor CRE1 from *Sclerotinia sclerotiorum* is functionally related to CREA from *Aspergillus nidulans* but not to the Mig proteins from *Saccharomyces cerevisiae. FEBS Lett.*, 453: 54–58.

130. Verdoes, J.C., Punt, P.J. and van den Hondel, C.A.M.J.J., 1995. Molecular genetic strain improvement for the overproduction of fungal proteins by filamentous fungi. *Appl. Microbiol. Biotechnol.*, 43: 195–205.

131. Wada, M., Beppu, T. and Horinouchi, S., 1996. Integrative transformation of the zygomycete *Rhizomucor pusillus* by homologous recombination. *Appl. Microbiol. Biotechnol.*, 45: 652–657.

132. Wang, J., Holden, D.W. and Leong, S.A.., 1988. Gene transfer system for the phytopathogenic fungus *Ustilago maydis. Proc. Natl. Acad. Sci., USA*, 85: 865–869.

133. Ward, M., Kodama, K.H. and Wilson, L.J., 1989. Transformaton of *Aspergillus awamorii* and *Aspergillus niger* by electroporation. *Exp. Mycol.*, 13: 289–293.

134. Weidner, G., d'Enfert, C., Koch, A., Mol, P.C. and Brakhage, A.A., 1998. Development of a homologous transformation system for the human pathogenic fungus *Aspergillus fumigatus* based on the pyrG gene encoding orotidine 50–monophosphate decarboxylase. *Curr. Genet.*, 33: 378–385.

135. Winans, SC., 1992. Two-way chemical signaling in *Agrobacterium* plant interactions. *Microbiol. Rev.*, 56: 12–31.

136. Yaver, D.S., Lamsa, M., Munds, R., Brown, S.H., Otani, S., Franssen, L., Johnstone, J.A. and Brody, H., 2000. Using DNA-tagged mutagenesis to improve heterologous protein production in *Aspergillus oryzae. Fungal Genet. Biol.*, 29: 28–37.

137. Ye, G.N., Daniell, H. and Sanford, J.C., 1990. Optimization of delivery of foreign DNA into higher-plant chloroplasts. *Plant Mol. Biol.*, 15: 809–819.

138. Yelton, M.M., Hamer, J.E. and Timberlake, W.E., 1984. Transformation of *Aspergillus nidulans* by using a trpC plasmid. *Proc. Natl. Acad. Sci., USA*, 81: 1470–1474.

139. Yoder, W.T. and Lehmbeck, J., 2004. Heterologous expression and protein secretion in filamentous fungi. In: (Eds.) *Advances in Fungal Biotechnology for Industry, Agriculture and Medicine*, (Eds.) J.S. Tkacz and L. Lange. Springer, pp. 210– 219.

140. Zambryski, P.C., 1992. Chronicles from the *Agrobacterium*–plant cell DNA transfer story. *Ann. Rev. Plant Physiol. Plant Mol. Biol.*, 43: 465–490.

141. Zhang, S., Fan, Y., Xia, Y.X. and Keyhani, N.O., 2010. Sulfonylurea resistance as a new selectable marker for the entomopathogenic fungus *Beauveria bassiana Appl. Microbiol. Biotechnol.*, DOI 10.1007/s00253–010–2636–x.

142. Zeilinger, S., Galhaup, C., Payer, K., Woo, S.L., Mach, R.L., Fekete, C., Lorito, M. and Kubicek, C.P., 1999. Chitinase gene expression during mycoparasitic interaction of *Trichoderma harzianum* with its host. *Fungal Genet. Biol.*, 26: 131–140.

Biotechnology: An Overview (2015)   Pages 223–278
Editors: Rajan Kumar Gupta, Nasim Akhtar and Deepak Vyas
Published by: DAYA PUBLISHING HOUSE, NEW DELHI

Chapter 16

Plant Cell Factory for Molecular Farming of Therapeutic Proteins

Nasim Akhtar

Department of Biotechnology, GITAM Institute of Technology,
GITAM University, Gandhi Nagar Campus, Rushikonda,
Visakhapatnam – 530 045, A.P.

ABSTRACT

Plant biotechnology has been developed at a faster pace during the last quarter of the 20th century. Plants are now considered as the bioreactor for molecular farming of compounds for pharmaceutical and health application. There is a growing demand for protein therapeutics and diagnostics, but the production system to meet those demands using established facilities is lacking at present. The success of molecular farming in plants depends on technology, economic considerations and public acceptance. A small number of plant-derived biologics are approaching commercialisation, but these are the minority that have met the technological challenges, cleared the regulatory hurdles and crossed the inertia of biotechnology industry. The cost reduction is the expected and most signicant driver of the plant biopharming industry. Although, plants have offered many advantages over established production system for the large-scale expression of recombinant therapeutic proteins, but several challenges remain to be addressed in terms of improving quality, yield and economic faesibility.

Introduction

The application of engineering principles for manipulation of biological system to improve products or process for the benefit of mankind is one of the applied aspects of biotechnology. As per the convention on biological diversity, biotechnology is defined as '*any technological application that uses biological systems, living organisms, or derivatives thereof, to make or modify products or processes for specific use.*' The synthesis of compound with biopharmaceutical, medical and health application in plants and

---

* *Corresponding Author:* E-mail: nasimakhtar01@yahoo.co.in

their cultivation is described as 'Molecular Pharming' or 'Molecular Farming'. It exploits the metabolic properties of living organisms for the production of valuable products with therapeutic applications. Thus, biotechnological drugs are those therapeutic products that are produced through the use of living organisms. The intact organism, their biomass or part of the body, isolated cells or the enzymes and other metabolites or the biotransformation of endogenous or exogenous substrates etc. constitute the biotechnology products which are important for agriculture, chemicals, food products, medicine, pharmaceuticals, and various other industrial and health applications.

Initially, microbes have been exploited for biotechnological products. Plant as bioreactor has been developed significantly in enhancing the basic and applied knowledge as well as in establishing production system. Animal model are also valuable systems, especially for the products such as antibodies, vaccines etc. There has been fast development in plant biotechnology during the last quarter of the 20th century. Genetically engineered plants with improved traits, pest and herbicide resistance, etc., have produced significant results since the first product 20 years ago. In contrast, the development of transgenic plants as cell factory to produce molecules and compounds for pharmaceutical and health application is still in its developmental phase.

One of the most important goals in molecular farming is the development of new drugs and vaccines for important diseases. Four types of products *viz.* growth factors (27 products), interferons (26 products), hormonal therapies (24 products) and interleukins (21 products) constitute the biotechnological drugs. The immediate therapeutic uses of these protein drugs shows that oncology comes first with 41 products followed by diabetes and endocrinology with 38, infectious diseases with 21 and haematology with 17 products. The market value of recombinant therapeutic proteins is expected to cross the mark of $50 billion by the end of present decade with a growth rate of 7 per cent. Talukder (2006) has indicated that this market value of the recombinant therapeutic proteins is only an indicator of the strategic importance of such commodities but their socio-political impacts extend far from realization.

Since, proteins are essential components of the living cells; they play a vital role as passive structural elements as well as active functional units. Proteins constitute more than half of the cells dry weight, comprise fundamental components of the cell structure and at the same time serve as the main instruments of molecular recognition and catalysis. There is ambiguity regarding the distinction between peptide and protein therapeutics. Any polymer of two or more amino acids linked by peptide bonds is considered a peptide. However, there must be a minimum number of amino acid residues to form a protein which is highly variable. Lien and Lowman (2003) have suggested the cutoff value as 100, so that any chain of fewer than 100 amino acids is considered a peptide and not a protein. Initially proteins and peptides are considered poor drug candidates because of their inability to cross biological membranes and cellular barriers efciently.

The polypeptides or proteins are important molecules involved in mediating large numbers of key biological functions. The unique properties of proteins, in terms of specicity, selectivity, efcacy, and low toxicity, makes them highly attractive biomolecule applied as a reagent for diagnostic, prophylactic, therapeutic as well as for various health complecations such as allergy, cancer, cardiovascular disease, gastrointestinal dysfunction, immunological disorders, and other infectious diseases.

A wide range of protein products have already been identified as target for molecular pharming in plants are antibodies, vaccines or biopharmaceuticals aimed at treatment of human and animal diseases (Tables 16.1 and 16.2). Plant made pharmaceutical drug products are in various development phases for almost all major human diseases and conditions, including Alzheimer's disease, Anthrax, Arthritis, Cancer, Chronic obstructive pulmonary disorder (COPD), Crohn's disease, Cystic Fibrosis, Diabetes, Heart Disease, Hepatitis B, Hepatitis C, HIV, Influenza, Iron deficiency, Kidney Disease, Measles, Multiple sclerosis, Non-Hodgkin's Lymphoma, Obesity, SARS and Spinal Cord Injuries. Animal and poultry vaccines and other medicines are also in development phase for diseases like Avian flu, Coccidiosis, Fowl pest, Parvovirus and Rabies. It is projected that even with 40 per cent approved plant made pharmaceutical drugs will have about $ 1bn market each with less than 50 farms spread over 40,000 acres by the end of current decade.

Presently, therapeutic proteins have been derived from three sources: (1) natural or bioactive peptides produced by plants or animals (including humans); (2) peptides isolated from libraries of recombinant cells; and (3) peptides discovered in chemical libraries. The chemical synthesis of peptides is very expensive and difcult to scale-up, resulting in the development of several approaches that allow eukaryotic cells (plants) to be used as peptide factories (Bray, 2003). The most promising and potentially beneficial developments in plant biotechnology are expected to take place in this area of molecular pharming.

This review is aimed at presenting the recent research work demonstrating the potential of plants cell to synthesize and assemble complex proteins and other molecules suitable for human therapeutic applications.

The Production System

A number of production system are available for therapeutic proteins such as the yeast, the insect cell culture, the mammalian cell culture, the bacterial system as well as the transgenic animal and plants. The most common large scale production systems for proteins are genetically engineered bacteria and yeast due to relatively ease of manipulation rapid growth cycle. Transgenic plants provide an attractive expression system for the production of foreign proteins compared with other production systems based on fermenter technologies, such as bacteria, virus, yeast, and mammalian cells. The advantages of because of plant based expression system are the speed of production in transient systems, the absence of contamination by pathogens, and the availability of eukaryotic protein processing systems for post-translational modication of proteins, the safety resulting from the inability of plants to support the replication of human viruses, and the ability of plants to synthesize peptides that are structurally and functionally identical to their native counterparts (Obembe *et al.*, 2011; Rybicki 2010).

Table 16.1. Status of plant-derived therapeutic product near to commercialization.

Company/Institution	Plant Used	Product	Indication	Clinical Stage
Agragen	Flax	Human serum albumin	Maintenance of blood plasma pressure	Pre-clinical
Applied Biotech. Institute and partner	Maize	Undisclosed	Undisclosed	Animal trials underway
Arizona State Univ.	Potato	Vaccine	E.Coli	Phase 1
	Potato	Vaccine	Hepatitis "B"	Phase 1
	Potato	Vaccine	Norwark virus	Phase 1
	Tobacco	Vaccine	Norwark virus	Phase 1/2
	Undisclosed	Oral Vaccine	Undisclosed	Phase 1 planned start mid '07
Arntzen group (Tacket et al., 2000)	Potato	Vaccine- Norwalk virus capsid protein	Norwalk virusinfection	Phase I
Arntzen group (Tacket et al., 1998)	Maize	E. coli heat- labile toxin	Diarrhoea	Phase I
Arntzen group (Richter et al., 2000)	Potato	Hepatitis B virus surface antigen	Hepatitis B	Phase I
Bayer Innovation	Tobacco	Anti-idiotype IgG antibodies	Non-Hodgkin's lymphomas	Phase 1
Biolex	Lemna	Alpha Interferon	Hepatitis "B" and "C" and Cancer	Phase 2b (x2 trials - "480" and "formulated dose")
	Lemna	Recombinant plasmin	Fibrinolytic "Clot-buster"	Phase 1 ready
	Lemna	Anti-CD20 mAb	Non-Hodgkin's lymphomas	Pre-clinical
Chlorogen and Partner	Tobacco	a TGF-Beta protein	Ovarian Cancer	Advanced animal trials
Cobento AS (formerly Cobento Biotech AS)	Aribidopsis	Human Intrinsic Factor	Vitamin B12 deficiency	Approved Coban product launched Successful cGMP production certified
	Aribidopsis	Transcobalamin	Diagnostic/research	Available from company

Contd...

Table 16.1–Contd...

Company/Institution	Plant Used	Product	Indication	Clinical Stage
CIGB (Cuba)	Tobacco	Recombinant Monoclonal (Mab)	Purification re-agent in Hep."B" vaccine	Approved mid 2006 in Cuba
Dow Agrisciences	Plant cell non-nicotine tobacco	Vaccine	Newcastle disease in poultry	**USDA Approved in Feb.06**
	Plant cell - tobacco	Vaccines	"Diseases of horses, dogs and birds"	Undisclosed
D. Yusibov and others	Spinach	Vaccine	Rabies	Phase 1 successful in 2002
Farmacule	Tobacco	Virtonectin	Research use	Available from company mid 2007 Distribution under negotiation
Guardian Biosciences	Canola	Edible vaccine	Coccidiosis in poultry	CFIA phase 2
LSBC (in Ch.11 bankruptcy)	Tobacco	Vaccine	Non-Hodgkin's lymphoma	Phase 1 successful in 2002
	Tobacco	Vaccine	Feline Parvovirus	"very advanced" - Status confidential
	Tobacco	Aprotinin	Non-clinical use	Was in Sigma catalogue in 2005
Medicago	Tobacco	Vaccine	Pandemic and Seasonal Influenza	Phase 2
Meristem Therapeutics (www.meristem-therapeutics. com)	Maize	Gastric lipase	Cystic fibrosis	Phase 2 (x2 "unformulated" and "formulated")
	Maize	Lactoferrin	Gastrointestinal disorders	Phase 1
	Maize	Collagen	Reconstructive surgery	Pre-clinical
Meristem Therapeutics and Solvay Pharmaceuticals (www.solvay.com)	Maize	Gastric lipase	Cystic fibrosis, pancreatitis	Phase II
	Tobacco	ICAM1/Rhinovirus prophylactic	Receptor for common cold	Phase 2

Contd...

Table 16.1–Contd...

Company/Institution	Plant Used	Product	Indication	Clinical Stage
NeoRx/Monsanto	Maize	Antibody	Cancer	Phase 2 (failed)
Nexgen Biotech	Oriental Melon	Thyroid stimulating hormone receptor	Diagnosis of Graves disease	Available from company
	Oriental Melon	Viral antigens	Rapid detection of Hantaan and Puumala	Available from company
Pharma-planta consortium	Tobacco	2G12 IgG	HIV prophylactic	Phase 1
Planet Biotechnology Inc. (www.planetbiotechnology.com)	Tobacco	sIgA "CaroRx"	Prevention of tooth decay	Phase 2 but already granted an EU licence as a medical device. Seeking EU. distribution partner.
Plantechno srl	Tobacco	B-Glucosidase	Gaucher's disease	Pre-clinical
	Rice	Apo-A1 (Milano)	Cardiovascular disease	Preclinical. Patent protection in the USA
	Soya	Lactoferrin	Infant formula enhancer	Preclinical.
Prodigene Inc. (www.prodigene.com/)	Maize	Lt-B vaccine	Traveller's diarrhoea	Phase 1 complete - not ongoing
	Maize	TGE vaccine	Piglet gastroenteritis	Phase 1 complete - not ongoing
	Maize	Avidin	Diagnostic use	Available in Sigma catalogue '07
	Maize	Trypsin	Wound care/insulin manufacture	Available in Sigma catalogue '07
	Maize	Aprotinin	Non clinical use	No longer in Sigma catalogue
	Maize	GUS	Non clinical use	No longer in Sigma catalogue

Contd...

Table 16.1–Contd...

Company/Institution	Plant Used	Product	Indication	Clinical Stage
Protalix	Carrot cell	Glucocerebrosidase	Gaucher's disease	**On Sale - named patient basis**
	Carrot cell	Acetylcholinesterase	Biodefense	Phase 1
	Carrot cell	Alpha-galactosidase	Fabry's Disease	Pre-clinical
	Carrot cell	Antitumor necrosis factor	Arthritis	Pre-Clinical
Sembiosys	Safflower	Insulin	Diabetes	Phase 3
	Safflower	Apolipoprotein AI	Cardiovascular	Phase 1 trial planned
Thomas Jefferson University/ Polish Academy of Science	Lettuce	Hepatitis B surface antigen	Hepatitis B	Phase 1
TransPharma srl	Tobacco	β-Glucosidase	Gaucher's disease	Phase 1
Ventria Biosciences	Rice	Lactoferrin	Infant formula enhancer	**On sale to infant formula makers**
	Rice	Lactoferrin	cell culture media	Available from company
	Rice	Lysozyme	Undisclosed	Undisclosed
	Rice	Lysozyme	For research purposes	Available from company

Modified from- http://www.molecularfarming.com/PMPs-and-PMIPs.html.

Table 16.2. Plant-derived recombinant proteins.

Origin of Recombinant Protein/ Epitope Source	Plant	Expression System	Expression Strategy	Production Level	Applications and Biological Assays	References
FMDV	Nicotiana benthamiana	PVX vector (Transient)	Synthetic gene with multiple epitopes	67 µg/g	Protection in challenge experiments	Andrianova et al. (2011)
Vibrio cholerae (humans)	Potato		Cholera toxin B-subunit	30 µg/g or 0.30 per cent TSP	Potein forms multimers, has receptor-binding activity and is immunogenic and protective when administered orally	Arakawa, T. et al. (1997) Arakawa, T. et al. (1998a, b)
PA of Bacillus anthracis	Tobacco	Nuclear (Stable)	HBcAg fusion	1 mg/kg	Epitope-specic antibody production	Bandurska et al. (2008)
Nef	Nicotiana tabacum	Nuclear (Stable, Transient)	Full length	0.2–0.7 per cent TSP	TA sequence enhances the protein accumulation in mice	Barbante et al. (2008)
Foot-and-mouth disease virus (agricultural domestic animals)	Arabidopsis		VP1	Not given	Immunogenic and protective when administered by injection	Carrillo, C. et al. (1998)
Rabbit haemorrhagic disease virus (rabbits)	Potato		VP60	0.30 per cent TSP	Immunogenic and protective when administered by injection	Castanon, S. et al. (1999)
Full length of Tat and GUS	Tomato	Nuclear (Stable, Transient)	Fusion protein	2–4 µ/mg TSP	Systemic immune responses when protein extracts applied by the intra-dermal route in mice	Cueno et al. (2010a, b)
Full length of Nef and Zein	Nicotiana tabacum	Nuclear (Stable, Transient)	Fusion protein	NR	Plant-derived antigen displays the antigenic determinants on Western blot analyses	De Virgilio et al. (2008)
Epitopes from gp41 fused to CPprotein from Cowpea mosaic virus (CPMV)	Cowpea	Nuclear (Transient)	Fusion protein	NR	Puried VLPs are immunogenic in mice inducing systemicimmune responses (IgG2a)	Durrani et al. (1998)

Contd...

Table 16.7–Contd...

Origin of Recombinant Protein/Epitope Source	Plant	Expression System	Expression Strategy	Production Level	Applications and Biological Assays	References
Trastuzumab-binding peptide	N. benthamiana	TMV-PVX (Transient)	TMV CVPs	0.6–1.0 mg/g	Recognition by specic antibodies	Frolova et al. (2010)
Transmissible gastro-enteritis coronavirus (pigs)	Arabidopsis		Glycoprotein S	0.06 per cent TSP	Immunogenic when administered by injection	Gomez, N. et al. (1998)
TBI epitopes from env and gag, and HBsAg as carrier	Nicotiana tabacum and Arabidopsis	Nuclear (Stable)	Fusion protein	Tobacco, 0.02 per cent TSP Arabidopsis, 0.026 per cent TSP	Plant-derived antigen from both tobacco and Arabidopsis display the antigenic determinants of both components	Greco et al. (2007)
Enterotoxigenic E. coli (humans)	Tobacco		Heat-labile toxin B-subunit	<0.01 per cent TSPa	Intact protein forms multimers and is immunogenic when administered orally	Haq, T.A. et al. (1995)
Enterotoxigenic E. coli (humans)	Potato		Heat-labile toxin B-subunit	0.19 per cent TSP	Receptor-binding activity and immunogenic and protective when administered orally	Haq, T.A. et al. (1995); Mason, H.S. et al. (1998) Tacket, C.O. et al. (1998)
Type II-collagen	Rice	Nuclear (Stable)	4x peptide fusion to glutelinA	0.98 µg/g seed	Oral tolerance studies in animal models	Hashizume et al. (2008)
Hepatitis B virus (humans)	Lupin (Lupinus spp.)		Envelope surface protein	<0.01 per cent FW	Immunogenic when administered orally	Kapusta, J. et al. (1999)
Tat	Nicotiana benthamiana and spinach	Nuclear (Transient)	Full length	300–500 µg/g fresh spinach leaf tissue	Induction of mucosal immune responses and neutralizing potential in mice	Karasev et al. (2005)

Contd...

Table 16.7–Contd...

Origin of Recombinant Protein/ Epitope Source	Plant	Expression System	Expression Strategy	Production Level	Applications and Biological Assays	References
V3 loop from gp120 and CTB as carrier protein	Potato	Nuclear (Stable)	Fusion protein	0.004 per cent TSP	The CTB–gp120 fusion protein is able to produce oligomer pentamer structures	Kim *et al.* (2004)
Retrocyclin-101 and Protegrin-1	Tobacco	Chloroplast (Stable)	Fusion of GFP	32–38 per cent TSP and 17–26 per cent TSP	Anti-microbial activity tested with plant pathogen	Lee *et al.* (2011)
NP of H1N1 Influenza A virus	*N. benthamiana*	PVX vector (Transient)	PVX CVPs	1.1 mg/g	Specific CD8 + T cells activated	Lico *et al.* (2009)
p24	*Arabidopsis thaliana*	Nuclear (Stable)	Full length	0.5 mg/g stems FW 0.2 mg/g leaves FW	Mucosal and systemic immune responses in mice	Lindh *et al.* (2008)
sIgA Anti-*S.* mutant	Tobacco			200-500 µg/g		Ma *et al.*, 1995, 1998)
Mouse GAD67	Potato			150 µg/g		Ma *et al.*, 1994
Ectodomain from gp41 and CTB as carrier protein	*Nicotiana tabacum*	Nuclear (Transient)	Fusion protein	NR	Puried VLPs are immunogenic in mice inducing high level of IgG and IgA antibodies	Marusic *et al.* (2001)
p25, p27 and Nef	*Nicotiana tabacum*	Nuclear (Stable)	Full length	0.7 per cent TSP	Proteins puried by His tag showing antigenic properties in mice	Marusic *et al.* (2007)
HBsAg	Tobacco			0.01 per cent TSP		Mason *et al.*, 1992; Thanavala *et al.*, 1995

Contd...

Table 16.7–Contd...

Origin of Recombinant Protein/ Epitope Source	Plant	Expression System	Expression Strategy	Production Level	Applications and Biological Assays	References
Hepatitis B virus (humans) HBsAg	Tobacco		Envelope surface protein	<0.01 per cent TSP	Virus-like particles form and extracted protein is immunogenic when administered by injection	Mason, et al. (1992) Thanavala, et al. (1995)
Norwalk virus capsid protein (humans)	Tobacco		Capsid protein	0.23 per cent TSP	Intact protein and virus-like particles form, immunogenic when administered orally	Mason, et al. (1996)
Norwalk virus capsid protein (humans)	Potato		Capsid protein	10–20 µg/g or 0.37 per cent TSP	Virus-like particles form and immunogenic when administered orally	Mason, et al. (1996) Tacket, et al. (2000)
P1 epitope from gp41 and CTB as carrier protein	*Nicotiana tabacum*	Nuclear (Transient)	Fusion protein	0.2 per cent TSP	Puried CTB-gp41 is immunogenic in mice by the i.n. route along with liposomes and CT	Matoba et al. (2004, 2009)
Gp41 of HIV	Tobacco	Nuclear (Stable)	CTB fusion	18.1 mg/kg (not puried)	G_{M1} binding assay and Epitope specic antibody production	Matoba et al. (2009)
Rabies virus (humans)	Tomato		Glycoprotein	1.00 per cent TSP	Intact protein	McGarvey, P.B. et al. (1995)
gp41 (731–752 aa)	Cowpea	Nuclear	(731–752 aa) (Transient)	NR	VLPs are immunogenic in mice by the s.c. route with aluminum hydroxide	McLain et al. (1996)
C4 and six different V3 domains from gp120	Lettuce	Nuclear (Transient)	Fusion protein	240 µg/g of leaf DW	Positive at systemic and mucosal levels	Rosales-Mendoza et al. (2012)
C4 and V3 domains from gp120	*Nicotiana tabacum*	Chloroplast (Transient)	Fusion protein	300 µg/g of leaf DW	Positive at systemic and mucosal levels	Rosales-Mendoza et al. (2012)

Contd...

Table 16.7–Contd...

Origin of Recombinant Protein/ Epitope Source	Plant	Expression System	Expression Strategy	Production Level	Applications and Biological Assays	References
M2 of H1N1 inuenza A virus	Vigna unguiculata	CPMV vector (Transient)	CPMV S-CP fusion	33 µg/g	Epitope-specic antibody production	Meshcheryakova et al. (2009)
Fulllength of bothp17 and p24	Nicotiana tabacum	Nuclear (Transient)	Fusion protein	220 µg/LFW (ER) 4800 µg/LFW (CH) 220 µg/LFW (C)	Mucosal and systemic immune responses by i.m. route in mice (DNA primary inoculation)	Meyers et al. (2008)
Fulllength of bothp17 and p24	Nicotiana tabacum	Nuclear (Stable)	Fusion protein	5–230 µg/LFW (CH) <1 µg/LFW (ER)	Mucosal and systemic immune responses in mice by the i.m. route	Meyers et al. (2008)
p24	Nicotiana tabacum	Nuclear (Transient)	Full length	0.01 µg/LFW (C) 3,691–16,148 µg/LFW (ER); 937–4,014 µg/LFW (CH)	Mucosal and systemic immune responses by i.m. route in mice (DNA primary inoculation)	Meyers et al. (2008)
p24	Nicotiana tabacum	Nuclear (Stable)	Full length	0.04 µg/LFW (C) 0.01–1.19 µg/LFW (ER) 636–2,994 µg/LFW (CH)	Mucosal and systemic immune responses by i.m. route in mice	Meyers et al. (2008)
NDV several antigens	N. benthamiana	PVX vector (Transient)	CMV CP fusion and VLPs generation	NR	Epitope-specic antibody production	Natilla and Nemchinov (2008)

Contd...

Table 16.7–Contd...

Origin of Recombinant Protein/ Epitope Source	Plant	Expression System	Expression Strategy	Production Level	Applications and Biological Assays	References
HCV R9 mimotope	Tobacco	CMV Vector (Transient)	CMV CVPs	10 mg/100 g	Recognition by sera of HCV infected patients and IFN$_7$ PBMC production	Nuzzaci et al. (2007, 2009)
Human insulin	A. thaliana	Nuclear (Stable)	Fusion to oleosin	0.75 per cent TSP	Insulin tolerance tests in animal models	Nykiforuk et al. (2006)
Full length p24	Nicotiana tabacum	Nuclear (Stable)	Fusedto a2–a3 IgA domains	1.4 per cent TSP	Systemic immune responses by s.c. route in mice.	Obregon et al. (2006)
VP2 of CPV	Tobacco	Chloroplast (Stable)	TDP fusion	0.4 mg/g (non-purified)	Epitope-specic antibody production	Ortigosa et al. (2010)
E6 and E7 of HPV	Tomato	Nuclear (Stable)	L1 fusion and VLPs generation	1 per cent TSP (non-purified)	Neutralizing antibodies and specic CD8 + T cells activated	Paz De la Rosa et al. (2009)
PA of B. anthracis	V. unguiculata	CPMV vector (Transient)	Inactivated CPMV CVPs	0.3 g/Kg	Recognition by specic antibodies	Phelps et al. (2007)
MUC1 tandem repeat tumor associated	N. benthamiana	Magnlcon system (Transient)	LTB fusion	88 µg/g	Epitope-specic antibody production	Pinkhasov et al. (2011)
Tat	Tomato	Nuclear (Stable)	Full length	900 µg/g of fresh fruit	Systemic and mucosal immune responses by protein in mice.	Rami'rez et al. (2007)
p24	Nicotiana benthamiana	Cytoplasmic (Transient)	Full length	3.23 mg/kg FW	Plant-derived protein isantigenic on an ELISA assay	Regnard et al. (2010)
Hepatitis B virus (humans)	Potato		Envelope surface protein	<0.01 per cent FWb	Immunogenic when administered orally	Richter et al. (2001)
p17	Nicotiana benthamiana	CH and C (Transient)	Full length	NR	Atigenic determinants detection on Western blot analyses	Scotti et al. (2009)
p24	Nicotiana benthamiana	CH and C (Transient)	Full length	NR	Antigenic determinants detection on Western blot	Scotti et al. (2009)

Contd...

Table 16.7–Contd...

Origin of Recombinant Protein/ Epitope Source	Plant	Expression System	Expression Strategy	Production Level	Applications and Biological Assays	References
Pr55Gag	*Nicotiana benthamiana*	ER, C, CH and AP (Transient)	Full length	NR	Antigenic determinants detection on Western blot	Scotti *et al.* (2009)
Pr55Gag	*Nicotiana benthamiana*	Nuclear and CH (Stable)	Full length	28 mg/kg FW (N) 38.6–338 mg/kg (CH)	Antigenic determinants detection on Western blot	Scotti *et al.* (2009)
Pr55Gag with a deletion of p17	*Nicotiana benthamiana*	CH and C (Transient)	Fusion Protein	NR	Antigenic determinants on Western blot	Scotti *et al.* (2009)
TCPA of *Vibrio cholerae*	Tomato	Nuclear (Stable)	CTB fusion	0.17 per cent TSP tomato fruit	G$_{M1}$ binding assay	Sharma *et al.* (2008)
HIV polyepitope derived from env and gag and the HBsAg as carrier	Tomato	Nuclear (Stable)	Fusion protein	7 ng/g FW	Plant material is immunogenic when administered by the i.p. and p.o. routes	Shchelkunov *et al.* (2006)
TBI and the HBsAg as carrier	*Lycopersicon esculentum*	Nuclear (Stable)	Fusion protein	0.3 ng/mg freeze-dried tomato	Plant material is immunogenic when administered by the i.p. and p.o. routes	Shchelkunov *et al.* (2006)
DPT polypeptide of *C. diphteriae*, *Bordetella pertussis*, *Clostridium tetani*	Tomato	Nuclear (Stable)	Synthetic gene with multiple epitopes	1-2.1 µg/g leaf, 2.25-6.8 µg/g fruit	Epitope specific antibody production in serum, gut and lungs	Soria-Guerra *et al.* (2007, 2011)
Enterotoxigenic *E. coli* (humans)	Maize		Heat-labile toxin B-subunit	Not given	Immunogenic and protective when administered orally	Streatfield *et al.* (2000)

Contd...

Table 16.7-Contd...

Origin of Recombinant Protein/ Epitope Source	Plant	Expression System	Expression Strategy	Production Level	Applications and Biological Assays	References
Transmissible gastro-enteritis coronavirus (pigs)	Maize		Glycoprotein S	0.01 per cent FW	Protective when administered by orally	Streatfield et al. (2000)
Der p 1 epitope of mite allergen	Rice	Nuclear (Stable)	—	4.5 µg/mg of seeds	Oral tolerance induction studies in animal models	Suzuki et al. (2011)
Human cytomegalo virus (humans)	Tobacco		Glycoprotein B	<0.02 per cent TSP	Immunologically related protein	Tackaberry, E.S. et al. (1999)
E. coli LT-B	Potato			3-4 µg/g		Tacket et al., 1998; Haq et al., 1995; Mason et al., 1998
3Crp T cell peptides string of cedar pollen	Rice	Nuclear (Stable)	Fusion to rice glutelin acidic subunit	35 µg/20 mg of grain	Digestibility and oral tolerance induction studies in animal models	Takagi et al. (2008)
Transmissible gastro-enteritis coronavirus (pigs)	Tobacco		Glycoprotein S	0.20 per cent TSP	Immunogenic when administered by injection	Tuboly et al. (2000)
Amyloid b peptides	Tobacco	CMV vector	CMV CVPs (Transient)	37 mg/10 g	Recognition by specic antibody and sub-cellular localization	Vitti et al. (2010)
Novokin epitope	Rice	Nuclear (Stable)	18X novokin	85 µg/g grain KDEL	Digestibility antihypertensive activity in animal models	Wakasa et al. (2011a)
Lactostatin of milk β-lactoglobulin	Rice	Nuclear (Stable)	Glutelin fusion to 6 repeated lactostatin	16.1 mg/g seed	Hypocolesterolemic activity after oral feeding	Wakasa et al. (2011b)
Foot-and-mouth disease virus (agricultural domestic animals)	Alfalfa		VP1	Not given	Immunogenic and protective when administered by injection or orally	Wigdorovitz et al. (1999)

Contd

Table 16.7–Contd...

Origin of Recombinant Protein/ Epitope Source	Plant	Expression System	Expression Strategy	Production Level	Applications and Biological Assays	References
VP1 of FMDV	Chenopodium quinoa	BaMV vector (Transient)	BaMV CVPs	0.2–0.5 mg/g	Antibodies, cell-mediated immunity activation and protection	Yang et al. (2007)
V3 loop from gp120 and the CP from Alfalfa mosaic virus	Nicotiana tabacum	Nuclear (Transient)	Fusion protein	NR	Puried VLPs are immunogenic in when administered by the i.p. route without the need of adjuvant.	Yusibov et al. (1997)
p24	Cucumber and Nicotiana benthamiana	Protoplast (Transient)	Full length	5 per cent TSP	Plant-derived protein displays the antigenic determinants on Western blot analyses	Zhang et al. (2000)
p24	Nicotiana tabacum	Nuclear (Stable)	Full length	3.5 mg/g TSP	Plant-derived antigen displays the antigenic determinants on Western blot analyses	Zhang et al. (2002)
VP1 of FMDV	Chenopodium amaranticolor	TNV vector (Transient)	TNV CVPs	330 µg/g	Epitope-specic antibody production (IgG, IgA and sIgA)	Zhang et al. (2010)
Nef and p24	Nicotiana tabacum	Choroplast (Transient)	Fusion protein	4–40 per cent TSP	Puried plant-derived antigen is immunogenic in mice by the s.c. route	Zhou et al. (2008) Gonzalez Rabade et al. (2011)

BaMV Bamboo mosaic virus; CMV Cucumber mosaic virus; CP coat protein; CPMV Cowpea mosaic virus; CPV Canine parvovirus; CTB cholera toxin B subunit; CVPs chimeric virus particles; Der p Dermatophagoides pteronyssinus; FMDV Foot and mouth disease virus; FW fresh weight; GFP green fluorescent protein; G_{M1} ganglioside M1; HBcAg Hepatitis B core antigen; HCV Hepatitis C virus; HBsAg surface antigen of the Hepatitis B virus; HIV-1 Human immunodeficiency virus type 1; HPV16 Human papillomavirus type 16; IFNc interferon gamma; LTB heat-labile enterotoxin B subunit of E. coli; MUC1 human epithelial mucin; NDV Newcastle disease virus; NP Nucleoprotein; PA protective antigen; PBMCs peripheral blood mononuclear cells; PRRSV Porcine reproductive and respiratory syndrome virus; PVX Potato virus X; TCPA toxin co-regulated pilus subunit A; TD tetramerization domain from human transcription factor p53; TMV Tobacco mosaic virus; TNV Tobacco necrosis virus; VLPs virus like particles; VP virus protein; per cent TSP percentage respect to total soluble protein; NR not reported, FW fresh weight, DW dry weight; CH chloroplast; C cytosol; N nuclear; LFW leaf fresh weight; ER endoplasmic reticulum; AP apoplast; s.c. subcutaneous; i.m. intramuscular; p.o. oral; i.p. intraperitoneal; i.n. intranasal.

The first recombinant plant-derived therapeutic protein, human serum albumin, has been produced in 1990 in transgenic tobacco and potato plants (Sijmons *et al.*, 1990). Twenty years on, of the numerous therapeutic proteins produced in plants only a limited number have made their way into clinical trials (Tables 16.1 and 16.2). The secretory immunoglobulin A (IgA) antibody comprising of four polypeptide chains that inhibits the major oral pathogen *Streptococcus* mutans from binding to teeth is the most advanced product in human clinical trials. The principle behind the development has been established for the production of many therapeutic proteins, including antibodies, blood products, cytokines, growth factors/hormones, recombinant enzymes and human and veterinary vaccines (Lico *et al.*, 2012; Twyman *et al.*, 2003). Additionally a few more products for the treatment of human diseases are approaching commercialization are listed in Table 16.1.

The use of *in vitro* plant cell culture systems has the advantage of providing sterile production conditions, which may be necessary for some pharmaceutical products. Moreover, all contact with the contaminating environment is eliminated during the cultivation of plant cells in a bioreactor system. The disadvantages of this system are the higher levels of technical expertise required (Rybicki 2010), higher production costs, low expression levels and relatively low protein contents (Hellwig *et al.*, 2004).

Cultivation of transgenic plants expressing therapeutic proteins in a greenhouse environment is advantageous. This is less expensive than cell culture systems, and the protein yield in specific plant storage tissues is higher than that resulting from cell culture. Moreover, greenhouses cultivation provide semi-sterile conditions, prevents contamination and can be considered a containment system that prevents the escape of transgene and hence, avoiding the risks to consumers and/or the environment. Further, greenhouse cultivation can reduce regulatory burdens, but it is costly. Feld cultivation of transgenic plants is considered as the least expensive but need to address the biosafety issues related to heath and environments (Mikschofsky *et al.*, 2011).

Besides many advantages of using plants to produce therapeutic proteins as protein bioreactors, there are only few current commercial applications (Table 16.1). This is because of the low yields of polypeptides produced in plants compared with those produced in microbial expression systems. There are inherent difficulties in maximizing peptide stability and accumulation, the best approach has to be optimized empirically for each peptide. The factors that affect recombinant protein yields in transgenic plants and other plant systems have recently been reviewed (Surzycki *et al.*, 2009). Low, medium and high level of expression have been demonstrated for the heterologous proteins in different tissues such as leaf (Alfaalfa, Arabidopsis, Lettuce, Tobacco) fruits (Cucumber, Potato, Tomato) and seeds (cowpea, maize, peas, rape, rice, soybean) (Table 16.2). The accumulation and stability of the nal peptide product are inuenced by several factors, including the strength of the promoter, intrinsic RNA and/or protein stability, subcellular localization, genetic background and host cell metabolism.

The Strategies

The expression of transgene from gene to protein requires construction of expression cassette and the vector, the tissue transformation and the regeneration of transgenic plants as well as testing of subsequent generations of seeds. Interaction between the multiple copies of the transgene can lead to co-supression of the expression of all of them. An ideal transformation experiment is the one uses a single copy of trangene as it expresses uniformly from one generation to another in a medndelian fashion. However, because of the long lasting effects of growth of plant cells in tissue culture, the expression of the transgene in the T0 transformant is not as reliable a basis for evaluation as expression in a seed-growing T_1 or T_2 plants. As a consequence transgene stability and inheritance have to be evaluated for several trangene generations. These processes take up to 2-3 years depending on the plant species, although milligram amounts of protein might be available after several months for initial testing (Hood *et al.*, 2002). Identification of transgenic material, containment of transgenes, control of recombinant protein, scalling up production as well as distribution and handling of genetically modified materials are the potential problems and must be addressed and taken in to consideration in order to get approval for the plant-based therapeutic products as drug.

Recombinant proteins have been produced following both stable transgene expression (Shinmyo and Kato 2010) and transient expressions systems (Komarova *et al.*, 2010) in many plant species (Table 16.2). It has been observed that transient expression produces higher yields and allows more rapid scale-up, whereas transgenic plants can provide an almost unlimited source of homogenous material expressing the target molecule (Lico *et al.*, 2008; Obembe *et al.*, 2011). The general technique of therapeutic protein production include the designing of chimeric expression cassette, components of plasmid vector, the marker (reporter and selectable), strength of promoter, use of enhancers, linkers and other sequences, condon usage, the target tissues, the delivery system, integration and transgene expression. Transgenic plants have been produced by *Agrobacterium* vector mediated as well as vectorless particle gun bombardment system of gene delivery for transformation. The low cost of experimentation, the ease of inoculation and delivery of target sequence as well as the ease of the characterization of trasngenge are the main advantages of *Agrobacterium* mediated transformation system. The particle bombardment is the physical process of DNA coating to an inner material and introduction to cells to produce transgenic plants. Expression of recombinant proteins have to be optimized with particular reference to codon preference of peptide sequences to match the preferences of the host plant (Hashizume *et al.*, 2008), targeting to subcellular compartments that allow peptides to accumulate in a stable form (Wakasa *et al.*, 2011a, b), the use of strong, tissue-specific promoters (Takaiwa *et al.*, 2007), and the testing of different plant species and systems. The highest peptide expression levels reported thus far have been achieved in transplastomic tobacco plants with peptide coding sequences fused to the p53 tetramerization domain, β-glucuronidase and green uorescent protein (GFP), respectively, resulting in fusion protein yields of 6–51 per cent TSP (Lee *et al.*, 2011; Lentz *et al.*, 2010; Ortigosa *et al.*, 2010).

The Plants

Plants produce large quantities of biomass without the need of expensive culture media and production can easily be scaled up according to market requirements (Lie´nard *et al.*, 2007). Transgenic plants, fruits and seeds have been selected as bioreactors for the production of various pharmaceutical proteins by virtue of their long-term stability over other expression systems. Plants offer economical advantages for certain applications and ease in scalability and avoid most safety issues concerning viral and toxin contaminants (Fischer *et al.*, 2004; Ma *et al.*, 2005). The development and refinement of plant genetic engineering and transformation techniques and an improving knowledge of plant molecular biology is continuously expanding the potential of plant biotechnology. The techniques of genetic transformation of plant cells, and regeneration of complete tramsgenic plants expressing heterologous genes for various traits have been perfected in many palnt species. A wide range of plant species for which transformation process have been reported include alfalfa, apple, asparagus, banana, barley, brassica, cabbage, canola, cantaloupe, carrot, cauliflower, cranberry, cucumber, eggplant, flax, grape, kiwi, lettuce, lupin, maize, melon, papaya, pea, peanut, pepper, plum, potato, raspberry, rice, service berry, soybean, squash, strawberry, sugar beet, sugarcane, sunflower, sweet potato, tomato, walnut, and wheat. Many diverse plant species have been transformed for production of recombinant proteins (Table 16.2).

The choice of plant species is crucial for successful plant biopharming for protein production since each plant species has unique characteristics that affect expression, product storage, downstream processes, and the quality of the final protein products. Most of the therapeutic proteins expressed to date have been in tobacco, although recently alfalfa, potatoes, soybean, tomato, lettuce, rice and wheat have also been used successfully (Table 16.2). Protein based pharmaceuticals have also been produced successfully in leaves, for example, via transplastomic expression, mostly in tobacco (Daniell, 2006; Daniell *et al.*, 2009) with a high expression levels of up to 40 per cent TSP. However, tobacco leaves are not well suited as an expression system due to their high content of phenolics and alkaloids. Other leafy crops that are interesting as platforms for the production of pharmaceuticals are alfalfa, spinach, lupins and lettuce (Rybicki 2010). The expression of pharmaceuticals in lettuce chloroplasts seems to be a good alternative. The major advantage of using green tissue (tobacco, alfalfa, soybean etc.) is their productivity. Both alfalfa and tobacco can support several crops (cuttings) per year, with potential annual biomass yields of 25 tonne ha^{-1} and >100 tonne ha^{-1}, respectively. By contrast, the maximum yields of wheat, rice and corn seed are ~3 tonne ha^{-1}, 6 tonne ha^{-1} 12–6 and 12 tonne ha, respectively. Other advantages of tobacco include its relative ease of genetic manipulation, production of large numbers of seeds (up to a million per plant) and an impending need to explore alternate uses for this hazardous crop. However, seeds are likely to have fewer phenolic compounds and a less complex mixture of proteins and lipids than green leaves, which might be an advantage in purification.

The Tissue

The plant species selected must efficiently concentrate bimass in the organ or tissue where the recombinant protein is expressed in order to achieve maximum production. The tissue chosen should be compatible for the stable accumulataion, correct processing and efficient recovery of the desired protein. There are two strategy *viz.* i. the whole plant expression (using constitutive promoter), and ii. the tissue specific expression system (using tissue specific/environmentally regulated promoter) for the production of foreign proteins in various plant species. Different tissues such as leaf (Alfaalfa, Arabidopsis, Lettuce, Tobacco) fruits (Cucumber, Potato, Tomato) and seeds (cowpea, maize, peas, rape, rice, soybean) (Table 16.2) have demonstrated low, medium and high level of expression of heterologous proteins. Among the various plant-based production platforms, seeds are very important because they are compact organs, provide a stable environment, and have a high protein content, which is the outcome of their inherent physiological function for producing and storing proteins (Stoger *et al.*, 2005; Lau and Sun, 2009; Boothe *et al.*, 2010; Peters and Stoger, 2011).

The seeds of transgenic plant have been proposed as bioreactors for the production but the low level of expression of foreign proteins in transgenic plant seeds is the major obstacle for industrial application. Cereal endosperm cells synthesize and accumulate storage proteins in special storage compartments called protein bodies (PBs). In order to achieve high level accumulation of foreign proteins in rice seed, Lico *et al.* (2009, 2012) have used two strategies: (1) targeting foreign polypeptides to PBs in endosperm, (2) the suppression of gene expressions of endogenous storage proteins (13 kDa prolamins and glutelins). The results of those studies have indicated that the signal peptide of storage proteins has a functional role in entering the ER and hGH polypeptides nally accumulated in PB-IIs via ER to PSV transport pathway in rice endosperm. Takagi *et al.* (2006) have indicated that the articial 7Crp peptide, seven major human T-cell epitopes derived from the Japanese cedar pollen allergens Cry j 1 and Cry j 2, in addition to a glutelin GluB-1 signal peptide is transported into both PB-I and PB-II. Furthermore, Masumura *et al.* (2006) has shown that human interferon-a with a 10 kDa prolamin signal peptide was also transported into both PB-I and PB- II in transgenic rice endosperm. Prolamins are aggregated to each other with ER lumen via disulde bonds, leading to the formation of PB-Is. Glutelins are synthesized in ER and form intramolecular disulde bonds, and then are transported to PB-IIs via the Golgi apparatus (Hsiung *et al.*, 1986; Vos de *et al.*, 1992; Leite *et al.*, 2000). Lico *et al.* (2012) have demonstrated the the human growth hormone accumulated to 470 µg/g at the maximum level in transgenic rice seeds. These data indicated that the signal peptide of the rice storage protein is an important factor for targeting foreign polypeptides to PB-I or PB-II in rice endosperm. Lettuce leaves are easy to be freeze-dried in comparison to tomatoes, bananas, tubers, or carrot. Moreover, though lyophilization is necessary, SHBsAg expression level in leaves is many times higher than in naturally dry seeds (Sunil Kumar *et al.*, 2006).

The Expression Pattern

The accumulation and stability of the nal peptide product are inuenced by several factors, including the strength of the promoter, intrinsic RNA and/or protein stability, subcellular localization, genetic background and host cell metabolism. It has been difcult to express single peptides in transgenic plants, probably because short heterologous peptides are unstable in plant cells. Heterologous peptide sequence has been engineered variously for improved stability. The major strategies have been used as: 1) the use of strong, tissue-specific promoters (Takaiwa *et al.*, 2007), 2) the codon optimization of peptide sequences to match the preferences of the host plant (Hashizume *et al.*, 2008), 3) the fusion of a peptide or repeated peptide sequence to the sequence of a carrier protein, similar to the endogenous or a heterologous plant protein, 4) the expression of peptide tandem repeats (usually 10–20 copies) (Yang *et al.*, 2012), 5) the production of recombinant viruses or virus-like particles that display the peptide on their surface (Komarova *et al.*, 2010), 6) the suppression of gene expressions of endogenous storage proteins by RNAi technology (Shigemitsu *et al.*, 2011), 7) the use of a linker sequence inserted between the peptide repeats or between the peptide and carrier protein to achieve correct peptide folding and release by enzymatic digestion in vitro (Lee *et al.*, 2011), 8) the targeting for correct glycosylation (Peter and Stoger, 2011; Yang *et al.*, 2012), 9) the sub-cellular localization in protein bodies (membrane boundaries around expreesed proteins to protect from proteases and desiccation during seed maturation) (Takagi *et al.*, 2010).

The Vector

A wide variety of *Agrobacterium* "disarmed" (non-tumorogenic) vector and vector system have been developed. Any T-DNA vector having left and right border sequences and is able to replicate in *Agrobacterium* on its own (binary vector) or be designed to recombine with a partner plasmid that does so (cointegrated vector) along with other chimeric sequences can be used to derive the foreign gene expression. Therapeutic protein expression follows similar principle to those described for plant cell transformation and transgenic plant generation. Transformation can be achieved using *Agrobacterium* (Circelli *et al.*, 2010; Horsch *et al.*, 1985; Koncz and Schell, 1986), particle bombardment (Christou, 1993, 1995), protoplast electroporation (Lindsey and Jones, 1987) or viral vectors (Porta and Lomonossoff, 1996).

Agrobacterium-mediated nuclear integration of recombinant T-DNA has been the standard method for the stable transformation of plants. Due to low expression of nuclear peptides, plastid transformation demonstrated high yields of heterologous peptides because multiple copies of the genome are present in each plastid and photosynthetic cells contains hundreds of plastids (Daniell *et al.*, 2002, 2009). Stable expression in transgenic plants obtained via *Agrobacterium*-mediated transformation has been the main system used for the expression therapeutic proteins. Both stable transgenic plants (Shinmyo and Kato 2010) and plant-based transient expressions systems (Komarova *et al.*, 2010) have been used to produce recombinant peptides. Typically, transient expression produces higher yields and allows more rapid scale-up, whereas transgenic plants can provide an almost unlimited source of homogenous material expressing the target molecule (Obembe *et al.*, 2011).

The earliest plant viral vector has been based on cauliflower mosaic virus (CaMV) a double stranded DNA virus replicates through an RNA intermediate. Application of plant viruses, such as alfalfa mosaic virus (AlMV) (Yusibov *et al.*, 2005), (BaMV) bamboo mosaic virus (Yang *et al.*, 2007), (ClYVV) clover yellow vein virus (Masuta *et al.*, 2000), (CPMV) cowpea mosaic virus (Phelps *et al.*, 2007), tobacco mosaic virus (TMV) (Frolova *et al.*, 2010), cucumber mosaic virus (Vitti *et al.*, 2010), (PVX) potato virus X (Lico *et al.*, 2009), (TNV) tobacco necrosis virus (Zhang *et al.*, 2010), (ZYMV) zucchini yellow mosaic virus (Arazi *et al.*, 2001) have been successfully used as vectors for expression and production of plant-based therapeutic proteins vaccines against rabies virus (Yusibov *et al.*, 1997; Modelska *et al.*, 1998), *Pseudomonas aeruginosa* (Brennan *et al.*, 1999a), hepatitis C virus (Nemchinov *et al.*, 2000), mink enteritis (Dalsgaard *et al.*, 1997), and *Staphylococcus aureus* (Brennan *et al.*, 1999b), among others. The ability to invade and replicate rapidly in plants, these engineered viruses can accumulate large amounts of viral proteins, including desired engineered antigenic proteins (Yusibov and Rabindran, 2008; Buetow and Korban, 2000). Moreover, fusing peptides to viral coat proteins will allow assembly of engineered antigenic protein genes into chimeric virus particles with enhanced immunogenicity of antigenic determinants (Yusibov and Rabindran, 2008). For example, CPMV virions contain 60 copies of each of the large and small coat protein subunits, and thus 60 copies of the antigenic protein can be expressed and displayed on the surface of each chimeric virus particle. This has been clearly demonstrated with expression in tandem of peptides 10 and 18 of the outer-membrane protein F of *Pseudomonas aeruginosa* in CPMV chimeric virus particles (Brennan *et al.*, 1999a). Moreover, using adjuvants such as QS-21, QuilA, Freund's complete adjuvant (FCA) and incomplete Freund's adjuvant (IFA) as carrier molecules for chimeric virus particles has been reported to enhance the immunogenicity of such plant-based vaccines (Brennan *et al.*, 1999a, b).

The Promoter

For the expression of transgene in plants initially the strongest available constitutive promoters have been used for molecular farming. Most commonly the gene expression has been achieved by constitutive promoters–a regular CaMV 35S promoter (Mason *et al.*, 1992; Kapusta *et al.*, 1999; Pniewski *et al.*, 2011, 2012; Ehsani *et al.*, 1997; Dogan *et al.*, 2000; Kong, *et al.*, 2001; Gao *et al.*, 2003) a 35S promoter with a dual enhancer (Mason *et al.*, 1992; Lam and Arntzen, 19965; Sunil-Kumar *et al.*, 2006; Shulga *et al.*, 2004; Richter *et al.*, 2000; Mason *et al.*, 2003; Joung *et al.*, 2004; Huang *et al.*, 2004, 2005), and a ubiquitin (Sunil-Kumar *et al.*, 2003, 2005, 2006; Srinivas *et al.*, 2008; Ganapathi *et al.*, 2007) or a hybrid promoter ocs-mas (Smith *et al.*, 2002). Some of the tissue specific promoter such as the tuber- (Shulga *et al.*, 2004; Richter *et al.*, 2000) the seed- (Hayden *et al.*, 2012a, b) or the fruit-specific (Shulga *et al.*, 2004; Srinivas *et al.*, 2008; Sunil-Kumar *et al.*, 2003, 2005, 2006) as well as the auxin-inducible promoters (Imani *et al.*, 2002) have been tried among non-constitutive promoters. Several novel promoters that can be induced by physical and chemical stimuli have been described recently (Padidam, 2003; Zuo and Chua, 2000). A peroxidase gene promoter isolated from sweet potato (*Ipomoea batatas*) has been used to drive the gusA reporter gene in transgenic tobacco producing 30 times more GUS activity than demonstrated by the cauliower mosaic virus (CaMV) 35S promoter following exposure to hydrogen peroxide, wounding or ultraviolet light (Kim *et al.*, 2003). A post-harvest induction of gene expression has been used by the CropTech mechanical gene activation (MeGA) system with the application of wound inducible promoter based on a tomato hydroxy3-methylglutaryl CoA reductase2 (HMGR2) gene.

The seed-specific promoter from the common bean (*Phaseolus vulgaris*) has resulted in excellent expression of a singlechain antibody in *Arabidopsis thaliana*. As compared to the CaMV 35S promoter, which resulted in antibody accumulation to 1 per cent TSP, the bean arc5-I promoter resulted in antibody levels in excess of 36 per cent TSP in homozygous seeds, and the antibody retained its antigen binding activity and afnity (De Jaeger, 2002). A trichome-specific promoter that might be useful for the secretion of recombinant proteins into the leaf guttation uid has also been described in tobacco (Wang *et al.*, 2002). Phytomedics Inc. (http://www.phytomedics.com), has commercialised another secretion system which involves the secretion of recombinant proteins into tobacco root exudates and the leaf guttation uid has been applied for the production of human secreted alkaline phosphatase and has been used for the secretion of recombinant antibodies (Drake *et al.*, 2003). A class I patatin promoter has been used to target the antigenic protein to potato tubers for expression of a synthetic LT-B gene. The amount of the recombinant LT-B protein in transgenic potato lines reached 17 mg g^{-1} fresh weight of tuber, corresponding to 1.3 mol kg^{-1} of monomeric recLT-B and 0.26 mol kg^{-1} of pentameric recLT-B.

The Enhancer

Several applications of transgenic technology require very high level of transgene expression. Further, the expression of a gene is the outcome of the interplay of multiple processes including transcription, splicing, mRNA transport mRNA stability, translation, protein stability and post translational modifications. Achieving high levels of transgene expression is one of the challenges for applications of transgenic technology. The use of strong promoters has been the main focus in such instances, 5′UTRs have also been shown to enhance transgene expression. Thus gene expression can be regulated not only by modulating transcriptional parameters but also by using post-transcriptional parameters like using optimal codons or by including 5′ and 3′ leader sequences. A 28 nt long synthetic 5′UTR (*synJ*) has been shown to enhance transgene expression under a strong promoter like 35S as well as under a weak promoter like NOS in dicotyledonous plants tobacco and cotton (Kanoria and Burma, 2012). Similarly, the 67 nt long 5′UTR derived from the Tobacco mosaic virus (TMV) RNA, widely called as 'ΩA leader' has been shown to enhance the translation of foreign gene transcripts both in vivo and in vitro (Fan *et al.*, 2012).

A trans-activation system has been described recently, that utilizes the viral vector for the expression of intracellular domain of the diabetes associated autoimmune antigen IA-2ic and anti-tetanus antibody 9F12, where it displayed tight control on transcription in transgenic plants (Hull *et al.*, 2005b). The strategy has been used successfully to enhance the expression level of reporter transgene (Yang *et al.*, 2001) by inclusion of a transcription factor REB (rice endosperm bZIP) in rice along with human lysozyme gene driven by rice globulin promoter where the presence of REB enhanced the expression level of lysozyme.

Endogenous proteolysis is a key determinant of the stability and yield of recombinant proteins in plants. The co-expression of recombinant protease inhibitors is another promising strategy for increasing recombinant protein yields in plants. The co-expression of tomato cystatin *SlCYS8*, an inhibitor of cystein (C1A) proteases, alongside C5-1 targeted to the cell secretory pathway has increased antibody yield by nearly 40 per cent after the usual 6-days incubation period, up to, 3 mg per tobacco plant. By contrast, C5-1 yield has been detected in greater by an additional 40 per cent following 8- to 10-days incubations in younger leaves with a high *SlCYS8* expression.

The presence of introns between the promoter and the coding sequences can significantly enhance the gene expression. Introns may act by stabilizing mRNA, thus leading to greater amount of translation products. Adding a 268 bp transcriptional enhancer from the pea plastocyanin gene (*PetE*) upstream of the CaMV 35S promoter containing the AMV leader further increased RSV-F gene expression up to 7.7-fold compared to the construct lacking both *AMV* leader and *PetE* transcriptional enhancer. Gene expression has also been intensified by viral transcription activators, as *TEV*- or *AlMV*- 5'UTR (Mason *et al.*, 1992, 2003; Richter *et al.*, 2000; Kong *et al.*, 2001; Joung *et al.*, 2004; Sojikul *et al.*, 2003; Huang *et al.*, 2005).

The Marker

Addition of a proper selectable marker gene is critical for recovery of transformed tissues allowing the proliferation of only the transformed cells in the presence of a selective agent while the non-transformed cells suppressed to divide. The *npt* II-encoding neomycin phosphotransferase is commonly used as a marker gene that determined resistance to the antibiotic hygromycin (Imani *et al.*, 2002). Sometimes kanamycin and analogues such as geneticin (G418) and paramomycin are also used with *npt* II gene. Similarly, either *hpt* or *bar* also determined resistance to hygromycin or herbicide glufosinate (Pniewski, *et al.*, 2011, 2012, Hayden *et al.*, 2012a, b), respectively. Pniewski (2013) has described the selection system based on the marker gene *bar* conferring resistance to strongly acting glufosinate, for the expression of S-HBsAg surface antigen in lettuce leaf, probably reduced the transformation rate, but, on the other hand, it guarantees high probability of regeneration of authentic transformants in comparison to kanamycin (Mohapatra *et al.*, 1999). For transient expression studies transformation efficiency is typically evaluated for GUS (β-glucuronidase) expression 2-3 days after transformation by immersing the tissues in a solution of 5-bromo-4-chloro-3-indolyl-β-D-glucuronic acid (x-gluc) for 16-24h. A modified green fluorescent protein (GFP) of jellyfish with alteration in chromophore has been developed as a very useful and sensitive reporter gene system. A β-glucuronidase and green uorescent protein (GFP) marker gene have been used in transplastomic tobacco plants with peptide coding sequences fused to the p53 tetramerization domain to achieve high level of expression resulting in fusion protein yields of 6–51 per cent TSP (Lentz *et al.*, 2010; Ortigosa *et al.*, 2010; Lee *et al.*, 2011).

The Linkers

The linker is the sequence which is usually inserted between the peptide repeats or between the peptide and carrier protein to achieve correct peptide folding. The linker includes an engineered proteolytic fusion site so that the native peptide can be released by enzymatic digestion in vitro (Lee *et al.*, 2011). Numerous linker has been used for epitope specific antibody production for example GPGP linker (Matoba *et al.*, 2009; Sharma *et al.*, 2008), PG_4SG_4P linker (Bandurska *et al.*, 2008), Xa protease cleavage site (Lee *et al.*, 2011), G_4S_2 linker (Andrianova *et al.*, 2011), GSP linker (Zhang *et al.*, 2010).

The Terminator

Besides commonly utilized NOS terminator, polyadenylation signals from plant genes are also used (Mason *et al.*, 1992, 2003, Richter *et al.*, 2000, Kong *et al.*, 2001, Joung *et al.*, 2004, Sojikul *et al.*, 2003, Huang *et al.*, 2005). It has been reported that the use of alternative polyadenylation signals to the commonly used nopaline synthase (NOS) 3' element in constructs, such as the 3' end of the soybean vegetative storage protein (*vspB*) gene and the 3' element from the potato *pinII* gene, has increased the amount of HBsAg mRNA (up to 16mg g^{-1} potato tuber) (Pniewski, 2013).

Protein Stability

Plant expressed proteins have demonstrated very low intrinsic stability due to the solubility of foreign proteins in water and lipids and susceptibility to degradation following pH variations. Further, naked peptides are degraded by proteolytic enzymes in the gastrointestinal tract and plasma, and are rapidly removed from circulation by hepatic and renal clearance (Vlieghe *et al.*, 2010). Several strategies have been applied for improvement of the therapeutic potential of peptides. These includes (*i*) Liposomes encapsulation or similar nanoparticles based on polylactic or polylactic-glycolic acid to achieve targeting specicity and controlled release (Ducat *et al.*, 2011); (*ii*) Conjugation with polyethylene glycol (PEG) to reduce renal and hepatic clearance; (*iii*) Substitution with non-natural counterparts such as D-aminoacids residues at predicted cleavage sites (Webb *et al.*, 2005); (*iv*) Circularization or modication of peptide bond to enhance bioavailability and selectivity (Craik *et al.*, 2002); (*v*) modication of terminal aminoacid by N-terminal acetylation or C-terminal amidation (Giuliani and Rinaldi 2011; Croft and Purcell 2011). In rice the low expression level of foreign proteins in transgenic plants has been over come by two strategies *viz.* (1) targeting foreign polypeptides to PBs in endosperm, (2) the suppression of gene expressions of endogenous storage proteins (13 kDa prolamins and glutelins) resulting in high level accumulation of foreign proteins (Shigemitsu *et al.*, 2011).

The Codon

Heterologous transgene from differernt organisms are processed differently with respect to their expression when compared to native plant gene. The non-plant DNA sources may contain sequences that are recognized by plants as signal for splicing or polyadenylation or read as codon that are rare in plants. For the efficient expression, the non-host sequences have to be modified to generate mRNA with more abundant plant codon that are efficiently recognized and read by the translational machinery of plants. Mason *et al.* (1998) have optimized the LT-B gene for plant mRNA processing by changing the 'AAT' codon coding for Asn to 'GTG' coding for Val, in order to accommodate a NcoI site surrounding the translation start site. The optimized synthetic gene, designated sLT-B, has been used in transforming potato plants, and found to accumulate 5–40-fold higher amounts of LT-B in leaf tissues than those lines carrying the bacterial (unmodified) LT-B gene (Mason *et al.*, 1998). Streatfield *et al.* (2001) have utilized a maize codon-optimized version of a barley α-amylase signal sequence, a cell secretion signal, to optimize antigenic protein accumulation in transgenic corn plants. Gene expression has also been intensified by codon usage of the S-HBsAg coding sequence according to a plant pattern (Mason *et al.*, 2003). Optimization of codon composition and regulatory elements for expression of human insulin like growth factor-1 in transgenic chloroplasts and evaluation of structural identity and functions have been reported by Daniell, *et al.* (2009).

Compartmentalization and Targetting

The purification of recombinant protein can be simplified through compartmentalization. Proteins that are tragetted to subcellular organelles play an important role in determining the yield of expressed foreign proteins because of the interrelated processes of folding, assembly and post-translational modication. By virtue of its physical separation from other proteins in the cell, purification of the desired proteins is achieved by fractionation to obtain an enriched subcellular fraction containing the recombinant protein. Proteins targetted to the secretory pathway using amino-terminal signal peptides usually accumulate to levels that are several orders of magnitude greater than those of protein expressed in the cytosol. Targeting is especially important if the recombinant protein is toxic to the host production system. The secretory pathway is more suitable for folding and assembly than the cytosol, and is therefore an advantageous site for high-level protein accumulation (Schouten *et al.*, 2002). Targetting of protein expression to endoplasmic reticulum (ER) is advantageous by providing an oxidizing environment and an abundance of molecular chaperones and only few proteases. Retention of expressed proteins in ER also inuences the structure of glycan chains (Ko *et al.*, 2003; Ramirez *et al.*, 2003). These features are likely to be the most important factors affecting protein folding and assembly. It has been shown recently that proteins that are targeted to the secretory pathway in transgenic plants interacts specically with the molecular chaperone BiP (Nuttall *et al.*, 2002). In the absence of further targeting information, proteins that accumulate in the secretory system are secreted to the apoplast. Depending on its size, the protein can be retained in the apoplast or might leach from the cell, with important implications for production systems that are based on cell-suspension cultures. The H/KDEL carboxy-terminal tetrapeptide tag has shown 2–10-fold greater and stable expression in the lumen of the ER compared to apoplast (Conrad and Fiedler, 1998; Schillberg *et al.*, 2000, 2003). The accumulation of avidin in the cytosol of transgenic tobacco plants is toxic, but plants can be regenerated successfully when this molecule is targeted to the vacuole (Murray *et al.*, 2002).

Targetting Sequence

Protein subunit oligomerization and cellular compartmentalization of the fusion proteins has been shown to improve the expression of recombinant proteins. A fusion protein of CTB with a human proinsulin gene (CTB-Pins) has expressed at up to 2.5 per cent TSP in lettuce (Ruhlman *et al.*, 2007). Similarly, the expression level of proinsulin [human proinsulin (A, B, C peptides)] containing three furin cleavage sites (CTB-PFx3)] has been enhanced up to 53 per cent of total leaf protein in old lettuce leaves (Boyhan and Daniell 2011). An endoplasmic reticulum (ER) retention signal SEKDEL, a six amino acids peptide peptide (Ser-Glu-Lys-Asp-Glu-Leu), is used in protein targeting constructs to transform both tobacco and potato plants

(Haq *et al.*, 1995). The LT-B transgene, driven by CaMV 35S, is flanked by the tobacco etch virus 5' untranslated region (*TEV* 5'-UTR) and the polyadenylation signal of a soybean vegetative storage protein (3' vspB) gene. The 5' *TEV* leader sequence served as a translational enhancer, and the 3' vspB also mediated the 3' end processing of the transcript. The recovered transgenic tobacco and potato plants expressing the transgene carrying the SEKDEL retention signal has shown significantly higher levels of the LT-B antigenic protein- a 1.2–2.8-fold increase in tobacco and 2–4-fold increase in potato (Korban, 2002).

Sandhu *et al.* (2000) have targeted the antigenic RSV-F protein to the fruit of tomato to develop a vaccine against the respiratory syncytial virus (RSV), which infects human particularly premature babies and infants, resulting in pneumonia and bronchiolitis. The targeting has been achieved by using a fruit ripening-specific promoter E-8 from tomato for the expression of the RSV-F gene in transgenic tomato plants. The amount of the recombinant RSV-F protein in transgenic tomato lines has reached 32.5mg g^{-1} fruit fresh weight. Earlier, Sandhu *et al.* (1999) reported that inserting the 5' untranslated leader sequence (37 bp) from alfalfa mosaic virus (*AMV*) RNA4 between the CaMV 35S promoter and the *RSV-F* gene increased transient viral expression in apple leaf protoplasts by 5.5-fold compared to the construct without this sequence.

Huang *et al.* (2001) have introduced three different constructs into tobacco, and found highest level of protein accumulation in transgenic plants containing the hemagglutinin (H) protein gene of measles virus, driven by CaMV 35S virus, with a 5' *TEV* leader sequence along with the ER retention sequence SEKDEL. Stoger *et al.* (2000) have expressed a single-chain Fv antibody (ScFv) against carcinoembryonic antigen (CEA), for targeting recombinant antibody either to the plant cell apoplast or to the ER by incorporating either a hexameric histidine (His6) tag or SEKDEL, respectively in both rice and wheat.

An articial 7Crp peptide, seven major human T-cell epitopes derived from the Japanese cedar pollen allergens Cry j 1 and Cry j 2, in addition to a glutelin GluB-1 signal peptide is transported into both PB-I and PB-II (Takagi *et al.*, 2006). Similarly, Masumura *et al.* (2006) have used a 10 kDa prolamin signal peptide for human interferon-α to be transported into both PB-I and PB- II in transgenic rice endosperm. While, a modied glucagon-like peptide 1 (mGLP-1) with rice chitinase signal peptide was sorted into the intercellular space (Yasuda *et al.*, 2006).

Post-Translational Modification and Glycosylation

Many eukaryotic proteins require extensive post-translational modifications to assume active confirmation. This is a major limitation of prokaryotic expression system. Plants utilize chaperons and protein disulfide isomerase to facilitate protein folding and to prevent large protein aggregate formation. An important aspect in expression of therapeutic protein is glycosylation as carbohydrate groups contribute to immunogenicity. The biological activity, immunogenicity, stability, correct targeting, and appropriate pharmacokinetic properties of many therapeutic proteins depends on correct N-glycosylation (Gomord *et al.*, 2010). N-glycosylation starts at first involving the attachment and modification of an oligosaccharide precursor in the endoplasmic reticulum (which is similar in all eukaryotes), followed by processing by enzymes resident in the golgi complex, (these enzymes are distinct in insects, plants, mammals and yeast), resulting in different oligosaccharide structures (Bardor *et al.*, 2009; Gomord *et al.*, 2010). Plant-derived glycoproteins carry specific β1,2-xylose and core α1,3-fucose residues instead of 1,6-fucose, β1,4-galactose and sialic acid, which are present in mammals. Recent researches are now focused not only on the humanization of plant glycans, but also the creation of specific glyco-forms for particular applications (Gomord *et al.*, 2010; Cox *et al.*, 2006; Rouwendal *et al.*, 2007).

Mucin-type *O*-glycosylation is another important post-translational modification that confers a variety of biological properties and functions to proteins. Yang *et al.* (2012) has established GalNAc *O*-glycosylation capacity in *N. benthamiana* plant cells by expression of a UDP-Glc(NAc) C4-epimerase and GalNAc-Ts, and have found *O*-glycosylation of the three different co-expressed secreted substrates.

A monoclonal antibody with human-like *N*-glycans in *Arabidopsis thaliana* has been produced through RNAi knocking out the genes encoding two enzymes responsible for the addition of plant-specific glycan residues (Schahs *et al.*, 2007). RNA interference has also been used to modulate the *N*-glycosylation machinery, by down regulating the endogenous genes responsible for the attachment of β1, 2-xylose and core α1, 3- fucose in *Lemna minor* and *Nicotiana benthamiana* (Cox *et al.*, 2006; Strasser *et al.*, 2008). Finally, Castilho *et al.* (2010) achieved protein sialylation in *N. benthamiana* by introducing the entire mammalian pathway for sialic acid synthesis. Hence, plants are now been used on an equal scale for glycoengineering as compared to yeast cells, where the modification of glycosylation enzymes is used to enhance product quality, *e.g.*, for antibodies and erythropoietin (Durocher and Butler, 2009; Elliott *et al.*, 2003; Li *et al.*, 2006).

Downstream Processing

In order to capitalize on the advantages of plant-based expression system in upstream production, it is necessary that the downstream purification of recombinant product be accomplished economically. An efcient recovery system of recombinant proteins need to be optimized for good yield for high-level expressed therapeutic proteins in plants. A secretory system is advantageous as plant cells need not necessary be disrupted during protein recovery without the interference of phenolic compounds. An afnity tag has been used to facilitate the recovery of proteins in order to avoid the istability of secretary recombinant protein in culture medium. The tag can be removed after purication to restore the native structure of the protein. The strategy has successfully been demonstrated in the oleosin fusion system (Moloney and van Rooijen, 1996). Oleosin fusion have been created with a number of different proteins varying in molecular weight from approximately 7-55 kDa, all of which are stably accumulated on the surface of oil bodies. A simple extraction procedure is used for recovery of fusion protein from oil bodies and the recombinant protein separated from its fusion partner by endoprotease digestion. Schillberg *et al.* (2000) have revised a strategy in which recombinant proteins are expressed as fusion constructs that contain an integral membrane-spanning domain derived from the human T- cell receptor, and are then puried from membrane fractions. Further, Leelavathi and Reddy (2003) have described the expression of His-tagged GUS-fusion proteins in tobacco chloroplasts. A foam fractionation system has been described for the extraction of Histidine tagged proteins recovery by Crofcheck *et al.* (2003). Production of a recombinant antimicrobial peptide in transgenic plants using a modied VMA intein expression system has been demonstrated by Morassutti *et al.* (2002). Additionally, fusing peptides to viral coat proteins will allow assembly of engineered antigenic protein genes into chimeric virus particles with enhanced immunogenicity of antigenic determinants (Yusibov and Rabindran, 2008).

Production and Storage System

The productivity and production costs of the different plant based therapeutic proteins are quite diverse and depend strongly on the specific region, choice of eld or greenhouse production and the plant tissue used. The different production conditions rendering a reliable assessment of inexpensive production systems are very difcult to optimize. It is estimated that in terms of the eld production of seed-based systems, in the USA, maize has been shown to have the highest productivity, with approximately 10t seed biomass/ha, and the lowest cost of production of $160 USD (per t), followed by rice, with 8t/ha and $370 USD/t, respectively (source FAO statistics for 2008; http://www.fao.org). Compared to this, pea seed biomass is relatively low (2.3 t/ha), and the production costs of $295 USD/t are much higher. However, assuming that the proportion and activity of recombinant proteins in pea seeds are equal to those in maize and rice, these disadvantages might be compensated by the high pea protein content (up to 40 per cent) compared with that of rice seeds (approximately 7 per cent) and maize seeds (8–10 per cent) (Boothe *et al.*, 2009, 2010).

The storage of plant-made therapeutic proteins is of great importance because the protein is harvested two or three times per year, and the isolation of the protein should be a continuous process. Hence, the stability of the transgene-encoded pharmaceutical harvested must be quite high for a complete decoupling of the cultivation cycle from the processing and purication of the desired protein (Boothe *et al.*, 2010). In green plant tissues, this can only be expected in freeze-dried material, but freeze-drying is a costly and energy intensive process. In contrast, pea seeds dry to a great extent on pea plants themselves. The residual humidity is less than 11 per cent in pea seeds after harvest compared to 12.5 per cent in maize and 13 per cent in rice (Boothe *et al.*, 2009). Under dry-seed conditions, the proteins are very stable. Mikschofsky *et al.* (2011) demonstrated that the amount of transgene-encoded CTB::VP60 was stable for more than 4 years in pea seeds. Hence, dry seeds permit relatively inexpensive storage at room temperature for years (Conrad and Fiedler 1998; Huang 2004; Stoger *et al.*, 2005).

Plant cell culture systems have the advantage of providing sterile production conditions, which may be necessary for some pharmaceutical products. Moreover, during the cultivation of plant cells in a fermenter, all contact with the environment is eliminated. The disadvantages of this system are the higher levels of technical expertise required (Rybicki 2010), higher production costs, low protein contents and relatively low expression levels (Hellwig *et al.*, 2004).

Transgenic plants producing pharmaceuticals is mostly cultivated in a greenhouse environment. This is because greenhouse cultivation is less expensive than cell culture systems, and the protein yield in specific plant storage tissues is higher than that resulting from cell culture. Additionally, greenhouses provide semi-sterile conditions that prevent contamination of the pharmaceuticals, and they can be considered a containment system that prevents the escape of transgenic plants that might pose a risk to consumers and/or the environment. Although greenhouse cultivation can reduce regulatory burdens, but it is costly and might lead to a greater variability in transgene expression compared with eld cultivation (Mikschofsky *et al.*, 2011).

Safety Issues

There is a considerable degree of public concdern in many parts of the world about the issues associated with genetically modifies plant production about potential impact of technology on food safety, human heath and the environment. Although, the field cultivation of transgenic plants is the least expensive methods to produce plant-derived pharmaceuticals. However, biosafety issues, including outcrossing potential, must be addressed if the genetically modified plants are to be cultivated in the field. The frequency of outcrossing from a transgenic line of eld-grown peas has been studied in 1997 and 1999 (Polowick *et al.*, 2002) using highly expressed β-glucuronidase as a marker of pollen transfer. Of the approximately 9000 offspring tested, only ve plants scored positive for the presence of the marker, representing a mean outcrossing rate of 0.07 per cent (Polowick *et al.*, 2002) indicating different pea lines expressing different pharmaceuticals can be grown directly side-by-side without the problem of cross-fertilisation. As an example, biosafety considerations related to outcrossing rates and seed dormancy prevent the application of canola-derived pharmaceuticals (Mallory-Smith and Zapiola 2008).

Other cross-pollinating plants, such as maize (Ramessar *et al.*, 2008) and safower, are subject to restrictions in terms of eld cultivation. One example is the cultivation of transgenic safower in Canada (SemBioSys Genetics, Calgary, AB, Canada), which only occurs miles away from other cultivated safower in the south of the USA Cross pollination is also a minor problem for tobacco, potato and carrot. Moreover, distances of 20 m are required for potato isolation to minimise the pollen-mediated escape of transgenes from potato (Conner and Dale 1996). Being a strict self-fertiliser, pea can be a superior system with respect to outcrossing and maintenance because seeds can not survive Northern European winters. However, pea is sensitive to several plant diseases and, therefore, plant protection might be more intensive compared to other species. Plant diseases are not a concern for greenhouse production in which pea can be cultivated easily and even higher harvests can be achieved (up to 5 t/ha) (Mikschofsky *et al.*, 2011). However, it remains to be ascertained whether this enhancement of yield also holds true for maize and rice.

AIDS/HIV

The acquired immune deciency syndrome (AIDS) caused by the human immunodeciency virus (HIV) is one of the most dreaded diseases worldwide. In 2008 the UNAIDS had estimated that 33.4 million people are HIV-positive and global spread of HIV-AIDS has reached pandemic proportions. The HIV belongs to the genus *Lentivirus* and the family Retroviridae. These RNA viruses are enveloped and produce characteristically slow and progressive infections. The introduction of highly active antiretroviral treatment (HAART) in industrialized countries had resulted in a considerably reduced disease progression to AIDS and has transformed HIV infection from a lethal disease to an effectively manageable chronic disease (Palella *et al.*, 1998; Walensky *et al.*, 2006).

The development of a plant-based vaccine on HIV has been explored over the last two decades and a gp120-based approache is performed by Rosales-Mendoza *et al.* (2012). The HIV antigens that constitute promising candidates on the development of vaccine are divided in three groups: early antigens, structural proteins and envelop proteins (Karasev *et al.*, 2005). Early antigen refers to those proteins that are expressed in the early stage of the virus life cycle. Six early proteins are present in the HIV: Tat, Rev, Vpu, Vif, Nef, and Vpr. These early components have been expressed in plants as full-length proteins and also as fusion proteins to improve their accumulation or enhancing immunogenicity (Table 16.2). Production of recombinant HIV-1/HBV virus-like particles has been achieved in *Nicotiana tabacum* and *Arabidopsis thaliana* plants for a bivalent plant-based vaccine development (Greco *et al.*, 2007). Kim *et al.* (2004) have reported the development of HIV-1 gp120 V3 cholera toxin B subunit fusion gene expression in transgenic potato. The human immunodeciency virus antigen Nef forms protein bodies in leaves of transgenic tobacco when fused to zeolin have been demonstrated by De Virgilio *et al.* (2008). The expression of full length HIV-1 Tat fusion protein in tomato proving the immunogenicity of the tomato-derived protein via oral, intraperitoneal, or intramuscular administration routes (Cueno *et al.*, 2010; Ramirez *et al.*, 2007). The Tat-based vaccines seem to be relevant as they have proven not only to be safe and immunogenic in preclinical models, but also effective in controlling virus replication and blocking the onset of the disease in monkeys (Goldstein *et al.*, 2000; Maggiorella *et al.*, 2004). Hence, plant-based vaccines have raised a new perspective which constitutes an alternative in conjunction with the traditional approaches, might facilitate the fight against HIV (Cueno *et al.*, 2010; Girard *et al.*, 2006a).

Allergies

Allergenic peptides have been expressed in transgenic rice seeds (Table 16.2) followed by an exhaustive investigation of the immunological mechanism activated after feeding in animal models (Takagi *et al.*, 2008; Suzuki *et al.*, 2011). It has been reported that subcutaneous or oral/sublingual administration of allergens is successful for a wide range of allergies including those to peanut (Jones *et al.*, 2009) and birch pollen (Mobs *et al.*, 2010). Suzuki and colleagues reported the production of a fragment (p45145) of the Der p 1 mite allergen, containing immunodominant human and mouse T cell epitopes. The fragment has been modied to achieve accumulation in type I endoplasmic reticulum-derived protein bodies (PB I) in rice seeds. Takagi *et al.* (2008) have found that endoplasmic reticulum targeted protein bodies (PB I) in a previous rice transformation experiments with cedar pollen allergenic peptides requires 20-fold lower dose of T cell epitopes for suppression of allergen-specific IgE in mice is about 20-fold lower compared to puried peptides. Furthermore, the ability of plant-derived peptides to suppress immune responses was specific to Der p 1, indicating that oral vaccines based on transgenic rice do not induce bystander immune suppression. Hence, oral delivery, which relies on the induction of oral tolerance, is the most intriguing delivery route for plant-derived allergens because it allows close control of dosage and administration frequency, and it has a lower patient impact than other routes (Levine 2010; Pajno and Barberi 2009).

Cancer

The plant-derived peptide has been demonstrated to induce systemic immunogenicity towards the tumor-associated antigen (Table 16.2). Pinkhasov *et al.* (2011) reported for the rst time the production in *N. benthamiana* plants of a breast tumor-associated antigenic peptide, the mucin-derived tandem repeat peptide (MUC1 TR). Cancer vaccines must break tolerance because this induces cytotoxic T cells that specically target the aberrant MUC1 variant (Reddish *et al.*, 1998). The plant-derived peptide was able to induce systemic immunogenicity towards the tumor-associated antigen. *N. benthamiana* plants has been infected with magnICON transient expression vectors containing a fusion construct comprising the MUC1 TR peptide and the *Escherichia coli* heat-labile enterotoxin B subunit (LTB), and the resulting peptide-based vaccine was tested for its ability to break tolerance in a human MUC1-tolerant mouse strain, MUC1.Tg (Pinkhasov *et al.*, 2011).

Cholera toxin B subunit (CT-B)

The two enteric diseases cholera and traveler's diarrhea, caused by *Vibrio cholerae* and enterotoxigenic strains of *Eschericia coli* (ETEC), respectively, are responsible for high mortality especially in young children in developing countries (Girard *et al.*, 2006). Oral vaccines, including plant-based formulations, are considered as alternatives or supplements for standard injection vaccines, since they have the potential for simplified vaccination procedures against pathogens penetrating through mucosal membranes or blood (Aziz *et al.*, 2007; Daniell, 2006, Girard *et al.*, 2006b; Mestecky *et al.*, 2008). A HIV-1 gp120 V3 cholera toxin B subunit fusion gene expression in transgenic potato has been reported by Kim *et al.* (2004). The transgenic expression of B subunit of cholera toxin (CT-B) from *Vibrio cholerae* in corn seeds has been explored to develop as a mucosal vaccine or vaccine component against cholera and traveler's diarrhea. CT-B has been expressed in a number of plant species (Table 16.2) including potato (Arakawa *et al.*, 1997, 1998a, b), carrot (Kim *et al.*, 2009), tomato (Jani *et al.*, 2002), tobacco (Daniell *et al.*, 2001; Jani *et al.*, 2004) and rice (Nochi *et al.*, 2007). Expression of the native cholera toxin B subunit gene and assembly of functional oligomers in transgenic tobacco chloroplasts has been reported by Daniell *et al.* (2001). Seeds can be superior site due to their natural protein storage ability that allow a high level of recombinant protein production compared to the costly high-level purication and presence of toxic alkaloid by-products in tobacco and other vegetable crops (Ma *et al.*, 2003). In addition, seeds can provide a stable dry environment for long-term storage and long distance transportation. The rice seed has been reported to be an adequate system for the production of CT-B (Nochi *et al.*, 2007). Karaman *et al.* (2012) have shown that maize derived CT-B can be immunogenic in mice. Since, the maize does not process the bacterial native signal peptide effectively, anti-CT-B antibodies induced by maize derived CT-B may be different than that induced by bacterial CT-B. Karaman *et al.* (2012) have established a combined vaccination system that can be effective for both CT-caused cholera and LT-caused traveler's diarrhea diseases. They found high level of antibody in the mice from the CTB5LTB5 combined treatment group than that of the CTB or LTB only groups but need further verication and improvement of the results. Further, repeating combined treatments with a higher dose of maize-derived CT-B would help to deduce more sound results for adjuvant/ synergistic actions of CT-B and LT-B. These results suggest that a synergistic action may be achieved using a CT-B and LT-B mixture that can lead to a more efficacious combined vaccine to target diarrhea induced by both cholera and enterotoxigenic strains of *Escherichia coli* (Karaman *et al.*, 2012).

Hepatitis B Surface Antigen

Potential plant-based oral vaccines require suitable antigen formulations and immunization protocols to ensure antigen stability through the alimentary tract and a balance between immune response and oral tolerance (Holmgren and Czerkinsky 2005; Mestecky *et al.*, 2007). Considerable progress has been made in prophylaxis against hepatitis B (HB) as one of the most effective and cheapest recombinant vaccines (Goldstein and Fiore 2001) (Table 16.2). However, one third of the worldwide human population is still suffering from HB and the number of hepatitis B virus (HBV) chronic carriers and patients suffering from HepB or hepatocellular carcinoma is still high (Kew 2010; Romano *et al.*, 2011), mainly in developing countries (Kao and Chen 2002).

Recombinant vaccines against the hepatitis B virus (HBV), based on the HBV small surface antigen (S-HBsAg) produced in yeasts, are introduced into hepatitis B (HepB) vaccination programmes in the early 1980s. This is a safe and effective vaccine demonstrated prophylaxis and control of HepB (Michel and Tiollais 2010). However, a tri-component vaccines has been suggested as more effective than vaccines based solely on S-HBsAg (Madalin'ski *et al.*, 2002; Rendi-Wagner *et al.*, 2006; Young *et al.*, 2001). The small HBV surface antigen (S-HBsAg) has been expressed in tobacco, potato, lettuce, and other plants, as well as in tissue and cell suspension cultures (Pniewski *et al.*, 2011, 2012). Anti-HBs immune response triggered by the oral administration of plant material containing SHBsAg has shown both in mice (Gao *et al.*, 2003; Joung *et al.*, 2004; Kapusta *et al.*, 1999, 2010; Kong *et al.*, 2001; Richter *et al.*, 2001) and in humans (Kapusta *et al.*, 1999, 2001; Mason *et al.*, 2003; Thanavala *et al.*, 2005). Pniewski (2013) has indicated that on the preparation of semi-products containing M/L-HBsAg, along with previously obtained plant-associated S-HBsAg constites a plant-derived oral tri-component vaccine against hepatitis B. Medium and large surface antigens of HBV have been expressed in lettuce leaf cells in the native and immunogenic forms. Accumulation of M/L-HBsAg, especially their VLP or aggregate forms, corresponded with the plant host. Moreover, both in lettuce and tobacco M-HBsAg is synthesised more intensively than L-HBsAg. It seems that the medium surface antigen is more durable in plant cells, similar to natural or recombinant expression systems (Bruss 2007). Lettuce is edible in the raw state, naturally free of harmful and anti-nutritional substances and for an orally administered vaccine the antigen concentration can be increased by lyophilisation (Pniewski *et al.*, 2011, 2012). Lettuce leaves are easy to be freeze-dried in comparison to tomatoes, bananas, tubers, or carrot. Moreover, though lyophilization is necessary, SHBsAg expression level in leaves is many times higher than in naturally dry seeds (Sunil Kumar *et al.*, 2006). Research on a plant-derived vaccine against hepatitis B virus (HBV) requires further detailed studies on mucosal and systemic immune response and possible tolerance induction. For the effective oral vaccination, parameters such as dosage, timing, adjuvants, antigen formulation, immunization routes, etc., need to be empirically tested.

Human Growth Hormone

Human Growth Hormone (hGH) is a well-characterized protein and is produced for medical supplies with the *Escherichia coli* expression system (Hsiung *et al.*, 1986). Rice seeds are potentially useful hosts for the production of hGH proteins (Table 16.2). Several proteins of therapeutic value, such as interferon β, erythropoietin, and epidermal growth factor have been accumulated to a level below 0.01 per cent of total soluble protein in plant tissues (Daniell *et al.*, 2001, 2009). It has been suggested that the low protein yield is caused by post-translational instability due mistargeting and proteolysis (Doran, 2006). Shigemitsu (2011) has expressed hGH fused to a 10 kDa prolamin signal peptide to target hGH to ER under the control of the endosperm-specific 10 kDa prolamin promoter. RNA interference (RNAi) is a gene-silencing technology that inhibits target gene expression specifically (Fire *et al.*, 1998; Montgomery *et al.*, 1998; Tuschl *et al.*, 1999). This has become a powerful tool to suppress target genes in plants (Monsoor *et al.*, 2006; McGinnis *et al.*, 2007; Kim *et al.*, 2008) in a bid to enhance the production of hGH proteins. Kuroda *et al.* (2010) have demonstrated that green fluorescent protein (GFP) is accumulated at a high level in transgenic rice seeds that suppressed rice endogenous storage protein (13 kDa prolamin, glutelin, and 26 kDa globulin) using RNAi approach. They found that the transgenic rice seeds accumulated hGH in PB-IIs and effectively suppressed the expression of endogenous glutelins or prolamins. The hGH accumulated to 470 µg/g at the maximum level approximately 0.7 per cent of total seed protein in transgenic rice (Table 16.2). The signal peptide of storage proteins has a functional role in entering the ER and hGH polypeptides nally accumulated in PB-IIs via ER to PSV transport pathway in rice endosperm. Masumura *et al.* (2006) has shown that human interferon-α with a 10 kDa prolamin signal peptide was also transported into both PB-I and PB-II in transgenic rice endosperm.

Prospects

Molecular farming is the application of biotechnology principles to pharmaceutical production. In the last few years, technical advances have improved the design, stability and delivery of therapeutic molecules, which have many advantages for the formulation of vaccines particularly when several peptides are combined to achieve precisely tuned immune responses. The major challenges in converting a plant-based therapeutic proteins in to a drug is equivalent to get an approval for an antibody produced from any other system. This is best described as good manufacturing practice (GMP) and the details compliance from monitoring the air to waste disposal is described by World Health Organization (WHO, 2007, 2009). Quite a few of the therapeutic protein to be used as drugs are in development and clinical trials phase (see www.phrma.org). The major transition of plant-derived pharmaceutical proteins to the clinic has already begun, with the recent development of several GMP-compliant production processes (Lai and Chen 2012) and industrial development of large-scale manufacturing infrastructure, such as the facilities commissioned by Medicago Inc., Kentucky BioProcessing and G-CON (Fischer *et al.*, 2011). These developments are likely to remove regulatory and economic hurdles that have until now prevented the use of plantbased production systems for pharmaceutical proteins. Despite the high costs and difficulties in the production, it seems that 'Molecular Farming' is one of the most promising options for large-scale therapeutic protein production in the future.

On the economic front the plant biopharming industry has demonstrated considerable growth and has attracted a large amount of investment in the current years. At present, it comprises, 100 small companies, each focusing on the development of a few products. Many companies such as Prodigene (http://www.prodigene.com/), Epicyte (http://epicyte.com), Large Scale Biology (http://www.lsbc.com/), SemBioSys Genetics (http://www.sembiosys.com/) and the European Meristem Therapeutics (http://www.meristem-therapeutics. com) are leading the market, each of which has tens of products in the pipeline. A number of productive collaborations and mergers among pharmaceutical companies, such as the collaboration of Meristem Therapeutics with Solvay Pharmaceuticals (http://www.solvay.com) for the development of plant-derived human gastric lipase have already taken place. Another product being CaroRxe, a recombinant antibody used for the prevention of dental caries (Planet Biotechnology; http://www.planetbiotechnology.com/products. html#carorx). The company Chlorogen (http://www.chlorogen.com/) has been founded recently, to exploit the chloroplast transgenic system for the commercial production of human therapeutics. Being highly technically sound, expertise demanding and cost intensive a few industries such as CropTech are collapsed, owing to its inability to achieve the expression levels required for commercial feasibility.

References

1.　Andrianova EP, Krementsugskaia SR, Lugovskaia NN, Mayorova TK, Borisov VV, Eldarov MA, Ravin NV, Folimonov AS, Skryabin KG (2011). Foot and mouth disease virus polyepitope protein produced in bacteria and plants induces protective immunity in guinea pigs. Biochemistry (Mosc). 76: 339–346.

2.　Arakawa T, Chong DKX and Langridge WHR (1998a). Efficacy of a food plant based oral cholera toxin B subunit vaccine. Nature Biotechnol. 16: 292–297 [PubMed: 9528012].

3.　Arakawa T, Chong DKX, Merrit JL, Langridge WHR (1997). Expression of cholera toxin B subunit oligomers in transgenic potato plants. Transgenic Res 6: 403–413.

4.　Arakawa T, Yu J, Chong DKX, Hough J, Engen PC and Langridge WHR (1998b). A plant–based cholera toxin B subunit – insulin fusion protein protects against the development of autoimmune diabetes. Nature Biotechnol. 16: 934–938.

5. Arazi T, Slutsky SG, Shiboleth YM, Wang Y, Rubinstein M, Barak S, Yang J, Gal-On A (2001). Engineering zucchini yellow mosaic potyvirus as a non–pathogenic vector for expression of heterologous proteins in cucurbits. J Biotechnol 87: 67–82.

6. Aziz MA, Midcha S, Waheed SM, Bhatnagar R (2007). Oral vaccines: new needs, new possibilities. BioEssays 29: 591–604.

7. Bandurska K, Brodzik R, Spitsin S, Kohl T, Portocarrero C, Smirnov Y, Pogrebnyak N, Sirko A, Koprowski H, Golovkin M (2008). Plant–produced hepatitis B core protein chimera carrying anthrax protective antigen domain–4. Hybridoma (Larchmt). 27: 241–247.

8. Barbante A, Irons S, Hawes C, Frigerio L, Vitale A, Pedrazzini E (2008). Anchorage to the cytosolic face of the endoplasmic reticulum membrane: a new strategy to stabilize a cytosolic recombinant antigen in plants. Plant Biotechnol J 6: 560–575.

9. Bardor M, Cabrera G, Stadlmann J, Lerouge P, Cremata JA, Gomord V, *et al.,* N–glycosylation of plant recombinant pharmaceuticals. Methods Mol Biol 2009; 483: 239–64.

10. Boothe J, Nykiforuk C, Shen Y, Zaplachinski S, Szarka S, Kuhlman P, Murray E, Morck D, Moloney, M M (2010). Seed–based expression systems for plant molecular farming. Plant Biotechnol J 8(5): 588–606.

11. Boothe JG, Nykiforuk CL, Kuhlman PA, Whelan H, Pollock WBRCS, Yuan S, Kumar R, Murray EW, Visser F, Martens K, Wu J, Pollock E, Given B, Szarka S, Zaplachinski S, Harry I, Keon R, Moloney MM (2009). Analytical characterization, safety and clinical bioequivalence of recombinant human insulin from transgenic plants. American Diabetes Association 69th Scientic Sessions Abstract 5–LB. American Diabetes Association, Alexandria.

12. Boyhan D, Daniell H (2011). Low–cost production of proinsulin in tobacco and lettuce chloroplasts for injectable or oral delivery of functional insulin and C–peptide. Plant Biotechnol J 9(5): 585–598. doi: 10.1111/j.1467–7652.2010.00582.x.

13. Bray BL (2003). Large–scale manufacture of peptide therapeutics. Nat Rev Drug 2: 587–593.

14. Brennan, F. R.; Bellaby, T.; Helliwell, S. M.; Jones, T. D.; Kamstrup, S.; Dalsgaard, K.; Flock, J.–I.; Hamilton, W. D. O. (1999a). Chimeric plant virus particles administered nasally or orally induce systemic and mucosal immune responses in mice. J. Virol. 73: 930–938.

15. Brennan, F. R.; Jones, T. D.; Gilleland, L. B.; Bellaby, T.; Xu, F.; North, P. C.; Thompson, A.; Staczek, J.; Lin, T.; Johnson, J. E.; Hamilton, W. D. O.; Gilleland, H. E. (1999b). *Pseudomonas aeruginosa* outer–membrane protein F epitopes are highly immunogenic in mice when expressed on a plant virus. Microbiology 145: 211–220.

16. Bruss V (2007). Hepatitis B virus morphogenesis. World J Gastroenterol 13: 65–73.

17. Buetow, D. E., Korban, S. S. Transgenic plants producing viral and bacterial antigens. AgBiotechNet 2, ABN 045, www.agbiotechnet.com; 2000.

18. Carrillo C, Wigdorovitz A, Oliveros JC, Zamorano PI, Sadir AM, Gomez N, Salinas J, Escribano JM and Borca MV (1998). Protective immune response to foot–and–mouth disease virus with VP1 expressed in transgenic plants. J. Virol. 72: 1688–1690.

19. Castanon S, Marin MS, Martin–Alonso JM, Boga JA, Casais R, Humara JM, Ordas RJ and Parra F (1999). Immunization with potato plants expressing VP60 protein protects against rabbit hemorrhagic disease virus. J. Virol. 73: 4452–4455.

20. Castilho A, Strasser R, Stadlmann J, Grass J, Jez J, Gattinger P, Kunert R, Quendler H, Pabst M, Leonard R, Altmann F, Steinkellner H (2010). In planta protein sialylation through overexpression of the respective mammalian pathway. J Biol Chem 285: 15923–15930.

21. Christou P (1993). Particle gun–mediated transformation. Curr. Opin. Biotechnol. 4: 135–141.

22. Christou P (1995). Strategies for variety–independent genetic transformation of important cereals, legumes and woody species utilizing particle bombardment. Euphytica 85: 13–27.

23. Circelli P, Donini M, Villani ME, Benvenuto E, Marusic C (2010). Efcient *Agrobacterium*–based transient expression system for the production of biopharmaceuticals in plants. Bioeng Bugs 1: 221–224.

24. Conner AJ, Dale PJ (1996). Reconsideration of pollen dispersal data from eld trials of transgenic potatoes. Theor Appl Genet 92(5): 505–508.

25. Conrad, U. and Fiedler, U. (1998). Compartment–specific accumulation of recombinant immunoglobulins in plant cells: an essential tool for antibody production and immunomodulation of physiological functions and pathogen activity. Plant Mol. Biol. 38, 101–109.

26. Cox K, Sterling J, Regan J, Gasdaska J, Frantz K, Peele C, *et al.*, Glycan optimization of a human monoclonal antibody in the aquatic plant *Lemna minor*. Nat Biotechnol 2006; 24: 1591–7.

27. Craik DJ, Simonsen S, Daly NL (2002). Thecyclotides: novel macrocyclic peptides as scaffolds in drug design. Curr Opin Discov Dev 5: 251–260.

28. Crofcheck C, Loiselle M, Weekly J, Maiti I, Pattanaik S, Bummer PM, Jayt M (2003). Histidine–tagged protein recovery from tobacco extract by foam fractionation. Biotechnol Prog, 19: 680–682.

29. Croft NP, Purcell AW (2011). Peptidomimetics: modifying peptides in the pursuit of better vaccines. Expert Rev Vaccines 10: 211–226.

30. Cueno ME, Hibi Y, Karamatsu K, Yasutomi Y, Imai K, Laurena AC, Okamoto T (2010). Preferential expression and immunogenicity of HIV–1 Tat fusion protein

expressed in tomato plant. Transgenic Res 19: 889–895.

31. Dalsgaard K, Uttenthal A, Jones TD, Xu F, Merryweather A, Hamilton WD, Langeveld JP, Boshuizen RS, Kamstrup S, Lomonossoff GP, Porta C, Vela C, Casal JI, Meloen RH, Rodgers PB (1997). Plant–derived vaccine protects target animals against a viral disease. Nat Biotechnol 15: 248–252.

32. Daniell H (2006). Production of biopharmaceuticals and vaccines in plants via the chloroplast genome. Biotechnol J 1: 1071–1079.

33. Daniell H, *et al.* (2009a). Optimization of codon composition and regulatory elements for expression of human insulin like growth factor–1 in transgenic chloroplasts and evaluation of structural identity and function. BMC Biotechnol; 9: 23. [PubMed: 19298646].

34. Daniell H, Khan MS and Allison L (2002). Milestones in chloroplast genetic engineering: an environmental friendly era in biotechnology. Trends Plant Sci. 7: 84–91.

35. Daniell H, Singh ND, Mason H, Streateld SJ (2009b). Plant–made vaccine antigens and biopharmaceuticals. Trends Plant Sci 14(12): 669–679. doi: 10.1016/j.tplants.2009.09. 009.

36. Daniell H, Streateld SJ, Wycoff K (2001a). Medical molecular farming, production of antibodies, biopharmaceuticals and edible vaccines in plants. Trends Plant Sci 6: 219–226.

37. Daniell, H. *et al.* (2001b). Expression of the native cholera B toxin subunit gene and assembly as functional oligomers in transgenic tobacco chloroplasts. J. Mol. Biol. 311: 1001–1009.

38. De Jaeger G, Scheffer S, Jacobs A, Zambre M, Zobell O, Goossens A, Depicker A, Angenon G (2002). Boosting heterologous protein production in transgenic dicotyledonous seeds using Phaseolus vulgaris regulatory sequences. Nat Biotechnol, 20: 1265–1268.

39. De Virgilio M, De Marchis F, Bellucci M, Mainieri D, Rossi M, Benvenuto E, Arcioni S, Vitale A (2008). The human immunodeciency virus antigen Nef forms protein bodies in leaves of transgenic tobacco when fused to zeolin. J Exp Bot 59: 2815–2829.

40. Dogan, B.; Mason, H.S.; Richter, L.; Hunter, J.B.; Shuler, M.L. (2000). Process options in hepatitis B surface antigen extraction from transgenic potato. *Biotechnol. Prog.*, 16: 435–441.

41. Doran PM (2006). Foreign protein degradation and instability in plants and plant tissue cultures. Trends Biotechnol 24: 426–432.

42. Drake, P.M.W. *et al.* (2003). Rhizosecretion of a monoclonal antibody protein complex from transgenic tobacco roots. Plant Mol. Biol. 52, 233–241.

43. Ducat E, Deprez J, Gillet A, Noel A, Evrard B, Peulen O, Piel G (2011). Nuclear delivery of a therapeutic peptide by long circulating pH–sensitive liposomes:

benets over classical vesicles. Int J Pharm 25 (Epub ahead of print). doi: 10.1016/ j.ijpharm.2011.08.034.

44. Durocher Y, Butler M. (2009). Expression systems for therapeutic glycoprotein production. Curr Opin Biotechnol; 20: 700–7.

45. Durrani Z, McInerney TL, McLain L, Jones T, Bellaby T, Brennan FR, Dimmock NJ (1998). Intranasal immunization with a plant virus expressing a peptide from HIV–1 gp41 stimulates better mucosal and systemic HIV–1–specific IgA and IgG than oral immunization. J Immunol Methods 220: 93–103.

46. Ehsani P, Khabiri A, Domansky NN (1997). Polypeptides of hepatitis B surface antigen produced in transgenic potato. Gene 190: 107–111.

47. Elliott S, Lorenzini T, Asher S, Aoki K, Brankow D, Buck L, *et al.* (2003). Enhancement of therapeutic protein in vivo activities through glycoengineering. Nature Biotechnology; 414–21.

48. Fan Q, Treder K, Miller WA (2012). Untranslated regions of diverse plant viral RNAs vary greatly in translation enhancement efficiency. BMC Biotechnol, 12(1): 22.

49. Fire A, Xu S, Montgomery MK, Kostas SA, Driver SE, Mello CC (1998). Potent and specific genetic interference by doublestranded RNA in Caenorhabditis elegans. Nature 391: 806–811.

50. Fischer R, Schillberg S, Hellwig S, Twyman RM, Drossard J (2011). GMP issues for plant–derived recombinant proteins. Biotechnol Adv. doi: 10.1016/ j.biotechadv.2011.08.007.

51. Fischer R, Stoger E, Schillberg S, Christou P, Twyman RM (2004). Plantbased production of biopharmaceuticals. Curr Opin Plant Biol 7: 152–158.

52. Frolova OY, Petrunia IV, Komarova TV, Kosorukov VS, Sheval EV, Gleba YY, Dorokhov YL (2010). Trastuzumab–binding peptide display by Tobacco mosaic virus. Virology 407: 7–13.

53. Ganapathi, T.R.; Sunil Kumar, G.B.; Srinivas, L.; Revathi, C.J.; Bapat, V.A. (2007). Analysis of the limitations of hepatitis B surface antigen expression in soybean cell suspension cultures. *Plant Cell Rep.,* 1575–1584.

54. Gao Y, Ma Y, Li M, Cheng T, Li S–W, Zhang J, Xia N–S (2003). Oral immunization of animals with transgenic cherry tomatillo expressing HBsAg. World J Gastroenterol 9: 996–1002.

55. Girard MP, Osmanov SK, Kieny MP (2006a). A review of vaccine research and development: the human immunodeciency virus (HIV). Vaccine 24: 4062–4081.

56. Girard MP, Steele D, Chaignat CL, Kieny MP (2006b). A review of vaccine research and development: human enteric infections. Vaccine 24: 2732–2750.

57. Giuliani A, Rinaldi AC (2011). Beyond natural antimicrobial peptides: multimeric peptides and other peptidomimetic approaches. Cell Mol Life Sci 68: 2255–2266.

58. Goldstein G, Manson K, Tribbick G, Smith R (2000). Minimization of chronic

plasmazviremia in rhesus macaques immunized with synthetic HIV–1 Tat peptides and infected with a chimeric simian/human immunodeciency virus (SHIV33). Vaccine 25: 2789–2795.

59. Goldstein ST and Fiore AE (2001). Toward the global elimination of hepatitis B virus transmission. J Pediatr 139: 343–345.

60. Gomez N, Carrillo C, Salinas J, Parra F, Borca MV and Escribano JM (1998). Expression of immunogenic glycoprotein S polypeptides from transmissible gastroenteritis corona virus in transgenic plants. Virology 249: 352–358.

61. Gomord V, Fitchette AC, Menu–Bouaouiche L, Saint–Jore–Dupas C, Plasson C, Michaud D, Faye L (2010). Plant–specific glycosylation patterns in the context of therapeutic protein production. Plant Biotechnol J 8: 564–587.

62. Gonzalez–Rabade N, McGowan EG, Zhou F, McCabe MS, Bock R, Dix PJ, Gray JC, Ma JK (2011). Immunogenicity of chloroplastderived HIV–1 p24 and a p24–Nef fusion protein following subcutaneous and oral administration in mice. Plant Biotechnol J 9: 629–638.

63. Greco R, Michel M, Guetard D, Cervantes–Gonzalez M, Pelucchi N, Wain–Hobson S, Sala F, Sala M (2007). Production of recombinant HIV–1/HBV virus–like particles in *Nicotiana tabacum* and *Arabidopsis thaliana* plants for a bivalent plant–based vaccine. Vaccine 25: 8228–8240.

64. Haq TA, Mason HS, Clements JD and Arntzen CJ (1995). Oral immunization with a recombinant bacterial antigen produced in transgenic plants. Science 268: 714–716.

65. Hashizume F, Hino S, Kakehashi M, Okajima T, Nadano D, Aoki N, Matsuda T (2008). Development and evaluation of transgenic rice seeds accumulating a type II–collagen tolerogenic peptide. Transgenic Res 17: 1117–1129.

66. Hayden, C.A.; Egelkrout, E.M.; Moscoso, A.M.; Enrique, C.; Keener, T.K.; Jimenez–Flores, R.; Wong, J.C.; Howard, J.A. (2012a). Production of highly concentrated, heat–stable hepatitis B surface antigen in maize. *Plant Biotechnol. J.*, 10: 979–984.

67. Hayden, C.A.; Streatfield, S.J.; Lamphear, B.J.; Fake, G.M.; Keener, T.K.; Walker, J.H.; Clements, J.D.; Turner, D.D.; Tizard, I.R.; Howard, J.A. (2012b). Bioencapsulation of the hepatitis B surface antigen and its use as an effective oral immunogen. *Vaccine*, 30: 2937–2942.

68. Hellwig S, Drossard J, Twyman RM, Fischer R (2004). Plant cell cultures for the production of recombinant proteins. Nat Biotechnol 22(11): 1415–1422. doi: 10.1038/Nbt1027.

69. Holmgren J, Czerkinsky C (2005). Mucosal immunity and vaccines. Nat Med 11[4 Suppl]: S45–53. doi: 10.1038/nm1213.

70. Hood, E.E. *et al.* (2002). Monoclonal antibody manufacturing in transgenic plants – myths and realities. Curr. Opin. Biotechnol. 13: 630–635.

71. Horsch RB, Fry JE, Hoffmann NL, Wallroth M, Eichholtz D, Rogers SG, Fraley

RT (1985). Simple and general method for transferring genes into plants. Science 227: 1229–1231.

72. Hsiung HM, Mayne NG, Becker GW (1986). High–level expression, efcient secretion and folding of human growth hormone in *Escherichia coli*. Nat Biotechnol 4: 991–995.

73. Huang N (2004). High–level protein expression system uses self pollinating crops as hosts. BioProcess Int 2: 54–59.

74. Huang X, Lu D, Ji G, Sun Y, Ma L, Chen Z, Zhang L, Huang J, Yu L (2004). Hepatitis B virus (HBV). vaccine–induced escape mutants of HBV S gene among children from Qidong area, China. Virus Res 99: 63–68.

75. Huang Z, Dry I, Webster D, Strugnell R and Wesselingh S (2001). Plant–derived measles virus hemagglutinin protein induces neutralizing antibodies in mice. Vaccine 19: 2163–2171.

76. Huang Z, Elkin G, Maloney BJ, Beuhner N, Arntzen CJ, Thanavala Y, Mason HS (2005). Virus–like particle expression and assembly in plants: hepatitis B and Norwalk viruses. Vaccine 23: 1851–1858.

77. Hull AK, Yusibov V, Mett V. (2005b). Inducible expression in plants by virus–mediated transgene activation. Transgenic Res;14: 407–16.

78. Imani, J.; Berting, A.; Nitsche, S.; Schaefer, S.; Gerlich, W.H.; Neumann, K.–H. (2002). The integration of a major hepatitis B virus gene into cell–cycle synchronized carrot cell suspension cultures and its expression in regenerated carrot plants. *Plant Cell Tiss. Org. Cult.*, 71: 157–164.

79. Jani D, Meena LS, Rizwan–ul–Haq QM, Singh Y, Sharma AK, Tyagi AK (2002). Expression of cholera toxin B subunit in transgenic tomato plants. Transgenic Res 11: 447–454.

80. Jani D, Singh NK, Chattacharya S, Meena LS, Singh Y, Upadhyay SN, Sharma AK, Tyagi AK (2004). Studies on the immunogenic potential of plant–expressed cholera toxin B subunit. Plant Cell Rep 22: 471–477.

81. Jones SM, Pons L, Roberts JL, Scurlock AM, Perry TT, Kulis M, Shrefer WG, Steele P, Henry KA, Adair M, Francis JM, Durham S, Vickery BP, Zhong X, Burks AW (2009). Clinical efcacy and immune regulation with peanut oral immunotherapy. J Allergy Clin Immunol 124: 292–300.

82. Joung YH, Youm JW, Jeon JH, Lee BC, Ryu CJ, Hong HJ, Kim HC, Joung H, Kim HS (2004). Expression of the hepatitis B surface S and preS2 antigens in tubers of *Solanum tuberosum*. Plant Cell Rep 22: 925–930.

83. Kanoria, S and Burma, PK (2012). A 28 nt long synthetic 50UTR (synJ). as an enhancer of transgene expression in dicotyledonous plants. BMC Biotechnology 2012, 12: 85, http://www.biomedcentral.com/1472–6750/12/85.

84. Kao JH, Chen DS (2002). Global control of hepatitis B virus infection. Lancet Infect Dis 2: 395–403.

85. Kapusta J, Modelska A, Figlerowicz M, Pniewski T, Letellier M, Lisowa O, Yusibov

V, Koprowski H, Plucienniczak A, Legocki AB (1999). A plant–derived edible vaccine against hepatitis B virus. FASEB J 13: 1796–1799.

86. Kapusta J, Modelska A, Pniewski T, Figlerowicz M, Jankowski K, Lisowa O, Plucienniczak A, Koprowski H, Legocki AB (2001). Oral immunization of human with transgenic lettuce expressing hepatitis B surface antigen. Adv Exp Med Biol 495: 299–303.

87. Kapusta J, Pniewski T, Wojciechowicz J, BociagP, Plucienniczak A (2010). Nanogram doses of alum–adjuvanted HBs antigen induce humoral immune response in mice when orally administered. Arch Immunol Ther Exp (Warsz). 58: 143–151.

88. Karaman, S., Cunnick, J. and Wang, K. (2012). Expression of the cholera toxin B subunit (CT–B). in maize seeds and a combined mucosal treatment against cholera and traveler's diarrhea. Plant Cell Rep.; 31: 527–537. DOI 10.1007/s00299–011–1146–3.

89. Karasev AV, Foulke S, Wellens C, Rich A, Shon KJ, Zwierzynski I, Hone D, Koprowski H, Reitz M (2005). Plant based HIV–1 vaccine candidate: Tat protein produced in spinach. Vaccine 7: 1875–1880.

90. Kew MC (2010). Epidemiology of chronic hepatitis B virus infection, hepatocellular carcinoma, and hepatitis B virus–induced hepatocellular carcinoma. Pathol Biol 58: 273–277.

91. Kim KY, Kwon SY, Lee HS, Hur Y, Bang JW, Kwak SS (2003). A novel oxidative stress–inducible peroxidase promoter from sweet potato: molecular cloning and characterization in transgenic tobacco plants and cultured cells. Plant Mol Biol, 51: 831–838.

92. Kim TG, Gruber A, Langridge WH (2004). HIV–1 gp120 V3 cholera toxin B subunit fusion gene expression in transgenic potato. Protein Expr Purif 37: 196–202.

93. Kim YS, Kim MY, Kim TG, Yang MS (2009). Expression and assembly of cholera toxin B subunit (CTB). in transgenic carrot (*Daucus carota* L.). Mol Biotechnol 41: 8–14.

94. Kim YS, Lee YH, Kim HS, Hahn KW, Ko JH, Joung H, Jeon JH (2008). Development of patatin knockdown potato tubers using RNA interference (RNAi). technology, for the production of human–therapeutic glycoproteins. BMC Biotechnol 8: 36.

95. Ko K, Tekoah Y, Rudd PM, Harvey DJ, Dwek RA, Spitsin S, Hanlon CA, Rupprecht C, Dietzschold B, Golovkin M, Koprowski H (2003). Function and glycosylation of plant–derived antiviral monoclonal antibody. Proc Natl Acad Sci USA,100: 8013–8018.

96. Komarova TV, Baschieri S, Donini M, Marusic C, Benvenuto E, Dorokhov YL (2010). Transient expression systems for plantderived bipharmaceuticals. Expert Rev Vaccines 9: 859–876.

97. Koncz C and Schell J (1986). The promoter of T*L*–DNA gene 5 controls the tissue specific expression of chimeric genes carried by a novel type of *Agrobacterium* binary vector. Mol. Gen. Genet. 204: 383–396.

98. Kong Q, Richter L, Yang YF, Arntzen CJ, Mason HS, Thanavala Y (2001). Oral immunization with hepatitis B surface antigen expressed in transgenic plants. Proc Natl Acad Sci USA 98: 11539–11544 [PubMed: 11553782].

99. Korban, SS (2002). Targeting and expression of antigenic proteins in transgenic plants for production of edible oral vaccines. In Vitro Cell Dev. Biol. Plants 38: 231–236.

100. Kuroda M, Kimizu M, Mikami C (2010). A simple set of plasmids for the production of transgenic plants. Biosci Biotechnol Biochem 74: 2348–2351.

101. Lai, H. and Chen, Q. (2012). Bioprocessing of plant–derived virus–like particles of Norwalk virus capsid protein under current Good Manufacture Practice regulations. *Plant Cell Rep.*, 31: 573–584. doi: 10.1007/s00299–011–1196–6.

102. Lam, D.M.–K. and Arntzen, C.J. (1996). Vaccines produced and administered through edible plants. U.S. Patent 5,484,719.

103. Larman HB, Zhao Z, Laserson U, Li MZ, Ciccia A, Gakidis MA, Church GM, Kesari S, Leproust EM, Solimini NL, Elledge SJ (2011). Autoantigen discovery with a synthetic human peptidome. Nat Biotechnol 29: 535–541.

104. Lau OS and Sun SSM (2009). Plant seeds as bioreactors for recombinant protein production. Biotechnol Adv 27(6): 1015–1022.

105. Lee SB, Li B, Jin S, Daniell H (2011). Expression and characterization of antimicrobial peptides Retrocyclin–101 and Protegrin–1 in chloroplasts to control viral and bacterial infections. Plant Biotechnol J 9: 100–115.

106. Leelavathi S, Reddy VS (2003). Chloroplast expression of His–tagged GUS fusions: a general strategy to overproduce and purify foreign proteins using transplastomic plants as bioreactors. Mol Bree; 11: 49–58.

107. Leite A, Kemper EL, da Silva MJ, Luchessi AD, Siloto RP, Bonaccorsi ED, El–Dorry HF, Arruda P (2000). Expression of correctly processed human growth hormone in seeds of transgenic tobacco plants. Mol Breed 6: 47–53.

108. Lentz EM, Segretin ME, Morgenfeld MM, Wirth SA, Dus Santos MJ, Mozgovoj MV, Wigdorovitz A, Bravo–Almonacid FF (2010). High expression level of a foot and mouth disease virus epitope in tobacco transplastomic plants. Planta 231: 387–395.

109. Levine MM (2010). Immunogenicity and efcacy of oral vaccines in developing countries: lessons from a live cholera vaccine. BMC Biol 8: 129.

110. Li H, Sethuraman N, Stadheim TA, Zha D, Prinz B, Ballew N, *et al.,* Optimization of humanized IgGs in glycoengineered *Pichia pastoris*. Nature Biotechnology 2006; 210–5.

111. Lico C, Chen Q, Santi L (2008). Viral vectors for production of recombinant proteins in plants. J Cell Physiol 216: 366–377.

112. Lico C, Mancini C, Italiani P, Betti C, Boraschi D, Benvenuto E, Baschieri S (2009). Plant–produced potato virus X chimeric particles displaying an inuenza virus–derived peptide activate specific CD8? T cells in mice. Vaccine 27: 5069–5076.

113. Lico, C., Santi, L, Twyman, RM, Pezzotti, M and Avesani, L (2012). The use of plants for the production of therapeutic human peptides Plant Cell Rep DOI 10.1007/s00299–011–1215–7.

114. Lien S and Lowman HB (2003). Therapeutic peptides. Trends Biotechnol 21: 556–562.

115. Lienard D, Sourrouille C, Gomord V, Faye L (2007). Pharming and transgenic plants. Biotechnol Annu Rev 13: 115–147.

116. Lindh I, Kalbina I, Thulin S, Scherbak N, Savenstrand H, Brave A, Hinkula J, Strid A, Andersson S (2008). Feeding of mice with Arabidopsis thaliana expressing the HIV–1 subtype C p24 antigen gives rise to systemic immune responses. APMIS 116: 985–994.

117. Lindsey K and Jones MGK (1987). Transient gene expression in electroporated protoplasts and intact cells of sugar beet. Plant Mol. Biol. 10: 43–52.

118. Lou X–M, Yao Q–H, Zhang Z, Peng R–H, Xiong A–S, Wang H–K (2007). Expression of the human hepatitis B virus large surface antigen gene in transgenic tomato plants. Clin Vaccine Immunol 14: 464–469.

119. Ma JK, Hiatt A, Hein M, Vine ND, Wang F, Stabila P, van Dolleweerd C, Mostov K and Lehner T (1995). Generation and assembly of secretory antibodies in plants. Science 268: 716–719.

120. Ma JK, Hikmat B, Wyco K, Vine M, Chargelegue D, Yu L, Hein M and Lehner T (1998). Characterization of a recombinant plant monoclonal secretary antibody and preventive immunotherapy in humans. Nature Med. 4: 601–606.

121. Ma JK–C, Barros E, Bock R, Christou P, Dale PJ, Dix PJ, Fischer R, Irwin J, Mahoney R, Pezzotti M, *et al.* (2005). Molecular farming for new drugs and vaccines: current perspectives on the production of pharmaceuticals in transgenic plants. EMBO Rep 6: 593–599.

122. Ma JK–C, Drake PM, Christou P (2003). The production of recombinant pharmaceutical proteins in plants. Nat Rev Genet 4: 794–805.

123. Ma JKC, Lehner T, Stabila P, Fux CI and Hiatt A (1994). Assembly of monoclonal antibodies with IgG1 and IgA heavy chain domains in transgenic tobacco plants. Eur. J. Immunol. 24: 131–138.

124. Madalinski K, Sylvan SP, Hellstrom U, Mikolajewicz J, ZembrzuskaSadkowska E, Piontek E (2002). Antibody responses to preS components after immunization of children with low doses of BioHepB. Vaccine 20: 92–97.

125. Maggiorella MT, Baroncelli S, Michelini Z, Fanales–Belasio E, Moretti S, Sernicola L, Cara A, Negri DR, Butto S, Fiorelli V, Tripiciano A, Scoglio A, Caputo A, Borsetti A, Ridol B, Bona R, ten Haaft P, Macchia I, Leone P, Pavone–Cossut MR, Nappi F, Ciccozzi M, Heeney J, Titti F, Cafaro A, Ensoli B (2004). Long–term

protection against SHIV89.6P replication in HIV–1 Tat vaccinated cynomolgus monkeys. Vaccine 22: 3258–3269.

126. Mallory–Smith C, Zapiola M (2008). Gene ow from glyphosate–resistant crops. Pest Manag Sci 64(4): 428–440.

127. Marusic C, Nuttall J, Buriani G, Lico C, Lombardi R, Baschieri S, Benvenuto E, Frigerio L (2007). Expression, intracellular targeting and purication of HIV Nef variants in tobacco cells. BMC Biotechnol 7: 12.

128. Marusic C, Rizza P, Lattanzi L, Mancini C, Spada M, Belardelli F, Benvenuto E, Capone I (2001). Chimeric plant virus particles as immunogens for inducing murine and human immune responses against human immunodeficiency virus type 1. J Virol 75: 8434–8439. doi: 10.1128/JVI.75.18.8434–8439.2001. [PMC free article] [PubMed].

129. Mason HS, Ball JM, Shi JJ, Jiang X, Estes MK, Arntzen CJ (1996). Expression of Norwalk virus capsid protein in transgenic tobacco and potato and its oral immunogenicity in mice. Proc Natl Acad Sci USA 93(11): 5335–5340.

130. Mason HS, Haq TA, Clements JD, Arntzen CJ (1998). Edible vaccine protects mice against Escherichia coli heat labile enterotoxin (LT): Potatoes expressing a synthetic LT–B gene. Vaccine 16: 1336–1343.

131. Mason HS, Lam DMK and Arntzen CJ (1992). Expression of hepatitis B surface antigen in transgenic plants. Proc. Natl. Acad. Sci. USA 89: 11745–11749.

132. Mason HS, Thanavala Y, Arntzen CJ, Richter E (2003). Expression of immunogenic hepatitis B surface antigen in transgenic plants. US Patent 6551 820 B1, 22.

133. Mason, H. S.; Ball, J. M.; Shi, J. J.; Jiang, X.; Extes, M. K.; Arntzen, C. J. (1996). Expression of Norwalk virus capsid protein in transgenic tobacco and potato and its oral immunogenicity in mice. Proc. Natl Acad. Sci. USA 93: 5335–5340.

134. Masumura T, Morita S, Miki Y, Kurita A, Morita S, Shirono H *et al.* (2006). Production of biologically active human interferon–a in transgenic rice. Plant Biotechnol 23: 91–97.

135. Masumura T, Morita S, Miki Y, Kurita A, Morita S, Shirono H *et al.* (2006). Production of biologically active human interferon–a in transgenic rice. Plant Biotechnol 23: 91–97.

136. Masuta C, Yamana T, Tacahashi Y, Uyeda I, Sato M, Ueda S, Matsumura T (2000). Development of clover yellow vein virus as an efcient, stable gene–expression system for legume species. Plant J 23: 539–546.

137. Matoba N, Kajiura H, Cherni I, Doran JD, Bomsel M, Fujiyama K, Mor TS (2009). Biochemical and immunological characterization of the plant–derived candidate human immunodeciency virus type 1 mucosal vaccine CTB–MPR. Plant Biotechnol J 7: 129–145.

138. Matoba N, Magerus A, Geyer BC, Zhang Y, Muralidharan M, Alfsen A, Arntzen CJ, Bomsel M, Mor TS (2004). A mucosally targeted subunit vaccine candidate

eliciting HIV–1 transcytosis–blocking Abs. Proc Natl Acad Sci USA 14: 13584–13589.

139. McGarvey PB, Hammond J, Dienelt MM, Hooper DC, Fu ZF, Dietzchold B, Koprowski H and Michaels FH (1995). Expression of the rabies virus glycoprotein in transgenic tomatoes. Biotechnology 13: 1484–1487.

140. McGinnis K, Murphy N, Carlson AR, Akula A, Akula C, Basinger H *et al*. (2007). Assessing the efciency of RNA interference for maize functional genomics. Plant Physiol 143: 1441–1451.

141. McLain L, Durrani Z, Wisniewski LA, Porta C, Lomonossoff GP, Dimmock NJ (1996). Stimulation of neutralizing antibodies to human immunodeciency virus type 1 in three strains of mice immunized with a 22 amino acid peptide of gp41 expressed on the surface of a plant virus. Vaccine 14: 799–810.

142. Meshcheryakova YA, Eldarov MA, Migunov AI, Stepanova LA, Repko IA, Kiselev CI, Lomonossoff GP, Skryabin KG (2009). Cowpea mosaic virus chimeric particles bearing the ectodomain of matrix protein 2 (M2E). of the inuenza A virus: production and characterization. Appl Mol Biol 43: 685–694.

143. Mestecky J, Nguyen H, Czerkinsky C, Kiyono H (2008). Oral immunization: an update. Curr Opin Gastroenterol 24: 713–719.

144. Mestecky J, Russell MW, Elson CO (2007). Perspectives on mucosal vaccines: is mucosal tolerance a barrier? J Immunol 179: 5633–5638.

145. Meyers A, Chakauya E, Shephard E, Tanzer FL, Maclean J, Lynch A, Williamson AL, Rybicki EP (2008). Expression of HIV–1 antigens in plants as potential subunit vaccines. BMC Biotechnol 8: 53.

146. Michel M–L, Tiollais P (2010). Hepatitis B vaccines: protective efcacy and therapeutic potential. Pathol Biol 58: 288–295.

147. Michelmore R, Marsh E, Seely S, Landry B (1987). Transformation of lettuce (Lactuca sativa). mediated by Agrobacterium tumefaciens. Plant Cell Rep 6: 439–442.

148. Mikschofsky H, Heilmann E, Schmidtke J, Schmidt K, Meyer U, Leinweber P, Broer I (2011). Greenhouse and eld cultivations of antigen–expressing potatoes focusing on the variability in plant constituents and antigen expression. Plant Mol Biol 76(1–2): 131–144. doi: 10.1007/s11103011–9774–0.

149. Mobs C, Slotosch C, Lofer H, Jakob T, Hertl M, Pfutzner W (2010). Birch pollen immunotherapy leads to differential induction of regulatory T cells and delayed helper T cell immune deviation. J Immunol 184: 2194–2203.

150. Modelska A, Dietzschold B, Sleysh N, Fu ZF, Steplewski K, Hooper DC, Koprowski H and Yushibov V (1998). Immunization against rabies with plant–derived antigen. Proc. Natl. Acad. Sci. USA 95: 2481–2485.

151. Moeller L, Gan Q, Wang K (2009). A bacterial signal peptide is functional in plants and directs proteins to the secretory pathway. J Exp Bot 60: 3337–3352.

152. Mohapatra U, McCabe MS, Power JB, Schepers F, van der Arend A, Davey MR

(1999). Expression of the bar gene confers herbicide resistance in transgenic lettuce. Transgenic Res 8: 33–44.

153. Moloney MM and van Rooijen GJH (1996). Recombinant proteins via oleosin partitioning. Inform; 7: 107–113.

154. Monsoor A, Amin I, Hussain M, Zafar Y, Briddon RW (2006). Engineering novel traits in plants through RNA interference. Trends Plant Sci 11: 559–565.

155. Montgomery MK, Xu S, Fire A (1998). RNA as a target of doublestranded RNA–mediated genetic interference in Caenorhabditis elegans. Proc Natl Acad Sci USA 95: 15502–15507.

156. Morassutti C, De Amicis F, Skerlavaj B, Zanetti M, Marchetti S (2002). Production of a recombinant antimicrobial peptide in transgenic plants using a modied VMA intein expression system. FEBS Let; 519: 141–146.

157. Murray C, Sutherland PW, Phung MM, Lester MT, Marshall RK, Christeller JT: Expression of biotin–binding protein, avidin and streptavidin, in plant tissues using plant vacuolar targeting sequences. Transgenic Res 2002, 11: 199–214.

158. Natilla A, Nemchinov LG (2008). Improvement of PVX/CMV CP expression tool for display of short foreign antigens. Protein Expr Purif 59: 117–121.

159. Nemchinov LG, Liang TJ, Rifaat MM, Mazyad HM, Hadidi A and Keith JM (2000). Development of a plant–derived subunit vaccine candidate against hepatitis C virus. Arch. Virol. 145: 2557–2573.

160. Nochi T, Takagi H, Yuki Y, Yang L, Masumura T, Mejima M, Nakanishi U, Matsumura A, Uozumi A, Hiroi T, Morita S, Tanaka K, Takaiwa F, Kiyono H (2007). Rice–based mucosal vaccine as a global strategy for cold–chain– and needle–free vaccination. Proc Natl Acad Sci USA 104: 10986–10991[PubMed: 17573530].

161. Nuttall J, Vine N, Hadlington JL, Drake P, Frigerio L, Ma JKC (2002). ER–resident chaperone interactions with recombinant antibodies in transgenic plants. Eur J Biochem 2002, 269: 6042–6051.

162. Nuzzaci M, Bochicchio I, De Stradis A, Vitti A, Natilla A, Piazzolla P, Tamburro AM (2009). Structural and biological properties of Cucumber mosaic virus particles carrying hepatitis C virusderived epitopes. J Virol Methods 155: 118–121.

163. Nuzzaci M, Piazzolla G, Vitti A, Lapelosa M, Tortorella C, Stella I, Natilla A, Antonaci S, Piazzolla P (2007). Cucumber mosaic virus as a presentation system for a double hepatitis C virusderived epitope. Arch Virol 152: 915–928.

164. Nykiforuk CL, Boothe JG, Murray EW, Keon RG, Goren HJ, Markley NA, Moloney MM (2006). Transgenic expression and recovery of biologically active recombinant human insulin from Arabidopsis thaliana seeds. Plant Biotechnol J 4: 77–85.

165. Obembe OO, Popoola JO, Leelavathi S, Reddy SV (2011). Advances in plant molecular farming. Biotechnol Adv 29: 210–222.

166. Obregon P, Chargelegue D, Drake PM, Prada A, Nuttall J, Frigerio L, Ma JK (2006). HIV–1 p24–immunoglobulin fusion molecule: a new strategy for plant–based protein production. Plant Biotechnol J 4: 195–207.

167. Ortigosa SM, Fernandez–San Millan A, Veramendi J (2010). Stable production of peptide antigens in transgenic tobacco chloroplasts by fusion to the p53 tetramerisation domain. Transgenic Res 19: 703–709.

168. Padidam M (2003). Chemically regulated gene expression in plants. Curr Opin Plant Biol 2003, 6: 169–177.

169. Pajno GB, Barberi S (2009). The history of sublingual immunotherapy. Int J Immunopathol Pharmacol 22: 1–3.

170. Palella FJ Jr, Delaney KM, Moorman AC, Loveless MO, Fuhrer J, Satten GA, Aschman DJ, Holmberg SD (1998). Declining morbidity and mortality among patients with advanced human immunodeciency virus infection. HIV Outpatient Study Investigators. N Engl J Med 338: 853–860.

171. Paz De la Rosa G, Monroy–Garcia A, Mora–Garcia Mde L, Pena CG, Hernandez–Montes J, Weiss–Steider B, Gomez–Lim MA (2009). An HPV 16 L1–based chimeric human papilloma virus–like particles containing a string of epitopes produced in plants is able to elicit humoral and cytotoxic T–cell activity in mice. Virol J 6: 2.

172. Peters, J. and Stoger, E. (2011). Transgenic crops for the production of recombinant vaccines and anti–microbial antibodies. Human Vaccines; 7(3): 367–374.

173. Phelps JP, Dang N, Rasochova L (2007). Inactivation and purication of cowpea mosaic virus–like particles displaying peptide antigens from Bacillus anthracis. J Virol Methods 141: 146–153.

174. Pinkhasov J, Alvarez ML, Rigano MM, Piensook K, Larios D, Pabst M, Grass J, Mukherjee P, Gendler SJ, Walmsley AM, Mason HS (2011). Recombinant plant–expressed tumour–associated MUC1 peptide is immunogenic and capable of breaking tolerance in MUC1.Tg mice. Plant Biotechnol J. doi: 10.1111/j.1467–7652.2011.00614.x.

175. Pniewski, T. (2013). The Twenty–Year Story of a Plant–Based Vaccine Against Hepatitis B: Stagnation or Promising Prospects? *Int. J. Mol. Sci.* 2013, *14*, 1978–1998; doi: 10.3390/ijms14011978.

176. Pniewski, T.; Kapusta, J.; Bociag, P.; Kostrzak, A.; Fedorowicz–Stronska, O.; Czyz, M.; Gdula, M.; Krajewski, P.; Wolko, B.; Plucienniczak, A. (2012). Plant expression, lyophilisation and storage of HBV medium and large surface antigens for a prototype oral vaccine formulation. *Plant Cell Rep.*, *31*, 585–595.

177. Pniewski, T.; Kapusta, J.; Bociag, P.; Wojciechowicz, J.; Kostrzak, A.; Gdula, M.; Fedorowicz–Stronska, O.; Wójcik, P.; Otta, H.; Samardakiewicz, S.; *et al.* (2011). Low–dose oral immunization with lyophilized tissue of herbicide–resistant lettuce expressing hepatitis B surface antigen for prototype plant–derived vaccine tablet formulation. *J. Appl. Genet.*, *52*, 125–136.

178. Polowick PL, Vandenberg A, Mahon JD (2002). Field assessment of outcrossing from transgenic pea (*Pisum sativum* L.). plants. Transgenic Res 11(5): 515–519.

179. Porta C and Lomonossoff GP (1996). Use of viral replicons for the expression of genes in plants. Mol. Biotech. 5: 209–221.

180. Ramessar K, Sabalza M, Capell T, Christou P (2008). Maize plants: an ideal production platform for effective and safe molecular pharming. Plant Sci 174(4): 409–419. doi: 10.1016/j.plantsci.2008.02.002.

181. Ramirez N, Rodriguez M, Ayala M, Cremata J, Perez M, Martinez A, Linares M, Hevia Y, Paez R, Valdes R *et al.* (2003). Expression and characterization of an anti–hepatitis B surface antigen glycosylated mouse antibody in transgenic tobacco plants, and its use in the immunopurication of its target antigen. Biotechnol Appl Biochem 2003, 38: 223–230.

182. Ramirez YJ, Tasciotti E, Gutierrez–Ortega A, Donayre Torres AJ, Olivera–Flores MT, Giacca M, Gomez–Lim MA (2007). Fruitspecic expression of the human immunodeciency virus type 1 tat gene in tomato plants and its immunogenic potential in mice. Clin Vaccine Immunol 14: 685–692.

183. Reddish M, MacLean GD, Koganty RR, Kan–Mitchell J, Jones V, Mitchell MS, Longenecker BM (1998). Anti–MUC1 class I restricted CTLs in metastatic breast cancer patients immunized with a synthetic MUC1 peptide. Int J Cancer 76: 817–823.

184. Regnard GL, Halley–Stott RP, Tanzer FL, Hitzeroth II, Rybicki EP (2010). High level protein expression in plants through the use of a novel autonomously replicating geminivirus shuttle vector. Plant Biotechnol J 8: 38–46.

185. Rendi–Wagner P, Shouval D, Genton B, Lurie Y, Rumke H, Boland G, Cerny A, Heim M, Bach D, Schroeder M, Kollaritsch H (2006). Comparative immunogenicity of a PreS/S hepatitis B vaccine in non– and low responders to conventional vaccine. Vaccine 24: 2781–2789.

186. Richter LJ, Thanavala Y, Arntzen CJ and Mason HS (2001). Production of hepatitis B surface antigen in transgenic plants for oral immunization. Nature Biotechnol. 18: 1167–1171 [PubMed: 11062435].

187. Romano L, Paladini S, Van Damme P, Zanetti AR (2011). The worldwide impact on the control and protection of viral hepatitis B. Dig Liver Dis 43S: S2–S7.

188. Rosales–Mendoza, S., Rubio–Infante, N., Govea–Alonso, D.O. and Moreno–Fierros, L. (2012). Current status and perspectives of plant–based candidate vaccines against the human immunodeficiency virus (HIV). Plant Cell Rep (2012). 31: 495–511 DOI 10.1007/s00299–011–1194–8.

189. Rouwendal G, Wuhrer M, Florack D, Koeleman C, Deelder A, Bakker H, *et al.* (2007). Efficient introduction of a bisecting GlcNAc residue in tobacco N–glycans by expression of the gene encoding human N–acetylglucosaminyltransferase III. Glycobiology 2007; 17: 334–44.

190. Ruhlman T, Ahangari R, Devine A, Samsam M, Daniell H (2007). Expression of cholera toxin B–proinsulin fusion protein in lettuce and tobacco chloroplasts–

oral administration protects against development of insulitis in nonobese diabetic mice. Plant Biotechnol J 5(4): 495–510. doi: 10.1111/j.1467–7652.2007.00259.x.

191. Rybicki EP (2010). Plant–made vaccines for humans and animals. Plant Biotechnol J 8: 620–637.

192. Salyaev RK, Stolbikov AS, Rekoslavskaya NI, Shchelkunov SN, Pozdnyakov SG, Chepinoga AV, Hammond RV (2010). Obtaining tomato plants transgenic for the preS2–S–HDEL gene, which synthesize the major hepatitis B surface antigen. Doklady Biochem Biophys 433: 187–190.

193. Sandhu JS, Krasnyanski SF, Domier LL, Korban SS, Osadjan MD and Buetow DE (2000). Oral immunization of mice with transgenic tomato fruit expressing respiratory syncytial virusF protein induces a systemic immune response. Transgenic Res. 9: 127–135 [PubMed: 10951696].

194. Sandhu, J. S.; Osadjan, M. D.; Krasnyanski, S. F.; Domier, L. L.; Korban, S. S.; Buetow, D. E. (1999). Enhanced expression of the human respiratory syncytial virus–F gene in apple leaf protoplasts. Plant Cell Rep. 18: 394–397.

195. Schahs M, Strasser R, Stadlmann J, Kunert R, Rademacher T, Steinkellner H. (2007). Production of a monoclonal antibody in plants with a humanized N–glycosylation pattern. Plant Biotechnol J; 5: 657–63.

196. Schillberg S, Fischer R, Emans N (2003). Molecular farming of recombinant antibodies in plants. Cell Mol Life Sci; 60: 433–445.

197. Schillberg S, Zimmermann S, Findlay K, Fischer R (2000). Plasma membrane display of anti–viral single chain Fv fragments confers resistance to tobacco mosaic virus. Mol Breed; 6: 317–326.

198. Schouten A, Roosien J, Bakker J, Schots A (2002). Formation of disulûde bridges by a single–chain Fv antibody in the reducing ectopic environment of the plant cytosol. J Biol Chem 2002, 277: 19339–19345.

199. Scotti N, Alagna F, Ferraiolo E, Formisano G, Sannino L, Buonaguro L, De Stradis A, Vitale A, Monti L, Grillo S, Buonaguro FM, Cardi T (2009). High–level expression of the HIV–1 Pr55gag polyprotein in transgenic tobacco chloroplasts. Planta 229: 1109–1122.

200. Sharma MK, Singh NK, Jani D, Sisodia R, Thungapathra M, Gautam JK, Meena LS, Singh Y, Ghosh A, Tyagi AK, Sharma AK (2008). Expression of toxin co–regulated pilus subunit A (TCPA). of *Vibrio cholerae* and its immunogenic epitopes fused to cholera toxin B subunit in transgenic tomato (*Solanum lycopersicum*). Plant Cell Rep 27: 307–318.

201. Shchelkunov SN, Salyaev RK, Pozdnyakov SG, Rekoslavskaya NI, Nesterov AE, Ryzhova TS, Sumtsova VM, Pakova NV, Mishutina UO, Kopytina TV, Hammond RW (2006). Immunogenicity of a novel, bivalent, plant–based oral vaccine against hepatitis B and human immunodeciency viruses. Biotechnol Lett 28: 959–967.

202. Shigemitsu, T, Ozaki, S, Saito, Y, Kuroda, M, Morita, S, Satoh, S and Masumura, T (2011). Production of human growth hormone in transgenic rice seeds: co–

introduction of RNA interference cassette for suppressing the gene expression of endogenous storage proteins. Plant Cell Rep; 31: 539–549. DOI 10.1007/s00299–011–1191–y.

203. Shinmyo A and Kato K (2010). Molecular farming: production of drugs and vaccines in higher plants. J Antibiot 63: 431–433.

204. Shulga NYa, Rukavtsova EB, Krymsky MA, Borisova VN, Melnikov VA, Bykov VA, YaI Buryanov (2004). Expression and characterization of hepatitis B surface antigen in transgenic potato plants. Biochemistry (Moscow). 69: 1158–1164.

205. Sijmons PC, Dekker BMM, Schrammeijer B, Verwoerd TC, van den Elzen PJM and Hoekema A (1990). Production of correctly processed human serum albumin in transgenic plants. Bio/Technology 8: 217–221.

206. Smith, M.L.; Mason, H.S.; Shuler, M.L. (2002). Hepatitis B surface antigen (HBsAg). expression in plant cell culture: Kinetics of antigen accumulation in batch culture and its intracellular form. *Biotechnol. Bioeng.* 80: 812–822.

207. Sojikul P, Buehner N, Mason HS (2003). A plant signal peptide hepatitis B surface antigen fusion protein with enhanced stability and immunogenicity expressed in plant cells. Proc Natl Acad Sci USA 100: 2209–2214.

208. Soria–Guerra RE, Rosales–Mendoza S, Marquez–Mercado C, LopezRevilla R, Castillo–Collazo R, Alpuche–Solis AG (2007). Transgenic tomatoes express an antigenic polypeptide containing epitopes of the diphtheria, pertussis and tetanus exotoxins, encoded by a synthetic gene. Plant Cell Rep 26: 961–968.

209. Soria–Guerra RE, Rosales–Mendoza S, Moreno–Fierros L, LopezRevilla R, Alpuche–Solis AG (2011). Oral immunogenicity of tomato–derived sDPT polypeptide containing Corynebacterium diphtheriae, Bordetella pertussis and Clostridium tetani exotoxin epitopes. Plant Cell Rep 30: 417–424.

210. Srinivas, L.; Sunil Kumar, G.B.; Ganapathi, T.R.; Revathi, C.J.; Bapat, V.A. Transient and stable expression of hepatitis B surface antigen in tomato (*Lycopersicon esculentum* L.). *Plant Biotechnol. Rep.*, **2008**, 2, 1–6.

211. Stoger E, Ma JKC, Fischer R, Christou P (2005). Sowing the seeds of success: pharmaceutical proteins from plants. Curr Opin Biotech 16(2): 167–173. doi: 10.1006/j.copbio.2005.01.005.

212. Stoger E, Vaquero C, Torres E, Sack M, Nicholson L, Drossard J, Williams S, Keen D, Perrin Y, Christou P, Fischer R (2000). Cereal crops as viable production and storage systems for pharmaceutical scFv antibodies. Plant Mol Biol 42(4): 583–590.

213. Strasser R, Stadlmann J, Schahs M, Stiegler G, Quendler H, Mach L, *et al.* (2008). Generation of glycoengineered *Nicotiana benthamiana* for the production of monoclonal antibodies with a homogeneous humanlike N–glycan structure. Plant Biotechnol J 2008; 6: 392–402.

214. Streateld SJ, Jilka JM, Hood EE, Turner DD, Bailey MR, Mayor JM, Woodard SL, Beifuss KK, Horn ME, Delany DE, Tizard IR and Howard JA (2001). Plant–based vaccines: unique advantages. Vaccine 19: 2742–2748.

215. Streateld SJ, Mayor JM, Barker DK, Brooks C, Lamphear BJ, Woodard SL, Beifuss KK, Vicuna DV, Massey LA, Horn ME, Delaney DE, Nikolov ZL, Hood EE, Jilka JM, Howard JA (2002). 2000 congress symposium on molecular farming: development of an edible subunit vaccine in corn against enterotoxigenic strains of Escherichia coli. *In vitro* Cell Dev Biol Plant 38(1): 11–17. doi: 10.1079/Ivp2001247.

216. Sunil Kumar GB, Srinivas L, Ganapathi TR, Bapat VA (2006). Hepatitis B surface antigen expression in transgenic tobacco (*Nicotiana tabacum*). plants using four different expression cassettes. Plant Cell Tissue Organ Cult 84: 315–323.

217. Sunil Kumar, G.B.; Ganapathi, T.R.; Bapat, V.A. (2007). Production of hepatitis B surface antigen in recombinant plant systems: An update. *Biotechnol. Prog.*, 23: 532–539.

218. Sunil Kumar, G.B.; Ganapathi, T.R.; Revathi, C.J.; Prasad, K.S.N.; Bapat, V.A. (2003). Expression of hepatitis B surface antigen in tobacco cell suspension cultures. *Prot. Exp. Purif.*, 32: 10–17.

219. Sunil Kumar, G.B.; Ganapathi, T.R.; Revathi, C.J.; Srinivas, L.; Bapat, V.A. (2005). Expression of hepatitis B surface antigen in transgenic banana plants. *Planta* 222, 484–493.

220. Surzycki R, *et al.* (2009). Factors effecting expression of vaccines in microalgae. Biologicals; 37: 133– 138. [PubMed: 19467445].

221. Suzuki K, Kaminuma O, Yang L, Takai T, Mori A, Umezu–Goto M, Ohtomo T, Ohmachi Y, Noda Y, Hirose S, Okumura K, Ogawa H, Takada K, Hirasawa M, Hiroi T, Takaiwa F (2011). Prevention of allergic asthma by vaccination with transgenic rice seed expressing mite allergen: induction of allergen–specic oral tolerance without bystander suppression. Plant Biotechnol J. doi: 10.1111/j.1467–7652.2011.00613.x.

222. Tackaberry ES, Dudani AK, Prior F, Tocchi M, Sardana R, Altosaar I and Ganz PR (1999). Development of biopharmaceuticals in plant expression systems: cloning, expression and immunological reactivity of human cytomegalovirus glycoprotein B (UL55). in seeds of transgenic tobacco. Vaccine 17: 3020–3029.

223. Tacket CO, Mason HS, Losonsky G, Clements JD, Levine MM and Arntzen CJ (1998). Immunogenicity in humans of a recombinant bacterial antigen delivered in a transgenic potato. Nature Med. 4: 607–609 [PubMed: 9585236].

224. Tacket CO, Mason HS, Losonsky G, Clements JD, Levine MM and Arntzen CJ (1998). Immunogenicity in humans of a recombinant bacterial antigen delivered in a transgenic potato. Nature Med. 4: 607–609 [PubMed: 9585236].

225. Tacket CO, Mason HS, Losonsky G, Estes MK, Levine MM and Arntzen CJ (2000). Human immune responses to a novel Norwalk virus vaccine delivered in transgenic potatoes. J. Infect. Dis. 182: 302–305 [PubMed: 10882612].

226. Takagi H, Hayashi Y, Jomori T, Takaiwa F (2006). The correlation between expression and localization of a foreign gene product in rice endosperm. Plant Cell Physiol 47: 756–763.

227. Takagi H, Hiroi T, Hirose S, Yang L, Takaiwa F (2010). Rice seed ER–derived protein body as an efcient delivery vehicle for oral tolerogenic peptides. Peptides 31: 1421–1425.

228. Takagi H, Hiroi T, Yang L, Takamura K, Ishimitsu R, Kawauchi H, Takaiwa F (2008). Efcient induction of oral tolerance by fusing cholera toxin B subunit with allergen–specic T–cell epitopes accumulated in rice seed. Vaccine 26: 6027–6030.

229. Takaiwa F, Takagi H, Hirose S, Wakasa Y (2007). Endosperm tissue is good production platform for articial recombinant proteins in transgenic rice. Plant Biotechnol J 5: 84–92.

230. Talukder, G. (2006). The market for bioengineered protein drugs, Business Communications Company Report, ID: BIO009E, BCC Inc., USA.

231. Thanavala Y, Mahoney M, Pal S, Scott A, Richter L, Natarajan N, Goodwin P, Arntzen CJ, Mason HS (2005). Immunogenicity in humans of an edible vaccine for hepatitis B. *Proc. Natl. Acad. Sci. USA 102*, 3378–3382. [PubMed: 15728371].

232. Thanavala Y, Yang YF, Lyons P, Mason HS and Arntzen CJ (1995). Immunogenicity of transgenic plant–derived hepatitis–B surface antigen. Proc. Natl. Acad. Sci. USA 92: 3358–3361.

233. Thanavala Y, Yang YF, Lyons P, Mason HS and Arntzen CJ (1995). Immunogenicity of transgenic plant–derived hepatitis–B surface antigen. Proc. Natl. Acad. Sci. USA 92: 3358–3361.

234. Torres AC, Cantliffe DJ, Laughner B, Bieniek M, Nagata R, Ashraf M, Ferl RJ (1993). Stable transformation of lettuce cultivar South Bay from cotyledon explants. Plant Cell Tissue Organ Cult 34: 279–285.

235. Tuboly T, Yu W, Bailey A, Degrandis S, Du S, Erickson L and Nagy E (2000). Immunogenicity of porcine transmissible gastroenteritis virus spike protein expressed in plants. Vaccine 18: 2023–2028.

236. Tuschl T, Zamore PD, Lehmann R, Bartel DP, Sharp PA (1999). Targeted mRNA degradation by double–stranded RNA in vitro. Genes Dev 13: 3139–3197.

237. Twyman, R. M., Stoger, E., Schillberg, S., Christou, P. and Fischer, R. (2003). Molecular farming in plants: host systems and expression technology TRENDS in Biotechnology, 21 (12): 570–578.

238. Vitti A, Piazzolla G, Condelli V, Nuzzaci M, Lanorte MT, Boscia D, De Stradis A, Antonaci S, Piazzolla P, Tortorella C (2010). Cucumber mosaic virus as the expression system for a potential vaccine against Alzheimer's disease. J Virol Methods 169: 332–340.

239. Vlieghe P, Lisowski V, Martinez J, Khrestchatisky M (2010). Synthetic therapeutic peptides: science and market. Drug Discov Today 15: 40–56.

240. Vos de AM, Ultsch M, Kossiakoff AA (1992). Human growth hormone and extracellular domain of its receptor: crystal structure of complex. Science 255: 306–312.

241. Wakasa Y, Tamakoshi C, Ohno T, Hirose S, Goto T, Nagaoka S, Takaiwa F (2011a). The hypocholesterolemic activity of transgenic rice seed accumulating lactostatin, a bioactive peptide derived from bovine milk beta–lactoglobulin. J Agric Food Chem 59: 3845–3850.

242. Wakasa Y, Zhao H, Hirose S, Yamauchi D, Yamada Y, Yang L, Ohinata K, Yoshikawa M, Takaiwa F (2011b). Antihypertensive activity of transgenic rice seed containing an 18–repeat novokinin peptide localized in the nucleolus of endosperm cells. Plant Biotechnol J 9: 729–735.

243. Walensky RP, Paltiel AD, Losina E, Mercincavage LM, Schackman BR, Sax PE, Weinstein MC, Freedberg KA (2006). The survival benets of AIDS treatment in the United States. J Infect Dis 194: 11–19.

244. Wang EM, Gan SS, Wagner GJ (2002). Isolation and characterization of the CYP71D16 trichome–specic promoter from *Nicotiana tabacum* L. J Exp Bot 2002, 53: 1891–1897.

245. Wang L, *et al.*, Immunogenicity of *Plasmodium yoelii* merozoite surface protein 4/ 5 produced in transgenic plants. Int. J. Parasitol 2008;38: 103–110. [PubMed: 17681344].

246. Wang ML, *et al.*, Production of biologically active GM–CSF in sugarcane: a secure biofactory. Transgenic Res 2005;14: 167–178. [PubMed: 16022388].

247. Wang X, *et al.*, A novel expression platform for the production of diabetes–associated autoantigen human glutamic acid decarboxylase (hGAD65). BMC. Biotechnol 2008;8: 87. [PubMed: 19014643].

248. Wang Y, *et al.*, Generation and immunogenicity of Japanese encephalitis virus envelope protein expressed in transgenic rice. Biochem. Biophys. Res. Commun 2009;380: 292–297. [PubMed: 19166811].

249. Wang Y, Wang W, Li N, Yu Y, Cao X (2002). Activation of antigenpresenting cells by immunostimulatory plant DNA: a natural resource for potential adjuvant. Vaccine 20: 2764–2771.

250. Webb AI, Dunstone MA, Williamson NA, Price JD, de Kauwe A, Chen W, Oakley A, Perlmutter P, McCluskey J, Aguilar MI, Rossjohn J, Purcell AW (2005). T cell determinants incorporating beta–amino acid residues are protease resistant and remain immunogenic *in vivo*. J Immunol 175: 3810–3818.

251. WHO (2007). Quality Assurance of Pharmaceuticals: a compendium of guidelines and related materials, Volume 2: good manufacturing practices and inspection. Second updated edition, 2007, 409pp.

252. WHO (2009). Expert Committee on Specifi cations for Pharmaceutical Preparations Forty–third report. WHO Technical Report Series, No. 953, 161pp.

253. Wigdorovitz, A. *et al.* (1999). Induction of a protective antibody response to foot and mouth disease in mice following oral or parenteral immunization with alfalfa transgenic plants expressing the viral structural protein VP1. *Virology* 255: 347–353.

254. Yang CD, Liao JT, Lai CY, Jong MH, Liang CM, Lin YL, Lin NS, Hsu YH, Liang SM (2007). Induction of protective immunity in swine by recombinant bamboo mosaic virus expressing foot–andmouth disease virus epitopes. BMC Biotechnol 7: 62.

255. Yang D, Wu L, Hwang YS, Chen L, Huang N. Expression of the REB transcriptional activator in rice grains improves the yield of recombinant proteins whose genes are controlled by a Reb–responsive promoter. Proc Natl Acad Sci USA 2001;98: 11438–43.

256. Yang Z, Drew DP, Jørgensen B, Mandel U, Bach SS, Ulvskov P, Levery SB, Bennett EP, Clausen H, and Bent L. Petersen BL (2012). Engineering Mammalian Mucin-type O–Glycosylation in Plants. The Journal of Biological Chemistry; 287(15): 11911–11923.

257. Yasuda H, Hayashi Y, Jomori T, Takaiwa F (2006). The correlation between expression and localization of a foreign gene product in rice endosperm. Plant Cell Physiol 47: 756–763.

258. Youm J–W, Won Y–S, Jeon JH, Ryu CJ, Choi Y–K, Kim H–C, Kim B–D, Joung H, Kim HS (2007). Oral immunogenicity of potatoderived HBsAg middle protein in BALB/c mice. Vaccine 25: 577–584.

259. Young, M.D.; Rosenthal, M.H.; Dickson, B.; Du, W.; Maddrey, W.C. (2001). A multi–center controlled study of rapid hepatitis B vaccination using a novel triple antigen recombinant vaccine. *Vaccine; 19,* 3437–3443.

260. Yusibov V, Mett V, Mett V, Davidson C, Muslychuk K, Gilliam S, Farese A, Macvittie T, Mann D (2005). Peptide–based candidate vaccine against respiratory syncytial virus. Vaccine 23: 2261– 2265.

261. Yusibov V, Modelska A, Steplewski K, Agadjanyan M, Weiner D, Hooper DC, Koprowski H (1997). Antigens produced in plants by infection with chimeric plant viruses immunize against rabies virus and HIV–1. Proc Natl Acad Sci USA 94: 5784–5788.

262. Yusibov V, Rabindran S. (2008). Recent progress in the development of plant derived vaccines. Expt. Rev. Vaccines; 7: 1173–1183. [PubMed: 18844592].

263. Zhang G, Leung C, Murdin L, Rovinski B, White KA (2000). In planta expression of HIV–1 p24 protein using an RNA plant virus–based expression vector. Mol Biotechnol 14: 99–107.

264. Zhang GG, Rodrigues L, Rovinski B, White KA (2002). Production of HIV–1 p24 protein in transgenic tobacco plants. Mol Biotechnol 20: 131–136.

265. Zhang Y, Li J, Pu H, Jin J, Zhang X, Chen M, Wang B, Han C, Yu J, Li D (2010). Development of Tobacco necrosis virus A as a vector for efcient and stable expression of FMDV VP1 to plants. Biotechnol. Appl. Biochem. 30: 101–108.

266. Zhou F, Badillo–Corona JA, Karcher D, Gonzalez–Rabade N, Piepenburg K, Borchers AMI, Maloney AP, Kavanagh TA, Gray JC, Bock R (2008). High–level expression of human immunodeciency virus antigens from the tobacco and tomato plastid genomes. Plant Biotechnol J 6(9): 897–913. doi: 10.1111/j.1467–7652.2008.00356.x.

267. Zuo, J. and Chua, N–H. (2000). Chemical–inducible systems for regulated expression of plant genes. Curr. Opin. Biotechnol. 11: 146–151.

Biotechnology: An Overview (2015)   *Pages 279–289*
Editors: Rajan Kumar Gupta, Nasim Akhtar and Deepak Vyas
Published by: DAYA PUBLISHING HOUSE, NEW DELHI

Chapter 17

Soil Remediation and Disposal

Madhuri Kaushish Lily*

*Department of Biotechnology, Modern Institute of Technology (MIT),
Dhalwala, Rishikesh – 249 201, Uttarakhand*

ABSTRACT

The selection of a suitable remediation process depends on the kind and concentration of pollutants, the soil type, the local availability of remediation processes, and economical aspects. Adequate treatment processes are available for all kinds of situations: biologically degradable pollutants should preferably be treated biologically.

Soils contaminated with non-biodegradable organic pollutants can be treated by thermal processes. To treat soils under controlled conditions, ex situ treatment, in which the soil is excavated and treated in specialized plants, should be preferred. In situ treatment avoids excavation and is therefore less costly, but it is often less effective and less controllable due to ubiquitous soil inhomogeneities. Additionally, it has to be assured that during in situ remediation no secondary pollution takes place and uncontrolled movement of the pollutants into uncontaminated areas is prevented. Therefore, extensive monitoring and securing measures may be necessary. This is especially true when natural attenuation processes are taken into consideration.

The excavation of significantly polluted soil and its disposal in landfills should be abolished. The possibilities for reuse of treated soils should be improved. It is essential that, as a first step in soil treatment, pre-investigations be performed to predict as far as possible the efficiency of the selected treatment process. This is especially true for biological soil treatment.

Keywords: Bioremediation, In situ remediation, Ex situ remediation, Pollutants, Soil.

* *Corresponding Author:* E-mail: mklily18@gmail.com

Introduction

Soil is continuously subjected to various types of contaminants such as pesticides, herbicides, insecticides, industrial effluents, polycyclic aromatic hydrocarbons (PAHs), toxic heavy metals and many more. For the treatment of contaminated sites, two types of methods are applicable including securing and remediation. Securing methods set up technical barriers for the protection of environment; however remediation achieves decontamination or reduction of pollutants. Securing measures often represent only a time restricted solution since the source of contamination remains in place and the technical barriers are subject to aging and environmental influences, therefore, future remediation activities become necessary.

On the basis their operation location and processing aspects, remediation methods are classified as *ex situ* and *in situ* processes on one hand, and thermal, chemical, physical, and biological processes on the other. The ex situ processes require excavation of the contaminated soil and soil treatment either at the site (on-site remediation) or at an external soil treatment plant (off-site remediation). In contrast, in situ treatment takes place at the site in the contaminated soil itself, without any soil excavation. Highly concentrated organic pollutants are treated by thermal processes however; they are suitable only to a small extent for the elimination of heavy metals. In soil scrubbing method, the coarse-grain fraction >63 μm is purified, transferring the pollutants into the water phase and/or into the fine-grain fraction. Since, this fine fraction is highly loaded with pollutants therefore it has to be treated and disposed of afterwards. For the large scale remediation purpose, the biopile process is applied for the effective treatment of biologically degradable pollutants such as mineral oil and its derivatives, aliphatic hydrocarbons, phenols, formaldehyde, and other soil contaminants.

Contaminated sites are remediated or secured, depending on the intended after-use, *e.g.*, housing, development of commercial or industrial facilities, or as recreational areas. If the contaminated site is not being used and so far no major dangerous contamination of the groundwater, surface water, etc. has occurred one relies more and more on natural attenuation processes in soil and groundwater (Dechema, 1992; Koning *et al.*, 2005).

Types of Remediation Processes

Thermal Processes

Thermal soil purification involves the use of thermal energy to transfer the pollutants from the soil matrix into the gas phase. The pollutants are released from the soil by vaporization and then burned followed by purification of the polluted gas. A variety of processing concepts are characterized for the thermal purification of contaminated soils by varying parameters such as temperature range, retention time for solids and waste gas in certain temperature zones, supply of oxygen, supply of reactive gases for gasification, supply of inert gas, kind of heat input, and optimum heat utilization, etc.

Thermal *Ex situ* Processes

Following processing steps are included in a thermal soil purification plant (Figure 17.1)

Soil Conditioning

In soil conditioning, the contaminated soil is made free of any interfering foreign matter such as scrap, plastics etc, thereafter broken, sieved, and homogenized by mechanical processes into particle size consistent with the technical requirements of the of the subsequent thermal treatment [<20 mm (fluidized bed) to 80 mm].

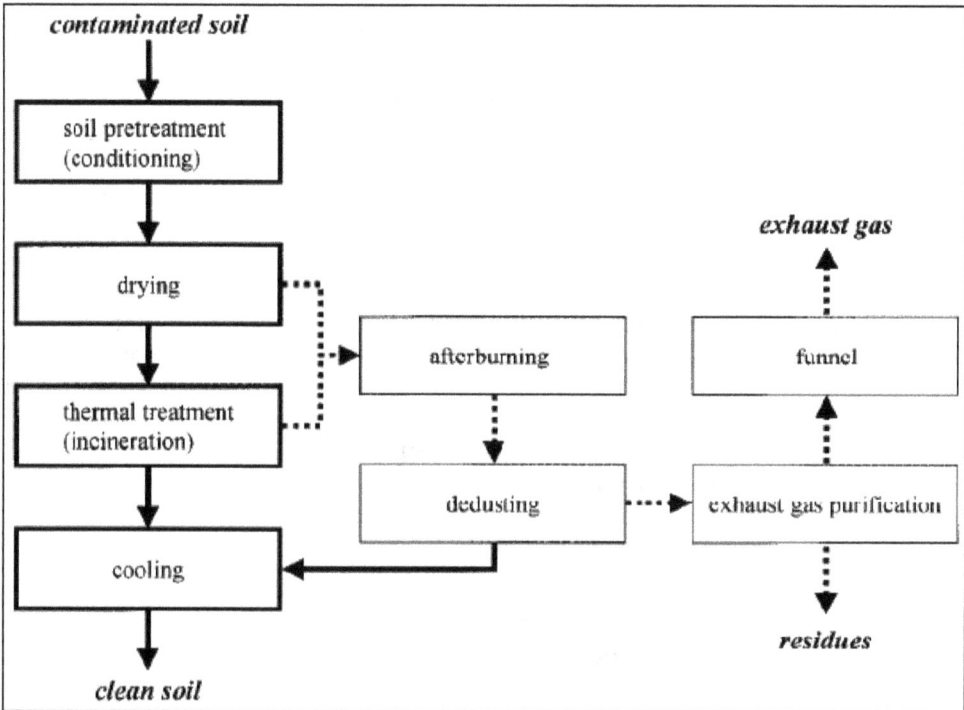

Figure 17.1. The thermal *ex situ* treatment process.

Thermal Treatment

During thermal treatment the soil is heated so that volatile pollutants are stripped from it. In the gas phase above the soil, the combustion of pollutants takes place, but in this phase complete destruction of volatile pollutants cannot be achieved. For this reason the gases are burned in an afterburner chamber at high temperature (~1200°C) for a certain retention time. Under these conditions dioxins are also destroyed.

Waste Gas Purification

Waste gas purification systems may vary and the selection of purification system is influenced by the local regulations on air emissions. The waste gas purification equipment mainly consists of three partial units including high-temperature afterburners, dedusters and flue gas purifiers. In high temperature afterburning at 900–1300°C, the organic pollutants (hydrocarbons) are completely oxidized. Dedusting involves the separation of fine soil particles from the gas stream by means of cyclones, hot gas filters, or combinations of different filter techniques specifically adapted to the process., flue gas purification minimizes the amounts of inorganic pollutants such as hydrogen chloride, hydrogen fluoride, and sulfur dioxide and brings the levels of dust and heavy metal emissions down to target values. Fundamentally, flue gas purification processes can be divided into a dry sorption process, a wet cleaning process (wet scrubbers), and a combination of both (semidry process). Special measures are required for the reduction of nitrogen oxides. To guarantee the target values of the waste gas, activated carbon filters are used as a final step in almost all processes.

The efficiency of the process depends not only on the effect of the temperature, but also on the physical properties of the pollutants. For these processes, the waste gas treatment normally includes no high-temperature afterburning of the waste gas. The purification is carried out via condensation and/or absorption (Fortmann and Jahns, 1996; ITVA, 1997; Koning *et al.*, 2005).

Thermal *In situ* Processes

Thermal in situ processes differ mainly in the kind of energy input for heating the soil matrix to transfer the pollutants into the gas phase. A combination of soil vapor extraction (SVE) and subsequent gas treatment is necessary to capture and treat the gas phase. In the steam-injection process, a hot steam–air mixture is passed into the unsaturated soil zone by steam–air injection (60–100°C). As a consequence, the volatile as well as the semivolatile compounds (Non aqueous phase liquids; NAPL) pass into the gas phase. The soil gas phase is extracted by means of gas extraction systems and is treated afterwards.

Mobilization of pollutants toward groundwater is prevented by specific temperature control and specific adjustment of the steam–air mixture. This process cannot be applied to soils with low permeability and for soils with very high impurities because these soils require long periods of heating time. In that case, the temperature within the unsaturated soil zone can be increased by imposing high-frequency electromagnetic fields by using electrodes (Jutterschenke, 1999; Koning *et al.*, 2001).

Application of Thermal Processes

Thermal processes can treat all kinds of pollutants that can be stripped from the soil under the influence of thermal energy. The operation temperatures and retention times depend on the type and concentration of the pollutants as well as on the intended use of the treated soil material.

Thermal ex situ processes are preferably employed when high initial concentrations of organic compounds are found and a high degree of purification is required. This process can remove mainly petroleum hydrocarbons (TPH), polycyclic aromatic hydrocarbons (PAH), volatile organic hydrocarbons (benzene, toluene, ethylbenzene, xylenes (BTEX)), phenolic compounds, cyanides, and chlorinated compounds such as polychlorinated biphenyls (PCB), pentachlorophenol (PCP), volatile halogenated hydrocarbons, chlorinated pesticides, polychlorinated dibenzodioxins (PCDD), and polychlorinated dibenzofurans (PCDF). Thermal ex situ processes can purify soil materials of all particle size distributions. A limitation on the proportion of silt (<30 per cent to 50 per cent) can become necessary for economic reasons. Furthermore, thermal ex situ processes can be used as a pre- or post-treatment step in conjunction with other ex situ processes. Some thermal processes having an exclusively thermal effect are mainly used for soils characterized by single substance class contamination with volatile pollutants.

On the other hand, because of their comparatively low generation of heat, in situ processes are suitable only for pollutants that can be stripped in the lower temperature range (*e.g.*, BTEX). The use of thermal in situ processes can also be restricted because of inhomogeneities or unsuitable soil water content.

The efficiency of thermal treatment processes for removing organic pollutants from contaminated soils approaches almost 100 per cent and is usually higher than the efficiency of biological or chemical/physical *ex situ* processes. However, to evaluate the application of various treatment processes, additional aspects, *e.g.*, the necessary energy input, technical requirements, treatment costs, possibilities for reuse of the treated soils, and other aspects have to be considered. Usually, the costs of thermal soil treatment are higher than the costs of biological or chemical/physical processes.

Chemical/Physical Processes

Chemical/physical soil treatment processes are mainly based on extraction and/ or wet classification processes. The principle of ex situ soil scrubbing technologies is to concentrate the contaminants in a small residual fraction by separation. In general, water (with or without additives) is used as an extracting agent. For the transfer of contaminants from the soil to the extracting agent, two mechanisms including strong shearing forces induced by pumping, mixing, vibration, high-pressure water jets and dissolution of contaminants by extracting agents are involved.

In situ extraction basically consists of percolation of an aqueous extracting agent into the contaminated soil. Percolation can be achieved by means of surface trenches, horizontal drains, or vertical deep wells. Soluble contaminants present in the soil dissolve in the percolate, which is pumped up and treated on-site.

Chemical/Physical *Ex situ* Processes

During soil scrubbing, the pollutants are detached from the soil particles by means of mechanical energy and/or solubilizing effects, often supported by surfactants. As a result, pollutants become concentrated in the liquid phase (water) and in the solid fine fraction of the soil containing pollutants sorbed onto the surface.

In general, soil scrubbing consists of the following steps (Figure 17.2):

1.Soil pretreatment 2. Soil washing 3. Separation by gravity (classification) 4. Separation of dispersed particles 5. Separation of process water and rinsing of the purified soil fraction 6. Process water recirculation 7. Wastewater purification 8. Waste gas purification.

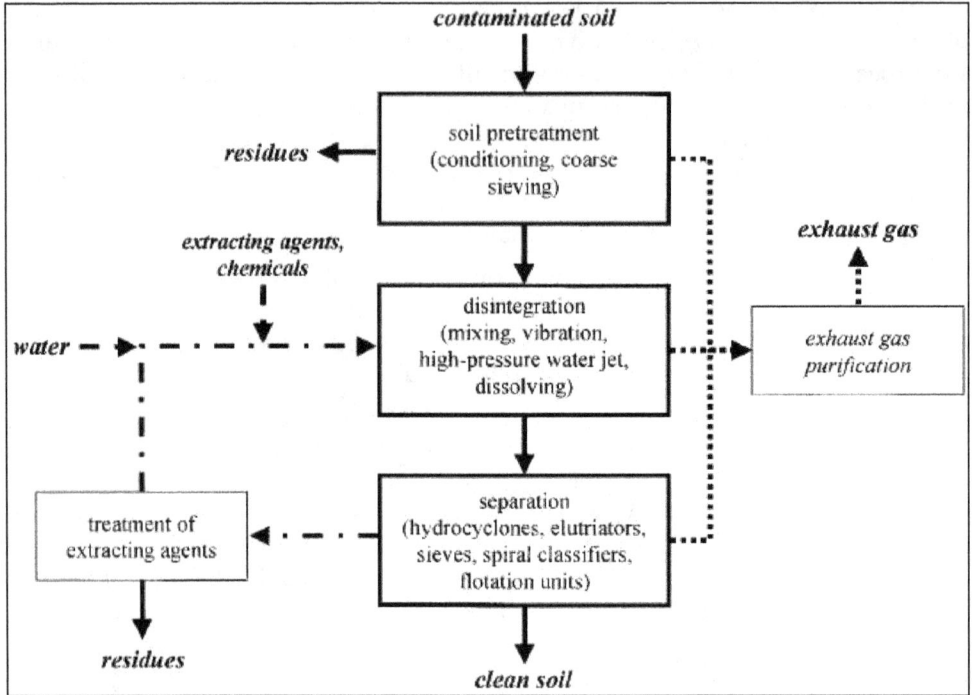

Figure 17.2. The chemical/physical *ex situ* process.

Chemical/Physical *In situ* Processes

During pump-and-treat processes, water is supplied to the soil so as to leach out the contaminants. The contaminated water is pumped to the surface and treated. Surfactants that increase the solubility of the pollutants may be added to the water. The extracted washing water is treated with standard wastewater treatment technologies. The possible applications of chemical/physical in situ processes are especially limited by the permeability of the soil. Soil vapor extraction (SVE) is an effective and economical process for decreasing highly volatile pollutants (*e.g.*, BTEX) in the unsaturated zone of permeable soils. Perforated pipes are placed in the contaminated soil area. The volatile pollutants are sucked out of the soil by using low vacuum blowers (ITVA, 1997). The extracted pollutants and condensates are treated on-site using activated carbon filters, compost filters, etc. The efficiency of the soil vapor extraction process is influenced by the characteristics of the soil (permeability, moisture content, temperature, homogeneity) and the kind of pollutants (vapor pressure).

Application of Chemical/Physical Processes

All kinds of pollutants that can be detached from soil particles and solubilized in the washing water can be treated by soil washing processes. Therefore, all kinds of pollutant groups have been treated with the ex situ soil scrubbing process: BTEX, TPH, PAH, PCB, PCP, PCDD and heavy metals etc (VDI, 2002; Koning *et al.*, 2001).

The actual purification of the polluted liquid phase and of the fine particle fraction can take place outside the soil washing plant, in separate treatment facilities. But the polluted fine fraction is usually dumped in a landfill, meaning that no actual treatment, but only separation, is achieved.

Biological Processes

The biological treatment involves the use of soil micro-organisms to convert organic pollutants (*e.g.*, hydrocarbons) into mainly CO_2, water, and biomass. Degradation may take place under aerobic as well as under anaerobic conditions. The aerobic process is predominantly used in soil remediation. For efficient biological treatment of contaminated soils, it is essential to optimize the environmental conditions for the micro-organisms (oxygen supply, water content, pH value, etc.). Under optimum conditions, the degradation processes are enhanced and the degree of degradation is improved, especially for high concentrations of readily degradable pollutants. Biological activity can be stimulated by doing soil homogenization, active aeration, moistening or drying, heating, addition of nutrients and substrates, or inoculation with micro-organisms. However, in comparison to thermal or chemical/physical treatment processes, less energy input but longer treatment periods are generally required.

Biological *Ex situ* Processes

Biological ex situ processes include 3 treatment steps (Figure 2.3) including 1. Mechanical pre-treatment 2. Addition of water, nutrients, substrates and micro-organisms 3. Biological treatment.

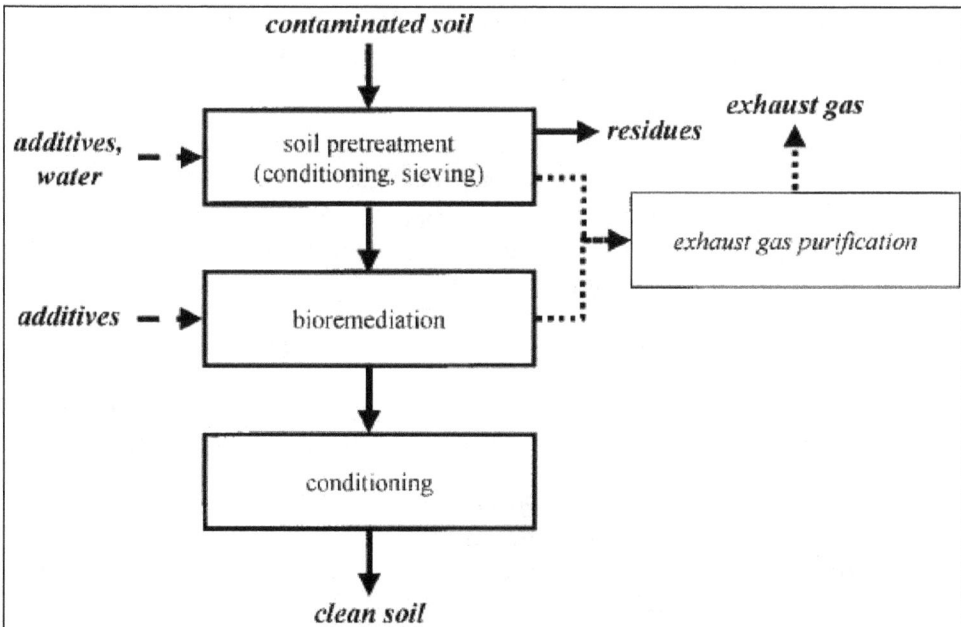

Figure 17.3: The biological *ex situ* process.

After excavation, the contaminated soil is mechanically broken up and sieved to remove disruptive material, homogenize the soil material, and loosen the soil structure. In addition, oxygen supply to the soil particles is improved. Mineral components separated during the sieving process can be crushed and later added to the contaminated soil. In order to activate biological degradation of the contaminants, water, nutrients, and substrates are added. Organic additives like compost, bark, wood chips, or straw serve as co-substrates or nutrient sources for the micro-organisms and as structural material.

Biological treatment of contaminated soils is done in thin soil layers (landfarming process), in biopiles, or in bioreactors (dry and slurry reactors).

1. In the landfarming process, the contaminated soil is treated in thin layers of up to 0.4 m thickness. Therefore, large treatment areas are required if large amounts of contaminated soil need to be treated. The pretreated soil is placed on foil, concrete, or a clay layer. Oxygen supply as well as mixing is enhanced by plowing, harrowing, or milling at regular intervals (Cookson, 1995; Koning *et al.*, 2001).

2. Biopiles for soil remediation are constructed in rectangular, oblong, or pyramidal forms are used. The height of the biopiles is usually between 0.8 and 3.0 m (Koning *et al.*, 2001; 2005). The biopile process can be carried out at water contents below and above the maximum water holding capacity (dry and wet systems). Dry systems can be operated with or without agitation (dynamic or static biopile process) whereas wet systems are all static. The dynamic biopile process involves decomposition of the soil by repeatedly plowing and turning the biopiles. If necessary, water and nutrients are added during the turning process. This increases the bioavailability of the pollutants, and the contamination is brought into close contact with micro-organisms, nutrients, water, and air. The biopile process is predominantly done dry and is used mainly for aerobic degradation processes. Oxygen contents >1 vol. per cent is ensured in all parts of the biopiles to achieve unrestricted aerobic degradation processes.

3. In the bioreactor process, soils are treated in the solid or slurry phase. The principle of solid-phase reactors is mechanical decomposition of the soil by abrasion and by intensive mixture of the components in a closed container ensuring permanent contact of contamination, micro-organisms, nutrients, water, and air. The soil can be aerated with an active aeration system or via exhaustion. However, the exhaust air has to be purified, for example, by the use of activated carbon filters or biofilters. Soils that is not suitable for solid bioreactors (clayey and silty soils) can be treated as slurry in suspension bioreactors. Suspension bioreactors are also suitable for treating the residual fine-particle fractions from the soil scrubbing process, which are highly loaded with contaminants. After treatment, the slurry is dewatered, and the contaminated water fraction is purified. Usually much of the water is recirculated (ITVA, 1994). Bioreactors offer better conditions for process control; therefore, usually shorter treatment periods are required as compared to landfarming and biopile.

Biological *In situ* Processes

Intrinsic bioremediation processes exploits the ability of micro-organisms to degrade organic pollutants under environmental conditions naturally present in the field. These processes can contribute to decreasing the levels of organic pollutants in soil over long-term periods (Newman and Barr, 1997; Koning *et al.*, 2005). In biological in situ treatment, the environmental conditions for the biological degradation of organic pollutants are optimized as far as possible. Oxygen usually has to be supplied, which can be done by artificial aeration or by addition of electron acceptors such as nitrate or oxygen-releasing compounds. Sometimes also H_2O_2 or O_3 dissolved in water is added. This accelerates the degradation of organic contaminants by the authochtonous microflora adapted to the present contaminants and addition of cultured micro-organisms is not necessary.

Application of Biological Processes

The biological turnover of organic pollutants depends mainly on the bioavailability and biodegradability of the contaminants, as well as on the environmental conditions for the degrading micro-organisms. Therefore, the degree of biological degradation achieved in a technical process is influenced by many factors (*e.g.*, type, concentration, and physical state of the contaminants, soil type, content of organic substances, adjustment of the environmental conditions) and can be limited for biological, physicochemical, or technical reasons (Dechema, 1991). Bioremediation methods are used for several different contaminants such as TPH, TNT, BTEX, phenols, PAHs, PCBs, etc. (Kastner, 2000; Koning *et al.*, 2005; Lily *et al.*, 2009; 2010).

Disposal

Due to economical reasons, the excavation and disposal of contaminated soil at landfills is most preferred alternative in redevelopment projects for contaminated sites. Although, the contaminated soil is removed from the original site, the contaminated soil remains usually untreated and therefore represents a potential source of environmental risk at the landfill site. Therefore, the landfilling of contaminated soil can be regarded as a securing measure with future remediation activities required from technical point of view. In order to save landfill capacity for other wastes and to promote the utilization of decontaminated soil, preference should be given to remediation measures.

Utilization of Decontaminated Soil

A major aspect in ex situ soil remediation is the reuse of decontaminated soil. During the various treatment processes, the soil materials change their chemical and physical properties in different ways. The reuse of biologically treated soil can be restricted by the residual concentrations of contaminants and of organic materials originating from organic additives (*e.g.*, compost, bark, wood chips). Such soil is not normally suitable for reuse as filling material or in agriculture; therefore, it is often used in landscaping. Thermally treated soil and soil from wet scrubbers can be used as filling material (*i.e.*, refill where excavated) but is not suitable for vegetation due to its inert nature. It is quite tedious to find adequate possibilities for utilization of the treated soil.

A crucial factor for the reuse of decontaminated soils is toxicological/ ecotoxicological assessment. For this purpose bioassays are conducted to measure the possible impacts of treated soils. Bioassays should be an appropriate tool if treated soils have to be tested with regard to their hazard potential since they integrate the effects of all relevant substances, including those not considered or recorded in chemical analyses (Dechema, 1995; Klein, 1999; Koning *et al.*, 2005).

References

1. Cookson, J.T., Jr., 1995. *Bioremediation Engineering: Design and Application.* McGraw-Hill, New York,

2. Dechema, E.V., 1995. *Bioassays for Soils.* Frankfurt/Main Dechema.

3. ITVA, 1994. Arbeitshilfe H 1–1 Dekontamination durch Bodenwaschverfahren. Berlin, IVTA.

4. ITVA, 1997. Arbeitshilfe H 1–6 Thermische Verfahrenur Bodendekontamination. Berlin, IVTA.

5. Jütterschenke, P., 1999. Thermische *in situ* Bodensanierung unter Einsatz hochfrequenter elektromagnetischer Felder: eine innovative Methode zur Reinigung kontaminierter Böden, in: Innovative Techniken der Bodensanierungein Beitrag zur Nachhaltigkeit, Deutsche Bundesstiftung Umwelt, (S. Heiden ed.), pp. 153–171. Heidelberg 1999: Spektrum Akademischer Verlag Fortmann, J., Jahns, P., Thermische Bodenreinigung, in: Altlasten: Erkennen, Bewerten, Sanieren 3rd edn. (Neumaier, H., Weber, H. H., eds.), Berlin 1996: Springer–Verlag, 272–303.

6. Kästner, M., 2000. Degradation of aromatic and polyaromatic compounds. In: *Biotechnology, Vol. 11b: Environmental Processes II,* (Eds.) H.-J. Rehm, G. Reed, A. Pühler and P. Stadler. Weinheim Wiley-VCH, pp. 211–239.

7. Klein, J., 1999. Biological soil treatment: status, development and perspectives. Bioremediation 1999: State of the art and future perspectives, *Proc. 9th Eur. Congr. Biotechnol ECB9,* July 11–15, Brussels, Belgium

8. Koning, M., Cohrs, I. and Stegmann, R., 2001. Development and application of an oxygen-controlled high-pressure aeration system for the treatment of TPH-contaminated soils in high biopiles: A case study. In: *Treatment of Contaminated Soil: Fundamentals, Analysis and Applications,* (Eds.) R. Stegmann, G. Brunner, W. Calmano and G. Matz. Springer-Verlag, Berlin, pp. 399–414.

9. Lily, M.K., Bahuguna, A., Dangwal, K. and Garg, V., 2009. Degradation of Benzo[a]pyrene by a novel strain *Bacillus subtilis* BMT4i (MTCC 9447). *Brazilian Journal of Microbiology,* 40(4): 884–892.

10. Lily, M.K., Bahuguna, A., Dangwal, K. and Garg, V., 2010. Optimization of an inducible chromosomally encoded nenzo[a]pyrene (BaP) degradation pathway in *Bacillus subtilis* BMT4i (MTCC 9447). *Annals of Microbiology,* 60(1): 51–58.

11. Koning, M., Hupe, K. and Stegmann, R., 2005. Soil remediation and disposal. In: *Environmental Biotechnology: Concepts and Applications*, (Eds.) H.-J. Jördening and J. Winter Copyright. Wiley–VCH Verlag GmbH and Co. KGaA, Weinheim.

12. Newman, A.W. and Barr, K.D., 1997. Assessment of natural rates of unsaturated zone hydrocarbon bioattenuation. In: *In situ and onsite Bioremediation*, Vol. 1, (Eds.) B.C. Alleman and A. Leeson. pp. 1–5. Columbus, OH, Battelle.

13. VDI, 2002. Emission control: plants for physical and chemical, thermal and biological soil treatment; immobilisation methods (VDI Guideline 3898). VDI/ DIN–Handbuch Reinhaltung der Luft (Air Pollution Prevention), Düsseldorf, VDI.

Biotechnology: An Overview (2015)
Editors: Rajan Kumar Gupta, Nasim Akhtar and Deepak Vyas
Published by: DAYA PUBLISHING HOUSE, NEW DELHI

Pages 291–300

Chapter 18

Application of Oxygen-Vectors in Improving the Performance of Submerged Aerobic Fermentation for the Production of Industrially Important Molecules in Bioreactors

Umesh K Narta[1]*, Rajesh Azad[1] and Wamik Azmi[2]

*[1]Government Post Graduate College,
Seema (Rohru) – 171 207, Himachal Pradesh
[2]Departmant of Biotechnology, H.P. University,
Shimla – 171 005, Himachal Pradesh*

ABSTRACT

The efficiency of oxygen transfer into the fermentation broths could be enhanced by adding in to the broths, the oxygen-vectors, such as hydrocarbons or fluorocarbons, without increasing the energy consumption for mixing or aeration. The media supplemented with oxygen-vectors by maintaining higher fraction of dissolved oxygen enhance the cell mass production along with the production of the molecules of the industrial importance and hence improve the overall performance of the aerobic fermentation economically. In the present article the use of oxygen-vectors as a simple and useful strategy of enhancing the oxygen solubility in the aqueous culture media during submerged fermentation is being discussed with a variety of microbial systems and bioprocesses engaged in large-scale production of the molecules of high industrial importance. The option of oxygen-vectors has been found to be economical, simple, reusable and non-toxic for the aerobic fermentation.

Keywords: Aerobic fermentation, Bioreactor, Oxygen-vector, Oxygen transfer rate.

----
* Corresponding Author: E-mail: umeshnarta@yahoo.co.in

Introduction

The importance of oxygen supply for microbial growth and product formation by micro-organism in submerged aerobic fermentation processes is well documented in the literature (Richards 1961; Lockhart and Squires, 1963; Narta *et al.*, 2011). Oxygen transfer from the gaseous phase to liquid media is critically important for successful bioreactor design and scale-up. This is mainly because oxygen is sparingly soluble in aqueous solutions and a continuous supply of oxygen is required for the growing micro-organisms. Even a brief interruption of oxygen supply can cause serious irreversible damage to the performance of aerobic fermentation. Achieving the sufficient oxygen supply is therefore a major challenge in economic large-scale liquid fermentations of aerobic micro-organisms. The aeration efficiency depends on oxygen solubilisation and diffusion rate into the broths, respectively and on the bioreactor capacity to satisfy the oxygen demand of microbial population. Generally, the availability of oxygen to the growing micro-organisms in a culture broth can be easily improved by increasing the stirrer speed and the aeration rate, but it leads to excessive power consumption and foam formation in the reactor, which is therefore neither effective nor economical. Further, the sensitivity of the growing micro-organisms to shear stress is rather high. One alternative approach is to generate oxygen chemically by introducing hydrogen peroxide to the medium, in which hydrogen peroxide will be converted to liquid oxygen by catalase inside the cells (Schlegel, 1997; Sonnleitner and Hahnemann, 1997). However, this approach may also necessitate simultaneous addition of catalase if the cells cannot produce sufficient catalase. In addition, hydrogen peroxide is toxic to cells, causing lower growth rates and biomass yields. Another simple and effective approach may be the addition of some organic solvents with high oxygen solubility to the fermentation broth as oxygen vectors, such as n-hexadecane (Ho *et al.*, 1990; Lee and Kim 2004; Li and Wu 2006), n-dodecane (Rols *et al.*, 1990; Wang 2000) and liquid paraffin (Narta *et al.*, 2011).

The addition of such non-aqueous solvents called oxygen-vectors in which oxygen solubilization is more than that in water (15-20 times) to the cultivation medium emerges out as one of the most effective schemes of improving oxygen transfer rate from gaseous phase to the micro-organisms growing under submerged conditions in aerobic fermentations. Higher oxygen solubility, non-toxicity to microbes, antifoaming action, lowering in energy consumption otherwise required for mixing intensifications makes hydrocarbons, perfluorocarbons and various oils as most sought after oxygen-vectors to be used in aerobic fermentation (Rols and Goma, 1989; Rols *et al.*, 1990; Rols and Goma, 1991; Jianlong, 2000; Xu *et al.*, 2007). The advantage of using oxygen-vectors in aerobic fermentation is that they increase oxygen transfer rate from the gas phase to the micro-organisms without the need for extra energy supply.

Mechanism of Action

The mechanism of the enhancement of the oxygen transfer by oxygen-vectors can be understood on the basis of the general equation of oxygen transfer rate (OTR) in a liquid culture which is expressed as:

$$OTR = K_L a(C^* - C_L)$$

The enhancement of oxygen transfer by the oxygen-vectors may have been achieved through their effect on the mass transfer coefficient K_La and/or the mass transfer driving force $(C^* - C_L)$, which is directly dependent on the oxygen solubility. Increase in $(C^* - C_L)$ by an oxygen-vector is possible because the higher oxygen solubility in these organic liquid called oxygen-vector than in the aqueous medium (Ho et al., 1990, Narta et al., 2011). The addition of oxygen-vectors determines the appearance of four phases in the bioreactor: the gas phase (air), the aqueous phase, the liquid organic phase and the solid phase (biomass), with the formation of new interfacial areas between the gas and liquid phases. Authors have suggested in these systems, five possible routes of the oxygen transfer from air to micro-organisms (Rols and Goma, 1989; Rols et al., 1990). As it can be seen in Figure 18.1, the oxygen transfer could occur directly to cells, or through oxygen-vectors adsorbed or not to bubble surface. Furthermore, the cells could be adsorbed to bubbles surface or to vector droplets. Among the mechanisms of oxygen transfer presented and analyzed in

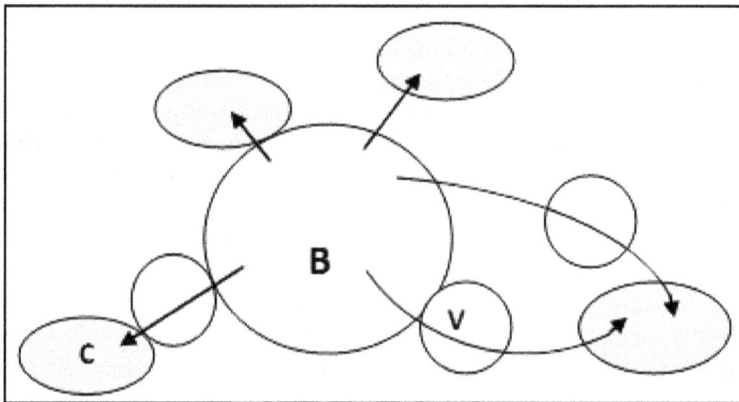

Figure 18.1. Possible mechanisms of oxygen transfer in presence of oxygen-vectors. (B: air bubble, C: cell, V: oxygen-vector droplet).

literature for these systems, the most plausible assumes that the hydrocarbon is adsorbed to bubbles surface, with or without the formation of a continuous film, the oxygen diffusion from air to micro-organisms occurring through oxygen-vector and then through aqueous phase or directly to the cells adsorbed to hydrocarbon droplets or film surface (Yoshida et al., 1970; Rols and Goma, 1989; Rols et al., 1990; Cascaval, 2006). The experiments indicated that the main resistance to oxygen transfer is due to the diffusion through aqueous boundary layer from hydrocarbon-aqueous phase interface, its negative influence being counteracted both by increasing the interfacial area of oxygen transfer and by accumulation of oxygen in organic phase, which acts as an oxygen reservoir. For this reason, the oxygen mass transfer coefficient corresponding to aqueous film, K_La, can also be used to describe the oxygen transfer in these systems, a being the gas–liquid interfacial area (Rols and Goma 1989; Rols et al., 1990; Cascaval, 2002).

Application of Oxygen-Vectors

The main oxygen-vectors used in biotechnology are hydrocarbons (MacLean, 1977; Rols and Goma 1989; Rols *et al.*, 1990; Ho *et al.*, 1990) and perfluorocarbons (Wei and Liu, 1998; Narta *et al.*, 2011) as well as oil added as antifoam agent (Yoshida *et al.*, 1970; Rols and Goma, 1991). The oxygen solubility in these compounds is manifolds higher than in water. The most frequently used oxygen-vectors along with their respective physical properties, which are in fact the determinants of their efficacy in improving the performance of the submerged aerobic fermentation, are enlisted in Table 18.1. The scope of their application is very wide and includes the hyper-production of the industrially important enzymes, therapeutic enzymes, antibiotics, food supplements, diagnostic molecules, metabolic intermediates and other biomolecules of immense economical and academic importance. The oxygen-vectors find application in a whole range of microbial system including bacteria, fungi, yeasts, actinomycetes and their suitability depends on their physical properties, which eventually decide their compatibility with a particular microbial system (Table 18.2). Wei and Liu (1997), and Narta *et al.* (2011) screened a number of oxygen vectors and ultimately selected n- dodecane and liquid paraffin, respectively in the optimisation studies on the production of L-asparaginase, an anticancer enzyme drug in two different microbial systems *viz. E. coli and Bacillus brevis*. Both of these studies reported a remarkable improvement in the oxygen solubility in the culture media, biomass production and L-asparaginase synthesis vis-à-vis the control experiment. Similar improvement has also been observed in the production of another industrially important enzyme, glucose-oxidase by Li and Chen (1994) with a variety of oxygen-vectors *viz.* n-dodecane, n-hexadecane and soyabean oil in *Asperigillus niger* fermentation. In one of the earliest investigations on the enhancement of the oxygen transfer in hybridoma cells culture, Cho and Wang (1988) used perfluorocarbons as the effective oxygen carriers. The application of oxygen-vectors has also proved to be effective even in the enhancement of antibiotics production by *Penicillium chrysogenum* fermentation in the n-hexadecane supplemented culture medium. The addition of n-hexadecane during penicillin fermentation resulted in significant increase in biomass

Table 18.1. Comparison of the physical properties of some frequently used oxygen-vectors with water.

Oxygen-Vectors	Density(Kg/L)	Surface Tension (N/m)	O_2–Solubility (mg/L)
n-Dodecane	0.749	22.60×10^{-3}	54.90
n-Hexane	0.66	18.43×10^{-3}	NA
n-Hexadecane	0.773	27.47×10^{-3}	50.40
Liquid paraffin	0.743	24.70×10^{-3}	39.50
Silicone oil	0.963	20.10×10^{-3}	NA
Perfluorocarbon	1.950	24.50×10^{-3}	118
Alkane mixture	0.742	45.20×10^{-3}	39.48
Water	1.000	71.20×10^{-3}	6.30

Table 18.2. Summary of the use of oxygen-vectors in different microbial systems for the production of molecules of industrial importance.

Sl.No.	O_2-Vectors	Microbial System	Commercial Product	Reference(s)
1.	n-Dodecane	*Aspergillus niger*	Citric acid	Jianlong, 2000
2.	Liquid paraffin	*Bacillus brevis*	L-asparaginase	Narta *et al.*, 2011
3.	n-Dodecane	*E.coli*	L-asparaginase	Wei and Liu, 1998
4.	n-Hexadecane	*Penicillium chrysogenum*	Penicillin	Ho *et al.*, 1990
5.	n-Hexadecane	*Acetobacter suboxydans*	L-sorbose	Giridhar and Sirivastava, 2000
6.	n-hexane and n-dodecane	*Blakeslea trispora*	Lycopene and β-carotene	Xu *et al.*, 1990
7.	silicone oil	*Aspergillus niger*	Citric acid	Ates *et al.*, 2002
8.	n-hexadecane, Silicone oil and n-dodecane	*Aspergillus niger*	Glucose-oxidase	Li and Chen, 1994
9.	n-hexadecane, n-tetradecane and n-dodecane	*Aspergillus terreus*	Lovastatin	Lai *et al.*, 2002
10.	N-Hexadecane	*Phaffia rhodozyma*	Carotenoid	Liu and Wu, 2006

and penicillin production, and also reduced the formation of mycelial pellets. (Ho *et al.*, 1990). Besides this a number of studies have reported the successful use of vegetable oils as effective oxygen-vectors in improving the performance of fermentation particularly the antibiotic fermentation, and the vegetable oils in such cases succeeded in extending the stationary phase during the course of the fermentation thereby maintaining high antibiotic production rates (Ryu and Hospodka, 1980; Cho and Wang, 1988).

Effect of Oxygen-Vectors

Addition of oxygen-vectors in the culture media has been found to have bearings on a number of fermentation parameters *e.g.* dissolved oxygen profile, cell mass production and the production of different metabolites and enzymes. Given the higher solubility of oxygen in these vectors they enhance the dissolved oxygen concentration in the media in which they are added and hence improve the overall performance of the aerobic fermentation for which the dissolved oxygen is a critical pre-requisite.

Effect on Dissolved Oxygen

As evident from the foregone discussion, solubility of oxygen is poor in the aqueous culture media and hence the aeration of culture media for aerobic fermentation and biological waste water treatment has been one of the bottlenecks of the conventional fermentation technology and continues to be so unless the strategies like use of oxygen-vectors are adopted, The latter improves the solubilisation of oxygen in the culture linearly with their increasing fraction unless the viscosity becomes hurdle with the increased concentration of these oxygen-vectors. The selected oxygen-vectors need to be optimised in terms of their concentration for a particular microbial system and if used in optimised condition, they significantly enhance the dissolved oxygen content of the culture media (Table 18.3). Cascaval *et al.* (2006) studied the influence of *n*-dodecane addition on oxygen transfer in stirred bioreactor, for different fermentation broths (simulated fermentation broths with a large domain of apparent viscosity variation, *Propionibacterium shermanii* and *Saccharomyces cerevisiae* cultures) and established some mathematical correlation for oxygen transfer in presence of *n*-dodecane. The proposed equations have been found to be useful for optimization or scaling-up of the fermentation in presence of oxygen-vectors and clearly describes the significant enhancement in the oxygen transfer rate and dissolved oxygen concentration vi-a-vis the simple broths. Jia *et al.* (1996) in their study on the enhancement of oxygen transfer in the tower bioreactor observed that addition of liquid hydrocarbons 4 per cent (v/v) as oxygen-vectors was capable of maintaining the dissolved oxygen well above 20 per cent throughout the course of the fermentation, which resulted in 1.3 fold increase in the yeast production rate. However, in the broth devoid of such oxygen-vectors, the dissolved oxygen became limiting after 8th h of the fermentation. During the production of lycopene and β-carotene by *Blakeslea trispora,* the addition of 1 per cent n-dodecane resulted in 24 per cent increase in relative dissolved oxygen concentration in the culture broth accompanied by the improved production of the desired metabolites (Xu *et al.,* 2007). Wei and Liu (1997) in an attempt to promote the production of antileukemic enzyme, L-asparaginase by the addition of n-dodecane (6 per cent v/v) were able to achieve above 80 per cent dissolved oxygen in the production broth of *E. coli* during the whole fermentation process, whereas in the normal broth the levels of the dissolved oxygen was as low as 40 per cent by the 14th h.

Table 18.3. Effect of oxygen-vector on dissolved oxygen status of the fermentation broth.

O_2-vectors	Microbial System	Dissolved O_2 (per cent)		Reference (s)
		(-)OV	(+)OV	
Liquid paraffin	*Bacillus brevis*	6.00	70.00	Narta *et al.*, 2011
n-Dodecane	*Aspergillus niger*	70	140	Jianlong, 2000
n-Dodecane	*E. coli*	54	88	Wei and Liu, 1997
n-Hexane	*Blakeslea trispora*	100	124	Xu *et al.*, 2007
Alkanes*	*Saccharomyces cerevisiae*	20	100	Jia *et al.*, 1996
Perfluorocarbon	*Saccharomyces cerevisiae*	17	30	Jia *et al.*, 1997

* Mixture of alkanes ranging between C_{12}-C_{16}; (-): without oxygen-vector, (+): with oxygen vector.

Effect on Biomass Production

In the typical aerobic fermentation, the positive effect of the oxygen-vector supplementation to the culture medium is also exhibited by multiplication rate of micro-organisms being cultivated in the vessel, which ultimately leads to higher biomass production. The biomass production assumes special importance in those cases where enzymes or metabolites are either expressed intracellularly or accumulate inside the microbial cells. Therefore the amount of biomass obtained during the fermentation assures higher amount of the intracellular enzyme or other metabolite which need to be released by cell disruption. The extent of improvement in biomass production however depends on the generation time, compatibility of the oxygen-vectors with the microbial system and other conditions. Their addition however demonstrates a profound improvement in biomass production as evident from the data in Table 18.5.

Table 18.4. Effect of oxygen-vector on the production of industrially important molecules through aerobic fermentation.

O_2-Vectors	Microbial System	Product	Production (g/L) (-)OV	(+)OV	Reference (s)
Liquid paraffin	*Bacillus brevis*	L-asn	1.50*	2.04*	Narta *et al.*, 2011
n-Hexadecane	*Acetobacter suboxydans*	L-Sorbose	64.83	82.12	Giridhar and Srivastava, 2000
n-Dodecane	*Aspergillus niger*	Citric acid	60	83	Jianlong, 2000
n-Dodecane	*E. coli*	L-asn	52*	62*	Wei and Liu, 1997
n-Hexane	*Blakeslea trispora*	β-carotene	0.335	0.481	Xu *et al.*, 2007
n-Hexane	*Blakeslea trispora*	Lycopene	0.254	0.375	Xu *et al.*, 2007
n-Hexadecane	*Penicillium chrysogenum*	Penicillin	0.20	0.307	Ho *et al.*, 1990

* Represent enzyme activities in IU/mL; L–asn: L–asparaginase

Table 18.5. Effect of oxygen-vectors on biomass production during the aerobic fermentation.

O_2-vectors	Microbial System	Biomass (g/L) (-)OV	(+)OV	Reference (s)
Liquid paraffin	*Bacillus brevis*	106	146	Narta *et al.*, 2011
n-Dodecane	*Aspergillus niger*	100	115	Jianlong, 2000
n-Dodecane	*E. coli*	400	450	Wei and Liu, 1997
n-Hexadecane	*Acetobacter suboxydans*	547	716	Giridhar and Srivastava, 2000
n-Hexane	*Blakeslea trispora*	53	55	Xu *et al.*, 2007
n-Hexane	*Blakeslea trispora*	55	63	Xu *et al.*, 2007
Perfluorocarbon	*Saccharomyces cerevisiae*	22	26	Jia *et al.*, 1997
Silicone oil	*Bacillus brevis*	106	117	Narta *et al.*, 2011

Effect on the Production of Metabolites

The aerobic fermentative processes besides other things are dependent on the availability of the oxygen to the microbes being cultivated in the bioreactor or the vessel. The conventional method makes use of the shaking for maintaining the requisite levels of the dissolved oxygen. But there is a limitation to shaking and agitation as it cannot meet the ever increasing requirement of rapidly multiplying microbial cells. The availability of the dissolved oxygen control the growth and multiplication of the micro-organism being cultivated in the vessel for the production of useful biomolecules by exerting effect on the metabolism of the microbes. The oxygen levels exert a positive pressure on the physiology and biochemical machinery of the microbes responsible for the synthesis of the molecules of the interest. Addition of oxygen vectors has been reported to enhance the production of a number of industrially useful molecules. Giridhar and Srivastava (2000) were able to improve sorbose accumulation and productivity in shake flask, batch, and fed-batch fermentation. The former is an intermediate in the process of Vitamin-C synthesis involving the microbial oxidation of D-sorbitol to L-sorbitol by *Acetobacter suboxydans*. They observed that 4 per cent addition of n-hexadecane in shake flasks resulted in a maximum accumulation of 82.12 Kg/m³ sorbose in 24 h, compared to 64.83 Kg/m³ without n-hexadecane. A net 15.9 Kg/m³ enhanced productivity accompanied by a total accumulation of 311.68 Kg/m³ sorbose indicate the economic nature of the process and make it more attractive option. The potential use of n-dodecane as an oxygen-vector for improvement of citric acid production by *Aspergillus niger* has been worked out by Jianlong (2000). Citric acid accumulation by a factor of 1.4-fold accompanied by 15 per cent increase in the dry mycelial weight over the control conditions was achieved when n-dodecane was used in the fermentation broth. Another study reports the successful use of silicone oil as oxygen-vector for increasing citric acid production by immobilised and freely suspended cells of *Aspergillus niger* by a factor of 2 and 1.6 fold respectively (Ates, 2002). An increase of significance to the tune of 51-78 per cent and 44-65 per cent, respectively was observed in lycopene and β-carotene production when oxygen-vectors, n-hexane and n-dodecane were added to cultures of *Blackeslea trispora* (Xu et al., 2007). N-hexadecane turned out to be the most beneficial for the yeast growth and carotenoid production. The addition of 9 per cent (v/v) n-hexadecane to the liquid medium at the time of inoculation was found to be optimal, increasing the carotenoid yield by 58 per cent and the oxygen transfer rate (OTR) by 90 per cent. The addition of n-hexadecane to shake-flask cultures of *Phaffia rhodozyma* significantly improved the oxygen transfer in culture, which led to the increase in carotenoid production (Liu and Wu 2006). Penicillin fermentation was equally improved by the addition of n-hexadecane as oxygen solubility enhancer. A significant increase in the cell growth in addition to the higher production rate of the lifesaving antibiotic penicillin in a 12.5 L capacity fermenter was attributed to an increased oxygen supply to *P. chrysogenum* due to higher oxygen solubility of fermentation media achieved by the addition of n-hexadecane (Ho et al., 1990).

Conclusion

The review of different studies with a variety of oxygen-vectors tried with diverse microbial system for the aerobic fermentation have shown that the addition of these nontoxic organic liquid at suitable volume fractions to liquid cultures of fermentation media not only significantly improves the production of the molecules of commercial interest, but have also enhanced the cell growth of the aerobic micro-organism. As oxygen supply is a major limiting factor for cell growth and metabolism of the aerobic micro-organism, the constant supply in an economical manner is of prime importance for a successful aerobic fermentative process. The inertness of oxygen-vectors allows the enhancement of OTR in the cultivation media without interfering with the rate of multiplication of the micro-organisms. The antifoaming action and the reusability are the additional attributes that make this strategy indispensable in the modern context. Therefore, the use of biocompatible and reusable oxygen-vectors is an efficient and economical option for improving the performance of large-scale aerobic fermentation processes. The optimisation of the concentration, the time of addition and suitability with a particular microbial system needs to be worked out for a successful scale-up of a submerged fermentative process.

References

1. Ates, S., Dingil, N., Bayraktar, E. and Mehmetoglu U., 2002. Enhancement of citric acid production by immobilized and freely suspended *Aspergillus niger* using silicone oil. *Process Biochemistry*, 38: 433–436.

2. Cascaval, D., Galaction, A.-I., Folescu, E. and Turnea, M., 2006. Comparative study on the effects of *n*–dodecane addition on oxygen transfer in stirred bioreactors for simulated, bacterial and yeasts broths. *Biochemical Engineering Journal*, 31: 56–66.

3. Cho, M.H. and Wang, S.S., 1998. Enhancement of oxygen transfer in hybridoma cell culture by using perfluorocarbon as an oxygen carrier. *Biotechnology Letters*, 10(12): 855–868.

4. Giridhar, R. and Srivastava, A.K ., 2000. Productivity enhancement in L-sorbose fermentation using Oxygen vectors. *Enzyme and Microbial Technology*, 27: 537–541.

5. Ho, C.S., Ju, L.K. and Baddour, R.F., 1990. Enhancing penicillin fermentations by increased oxygen solubility through the addition of *n*–hexadecane. *Biotechnology and Bioengineering*, 36: 1110–1118.

6. Jia, S., Park, Y.S. and Okabe, M., 1996. Enhanced oxygen transfer in tower bioreactor on addition of liquid hydrocarbons. *Journal of Fermentation and Bioengineering*, 82(2): 191–193.

7. Jianlong, W ., 2000. Enhancement of citric acid production by *Aspergillus niger* using *n*-dodecane as an oxygen-vector. *Process Biochemistry*, 35: 1079–1083.

8. Lai, L.-S.T., Tsai, T.-H. and Wang, T.C., 2002. Application of oxygen vectors to *Aspergillus terreus* cultivation. *Journal of Bioscience and Bioengineering*, 94(5): 453–459.

9. Lee, B.S. and Kim, E.K., 2004. Lipopeptide from *Bacillus sp.* GB 16 using a novel oxygenation method. *Enzyme and Microbial Technology*, 35: 639–647.

10. Li, T.H. and Chen, T.L., 1994. Enhancement of glucose oxidase fermentation by addition of hydrocarbons. *Journal of Fermentation and Bioengineering,* 78(4): 298–303.

11. Liu, Y.S. and Wu, J.Y., 2006. Use of *n*-hexadecane as an oxygen vector to improve *Phaffia rhodozyma* growth and carotenoid production in shake-flask cultures. *Journal of Applied Microbiology,* 101: 1033–1038.

12. Lockhart, W.R. and Squires, R.W., 1963. Aeration in the laboratory. *Advances in Applied Microbiology,* 5: 157–187

13. MacLean, G.T., 1977. Oxygen diffusion rates in organic fermentation broth. *Process Biochemistry,* 12: 22–28.

14. Mattiasson, B. and Adlercreutz, P., 1983. Use of perfluorochemicals for oxygen supply to immobilised cells. *Annal of New York Academy of Sciences,* 413: 545–553.

15. McMillan, J.D. and Wang, D.I.C., 1987. Enhanced oxygen transfer using oil-in-water dispersions. *Annal of New York Academy of Sciences,* 506: 569–572.

16. Narta, U., Roy, S., Kanwar, S.S. and Azmi, W., 2011. Improved production of L-asparaginase by *Bacillus brevis* cultivated in the presence of oxygen-vectors. *Bioresource Technology,* 102: 2083–2085.

17. Richards, J.W., 1961. Studies in aeration and agitation. *Progresses in Industrial Microbiology,* 3: 141–172.

18. Rols, J.L., Condoret, J.S., Fonade, C. and Goma, G., 1990. Mechanism of enhanced oxygen transfer in fermentation using emulsified oxygen vectors. *Biotechnology and Bioengineering,* 35: 427–435.

19. Rols, J.L. and Goma, G., 1989. Enhancement of oxygen transfer rates in fermentation using oxygen-vectors. *Biotechnology Advances,* 7: 1–8.

20. Ryu, D.D.Y. and Hospodka, J., 1980. *Biotechnology and Bioengineering,* 26: 148–298.

21. Schlegel, H., 1997. Aeration without air oxygen supply by hydrogen peroxide. *Biotechnology and Bioengineering,* 81: 257–266.

22. Sonnleitner, B. and Hahnemann, U., 1997. Robust oxygen supply by controlled addition of hydrogen peroxide to microbial cultures. *Bioprocess Engineering,* 17: 215–219.

23. Wang, J., 2000. Enhancement of citric acid production by *Aspergillus niger* using *n*-dodecane as an oxygen-vector. *Process Biochemistry,* 35: 1079–1083.

24. Wei, D.z.H. and Liu, H., 1998. Promotion of L-asparaginase production by using *n*-dodecane. *Biotechnology Techniques,* 12(2): 129–131.

25. Xu, F., Yuan, Q.P. and Zhu, Y., 2007. Improved production of lycopene and β-carotene by *Blakeslea trispora* with oxygen-vectors. *Process Biochemistry* 42: 289–293.

26. Yoshida, F., Yamane, T. and Miyamoto, M., 1970. Oxygen absorption into oil-in-water emulsions. A study of hydrocarbon fermenter. *Ind. Eng. Chem. Proc. Des. Dev.,* 9: 570–576.

Biotechnology: An Overview (2015)
Editors: Rajan Kumar Gupta, Nasim Akhtar and Deepak Vyas
Published by: DAYA PUBLISHING HOUSE, NEW DELHI

Pages 301–316

Chapter 19

Pesticides and their Degradation

Ashutosh Bahuguna *

*Department of Biotechnology and Microbiology,
Modern Institute of Technology (M.I.T.) Dhalwala,
Rishikesh – 249 201, Uttarakhand*

Introduction

Pesticides constitute a heterogeneous category of chemical substance or mixture used for preventing, destroying, repelling, or mitigating any pest. A pest is any harmful, destructive, or troublesome animal, plant or micro-organism.

Pesticides are defined by the Environmental protection agency (EPA) as any substance or mixture of substances that interferes with a pest. These can be toxins, attractants, mating disruptors, hormones or hormone disrupters or repellents.

Their application is still the most effective and accepted means for the protection of plants from pests and has contributed significantly to enhanced agricultural productivity and crop yields. A total of about 890 active ingredients are registered as pesticides in the USA and currently marketed in some 20,700 pesticide products.

The intensive use of pesticides in agriculture is also well known to be coupled with the "green revolution". Green revolution was a worldwide agricultural movement that began in Mexico in 1944 with a primary goal of boosting grain yields in the world that was already in trouble with food supply to meet the demand of the rapidly growing human population. The green revolution involved three major aspects of agricultural practices, among which the use of pesticides was an integral part. Following its success in Mexico, green revolution spread over the world. Pest control has always been important in agriculture, but green revolution in particular needed more pesticide inputs than did traditional agricultural systems because, most of the high yielding varieties were not widely resistant to pests and diseases and

---

* *Author:* E-mail: ashubahuguna@gmail.com

partly due to monoculture system. Each year pests destroy about 30-48 per cent of world's food production. For example, in 1987 it was reported that, one third of the potential world crop harvest was lost to pests. A further illustration to the pest problem in the world is shown in Table 19.1.

Table 19.1. Estimated per cent losses caused by pests in some world's major crops per year (Hellar, 2002).

Crop	Estimated Per cent Losses			
	Insect	Diseases	Weeds	Total
Rice	26	9.6	10.8	46.4
Maize	12.4	9.4	13.0	34.8
Wheat	5.0	9.1	9.8	23.9
Potatoes	6.5	21.8	4.0	32.3
Soybeans	4.5	11.1	13.5	29.1
Sugarcane	9.2	10.7	25.1	45.0

Insect pests and rodents also account for a big loss in stored agricultural products. Internally feeding insects feed on grain endosperm and the germ the result of which is the loss in grain weight, reduction in nutritive value of the grain and deterioration in the end use quality of the grain. Externally feeding insects damage grain by physical mystification and by excrement contamination with empty eggs, larval moults and empty cacoons. A common means of pest control in stored agricultural products has always been the use of insecticides such as malathion, chlorpyrifos-methyl or deltamethrin impregnated on the surfaces of the storage containers. It is therefore quite apparent that, the discovery of pesticides was not a luxury of a technical civilization but rather was a necessity for the well being of mankind.

History of Pesticides

The historical background of pesticides use in agriculture is dated back to the beginning of agriculture itself and it became more pronounced with time due to increased pest population paralleled with decreasing soil fertility. However, the use of modern pesticides in agriculture and public health is dated back to the 19th century.

By the 15th century, toxic chemicals such as arsenic, mercury and lead were being applied to crops to kill pests. In the 17th century, nicotine sulfate was extracted from tobacco leaves for use as an insecticide. The 19th century saw the introduction of two more natural pesticides, pyrethrum, which is derived from chrysanthemums, and rotenone, which is derived from the roots of tropical vegetables and use as an insecticide. An important milestone was introduction of phenyl mercury in 1913 for the protection of seed (fungicide).

The first generation of pesticides involved the use of highly toxic compounds, arsenic (calcium arsenate and lead arsenate) and a fumigant hydrogen cyanide in 1860's for the control of such pests like fungi, insects and bacteria. Other compounds included Bordeaux mixture (copper sulphate, lime and water) and sulphur. Their use was abandoned because of their toxicity and ineffectiveness.

The second generation involved the use of synthetic organic compounds. The first important synthetic organic pesticide was dichlorodiphenyltrichloroethane (DDT) first synthesized by a German scientist Ziedler in 1939. In its early days DDT was hailed as a miracle because of its broad-spectrum activity, persistence, insolubility, inexpensive and ease to apply. DDT in particular was so effective at killing pests and thus boosting crop yields and was so inexpensive to make its use quickly spread over the globe.

Problems Associated with Pesticides

1. Most of the pesticides are violent poisons and their handling is hazardous. The selective action of pesticides is never perfect and many non-target organisms are affected by their toxicity some of which may be useful organisms. For example elimination of some insects and bees which vitally aid in pollination of many plants could cause considerable damage to agricultural productivity.These non-target organisms may also include domestic animals, live-stock, poultry etc. Accidental exposures may lead to human casualties as well.

2. Most of these synthetic organics or their decomposition product persists in toxic state in the environment for long durations. Thus, once applied they continue to harm the non-target organisms for long periods of time. For example DDT may be eliminated within two years only, under favorable conditions while its persistence for period ranging 20-25 years has also been reported. The persistence of the pesticides in the environment depends upon their chemical and physical properties, formulation (*e.g.* liquid, powder or granules etc), type of soil, its moisture content, temperature, physical properties of soil, composition of soil microflora and the plant species present.

3. Most of the pesticides are lipophilic, as a result they show bio-accumulation and bio-magnification, which cause problems at higher trophic levels in an ecosystem. Pesticides even though present in extremely low concentration in the soil or the surrounding water are taken up by various microbes, plants and animals which may accumulate and concentrate them several thousand times.

   For example bio-magnification of DDT from water by continued absorption through plankton and subsequent passage of plankton to fishes through food chain and finally to sea-eagle. The concentration builds up as we move up in the food chain *i.e.* plankton to small fishes, small fishes to large fishes, and large fishes to sea eagle (Table 19.2). In case of DDT, a 10^5 fold increase occurs in sea eagles as compared to the concentration present in the aqueous environment.

   Toxaphene may occur in lake water only in the concentration range of 0.0002-0.0006 mg per liter but the water plants growing in the lake may contain as much as 0.2-0.4 mg per kg, invertebrates 0.5-1.5 mg per kg and trout and salmon may contain as much as 3.0-6.0 mg per kg. Toxaphene is decomposed very slowly and even after a period of six or seven years there is little significant change in its concentration. Similar bioaccumulation and biomagnifications have been recorded for a number of other pesticides as well.

Table 19.2. Bio-magnification of DDT.

Organism Environment	DDT Concentration	Fold Increase Over Concentration in Water
Water	0.3 ppb	–
Plankton	30 ppb	$100(=10^2)$
Small fishes	0.3 ppm	$1,000(=10^3)$
Large fishes	3.0 ppm	$10,000(=10^4)$
Sea-eagle	30 ppm	$100,000(=10^5)$

4. Differential toxic action of pesticides on different species results in an unbalanced biotic community. Species susceptible to the toxic action of the pesticide are eliminated while the resistant ones multiply without competition. Organisms on which the eliminated species depended for its food requirement are also affected. Absence of predation causes these forms to multiply in numbers. Thus a chain of events is initiated which alter the species composition of the ecosystem beyond recognition. Insignificantly pest may become so numerous as to cause another pestilence.

5. Pesticides exert a remarkable influence on microbial community in an eco system. Moderate concentration of most of the pesticides; to begin with depress the activity of microbes. Since microbes function as co-ordinate entity with each of its components performing certain specific task, selective action of pesticides on different components of microbial population causes changes in over-all activity of the community which is no longer capable of carrying out its normal function.

Pesticides may alter the soil microbial population, both qualitatively and quantitatively, in several ways. The most obvious effect is that of the direct toxicity of applied pesticide to the susceptible microbial species. Other micro-organisms become resistant to the pesticide and can increase their biomass because of decreased competition.

Pesticides reaching the soil in significant quantities have direct effect on soil microbiological aspects, which in turn influence plant growth. Some of the most important effects caused by pesticides are, alterations in ecological balance of the soil microflora, continued application of large quantities of pesticides may cause ever lasting changes in the soil microflora, adverse effect on soil fertility and crop productivity, inhibition of nitrogen fixing soil micro-organisms such as *Rhizobium, Azotobacter, Azospirillum* etc. and cellulolytic and phosphate solubilizing micro-organisms, suppression of nitrifying bacteria, *Nitrosomonas* and *Nitrobacter* by soil fumigants ethylene bromide, telone, and vapam have also been reported, alterations in nitrogen balance of the soil, interference with ammonification in soil, adverse effect on mycorrhizal symbioses in plants and nodulation in legumes, and alterations in the rhizosphere microflora, both quantitatively and qualitatively.

Potential Health Effects of Pesticides

People can be exposed to pesticides in three ways:

1. Breathing (inhalation exposure).
2. Getting it into the mouth or digestive tract (oral exposure).
3. Contact with the skin or eyes (dermal exposure).

Pesticides can enter the body by any one or all three of these routes. Inhalation exposure can happen if one breathes air containing pesticide as a vapor, as an aerosol, or on small particles like dust. Oral exposure happens when one eat food or drink water containing pesticides. Dermal exposure happens when ones skin is exposed to pesticides. This can cause irritation or burns.

Pesticide Toxicity and Exposure

Hazard, or risk, of using pesticides is the potential for injury, or the degree of danger involved in using a pesticide under a given set of conditions. Hazard depends on the toxicity of the pesticide and the amount of exposure to the pesticide and is often illustrated with the following equation:

Hazard = Toxicity x Exposure

The toxicity of a pesticide is a measure of its capacity or ability to cause injury or illness. The toxicity of a particular pesticide is determined by subjecting test animals to varying dosages of the active ingredient (a.i.) and each of its formulated products. Toxicity caused by pesticides are categorize as acute Toxicity, Chronic Toxicity and Chronic Effect.

Acute Toxicity

Acute toxicity of a pesticide refers to the chemical's ability to cause injury to a person or animal from a single exposure, generally of short duration. The harmful effects that occur from a single exposure by any route of entry are termed "acute effects". Acute toxicity is determined by examining the dermal toxicity, inhalation toxicity, and oral toxicity of test animals. Acute toxicity is measured as the amount or concentration of a toxicant the active ingredient (a.i.) required to kill 50 per cent of the animals in a test population. This measure is usually expressed as the LD_{50} (lethal dose 50)or the LC_{50} (lethal concentration 50). LD_{50} and LC_{50} values are based on a single dosage and are recorded in milligrams of pesticide per kilogram of body weight (mg/kg) of the test animal or in parts per million (ppm). LD_{50} and LC_{50} values are useful in comparing the toxicities of different active ingredients and different formulations containing the same active ingredient. The lower the LD_{50} or LC_{50} value of a pesticide product, the greater its toxicity to humans and animals.

Chronic Toxicity and Chronic Effects

The chronic toxicity of a pesticide is determined by subjecting test animals to long-term exposure to the active ingredient. Any harmful effects that occur from small doses repeated over a period of time are termed "chronic effects". Suspected chronic effects from exposure to certain pesticides include birth defects (teratogenesis), toxicity to a fetus, production of benign or malignant tumors (carcinogenic/ oncogenesis), genetic changes (mutagenic), blood disorders, nerve disorders, endocrine disruption, and reproduction effects.

Pesticide Signal Words

Pesticides products are categorized on the basis of their relative acute toxicity (their LD_{50} or LC_{50} values). According to the toxicity exert by pesticides they are labeled as danger and poison, warning and caution symbol.

Pesticides that are classified as highly toxic (Toxicity Category I) on the basis of either oral, dermal, or inhalation toxicity must have the signal words danger and poison printed in red displayed on the front panel of the package label. The acute (single dosage) oral LD_{50} for pesticide products in this group ranges from a trace amount to 50 mg/kg.

Pesticide products considered moderately toxic (Toxicity Category II) must have the signal word warning and displayed on the product label. In this category, the acute oral LD_{50} ranges from 50 to 500 mg/kg.

Pesticide products classified as either slightly toxic or relatively nontoxic (Toxicity Categories III and IV) are required to have the signal word caution on the pesticide label. Acute oral LD_{50} values in this group are greater than 500 mg/kg (Table 19.3) summarizes the LD_{50} and LC_{50} values for each route of exposure for the four toxicity categories and their associated signal word.

Table 19.3. Toxicity category of pesticides.

Routes of Exposure	Toxicity Category			
	I	*II*	*III*	*IV*
Oral LD_{50}	Up to and including 50 mg/kg	50–500 mg/kg	500–5,000 mg/kg	>5,000 mg/kg
Inhalation LC_{50}	Up to and including 0.2 mg/l	0.2–2 mg/l	2–20 mg/l	>20 mg/l
Dermal LD_{50}	Up to and including 200 mg/kg	200–2,000 mg/kg	2,000–20,000 mg/kg	>20,000 mg/kg
Eye Effects	Corrosive corneal opacity not reversible within 7 days	Corneal opacity reversible within 7 days; irritation persisting for 7 days	Corneal opacity irritation reversible within 7 days	No irritation
Skin Effects	Corrosive	Severe irritation at 72 hours	Moderate irritation at 72 hours	Mild or slight irritation at 72 hours
Signal Word	Danger Poison	Warning	Caution	Caution

Classification of Pesticides

Synthetic pesticides are classified in different ways; there are three most popular ways of classifying pesticides which are:

1. Classification based on the mode of action.
2. Classification based on the targeted pest species.
3. Classification based on the chemical composition of the pesticide.

Classification of Pesticides Based on the Mode of Action

Under this type of classification, pesticides are classified based on the way in which they act to bring about the desired effect. In this way pesticides are classified into two groups.

I) Contact (non-systemic) pesticides.

II) Systemic pesticides.

The non-systemic pesticides are those that do not appreciably penetrate plant tissues and consequently not transported within the plant vascular system. The non systemic pesticides will only bring about the desired effect when they come in contact with the targeted pest, hence the name contact pesticides. Examples of contact pesticides are paraquat and diquatdibromide.

The systemic pesticides are those which effectively penetrate the plant tissues and move through the plant vascular system in order to bring about the desired effect. Examples of systemic pesticides include 2, 4-D and glyphosate Rodenticides (Stomach poisons that bring about the desired effect after being eaten) and fumigants (those pesticides that produce vapor which kills the pests) are also comes under this classification.

Classification of Pesticides Based on the Targeted Pest Species

In this type of classification, pesticides are named after the name of the corresponding pest in target as shown in Table 19.4.

Table 19.4. Classification of pesticides based on the target organisms.

Type of Pesticide	Target Organism/Pest
Insecticides	Insects
Herbicides	Weeds
Rodenticides	Rodents
Fungicides	Fungi
Acaricides and Miticides	Arachnids of the order Acarina such as ticks and Mites
Molluscicides	Mollusks
Bactericides	Bacteria
Avicides	Bird pests
Virucides	Virus
Algicides	Algae

Classification of Pesticides Based on the Chemical Composition

In this category pesticides are grouped according to the chemical nature of the active ingredients. It is the most useful classification as it gives idea about effectiveness, physical and chemical properties of pesticides. According to this classification pesticides can be classified into four categories;

1. Organochlorines
2. Organophosphorus
3. Carbamates
4. Pyrethroids and miscellaneous chemicals.

Organochlorines

Organochlorines pesticides are organic compounds with five or more chlorine atoms. They are first synthetic pesticides used in agriculture, most of the members of this group used as insecticides for the control of wide range of insects. These pesticides are having long lasting effect into environment as they are resistant to most chemical and microbial degradation. Organochlorine insecticides act as nervous system disruptors leading to convulsions and paralysis of the insect and its eventual death. Some of the commonly used representative examples of organochlorine pesticides are DDT, lindane, endosulfan, aldrin, dieldrin and chlordane (Figure 19.1).

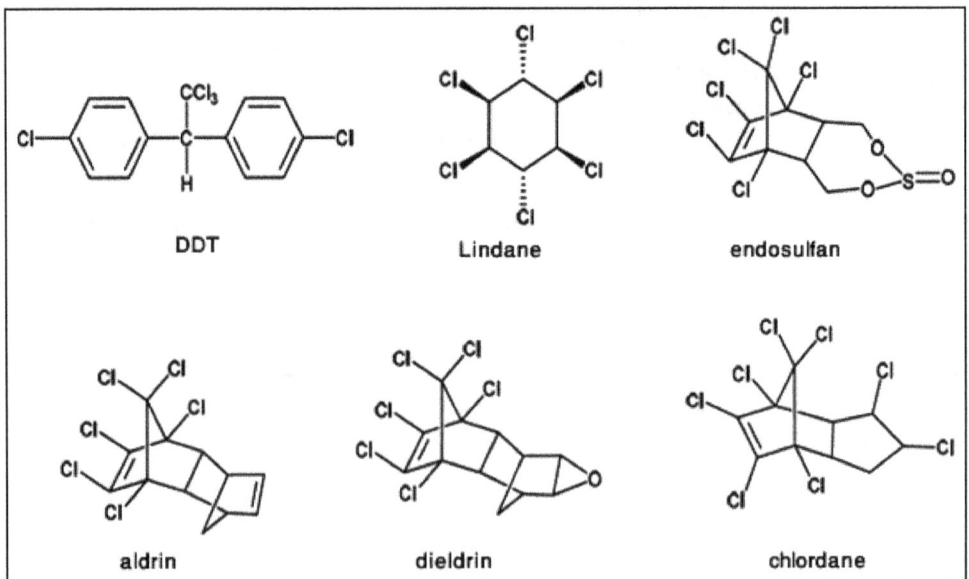

Figure 19.1. Structure of various organochlorines pesticides [*Source*: Zacharia (2011)].

Organophosphorus

Organophosphorus insecticides on the other hand contain a phosphate group as their basic structural framework. Organophosphorus insecticides implicate the nervous system of the target organism. These pesticides inhibit the activity of acetyl choline esterase an enzyme which catalyze the removal of acetyl choline from synaptic cleft. In contrast to organochlorine pesticides organophosphorus are easily degraded in the environment by various chemical and biological reactions. Some of the common used organophosphorus insecticides include parathion, malathion, diaznon and glyphosate (Figure 19.2).

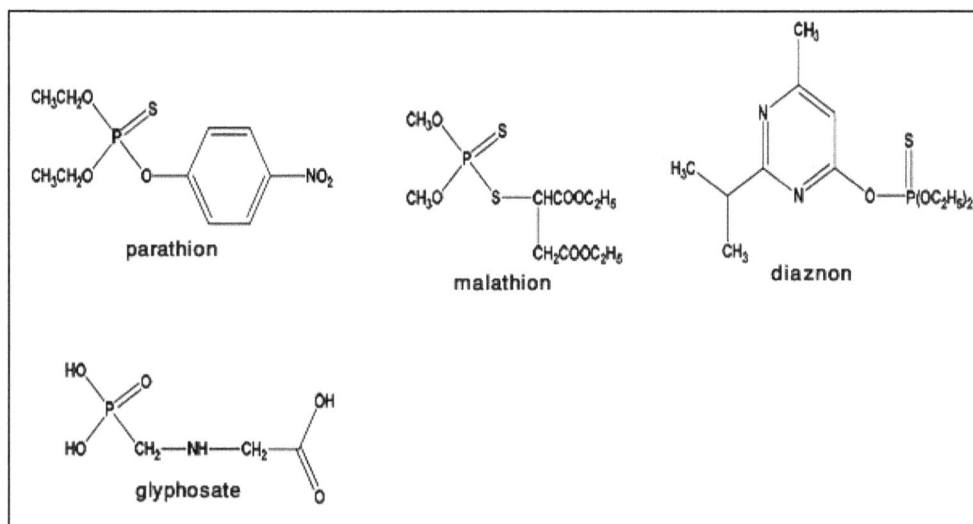

Figure 19.2: Structure of various organophosphorus pesticides [Source: Zacharia (2011)].

Carbamates

Carbamates are organic pesticides derived from carbamic acid. In compare to organophosphorous pesticides these pesticides have lower toxicity to mammalian system. Some of the widely used insecticides under this group include carbaryl, carbofuran and aminocarb (Figure 19.3).

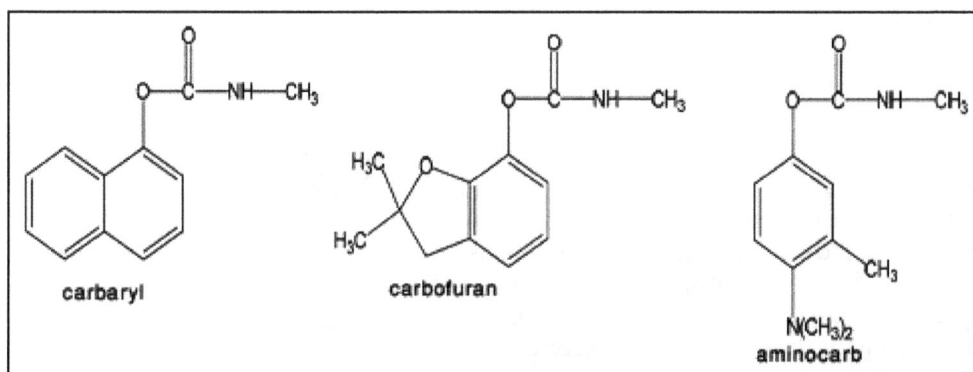

Figure 19.3. Structure of various carbamates pesticides [Source: Zacharia (2011)].

Pyrethroids

Pyrethroids are synthetic analogues of the naturally occurring pyrethrins, a product of flowers from pyrethrum plant (*Chrysanthemum cinerariaefolium*). Pyrethroids are of immense importance due to their fast knocking down effect against insect, minimum mammalian toxicity and easy to biodegrade. Although the naturally occurring pyrethrins have good pesticidal activity but still its direct use is impractical as it undergo fast photochemical degradation The synthetic analogues of the naturally occurring pyrethrins (pyrethroids) were developed by the modification of pyrethrin structure by introducing a biphenoxy moiety and substituting some hydrogens with halogens in order to confer stability at the same time retaining the basic properties of pyrethrins. The most widely used synthetic pyrethroids include permethrin, cypermethrin and deltamethrin (Figure 19.4).

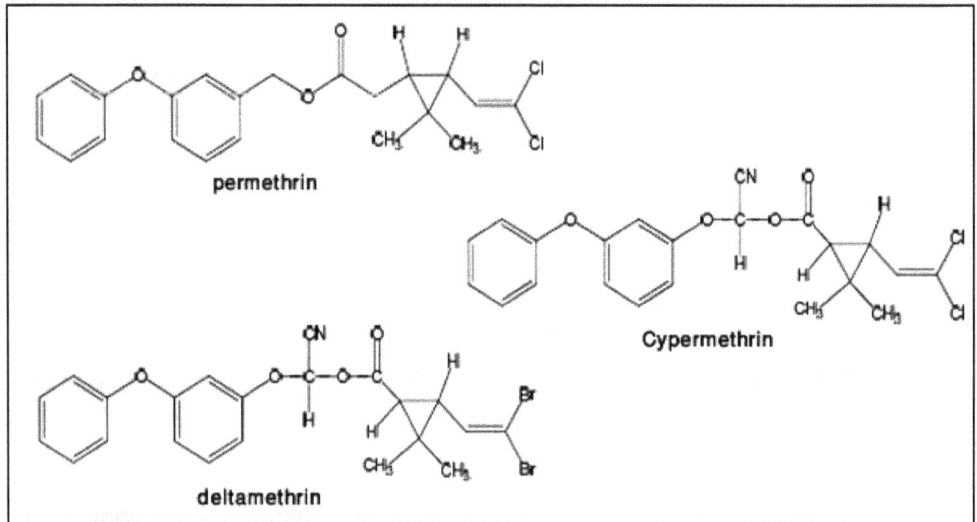

Figure 19.4. Structure of various pyrethroids pesticides [Source: Zacharia (2011)].

Other miscellaneous groups of pesticides that are worth mentioning include among others phenoxyacetic acid under which the herbicide 2,4-D belongs and bipyridyls under which the herbicides paraquat and diquat belong (Figure 19.5).

Degradation of Pesticides

About 30 per cent of agricultural produce is lost due to pests. Hence, the use of pesticides has become indispensable in agriculture. The abusive use of pesticides for pest control has been widely used in agriculture. However, the excessive use of pesticides causes serious harm and problems to humans as well as to the environment. Commonly used treatment methods for remediation of site contaminated with pesticides are expensive, some time insufficient and have some adverse effect on environment. Hence there is a strong need to develop some new technologies to reduce the pesticide contaminants from environment.

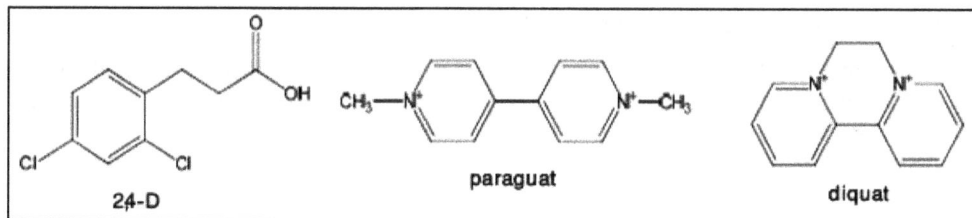

Figure 19.5. Structure of few miscellaneous groups of pesticides [Source: Zacharia (2011)].

One possible approach is use of micro-organisms to remove pollutants from contaminated sites, many soil micro-organisms have the ability to act upon pesticides and convert them into simpler non-toxic compounds. This process of degradation of pesticides and conversion into non-toxic compounds by micro-organisms is known as "biodegradation". This process is effective, minimally hazardous, economical, versatile and environment-friendly.

Further more the biochemical and genetic basis of microbial degradation catch much attention. Many genes/Enzymes that provide ability to degrade these compounds have been identified; genetic engineering may be used to enhance the performance of such micro-organisms that have the essential properties for biodegradation

The chemical reactions leading to biodegradation of pesticides fall into several broad categories.

Detoxification

Conversion of the pesticide molecule to a non-toxic compound. Detoxification is not synonymous with degradation. Since a single change in the side chain of a complex molecule may render the chemical non-toxic.

Degradation

The breaking down/transformation of a complex substrate into simpler products leading finally to mineralization. Degradation is often considered to be synonymous with mineralization, *e.g.* Thirum (fungicide) is degraded by a strain of *Pseudomonas* and the degradation products are dimethlamine, proteins, sulpholipids, etc.

Conjugation (Complex Formation or Addition Reaction)

In which an organism make the substrate more complex or combines the pesticide with cell metabolites. Conjugation or the formation of addition product is accomplished by those organisms catalyzing the reaction of addition of an amino acid, organic acid or methyl group to the substrate, for *e.g.*, in the microbial metabolism of sodium dimethlydithiocarbamate, the organism combines the fungicide with an amino acid molecule normally present in the cell and there by inactivate the pesticides/chemical.

Practical Approach of Biodegradation

Biodegradation of pesticides/herbicides is greatly influenced by the soil factors like moisture, temperature, pH and organic matter content, in addition to microbial population and pesticide solubility. Optimum temperature, moisture and organic matter in soil provide congenial environment for the break down or retention of any pesticide added in the soil.

For the successful biodegradation/bioremediation of a contaminant following strategies are needed.

Biostimulation

Addition of nutrients nitrogen and phosphorus to stimulate indigenous micro-organisms in soil.

Bioventing

Process/way of biostimulation by which gases stimulants like oxygen and methane are added or forced into soil to stimulate microbial activity.

Bioaugmentation

It is the inoculation/introduction of micro-organisms capable of degrading contaminat in the contaminated site/soil to facilitate biodegradation.

Microbial Degradation of Organochloride Pesticides

The fate of pesticides in the environment is determined by both biotic and abiotic factors. The microbial culture able to degrade pesticides are usually isolated from the site contaminated with these pesticides. DDT-metabolising microbes have been isolated from a range of habitats, including animal feces, soil, sewage, activated sludge, and marine and freshwater sediments. Under reducing conditions, reductive dechlorination is the major mechanism for the microbial conversion of both the o, p'-DDT and p, p'-DDT isomers of DDT to DDD (dichlorodiphenyldichloroethane). The reaction involves the substitution of an aliphatic chlorine for a hydrogen atom. The suggested pathway for the transformation of DDT by bacteria is shown in Figure 19.6. Degradation proceeds by successive reductive dechlorination reactions of DDT to yield 2,2-bis(p-chlorophenyl)ethylene (DDNU), which is then oxidised to 2,2- bis(p-chlorophenyl)ethanol (DDOH). Further oxidation of DDOH yields bis($p$chlorophenyl) acetic acid (DDA) which is decarboxylated to bis(p-chlorophenyl)methane (DDM). DDM is metabolized to 4,4'dichlorobenzophenone (DBP) or, alternatively, may undergo cleavage of one of the aromatic rings to form p-chlorophenylacetic acid (PCPA).

Matsumura *et al.* (1968) showed the breakdown of dieldrin in the soil by a *Pseudomonas* sp. which was isolated from a soil sample of the dieldrin factory yards. Later, authors showed the biodegradation of aldrin, endrin and DDT with bacteria that were shown to be able to degrade dieldrin .

Microbial Degradation of Organophosphate Pesticides

Methyl parathion (O, O-dimethyl-O-(p-nitro-phenylphosphorothioate) is one of the most used organophosphorus pesticides. This product is widely used throughout

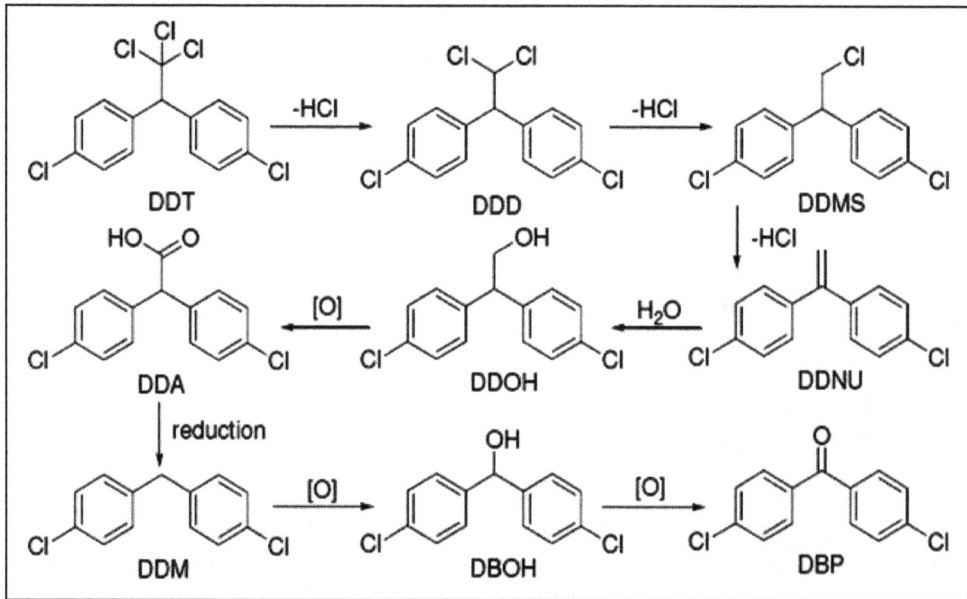

Figure 19.6. Proposed pathway for bacterial metabolism of DDT [Source: Andreì (2011)].

the world and its residues are regularly detected in a range of fruits and vegetables. Bacteria with the ability to degrade methyl parathion have been isolated. Methyl parathion hydrolase gene, *mpd*, which is responsible for hydrolyzing methyl parathion to *p*-nitrophenol and dimethyl phosphorothioate, has also been cloned from these strains.

A fenpropathrin-degrading bacterium, *Sphingobium* sp. JQL4-5, was isolated and characterized. A stable, genetically engineered strain, JQL4-5-*mpd*, capable of simultaneously degrading fenpropathrin and methyl parathion was constructed by random insertion of the methyl parathion hydrolase gene (*mpd*) into the chromosome of strain JQL4-5. Soil treatment results indicated that JQL4-5-*mpd* is a promising genetically engineered micro-organism (GEM) in the bioremediation of multiple pesticide contaminated environments. Organophosphorus hydrolase (OPH), isolated from both *Flavobacterium* sp. ATCC 27551 and *Pseudomonas diminuta* MG, is capable of hydrolyzing a wide range of oxon and thion organophosphorus pesticides (Ops). Subsequently, a dual-species consortium comprising engineered *E. coli* (with *mpd* gene) and a natural *p*-nitrophenol (PNP) degrader *Ochrobactrum* sp. strain LL-1 for complete mineralization of dimethyl OPs was studied. The dual-species consortium possesses the enormous potential to be utilized for complete mineralization of *p*-nitrophenol (PNP) substituted OPs in a laboratory-scale bioreactor. These studies demonstrated that methyl parathion (MP) could be degraded via the methyl parathion (MP) to *p*-nitrophenol (PNP) to hydroquinone to Krebs cycle (Figure 19.7) by the dual species consortium. The data confirm that the mineralization process of methyl parathion (MP) is initiated by hydrolysis leading to the generation of PNP and

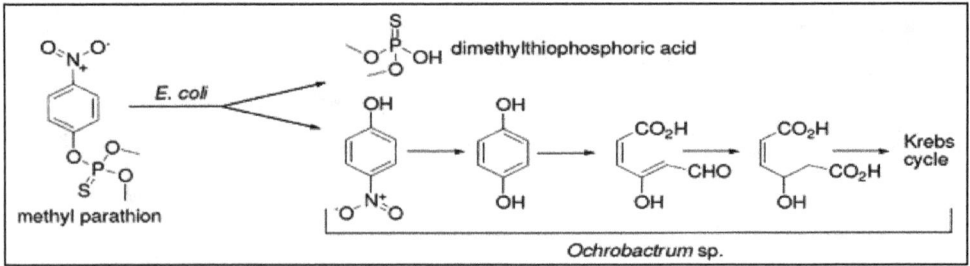

Figure 19.7: Proposed pathway for the biodegradation of MP by microbial consortium [Source: Andreì *et al.* (2011)].

dimethylthiophosphoric acid, and PNP degradation, then, proceeds through the formation of hydroquinone.

Microbial Degradation of Carbamate Pesticides

The biodegradation of carbamates has been investigated by different micro-organisms that metabolize carbamate pesticides. Carbofuran is one of the pesticides belonging to the *N*-methylcarbamate class used extensively in agriculture. It exhibits high mammalian toxicity and has been classified as highly hazardous. A number of bacteria capable of degrading carbofuran (*Pseudomonas, Flavobacterium, Achromobacterium, Sphingomonas, Arthrobacter*) have been isolated and characterized.

Carbofuran was degraded first to carbofuran phenol and the resultant was degraded to 2-hydroxy-3-(3-methylpropan-2-ol) phenol by *Sphingomonas* sp. and *Arthrobacter* sp. (Figure 19.8).

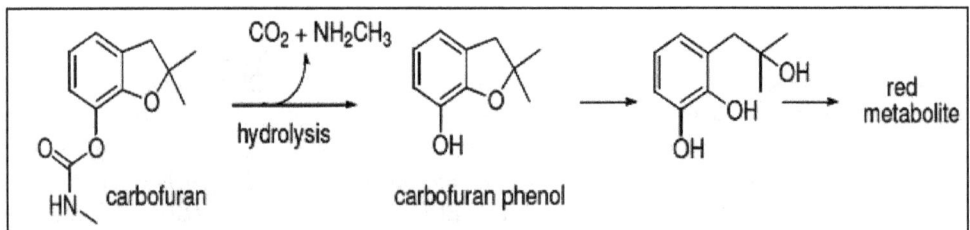

Figure 19.8: Biodegradation of carbofuran by *Sphingomonas* sp.[Source: Andreì *et al.* (2011)].

References

1. Andreì Luiz Meleiro Porto, Gliseida Zelayaraìn Melgar, Mariana Consiglio Kasemodel and Marcia Nitschke, 2011. Biodegradation of pesticides. In: *Pesticides in the Modern World: Trends in Pesticides Analysis*, (Ed.) Margarita Stoytcheva.

2. Asthana, D.K. and Asthana M., 2003. *Environment: Problem and Solutions.* S. Chand and Company Limited, New Delhi.

3. Bolognesi, C., 2003. Genotoxicity of pesticides: A review of human biomonitoring studies. *Mutation Research*, 543: 251–272.

4. Buchel, K.H., 1983. *Chemistry of Pesticides*. John Wiley and Sons, Inc. New York, USA.

5. De Schrijver, A. and De Mot, R., 1999. Degradation of pesticides by actinomycetes. *Critical Review in Microbiololy*, 25(2): 85–119.

6. Drum, C., 1980. *Soil Chemistry of Pesticides*. PPG Industries, Inc., USA.

7. Gruzdyev, G.S., Zinchenko, V.A., Kalinin, V.A. and Slovtsov, R.I., 1988. *The Chemical Protection of Plants*, (Ed.) G.S. Gruzdyev. MIR Publishers, Moscow (English translation.).

8. Gavrilescu, M., 2005. Fate of pesticides in the environment and its bioremediation. *Engineer in Life Science*, 5(6): 497–526.

9. Hong, L., Zhang, J.J., Wang, S.J., Zhang, X.E. and Zhou, N.Y., 2005. Plasmid-borne catabolism of methyl parathion and *p*-nitrophenol in *Pseudomonas* sp. strain WBC-3. *Biochemistry and Biophysics Research Communications*, 334(4): 1107–1114.

10. Hellar, H., 2002. Pesticides residues in sugarcane plantations and environs after long-term use: The case of TPC Ltd, Kilimanjaro Region, Tanzania.

11. Johnsen, R.E., 1976. DDT metabolism in microbial systems. *Pesticide Reviews*, 61: 1–28.

12. Keneth, M., 1992. *The DDT Story*. The British Crop Protection Council, London, UK.

13. Kim, I.S., Ryu, J.Y., Hur, H.G., Gu, M.B., Kim, S.D. and Shim, J.H., 2004. *Sphingomonas* sp. strain SB5 degrades carbofuran to a new metabolite by hydrolysis at the Furanyl Ring. *Journal of Agricultural and Food Chemistry*, 52(8): 2309–2314.

14. Lal, R. and Saxena, D.M. 1982. Accumulation, metabolism and effects of organochlorine insecticides on micro-organisms. *Microbiological Reviews*, 46(1): 95–127.

15. Loganathan, P.K. and Natarajan, P., 2008. Toxic effects of pesticides: A review on cytogenetic biomonitoring studies. *Facta Universitatis Medicine and Biology*, 15(2): 46–50

16. Liu, Z., Hong, Q., Xu, J.H., Wu, J., Zhang, X.Z., Zhang, X.H., Ma, A.Z., Zhu, J. and Li, S.P., 2003. Cloning, analysis and fusion expression of methyl parathion hydrolase. *Acta Genetica Sinica*, 30(11): 1020–1026.

17. Matsumura, F., Boush, G.M. and Tai, A., 1968. Breakdown of dieldrin in the soil by a micro-organism. *Nature*, 219(5157): 965–967.

18. McFarlane, J.A., 1989. Guidelines for pest management research to reduce stored food losses caused by insects and mites. *Overseas Development and Natural Institute Bulletin* No. 22, Chatham, Kent, UK.

19. Mulbry, W.W. and Karns, J.S., 1989. Parathion hydrolase specified by the *Flavobacterium* opd gene: Relationship between the gene and protein. *Journal of Bacteriology*,171(12): 6740–6746.

20. Muir, P., 2002. *The History of Pesticides Use*. Oregon State University Press, USA.

21. Pal, R., Chakrabarti, K., Chakraborty, A. and Chowdhury, A., 2006. Degradation and effects of pesticides on soil microbiological parameters: A review. *International Journal of Agricultural Research*, 1: 240–258.

22. Patil, K.C., Matsumura, F. and Boush, G.M., 1970. Degradation of endrin, aldrin, and DDT by soil micro-organisms. *Journal of Applied Microbiology*, 1(5): 879–881.

23. Rochkind-Dubinsky, M.L., Sayler, G.S. and Blackburn, J.W., 1987. Microbiological decomposition of chlorinated aromatic compounds. In: *Microbiology Series*, 18: 153–162, New York, USA.

24. Schroll, R., Brahushi, R., Dorfler, U., Kuhn, S., Fekete, J. and Munch, J.C., 2004. Biomineralisation of 1,2,4–trichlorobenzene in soils by an adapted microbial population. *Environmental Pollution*,127(3): 395–401.

25. Singh, B.D., 2005. *Biotechnology*. Kalyani Publisher, New Delhi.

26. Serdar, C.M., Murdock, D.C. and Rohde, M.F., 1989. Parathion hydrolase gene from *Pseudomonas diminuta* MG subcloning, complete nucleotide sequence, and expression of the mature portion of the enzyme in *Escherichia coli. Nature Biotechnology*, 7: 1151–1155.

27. Vocke, G., 1986. *The Green Revolution for Wheat in Developing Countries*, US Department of Agriculture, USA.

28. Yuanfan, H., Jin, Z., Qing, H., Qian, W., Jiandong, J. and Shunpeng, L., 2010. Characterization of a fenpropathrin-degrading strain and construction of a genetically engineered micro-organism for simultaneous degradation of methyl parathion and fenpropathrin. *Journal of Environmental Management*, 91(11): 2295–2300.

29. Zacharia, Tano J., 2011. Identity, physical and chemical properties of pesticides. In: *Pesticides in the Modern World: Trends in Pesticides Analysis*, (Ed.) Margarita Stoytcheva.

Biotechnology: An Overview (2015)
Editors: Rajan Kumar Gupta, Nasim Akhtar and Deepak Vyas
Published by: DAYA PUBLISHING HOUSE, NEW DELHI

Pages 317–351

Chapter 20

Microbial Megacell in Advanced Biological Processes

Rajesh K. Srivastava*

Department of Biotechnology, GITAM Institute of Technology,
GITAM University, Gandhi Nagar Campus, Rushikonda,
Visakhapatnam – 530 045, A.P.

Introduction

Filamentous growth is a fungal differentiation behaviour and fungal scavenging response, occurred to extra-cellular stimuli such as nutrient limitation (Bharucha, *et al.*, 2008). Megacell or filaments is found in range of size from 0.8 to 5 μm in width and from 5 to > 500 μm in length. Growth morphology of filamentous fungi and some filamentous nature of bacteria is an important parameter for productivity of several industrial processes (Villena and Gutierrez-Correa, 2007; Srivastava *et al.*, 2009). Exposure (*i.e.*15–30 min) of cells harboring synthetic genetic circuit to small molecule signals (*i.e.* anhydrotetracycline or IPTG), triggered long-term and uniform cell elongation, with cell length being directly proportional to the time elapsed following a brief chemical exposure (Gardner *et al.*, 2000). Many different species including plant and animal pathogens (such as some bacteria and fungi), and yeasts like the baker's (or budding) yeast *Saccharomyces cerevisiae* undergo filamentous growth (Paul and George, 2012). Controlling of mycelial morphology is important factor or key parameter for attaining the optimum metabolic stage for maximum production of metabolites in industrial application. Engineering studies of filamentous organism is used to design and operation for industrial fermentation processes. Filamentous fungi are grown by the polar extension of hyphae. Polar growth requires the specication of sites of germ tube or branch emergence which is then followed by the recruitment of the morphogenetic machinery to those sites for localized cell wall

---

* *Corresponding Author:* E-mail: rajeshksrivastava73@yahoo.co.in

deposition. Filamentous growth in yeast is found with major changes such as an increase in cell length, a reorganization of polarity and enhanced cell–cell adhesion. (Wendland and Philippsen, 2000; Sudbery, 2001). Yeast cells could drastically change their morphology, forming the shmoo tip (pear-like cell morphology) and pseudohyphae (lamentous cell morphology) or ascospores (four cells within a cell) under starvation conditions (Lynn and Magee, 1970; Trueheart *et al.*, 1987; Mosch and Fink, 1997).

Bacteria occur singly, in small chains or clumps. Typically bacteria have cell size in few micrometers with variation in shapes, ranging from spheres to rods and spirals. Bacterial cell size depends on the balance between biomass synthesis and cell division. Unicellular organism has tendency to increase in the number of individuals in the population during its cell division. Bacteria are unicellular and most multiply by binary fission. (Errington *et al.*, 2003).

Megacells are those bacterial cells that sizes are larger than normal size (*i.e.* 2-5 µm in length). It meant that megacells are larger in their length than normal length. Normally, bacteria multiply by elongating at a certain stage of growth with beginning the formation of a septum at mid-cell, and finally completing the septum and dividing into two cells (Wu and Errington, 2004; Hu and Lutkenhaus, 1999). Megacell would be formed when cell elongation process will start in bacteria. It is an anomalous growth, found in certain bacterial species, such as *E. coli, Helicobacter pylori, Caulobacter crescentus* and some species of *Bacillus* (Everis and Betts, 2001). Induction of megacell formation could be occurred as resultant of delayed cell division protein expression. In bacterial megacell, cell has tendency to elongate continue, without division and septum formation. Megacells are long strands of non-dividing bacteria which could contain reduced or enhanced quantities of metabolites or endotoxin. (Dofferhoff and Buijs 1996; Buijs *et al.*, 2006). Megacell formation could be found in bacteria responding to various conditions such as stresses, (*i.e.* DNA damage or inhibition of replication), pH change, temperature change effect and metabolic alteration (change of metabolites concentration) (Norton *et al.*, 1993; Srivastava *et al.*, 2009). Binding of β-lactam agents to penicillin-binding-proteins (PBP-3) causes the formation of megacells/filaments which are due to the inability of bacteria to septate after doubling of their cell mass. Extensive DNA damage through the SOS response system and nutritional changes may also cause bacterial megacell formation. It has seen that key genes such as sulA and minCD has involved in filamentation in *E. coli* (Bi, and Lutkenhaus, 1993). Here we shall be review the effects of morphological changes of some micro-organism in biological process which has many applications in food or pharmaceutical industries and medical field.

The Fungi

Multi-cellular Fungi

Filamentous fungi are eukaryotic and multi-cellular organisms, used in industrial metabolites production (*i.e.* in secreting enzymes or proteins, organic acid, antibiotics or low mol.wt sugars) due to their metabolic versatility and to their capability (Villena and Gutierrez-Correa, 2007). Biofilms are morphological structure of filamentous fungi and could form as a result of antibiotic-resistant mode of microbial life, found

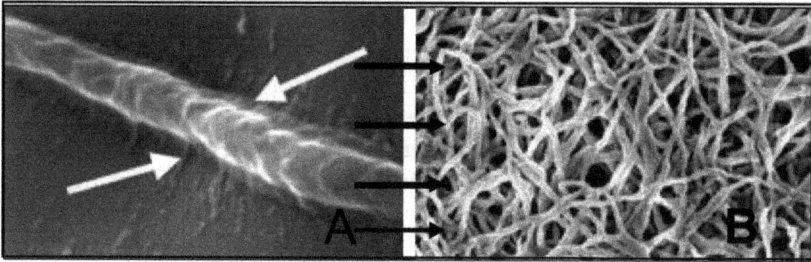

Figure 20.1. Cryo-scanning electron microscope microphotographs of *Aspergillus niger* ATCC 10864, biofilm development on polyester cloth (A) Hyphal elongation at 8 h (B) Biofilm formation at 72 h (Villena and Gutierrez-Correa, 2007).

in natural and industrial settings. In industrial application, study of morphological patterns of biofilms of filamentous fungi such as *Aspergillus niger* ATCC 10864 is useful for analysis of differential physiological behaviour of the strain and it provides the information for enhancement of lignocellulolytic enzyme productivity in submerged cultures (SC). Hyphae elongation at 8 h of this strain could be seen in Figure 20.1A and this hyphe elongation causes the biofilm formation at 72 h (Papagianni, 2004; Villena and Gutierrez-Correa, 2007). Fungal biofilms are morphological efficient systems for metabolites production and share favourable physiological aspects also (shown in Figure 20.1B). The surface-associated communities of bacterial cells are encapsulated in an extracellular polymeric substances (EPS) matrix. Fungal morphology of *A. niger* could be affected by spore inoculum level in fermentation of citric acid. Morphological features of *Aspergillus niger* in submerged citric acid fermentation is studied by digital image analysis and it has been classified in to four main object types such as globular or elongated pellets, clumps or free mycelial trees by using an artificial neural network (ANN) (Papagianni and Mattey, 2006). The Morphology number could be used to good characterization of fungal morphology. It helps to distinguish between different pellet and clump morphologies as a particle like and mycelial morphology. Determination of the fractal dimension and macro morphologic approach could be a promising method for holistic characterization of mycelial morphology (Wucherpfennig *et al.*, 2011).

$$\text{Morphology Number} = \frac{2 \cdot \sqrt{A} \cdot S}{\sqrt{\pi} \cdot D \cdot E}$$

Where A is the projected area, S is the image analysis parameter solidity, D is the maximal diameter of the pellet and E is the elongation (aspect ratio) of the particle.

Butyrolactone I is another secondary metabolite, produced by *A. terreus*, inhibits eukaryotic cyclin-dependent kinases via controlling cell cycle progression. The cyclin-dependent kinases contain serine, threonine, or tyrosine residues and are regulated by its residues phosphorylation and dephosphorylation process. Compound butyrolactone I have induced the morphological and sporulation changes in *A. terreus* and also enhanced secondary metabolite (*i.e.* lovastatin) production. It is due to physiological changes with alteration in abundance and complexity of nutrient sources (Schimmel *et al.*, 1998). Morphological changes of *A. terreus* NRRL 255 has been occurred from third days to ten days of fermentation process for lovastatin. Early growth phase (trophophase) is followed by late lovastatin production phase (idiophase) starts after 150 h and extends till 240 h of fermentation, shown in Figures 20.2A to 20.2B (Gupta *et al.*, 2007). In trophophase phase of fermentation, cell filamentation process is found more due to hyphe extension than idiophase. But we could see the more clumping of filamentous structure with weak cell wall structures and multinucleate followed by some conidia formation in idiophase of *A. terreus* NRRL 255.

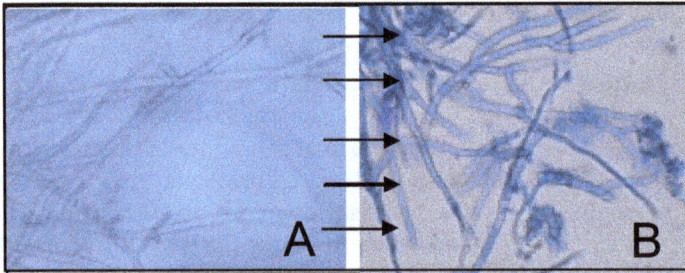

Figure 20.2. Morphological changes of *A. terreus* during 10 days fermentation process for lovastatin (A) 3rd day of lovastatin production at less productivity (B) 9th day of lovastatin production in idiophase with enhanced productivity (Gupta *et al.*, 2007).

Rhizopus chinensis CCTCC M201021 is a typical filamentous fungus and relationship between the change of morphology (*i.e.* effects of micromorphology and macromorphology) and the mycelium-bound lipase production by *R. chinensis* has been studied. Mycelia from fully entangled filaments (shown in Figure 20.3B) has exhibited high biosynthetic activity by rapid maltose consumption high enzyme production, and the situation is reverse for mycelia from dispersed mycelia shown in Figure 20.3A (Tang *et al.*, 2009). Hypha from the fully entangled filaments is asymmetric, no dissepiments formed and tips of mycelia are clear. Mycelial fragmentation is common in fungal fermentations. Similar effect for the mutant strain *R. chinensis* Y92 M on the mycelium-bound lipase production has been also reported (Teng and Xu, 2008).

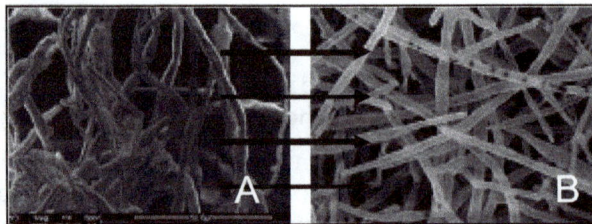

Figure 20.3. Electron microscopic picture of morphology for the (A) dispersed mycelia and (B) the fully entangled filaments (Tang *et al.*, 2009).

Mechanism of Filamentous Growth in Multi-Cellular Fungi

☆ Filamentous growth is dependent on the position of germ tube emergence, the relaying of positional information via RhoGTPase modules, and the recruitment of morphogenetic machinery components including cytoskeleton, polarisome and ARP2/3 complexes, and the vesicle tracking system (Wendland and Philippsen, 2000; Weinzierl *et al.*, 2003).

☆ Polarity establishment could be occurring at two distinct points. First point is primary germ tube emergence from the spore and second point is branch emergence from the hypha. Polarity establishment requires the specication of sites. Polarity maintenance is dependent upon the sustained localization of the morphogenetic machinery at the tips of extending hyphae. *Aspergillus nidulans* and *Neurospora crassa* have uncovered several "pioneer" polarity proteins along with others whose importance in polarity (Seiler and Plamann, 2003; Shi, *et al.*, 2003).

☆ Actin filaments are required for the establishment and maintenance of hyphal polarity in filamentous fungi which has been confirmed by use of actin-depolymerizing agents (Heath, 1994). Role for actin filaments in polarized morphogenesis is to tracks the myosin-based transport of vesicles to polarization sites. And it could direct localized vesicle transport from the apical vesicle cluster to the hyphal tip (Bourett and Howard, 1991; Roberson, 1991).

☆ Filamentous fungi use the cortical markers to specify positional information in a morphological structure system. *A. gossypii* markers (*i.e.* Bud3) are originally deposited at the hyphal tip and may be specifying sites of branch emergence (Knechtle *et al.*, 2003). Species-specic cell wall protein could mark the polarization sites as Bud8p and Bud9p found in *S. cerevisiae*. Sites of germ tube and branch emergence are specied by positional signals, emanating from occupied receptors. And filamentous fungi are capable of responding to pheromones during mating. Filamentous fungi could mark the polarization sites by receptor clustering in response to signals other than pheromones. For example, saprophytic fungi might use nutrient receptors to specify the polarization site, while plant pathogens might use receptors that bind to specic plant surface components (Sudbery, 2001).

☆ Positional information is relayed by the GTPase signaling modules to the morphogenetic machinery that remodels the cell surface. The bulk of the morphogenetic machinery is involved in the establishment and maintenance of hyphal polarity. These elements include the cytoskeleton and associated proteins, as well as the vesicle tracking complexes that direct the formation, transport, and fusion of secretory vesicles with target membranes. Rho GTPases and Cdc42p is function as a molecular on/off switch, relaying signals in its active GTP-bound form, but not in its inactive GDP-bound form. They have central role in transducing polarity information and Cdc42p accumulates at the plasma membrane in areas of polar growth (Richman *et al.*, 2002; Ziman *et al.*, 1993).

☆ The Cdc42 GTPase module is needed for germ tube emergence and is essential for those dimorphic and filamentous fungi most closely related to *S. cerevisiae* (*A. gossyippi* and *C. albicans*). In both fungi, localization studies and mutant phenotypes support the roles in polarity establishment of germ tubes and branches in addition to the expected role in septation. (Sharpless and Harris, 2002). In *A. nidulans*, the septin AspB localizes to emerging secondary germ tubes, branches, and conidiophore layers. Strains carrying a temperature-sensitive mutation in aspB show hyperbranching as well as conidiophore defects (Westfall and Momany, 2002)

☆ The polarisome directs the formation of actin cables at polar growth sites in filamentous fungi. Most polarisome components are conserved in *A. nidulans* (shown in table 1), and the formin SepA controls the assembly of actin cables at hyphal tips and septation sites (Sharpless and Harris, 2002). Similarly, the Arp2/3 components are conserved in *A. nidulans*. Signals emanating from Cdc42 GTPase modules are triggered by local organization of actin filaments via similar mechanisms in filamentous fungi and yeast. Microtubules have a critical role in regulating the direction of hyphal extension (Riquelme *et al.*, 1998).

Unicellular Fungi (The Yeast)

The dimorphic change of a single-cell budding yeast to a filamentous form enables the *Saccharomyces cerevisiae* to forage for nutrients where as opportunistic pathogen *Candida albicans* invades the human tissues and evades the immune system. Ryan *et al.*, have identified the genes, involved in morphologically distinct forms of filamentation such as haploid invasive growth, biofilm formation, and diploid pseudohyphal growth (Ryan *et al.*, 2013). It has been reported about core genes with general roles in filamentous growth such as MFG1 (YDL233w). Its product binds two morphogenetic transcription factors (*i.e.* Flo8 and Mss11) and functions as a critical transcriptional regulator of filamentous growth in both *S. cerevisiae* and *C. albicans* (Shively *et al.*, 2013, Ryan *et al.*, 2013). Contact of yeast cells with 10 μg of 5-fluorocytosine (5-FC) per ml, has marked the enlargement in cell volume of the *C. albicans* cells (shown in Figure 20.4B) whereas no further change was observed with

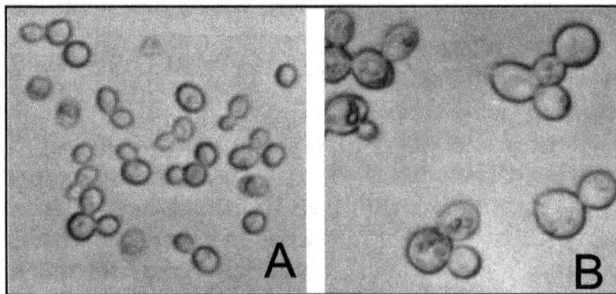

Figure 20.4. Enlargement of *C. albicans* cells induced by 5-FC treatment.

(A) Normal cell (B) induced cells after 12 h contact with 10 μg of 5-FC per ml (X 1,400) (Arai *et al.*, 1977).

Table 20.1. Homologues of Cdc42 module proteins from filamentous and dimorphic fungi.

Sl.No.	Fungus Gene name	Comments a	Reference
Rho GTPase: CDC42			
1.	*Ashbya gossypii* AgCDC42	Isotropic, delocalized, actin patches, no GTE	Wendland and Philippsen, 2001
2.	*Aspergillus nidulans* modA	OE dom hyperactive modA: delayed GTE, swollen hyphae, no conidia	Harris *et al.*, 1994
3.	*Candida albicans* CaCDC42	DDCaCDC42 +inducible CaCDC42 in repressing conditions	Ushinsky *et al.*, 2002
4.	*Suillus bovines* SbCDC42	RNA and protein expressed in vegetative and ectomycorrhiza-forming hyphae; co-localized with actin at hyphal tips	Gorfer *et al.*, 2001
Rho GTPase: RAC			
1.	*Aspergillus niger* racA	DracA: viable, accelerated 2°GTE and increased apical branching, compact, reduced conidiation, actin localization normal	Ram *et al.*, 2001
2.	*P. marneffei* cB	DcB: yeast cells normal, hyphae and conidiophores defective polarization, hyperbranch and tip branch. GFP::CB co-localized with actin at septa and hyphal tips	Boyce *et al.*, 2003
3.	*Yarrowia lipolytica* YlRAC1	DYlrac1: yeast round, pseudohyphae, no true hyphae; actin polarized; YlRAC1 RNA increases in yeast to hyphae transition	Hurtado *et al.*, 2000
GEF: CDC24			
1.	*A. gossypii* AgCDC24	Complements *S. cerevisiae* cdc24; DAgcdc24 [b] identical to DAgcdc42	Wendland and Philippsen, 2000
2.	*Ustilago maydis* don1	Don1 mutant and Ddon1: yeast separation and bud site selection defects; normal filamentous growth of dikaryon; interacts with UmCDC42 in 2 hybrid	Weinzierl *et al.*, 2002
GAP			
1.	*A. gossypii* AgBEM2	DAgbem2: swollen germ cells, possible extra GTE from germ cell, increased tip branching	Wendland and Philippsen, 2000
PAK: STE20			
1.	*C. albicans* CST20	DDcst20 reduced hyphae on some media; reduced virulence	Kohler and Fink, 1996

S. cerevisiae. Mechanism of action of 5-FC is to selectively inhibit DNA synthesis and that cause thymineless death in morphological changes of *C. albicans* (Arai *et al.*, 1977).

Filamentous growth of the budding yeast *Saccharomyces cerevisiae is* induced by overexpression of the Whi2. Whi2 is concentrated at the tips of the filaments. Gene overexpression of Whi2 from the GAL1 promoter causes the filamentous cells growth because of fail to complete cytokinesis and the budding pattern changes from axial to polar, cells become elongated and cell size increases threefold (Rahman, *et al.*, 1988). Cells show hyperpolarized growth, new cell wall synthesis is confined to apical cells, cells become elongated in shape, the axial ratio doubles, and volume increases threefold. New cell wall material is incorporated only into the tips of filaments. It has been confirmed by performing the A Concanavalin A pulse-chase experiment. Bud of *S. cerevisiae* (shown in 20.5A) could develop the pseudohyphae shown in Figure 20.5D (*i.e.* result from nitrogen-limited growth) and the filaments during the invasive growth of haploids (shown in Figures 20.5B and 20.5C). However, Whi2-induced filament formation is reduced, but not blocked, by mutations in STE7, STE12 or STE20 where as no blockage of pseudohypha formation (Radcliffe *et al.*, 1997, Cvrckova, *et*

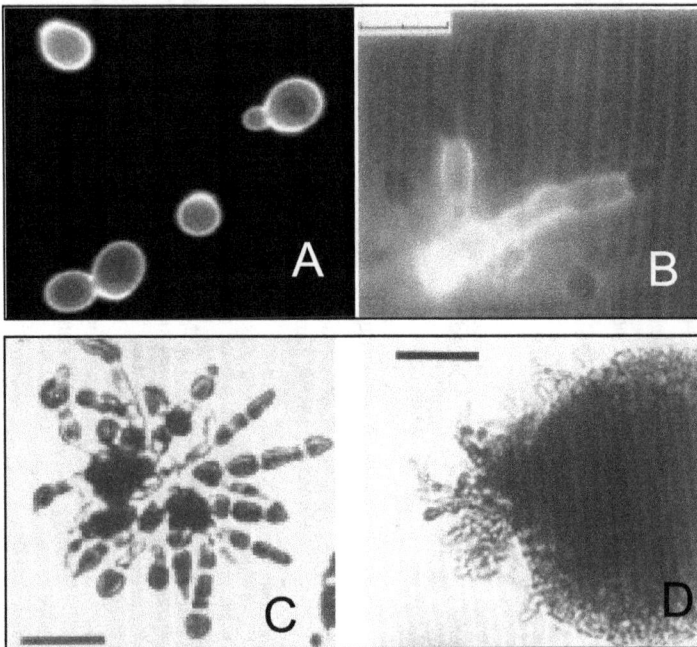

Figure 20.5. Morphological structures of *S. cerevisiae* in bioprocesses.

(A) Budding yeast *Saccharomyces cerevisiae* (B) New cell wall material is incorporated only into the tips of filaments via confirming to A Concanavalin A pulse-chase experiment, Bar is 10µm; (C) *Saccharomyces cerevisiae* strain L5366h-pWHIM5 cells were streaked onto the surface of a minimal medium plus galactose agar plate and examined by transmitted light microscopy. Bar, 20µm; (D) Pseudohypha formation is occurred on agar plates containing nitrogen-limiting medium Bar, 80µm (Girneno *et al.*, 1992; Radcliffe *et al.*, 1997).

al., 1995). Pseudohypha formation could be seen on agar plates containing nitrogen-limiting medium. Diploid cells are grown to form the pseudohyphae, in which budding switches from a bipolar to a unipolar pattern, cytokinesis is not completed and cell growth becomes highly polarized (Girneno *et al.*, 1992, Wright, *et al.*, 1993). In other study, changes in yeast morphology are caused by a high concentration of extracellular Ca^{2+}.

A calcium-sensitive cls4 mutant of *S. cerevisiae* stops dividing in the presence of 100 mM Ca^{2+}, producing large, rounded, and unbudded cells. The zds1 mutant shows defective growth in rich medium in the presence of 50 to 300 mM Ca^{2+}, with the cells forming an elongated bud and most of the elongated cells having a single nucleus at the mother/bud neck with characteristic of G delay. zds1 (zillion different screens) mutant elongated buds have signicantly increased parameter of C114_C (*i.e.* bud axis ratio of budded cell with two nuclei) after Ca^{2+} treatment of the five samples (P 0.01) (Bi and Pringle, 1996; Yu *et al.*, 1996). Morphological parameters (P) include a dispersion which is the variation of values caused by experimental errors. Dispersion (*e.g.*, standard deviation and quartile deviation) to the high-dimensional cluster analysis is obtained using the data obtained from replicated experiments. However, the bud morphology of the cls mutants with exception of class V cls mutants is coherently rounded rather than elongated in the presence of a high concentration of Ca^{2+} (Yu *et al.*, 1996; Ohnuki *et al.*, 2007). Microscopic examinations of cells are done to visualize the changes in cell shape and specic assays are used to examine the response in haploid and diploid cells. Both haploid and diploid cells can undergo filamentous growth, form biofilms, or enter stationary phase (quiescence) in response to nutrient (glucose or nitrogen) limitation. Diploid cells also sporulate in response to the limitation of carbon and nitrogen sources. Secreted alcohols act as autoinducers to stimulate filamentous growth.

The budding yeast *Saccharomyces cerevisiae* undergoes filamentous growth and is regulated by evolutionarily conserved signaling pathways such as mitogen activated protein kinase (MAPK), rat sarcoma/protein kinase A (RAS/PKA), sucrose nonfermentable (SNF), and target of rapamycin (TOR). These pathways regulate the cell differentiation to the filamentous type, which is characterized by changes in cell adhesion, cell polarity, and cell shape. Flamentous growth is a more complex and globally regulated behaviour. It could be pave the way for future investigations for study to eukaryotic cell differentiation behavior. The filamentation response could be varying among species, ranging from mycelial mat or hyphal formation in true filamentous fungi and in yeasts, it could be to subtle changes in cell shape. Budding yeast does not undergo true hyphal growth, but a pseudohyphal growth pattern could be formed in which cells fully separate by cytokinesis with non- multinucleate and remain attached to each other by proteins in the cell wall.

Mechanism of Yeast Filamentation

☆ In yeast two distinct multi-protein complexes (such as the polarisome and the Arp2/3 complex) the downstream of the Cdc42 GTPase module is to direct the localized assembly of actin filaments at polarization sites (Trueheart *et al.*, 1987; Harris and Momany, 2004). The polarisome regulates the formation of linear, unbranched actin filaments (actin cables). The key component of the polarisome is the formin Bni1p, which binds to the barbed ends of actin filaments and nucleates microfilament assembly (Pruyne *et al.*, 2002; Sagot *et al.*, 2002a; Sagot *et al.*, 2002b).

☆ The remaining polarisome components appear to regulate the timing and location of Bni1p activity (Sagot *et al.*, 2002b). The Arp2/3 complex regulates the formation of branched actin filaments, which unlike linear cables, form a fine meshwork that typically underlies the cell surface. In yeast, this complex regulates actin patch formation and key components formation include the WASP homologue Las17p/Bee1p, the actin-related proteins Arp2 and Arp3, and the class I myosin Myo3p (Lechler *et al.*, 2001).

☆ To specify the septation site, there are distinct differences in the mechanisms of budding and binary fission. Yeast cells re-localize the morphogenetic apparatus from the poles to this site in order to undergo cytokinesis. First, *S. cerevisiae* bud site markers Bud3p, Bud8p, and Bud9p are present in the proteome of *S. pombe* (Ziman *et al.*, 1993). Second, the conserved Cdc42p signaling module relays positional information to the morphogenetic machinery. Third, scaolding proteins, such as formins and septins, play an important role in organizing the morphogenetic machinery at sites of polarized growth (Pruyne *et al.*, 2002).

☆ Mating site position is specied by an occupied mating receptor complexed with a heterotrimeric G-protein. The conserved Cdc42 GTPase module transduces positional information to several effectors that promote localized organization of the morphogenetic machinery (Zhang *et al.*, 2001). Similarly, in *S. pombe*, cell ends are specied by cortical markers working with a Ras GTPase, and positional information is transduced through the Cdc42 GTPase module to the morphogenetic machinery (Chang and Peter, 2003).

☆ In *S. cerevisiae*, the cortical landmark proteins generate the positional signal (Pringle *et al.*, 1995). These include markers, define the axial budding pattern (Bud3p, Bud4p, and Axl2p) or the bipolar budding pattern (Bud8p, Bud9p, and Rax2p) (Wright *et al.*, 1993). Several of these markers (Axl2p, Bud8p, and Bud9p) are cell wall proteins whose delivery to the cell surface is tightly coordinated with cell cycle progression and changes in the pattern of localized secretion (Lord *et al.*, 2000; Schenkman *et al.*, 2002).

☆ In *S. cerevisiae*, pheromone dependent stabilization of an adaptor protein that interacts with the free Gb subunit and the Cdc42 GTPase module appears to override the cortical bud site markers (Shimada *et al.*, 2000).

Filamentous Growth of Yeast by Nutrient-Sensing Pathways

Filamentous growth is linked to a nutritional scavenging response and this morphological behaviour is related to nutrition in two important ways; first strains defective for ammonium utilization are hyperlamentous and shows the connection of filamentous growth with nitrogen levels. Lack of fermentable carbon source could trigger for filamentous growth. Second, the global nutrient regulatory GTPase Ras2 is found to be required for filamentous growth regulation. There are four signaling pathways such as rat sarcoma/protein kinase A (RAS/PKA), sucrose nonfermentable (SNF), target of rapamycin (TOR), and MAPK are found to regulate filamentous growth (Kataoka *et al.*, 1984). The signaling pathway triggers the invasive growth in haploid cells via utilizing the Ste20, Ste7, Ste11, and Ste12. Filamentous or pseudohyphal growth in diploid *S. cerevisiae* cells is characterized by cell elongation, amitotic delay, symmetric cell division, unipolar budding and persistent physical attachment of mother and daughter cells shown in Figure 20.6.

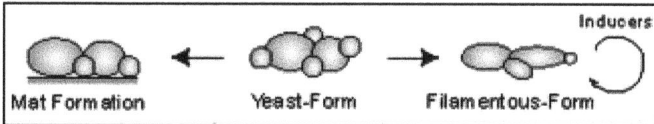

Figure 20.6. The life cycle of the budding yeast Saccharomyces cerevisiae. Both haploid and diploid cells can undergo filamentous growth, form biofilms, or enter stationary phase (quiescence) in response to nutrient (glucose or nitrogen) limitation (Paul and George, 2012).

RAS/PKA Pathway

RAS is involved in filamentous growth and it is components of the molecular pathway via playing an important role in growth habit. There are reports on yeast's RAS genes expression rates such as RAS1 and RAS2. TheRAS2 gene is expressed at higher levels than RAS1 and it is responsible for the majority of Ras function (Kataoka *et al.*, 1984). Ras2 activates adenylate cyclase produces the second messenger cyclic adenosine monophosphate (cAMP) (Toda *et al.*, 1985). Levels of cAMP are critical for the decision of whether or not cells undergo filamentous growth (Mosch *et al.*, 1996). Overexpression of the gene encoding the phosphodiesterase (Pde2) has dampened the filamentous growth and also suppressed the hyperlamentation, induced by activated RAS (Ward *et al.*, 1995). cAMP regulates the activity of a family of protein kinases, (*i.e.* protein kinase A) (PKA). Binding of cAMP to a regulatory subunit (*i.e.* yeast Bcy1) releases the PKA, activating its kinase activity. Budding yeast has three different PKAs (such as Tpk1, Tpk2 and Tpk3), which are 75 per cent homologous in their catalytic domains but differing in their N-terminal regions (Toda *et al.*, 1987). All three Tpks associate with the regulatory subunit Bcy1 with Ras2/cAMP activation of PKA is required for filamentous growth. The PKA pathway regulates the filamentous growth by regulating the transcription factor (*i.e.* Flo8), via Tpk2 with regulating the dual-specicity tyrosine-regulated kinase (DYRK) Yak1, via Tpk1.Yak1 has a positive role in regulating filamentous growth (Zhang *et al.*, 2001). The G-protein coupled receptor (GPCR) Gpr1 and its associated heterotrimeric G protein regulate the Ras2 GTPase activating proteins (GAPs) (such as Ira1 and Ira2). Ras2 regulates adenylate cyclase, which produces cAMP. cAMP can bind to Bcy1 and inactivates the protein via releasing the Tpk1, Tpk2, and Tpk3 which activate Flo8 and other targets. It contributes to nutrient-regulated filamentous growth (Harris and Momany, 2004).

Snf1 Pathway

Depletion of fermentable carbon sources, like glucose, could trigger the filamentous growth response. By removing and adding back various nutrients and examining the effects on cell and colony morphology, it has been confirmed that depletion of fermentable carbon sources, like glucose has resulted the induction of filamentous growth (Cullen and Sprague 2000). Snf1 operates in a separate pathway from Gpr1, by regulating the repressors Nrg1 and Nrg2 at the FLO11 promoter and this gene is required for filamentous growth. Nrg1 and Nrg2 are functioned by recruitment of the Cyc8–Tup1 complex to promoters (Kuchin *et al.*, 2002). Thus different glucose-sensing pathways such as Gpr1/Gpa2/Ras2/PKA and Snf1, regulate filamentous growth in yeast.

TOR Pathway

Limiting the xed nitrogen (specically ammonia) is a trigger of filamentous growth. Specically, mutants defective for ammonium transport are hyperfilamentous. It suggests that ammonium starvation might be a trigger for filamentous growth (Gimeno *et al.*, 1992). The high-affinity ammonium transporter (*i.e.* Mep2) is required for filamentous growth. The filamentation defect of the mep2 mutant arises apparently (Boeckstaens *et al.*, 2008). A specific role for transporter is in communicating a signal through a small region in its cytosolic domain. Nitrogen signals could be interpreted by the TOR pathway, an evolutionarily conserved nutrient-regulatory pathway (Rutherford *et al.*, 2008). The serine/threonine protein kinase TOR regulates cellular homeostasis by coordinating metabolic processes with cellular nutrient levels (Sengupta *et al.*, 2010). The TOR pathway regulates the transcription factor (*i.e.* Gcn4) and Gcn4 is a regulator of FLO11 expression (Heitman *et al.*, 1991). The TOR pathway regulates filamentous growth and it is apparently independent of the RAS/PKA and MAPK pathways. Glucose limitation induces filamentous growth in both haploid and diploid cells. Indeed, cells grown in nutrient-rich (high glucose) conditions do not produce pseudohyphae. Glucose/sucrose is required for filamentous growth in a Gpr1-dependent manner. Rme1p (a zinc-nger type transcriptional factor) promotes the mitotic/meiotic decision. Rme1p binds directly to the FLO11 promoter to induce cell–cell adhesion and invasive growth (van Dyk *et al.*, 2003). Rapamycin inhibits filamentous growth under nitrogen-limited conditions and this inhibition is mediated by the TOR pathway phosphatases Tap42 and Sit4 (Cutler *et al.*, 2001).

MAPK Pathway

A MAPK is phosphorylated and activated by a MAPK- Kinase (MAPKK), an upstream protein kinase. Each pathway is initiated by a distinct upstream regulator and individual MEKK-MEK-MAPK modules control mating, cell-wall integrity, pseudohyphal development and filamentous invasive growth. A MAPK pathway is composed of kinases and function in the mating or pheromone response pathway. Four proteins such as the p21-activated (PAK) kinase Ste20, the MAPKKK Ste11, the MAPKK Ste7 are required for mating in haploid cells. The transcription factor (*i.e.* Ste12) is also required for filamentous growth in diploids (Liu *et al.*, 1993). We find the reports that haploid cells utilize the "core module" of Ste20/Ste11/Ste7/Ste12 for mating, whereas diploid cells utilize same core module for filamentous growth regulation shown in Figure 20.7A. Cdc42 is an essential protein, required to establish the cell polarity. It has been shown that Cdc42 and Ste20 function in both the mating pathway and the filamentation pathway (Gao *et al.*, 2007; Tong *et al.*, 2007). Nutrient limitation induces filamentous growth shown in Figure 20.7B. Kss1 is the MAP kinase for the filamentation pathway ((Elion *et al.*, 1990). The kss1 mutant has a strong invasive growth defect and showed reduced activity of a filamentation response element (FRE) (Mosch *et al.*, 1996). Deleting FUS3 restored invasive growth to the kss1 mutant. ste7 fus3, kss1 triple mutants has invaded the agar as well as wildtype cells (Cook *et al.*, 1997). Fus3 and Kss1 (mainly Kss1 from genetic evidence) had an inhibitory function in filamentous growth. A dual role for Kss1 in MAPK regulation could be explained by changes in its phosphorylated (active) state. Unphosphorylated Kss1 functions as an inhibitor, whereas phosphorylated Kss1, catalyzed by Ste7, is function as an activator (Cook *et al.*, 1997).

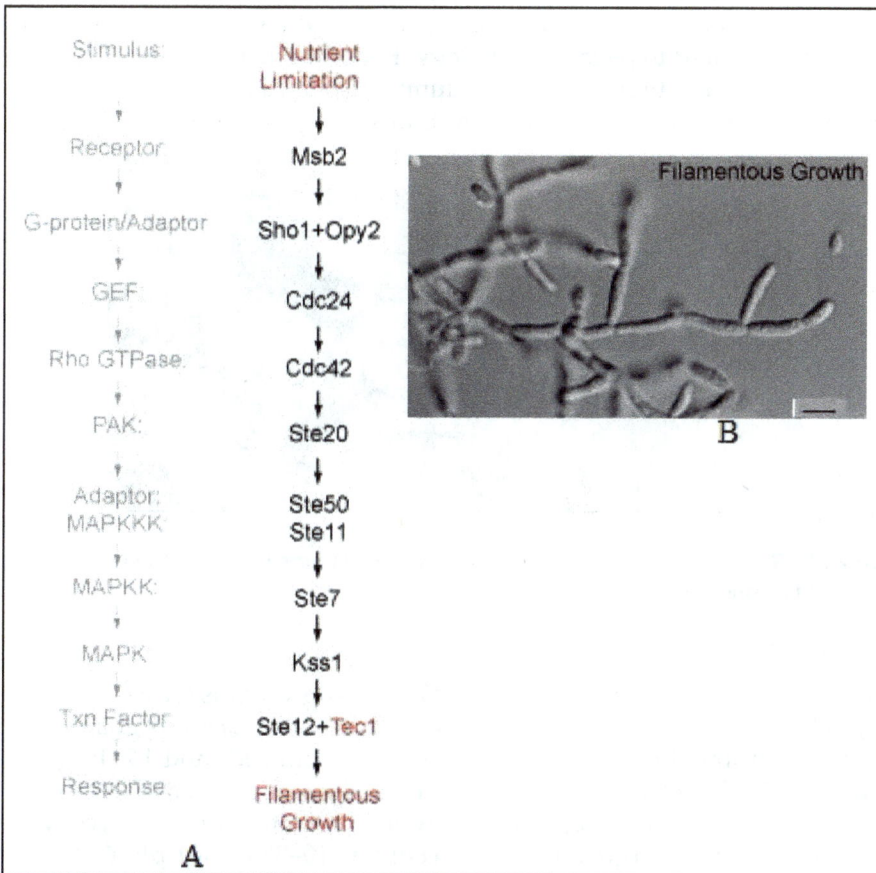

Figure 20.7. MAPK pathway is shown.

(A) Colored proteins represent pathway-specific factors; protein shown in black, function in this pathway. Nutritional signals feed into lamentous growth and pathway regulation is not well understood. (B) MAPK morphogenesis in yeast and the filamentous growth pathway induces lamentous growth, branched chains of elongated and connected cells. Bar, 5 μm (Paul and George, 2012).

Actinomycete

The effect of spore concentration on the morphology of filamentous fungi has been extensively reported and whereas there is less information for actinomycetes (Smith and Calam, 1980; Du *et al.*, 2003). *Saccharopolyspora erythraea* is an industrially important actinomycete used for the production of erythromycin. Submerged cultivation of *Saccharopolyspora erythraea*, at different initial spore concentrations was done to study the inter-relationship between inoculum concentration, morphology, rheology and erythromycin production. Pellet morphology was dominant at 10^3 and 10^4 spore/ml initial spore concentrations shown in Figure 20.8B whereas presence of clump morphology was at initial spore concentrations of 10^5-10^7 spore/ml shown in Figure 20.8A (Ghojavand *et al.*, 2011). For *Saccharopolyspora erythraea* cultivation, clump morphology is found more suitable for erythromycin production compared to pellet morphology. For other filamentous actinomycetes, there is also increase in the proportion of clump/dispersed growth morphology with an increase in initial spore concentration (ElEnshasy *et al.*, 2000).

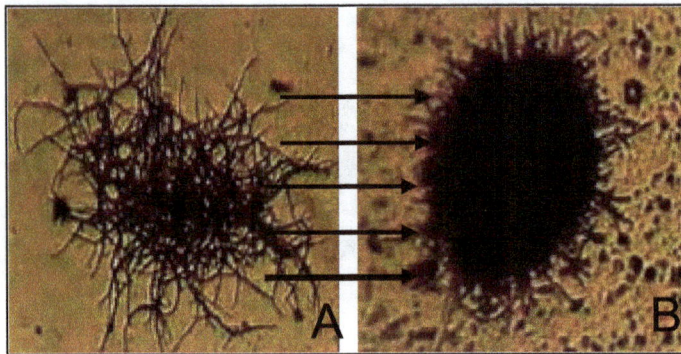

Figure 20.8. *Saccharopolyspora erythraea* **(NUR001) morphology (A) clump form; (B) pellet form (Ghojavand *et al.*, 2011).**

The Bacteria

Freshwaters bacterial species *e.g. Flectobacillus* spp. (such as *F. lacus, F. roseus*) are small, rod-shaped and can able to form filamentous cells. *F. lacus* is gram-negative, aerobic, almost straight rods, approximately 0.3–0.6µm wide and 4.7–10.0µm long (Chung *et al.*, 2006). Cells are nonmotile and do not possess flagella. Colonies on R2A agar are pale pink to rose in colour and are round (2–6 mm in diameter), convex and smooth with entire margins. Growth occurs at 10–35ºC and pH 6–9; optimal temperature and pH for growth are 25–30ºC and pH 7 respectively. Grows Induction of cell filamentation in freshwater bacteria is found due to defense mechanism against bacterivorous protests (Chung *et al.*, 2006).

Under starvation conditions bacterial rod shaped cells could change their size and become coccoid. *Proteus vulgaris* is one of marine bacteria and it causes the food poisoning and food borne disease. It is gram negative bacteria and found in aquatic environment. It has tendency to form biofilm with food contact surface. Here in case

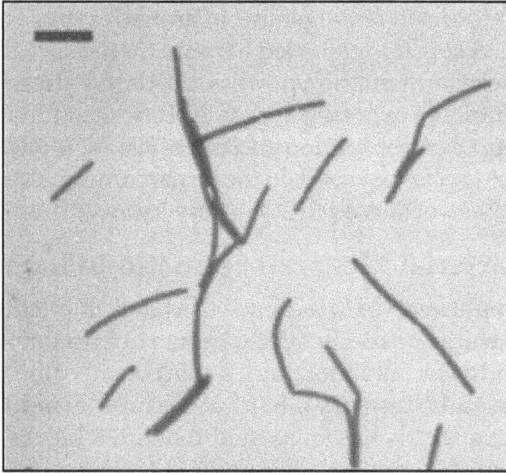

Figure 20.9. Transmission electron micrograph of negatively stained of *F. lacus* cells of strain CL-GP79ᵀ with bar 5 μm (Chung *et al.*, 2006).

of *Proteus vulgaris* under long term starvation condition, the cell length is found to shortening with decreased area (shown in Figure 20.10B) (Myszka and Craczyk, 2011). Similar results are found in case of *E. coli*.

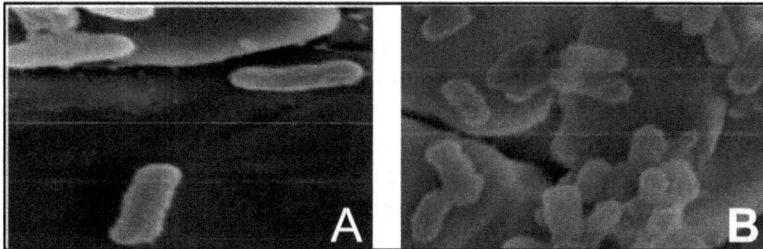

Figure 20.10. Scanning electron micrographs of *Proteus vulgaris* cell on stainless steel surface in 1000X. (A) Optimum nutrient available condition (B) starvation condition (Myszka and Craczyk, 2011).

Regular helical shape of *Helicobacter pylori* could develop the extended filamentous structures due to highly elongation with a delay in cell division and contained non-segregated chromosomes. It is regulated by peptidases enzymes that cause peptidoglycan relaxation and the coiled-coil-rich proteins (Ccrp) and also responsible for the helical cell shape *in vitro* as well as *in vivo*. *M. tuberculosis* and *Mycobacterium smegmatis* cell can also form filamentation.

Bacterial elongation could dependent on growth rate enhancement and cell division suppression. Both processes could depend on the quantity and quality of substrate. Filamentous or megacell phenotype can confer protection against lethal environments or induction of specific metabolite formation. Filamentous *E. coli* (size up to 70 μm) has been found in pathogenesis in human cystitis where as the length of *Bacillus pumilus* IFO13322 is also reported to more than 80 μm depending on media composition in D-ribose fermentation process. Oxidative stress, nutrient limitation, DNA damage and antibiotics exposure are some of stress conditions to which bacteria respond, altering their DNA replication and cell division. *Mycobacterium tuberculosis* also elongates after being phagocytized. In the urinary tract infection, rod morphology of *Escherichia coli* has been observed to SOS stress-induced filaments form.

Mechanism of Bacterial Megacell Formation/Filamentation

Blocking DNA replication in *E. coli* leads to a block in cell division and resulting in filamentation. There are two mechanisms, known to exist for blocking cell division. First, it involves the inducible inhibitor SulA, a component of the SOS response shown in Figure 20.11B (Dari and Huisman, 1983). SulA is rapidly induced following DNA damage. And genetic analysis has suggested that SulA blocks the cell division by interaction with FtsZ protein shown in Figure 20.11A. Recent biochemical and genetic data have supported this suggestion. SulA interacts with FtsZ and blocks FtsZ's function in division. FtsZ forms a ring at the division site, and induction of SulA is sufficient to abolish this localization of FtsZ (Bi and Lutkenhaus, 1993; Dai *et al.*, 1994). The binding of SulA to FtsZ prevents it from localizing to the division site. A MalE-SulA fusion blocks the binds to FtsZ in the presence of GTP but it does not affect the GTPase activity of FtsZ (Higashitani *et al.*, 1997; Higashitani *et al.*, 1995).

Second, FtsZ is limited by blocking replication of a plasmid carrying the ftsZ gene. This leads to the disappearance of FtsZ and causes cessation of cell division. Expression of antisense RNA under the control of the lac promoter reduces the level of FtsZ by about 80 per cent . This reduction in FtsZ is resulted to delay in division. However, the culture is achieved a steady state with division occurring at a larger cell size and level of FtsZ controls the frequency of division (Figure 20.11A). It has also been reported that decreasing the level of FtsZ affects nucleoid segregation (Tetart *et al.*, 1992). FtsZ is localized to the division site in *E. coli* in a pattern designated the Z ring. The analysis of dividing cells at different stages of septation has revealed that FtsZ is localized at the leading edge of the invaginating septum throughout division. Cells without an invagination have been rarely observed with FtsZ localization at midcell (Raychaudhuri and Park, 1992). In newborn cells, Z ring formation is initiated from a single spot to form a ring. This process is rapid, requiring less than 1 minute. During division the diameter of Z ring has decreased at the leading edge of the septum. *E. coli* has enough concentration of FtsZ protein to form (~20) Z-rings. FtsZ is contained a sequence motif, present in tubulin and is believed to involve its GTPase activity. This leads to an examination of the interaction of FtsZ with nucleotides (de Boer *et al.*, 1992; Mukherjee *et al.*, 1993; RayChaudhuri and Park, 1992). Furthermore, the GTPase activity has showed a dependence of the FtsZ concentration. In *Bacillus subtilis*, it was found that asymmetric septum formation divides the cell into two different-sized cells with different developmental fates. The small cell, or prespore, is destined to become the spore and the large or mother cell participates in formation of the spore but does not survive. This developmental regulation over the topology of cell division has made the *B. subtilis* an interesting organism for the study of cell

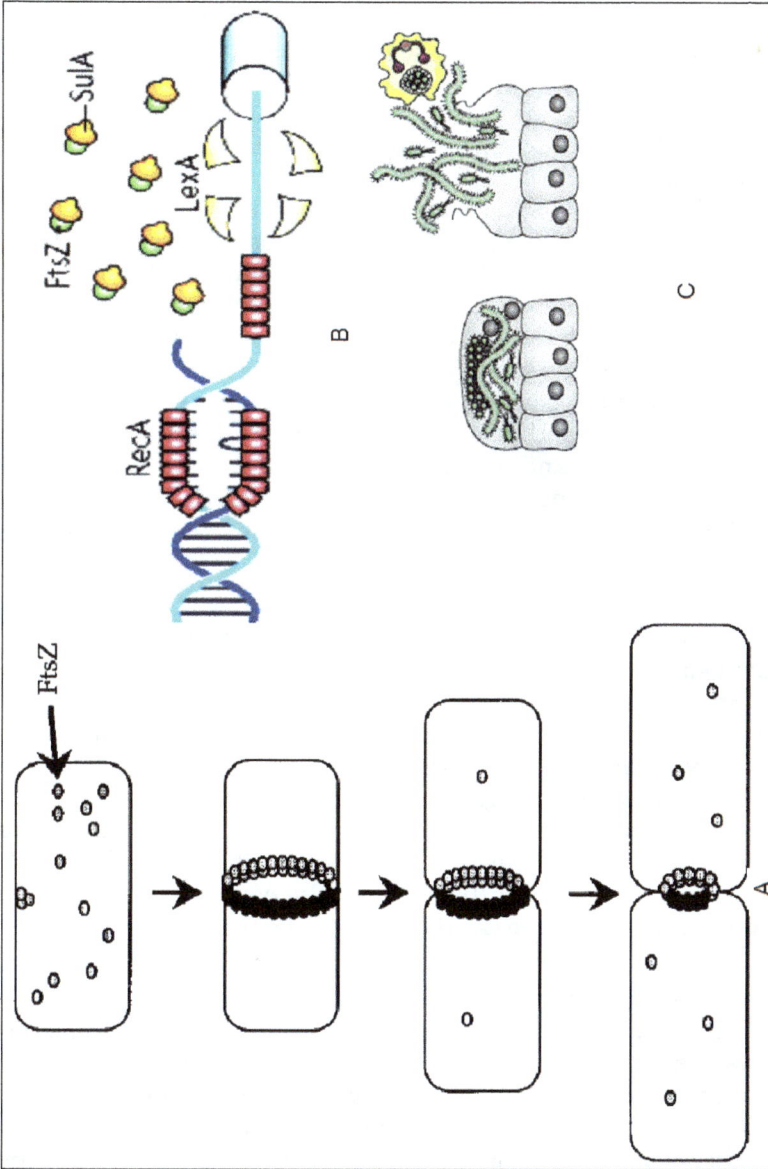

Figure 20.11. Z-ring formation at mid-cell, which results in the formation of non-septate bacterial filaments.
(A) FtsZ protein is expressed and localized at cell division sites (B) LexA regulon combine with, SulA, to prevent the transmission of mutant DNA to new daughter cells. SulA specifically inhibits polymerization of the division protein FtsZ by binding to the FtsZ monomer. This process blocks Z-ring formation at mid-cell, which results in the formation of non-septate bacterial filaments.
(C) Filamentation of Uropathogenic *Escherichia coli* (UPEC) reside intracellularly within bladder epithelial cells (Raychaudhuri and Park, 1992; de Boer *et al.*, 1992; Justics, 2004).

division. Consistent with this, the min locus has been found in B. *subtilis*, although it lacks a minE homologue.

A small number of intracellular bacteria have responded to the activation of host immune effectors by filamentation. Epithelial-cell death has accompanied the bacterial growth, as an exposure of filamentous, shown in Figure 20.11C. During vegetative septation, FtsZ is present at midcell at the leading edge of the invaginating septum; however, as cells enter sporulation. FtsZ is localized to both cell poles. Thus, it appears as though nucleation sites at both poles are unmasked, and they are preferred over the central site. If FtsZ is limiting, then Z ring would only form at one cell pole. Asymmetric septation will only occurred at one cell pole. In the absence of another early-acting sporulation gene product (SpoOH), blocks the contraction of the FtsZ ring with no septation. One morphological feature of the asymmetric septum is much thinner than the vegetative septum owing to lower cell wall content. SpoIIE, like FtsA in E. *coli* is recruited to the division site by direct interaction with the Z ring (Addinall and Lutkenhaus, 1996). In *Bacillus subtilis*: nucleoid segregation is unaffected over a wide range of FtsZ levels (Partridge and Wake, 1995). Surprisingly, blocks to division can be suppressed by increasing the level of FtsZ. This suggests that FtsZ or some other component of the division machinery normally depends on the protein folding machinery. Both the B. *subtilis* and E. *coli* FtsZs have low specific activity at concentrations below 50 µg ml^{-1}, and maximal specific activity at concentrations above 200–300 µg ml^{-1}. Observations are interpreted as a requirement for a protein-protein interaction for the GTPase activity. There are some more mechanisms of elongation or megacell formation.

☆ Base excision repair (BER) is a cellular mechanism that repairs the damaged DNA throughout the cell cycle. It is responsible primarily for removing small, non-helix-distorting base lesions from the genome. BER mechanism involves five sequential reactions: (i) base removal; (ii) incision of the resulting abasic site; (iii) processing of the generated termini at the strand break; (iv) DNA synthesis, and (v) ligation. Lack of Base Excision Repair (BER) activity via mutating the genes (such as gene of glycosylase or AP-endonucleases enzymes) in E. *coli causes the* strong filamentation structure via elongation or megacell formation. BER is responsible for repair of DNA damage by involving the bifunctional glycosylase and purinic/Apirimidinic (AP)-endonucleases (Fortin *et al.*, 2003).

☆ SOS gene products are involved in cytogenesis, DNA recombination, DNA replication, DNA damage repair, and segregation of chromosomes during cell division. In SOS gene, sulA is induced to inhibit and delay the cell division transiently leading to cell filamentation. *SulA/FtsZ* protein mediates filamentation process in bacteria by halting cell division and repair DNA mechanism (Dallo and Weitao, 2010). The single-stranded DNA regions induce mutations due to the action of different external cues. The major bacterial recombinase (RecA) binds to this DNA region and is activated by the presence of free nucleotide triphosphates. This activated RecA stimulates the autoproteolysis of the SOS transcriptional repressor LexA. The LexA regulon includes a cell division inhibitor, SulA that prevent the

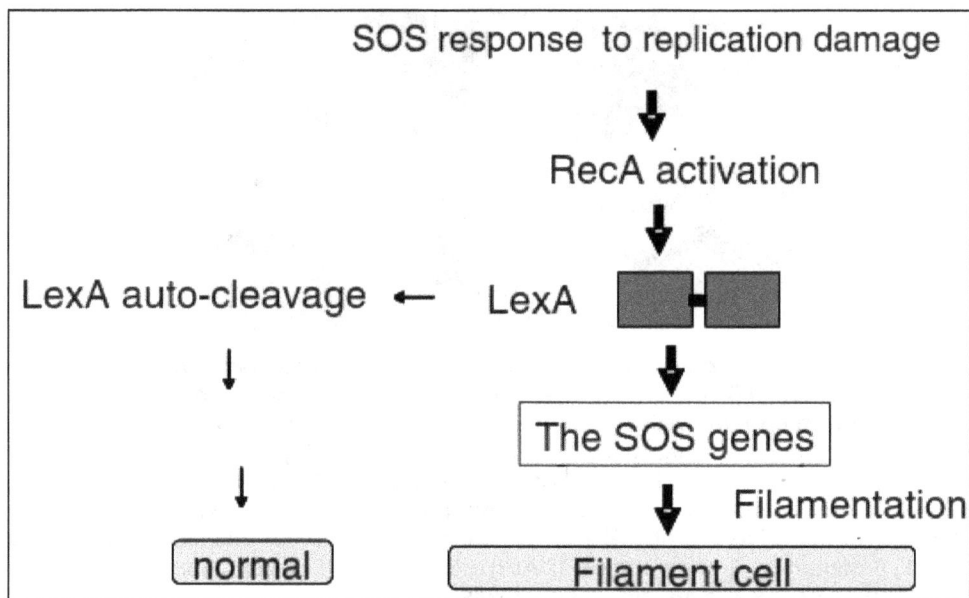

Figure 20.12. In a SOS responses, LexA and RecA control the SOS genes. LexA represses replication genes. Autocatalytic cleavage of LexA causes the SOS genes derepression and form normal cell where as complete LexA protein causes the SOS genes expressed for repair. Cell division is inhibited and delayed resulting filamentation to allow repair before cell division (Dallo and Weitao, 2010). FtsZ is limited by blocking replication of a plasmid carrying the ftsZ gene. sporulation gene product (SpoOH), blocks the contraction of the Z-ring with no septation.

transmission of mutant DNA to the daughter cells. SulA is a dimer, binds to FtsZ (a tubulin-like GTPase) in a 1:1 ratio and acts specifically on its polymerization process and forms the non-septated bacteria filaments. And this above process is crucial for the development to filamentous forms of bacteria.

☆ Inhibition of Z-ring formation and nucleoid separation is occurred in elongated bacterial cells. Filamentation of bacilli is reported and is inhibition of one of the steps of cell division mechanism. *Bacilli* show filamentous morphology when FtsZ assembly and consequently Z-ring formation is perturbed (Beuria *et al.*, 2005; Jaiswal *et al.*, 2007). Nucleoid staining is done to ascertain if Z-ring formation and nucleoid separation were affected in the elongated *Bacillus pumilus* ATCC 21951 cells. In LB broth cultures, the cells showed clear nucleoid separation and Z ring formation. This was expected as the unstained cells had showed a clear septation between adjacent cells in a chain and prominent constrictions at the middle of a single cell (Figure 20.13A). The cell growth appears to have continued without cell division thereby leading to an elongated cell with multiple copies of the chromosome (Figure 20.13B). Z ring formation, an

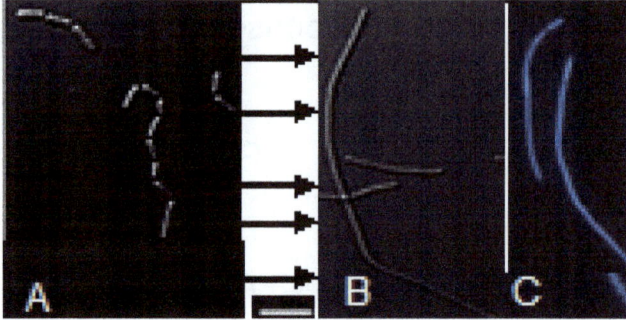

Figure 20.13. *Bacillus pumilus* **IFO 13322 cell are in megacell stage after 60-72 hours after fermentation of D-ribose production in fermentor media containing 200 g/l glucose and 9 g/l corn steep liquor with 5g/l ammoniums sulphate. Additional media components included calcium carbonate: 16.0 g/l, manganese sulfate: 0.5 g g/l, L-leucine: 0.5 g/l and tryptophan: 0.05g/l. (A) DIC images of cell in LB medium after 24 hours (B) DIC images of cells in production medium (details mentioned above) after 72 hours (C) Megacell shows DNA materials after DAPI stain (Srivastava *et al.*, 2009). Nucleoid was stained with 4', 6-diamidino-2-phenylindole (DAPI) and visualized at excitation and emission wavelengths of 359 and 461 nm, respectively.**

essential first step in bacterial cytokinesis, was delayed in glucose-rich media. Srivastava *et al.*, have analyzed the FtsZ and nucleoid localization in medium containing glucose, ammonium sulfate and CSL for growth in fermentor, which had demonstrated cell elongation.

☆ ATPase required for the correct placement of the division site. Cell division inhibitors MinC and MinD act in concert to form an inhibitor capable of blocking formation of the polar Z ring septums. Rapidly oscillates between the poles of the cell to destabilize FtsZ filaments that have formed before they mature into polar Z rings. Proteins that are involved in regulation of cell division and cell cycle progression remain undefined in *Mycobacterium tuberculosis*. Septum site determining protein (Ssd) in *E.coli* is MinD (*i.e.* 270 AA) and MinC (*i.e.* 231 AA). In *Escherichia coli*, the cell division site is determined by the cooperative activity of min operon products MinC, MinD, and MinE. MinC is a nonspecific inhibitor of the septum protein FtsZ, and MinE is the supressor of MinC. MinD plays a multifunctional role. It is a membrane-associated ATPase and is a septum site-determining factor through the activation and regulation of MinC and MinE (Sakai *et al.*, 2001). Increased expression of ssd in *M. smegmatis* and *M. tuberculosis* inhibited septum formation resulting in elongated cells devoid of septa. Ssd protien (is encoded by *rv3660c*) promotes the filamentation in *M. tuberculosis* in response to the stressful intracellular environment. It is similar to MinD. Ssd is a part of a global regulatory mechanism in *M. tuberculosis* and *Mycobacterium smegmatis* via promoting a shift into an altered metabolic state. The protein encoded by rv3660c is a member of the septum site determining protein family (England *et al.*, 2011).

☆ *Helycobacter pylori*, is a Gram-negative bacterium, highly motile, microaerophilic, spiral-shaped organism, which colonizes the stomachs of at least half of the world's human population and its filamentation mechanism are regulated the peptidases (*i.e.* cause peptidoglycan relaxation) and the coiled-coil-rich proteins (Ccrp) (*i.e.* responsible for the helical cell shape *in vitro* as well as *in vivo*). There are four Ccrps, having different multimerization and filamentation properties and different types of smallest subunits. Mre, is protein not exactly involved in the maintenance of cell shape but in the cell cycle. Mutant cells of Mre protein were highly elongated due to a delay in cell division and contained non-segregated chromosomes. Deletion of ccrp59 leads to the formation of 100 per cent straight cells. The percentage of straight cells in the population was 50 per cent in the ccrp58 mutant (n = 400) and 40 per cent in the ccrp1142 mutant (n = 130) (Specht *et al.*, 2011).

☆ The filamentation contributes to a pathogen's resistance due to antimicrobial agent. The induction of bacterial filamentation by antibiotics can alter bacterial virulence, which would have important implications for the pathogenesis and treatment of the disease as well. The ability of some β-lactam antibioticsto induce filamentation in gram negative bacteria is to inhibit penicillin-binding proteins (PBPs). This inhibition stops the assembly of peptidoglycan network in bacterial cell wall. PBP-1 inhibition leads to rapid cell death. PBP-2 and PBP-3 inhibition lead to morphological change turn normal cells to spheroplasts and filamentous cells (elongated rods) respectively.

☆ PBP-3 forms the septum in dividing bacteria by involving transpeptidase so inhibition of this PBP-3 protein leads to the incomplete formation of septa in dividing bacteria resulting in the formation of long strands of bacteria (filaments). The filamentation also depends on the expression levels of an efflux pump (MexAB-OprM) and the minimum inhibitory concentration of chloramphenicol.

☆ Bacterial elongation could occur due to nutritional stress. The filamentation is triggered by a limitation in the availability of one or more nutrients. Since the filament can increase a cell's uptake–proficiency surface without changing its surface-to-volume ratio appreciably, this may be enough reason for cells to be filament. Moreover, the filamentation benefits bacterial cells attaching to a surface because it increases specific surface area in direct contact with the solid medium. The filamentation may allows bacterial cells to access nutrients by enhancing the possibility that part of the filament will contact a nutrient-rich zone and pass compounds to the rest of the cell's biomass. For example, *Actinomyces israelii* grows as filamentous rods or branched in the absence of phosphate, cysteine, or glutathione.

Z Ring and Accessory Protein Required in Bacterial Division

Assembly of FtsZ is important for bacterial cytokinesis. FtsZ is a GTP-binding protein with GTPase activity. The effects of pH on the assembly and structural

Table 20.2. Example of filamentous bacteria developed due to different biological mechanisms (Srivastava *et al.,* 2009, Justice *et al.,* 2008)

Species	Niche or Disease	Filamentation Observed	Other Observations	Mechanism	Biological Role
Uropathogenic *Escherichia coli*	Urinary-tract infections	Mouse bladder and human urine (Justice et al., 2006)	None	SulA and, possibly, the SOS response	Protection against Phagocytosis
Caulobacter crescentus	Freshwater	Prolonged growth (Wortinger et al., 1998)	Resistance to heat, oxidative stress and pH	Decreased FtsZ Concentration	Survival
Mycobacterium tuberculosis	Tuberculosis	Intracellular(macrophages) (Chauhan et al., 2006)	None	Z-rings absent and, possibly, the SOS response	Unknown
Burkholderia pseudomallei	Melioidosis	Unknown	Intracellular macrophages (Jones et al., 1996)	RecA and, possibly, the SOS response	Intracellular survival
Bacillus pumilus IFO13322	Aerobic, fermentative	Delayed Z-protein expression and Z rings inhibition	Srivastava et al., 2009	Z-ring absent and metabolite alteration	D-ribose production
*Bacillus subtilis*168	Sanguinarine bind to Fts Z and as anti-bacterial agents	Inhibition of Z-ring	Beuria et al., 2005	Reduced the frequency of the occurrence of Z rings/micrometer	Drug development
Proteus mirabilis	Disease causes	Inhibition of Z-rings	(Allison et al., 1992; Rather 2005)	High levels of quorum-sensing molecules	Protection from the host immune response

properties of FtsZ were examined. FtsZ retained GTP binding ability but lost GTPase activity at pH 2.5. In the presence of GTP, FtsZ formed protofilaments at pH 7 while it formed aggregates instead of protofilaments at pH 2.5, indicating that GTP hydrolysis is important for the assembly of FtsZ into protofilaments (Santra and Panda 2007). The cell division protein FtsZ plays an important role during binary fission of bacteria (Addinall and Holland 2002; Bi and Lutkenhaus 1991; Bramhill 1997; Errington *et al.*, 2003).

In the early stage of cell division, FtsZ self-assembles to form an extremely dynamic cytokinetic Z-ring at the middle of the bacterial cell, which provides a platform for the recruitment of nine other proteins involved in the cytokinesis (Lutkenhaus and Addinall 1997; Rothfield *et al.*, 1999; Stricker *et al.*, 2002). Biochemical studies of the cell division process have initiated with studies of the GTPase activity and the in vitro polymerizations of FtsZ. Nine proteins have now been identified that are required for cytokinesis (FtsA, FtsI, FtsK, FtsL, FtsN, FtsQ, FtsW, FtsZ, and ZipA) shown in Figure 20.14. The "fts" stands for filamentation thermo sensitive but this terminology is now applied to most cell division genes. FtsZ is the most abundant of the known cell division proteins and it is present at 10,000–20,000 copies per cell in rapidly growing cultures of *Escherichia coli* (Bi and Lutkenhaus, 1991). In contrast, the other division proteins are present at less than 1–2 per cent of Fts Z concentration (Carson *et al.*, 1991; Guzman *et al.*, 1992; Wang and Gayda 1992). As shown in Figure 20.14, FtsZ is formed a membrane-associated ring that extended circumferentially around the cell cylinder at the site of septum formation. The ring appeared to be formed by FtsZ molecules located on the inner surface of the cytoplasmic membrane.

The Z-ring remained associated with the leading edge of the in-growing septum during the remainder of septal invagination and disappeared from the division site after completion of septation (de Boer *et al.*, 1992.; Mukherjee *et al.*, 1993; Rothfield and Zhao 1996). The ftsZ gene product acts prior to other known cell division proteins. This is originally inferred from the different phenotypes of the nonseptate filaments of fts mutants. The unusual concentration-dependent kinetics of the FtsZ GTPase resembles to self-associating systems such as the nucleotide-dependent assembly of microtubules and actin filaments, leading to the binding of GTP to FtsZ in-vitro system might be related to a multimerization step, leads to production of the FtsZ ring in vivo (de Boer *et al.*, 1992.). Tubulin and FtsZ both polymerize in GTP dependent manner hydrolyze GTP, form protofilaments in vitro and both share the common GTP binding signature motifs (Bramhill and Thompson 1994; Errington *et al.*, 2003; Lu *et al.*, 1998; Mukherjee and Lutkenhaus 1998). A protein has been identified in *Caulobacter* (CtrA) that regulates both the initiation of chromosome regulation and the transcription of ftsZ, and may play an important role in the coordination process. It has been shown that selection of the proper division site at midcell in *Escherichia coli* is required the cooperative action of the products of the minCDE gene locus. There are enough FtsZ molecules in the average *E. coli* cell to form several circumferential rings around the cell. Therefore, the Z-ring might consist of several parallel rings, each consisting of one or more FtsZ protofilaments held together by side-to-side interactions, that are available. *Chlamydia* undergoes cell division in the absence of FtsZ and it might be due to some protein involvement. The nearly universal

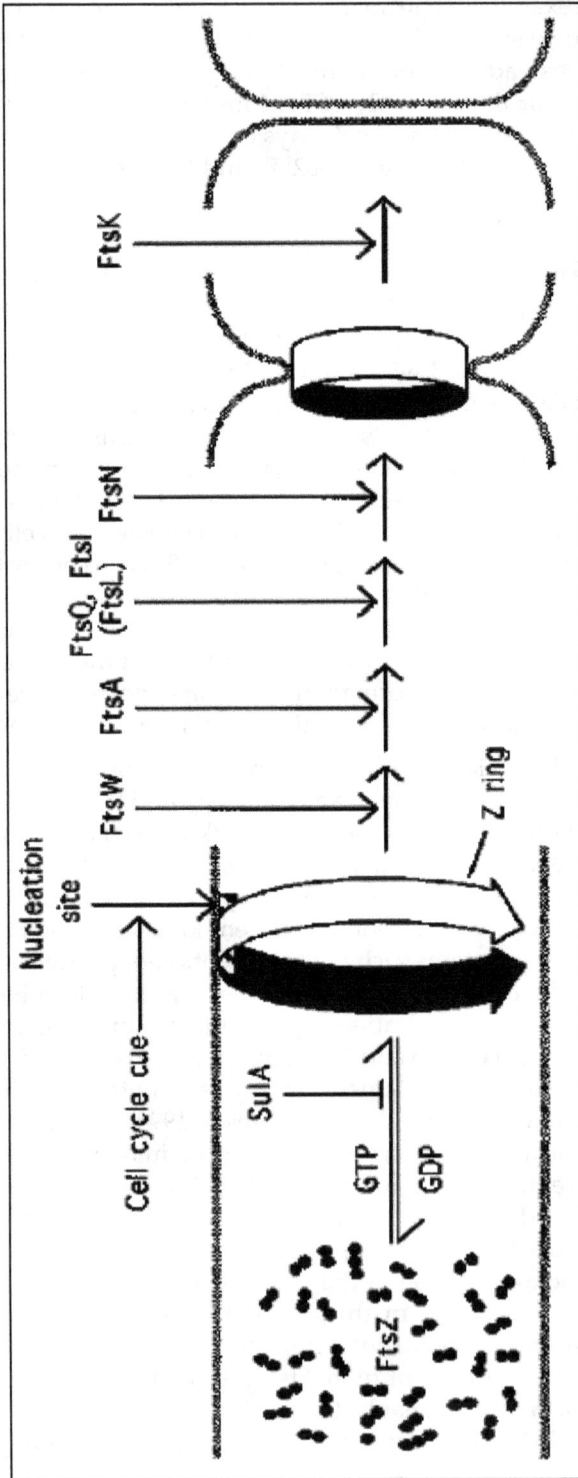

Figure 20.14. Diagramatic presentation of role of Z ring and accessory protein such as FtsA, FtsI, FtsK, FtsL, FtsN, FtsQ, FtsW, FtsZ, and ZipA in *E.coli* cell division process. This figure shows the involvement of Fts Z protein with GTPase activity in Z ring formation. Sul A also shows its role in GTP hydrolysis process (Lutkenhaus and Addinall, 1997).

distribution of FtsZ, with a marked degree of sequence conservation, implies that a similar mechanism exists for initiation of the division process in the bacterial and archaeal kingdoms. FtsZ appears to be the first component to appear in the ring, since none of the other proteins is capable of localizing to midcell in the absence of FtsZ. It is significant that although formation of the ZipA and FtsA rings requires FtsZ, ZipA and FtsA assemble independently of each other. Assembly of all of the other division proteins requires FtsA as well as FtsZ.

Nuclear Stain for Bacteria

4', 6-Diamidino-2-phenylindole (DAPI) stain is fluorescence dye and is used to visualize the nuclear DNA of both living and dead cells. Its molecular formula and weight is $C_{16}H_{15}N_5$ and 277.25 respectively. DAPI analogs, such as 4', 6-diamidino-2-phenylindole, are also known as aromatic diamidine. DAPI contains an imidazoline ring or tetrahydropyrimidine ring which has DNA binding capability. With DAPI analog, DNA interacts by intercalation, condensation or minor groove binding. Except DAPI analog, mithramycin is also used, known as fluorochrome and cellular DNA is also measured by microspectrophotometry. DAPI analog is used as biological and cytological tool and forms fluorescent complex with duplex DNA which in turn induced the triplex formation between the DNA and RNA hybrid. This triplet has potentially important applications in the design of novel antisense, antigene, antiviral, and diagnostic strategies (Zhitao *et al.*, 1997).

Future Prospectus

We have gone the studies of different micro-organisms with respect to change of morphological structures, involved in different advanced biological processes. We have found and observed the different physiological and metabolic behaviour of micro-organisms which has changed with change in macro and micro-morphological structure. Future prospectus of this megacell formation or filament structure study would be beneficial for designing the operation of industrial bioprocess, in pharmaceutical (in drug design) and medical field (curing of diseases such as cancer). More information would get via study of molecular biology, cellular biology and physiological aspects with respect to change of morphological structures that would be future prospectus of this study.

Conclusion

Filamentous growth of some fungi and bacteria is found an important parameter for productivity of several industrial processes. Megacell/filamentous cell are larger than normal cell. Controlling of mycelial morphology is important factor for biological processes. Megacell can be formed by short exposure of anhydrotetracycline or IPTG compounds via getting the small molecule signal. Yeast *Saccharomyces cerevisiae* undergo filamentous growth via major changes such as an increase in cell length, a reorganization of polarity and enhanced cell–cell adhesion. Polar growth requires the specication of sites of germ tube or branch emergence. Engineering studies of filamentous organism is used to design and operation for industrial fermentation processes. Forming the shmoo tip (pear-like cell morphology) and pseudohyphae (lamentous cell morphology) or ascospores (four cells within a cell) under starvation conditions are different morphological structures of yeast in different in biological processes. Filamentous growth in fungi is found to depend on the position of germ tube emergence with relaying of positional information via RhoGTPase modules, and the recruitment of morphogenetic machinery components including cytoskeleton, polarisome and ARP2/3 complexes, and the vesicle tracking system. *Aspergillus nidulans* and *Neurospora crassa* have uncovered several "pioneer" polarity proteins. Actin filaments are required for the establishment and maintenance of hyphal polarity in filamentous fungi. The polarisome direct the formation of actin cables at polar growth sites in filamentous fungi. The Cdc42 GTPase module is needed for germ tube emergence and is essential in those dimorphic and filamentous fungi. Microtubules have a critical role in regulating the direction of hyphal extension. Sites of germ tube and branch emergence are specied by positional signals emanating from occupied receptors. And filamentous fungi are capable of responding to pheromones during mating. The septin AspB localizes to emerging secondary germ tubes, branches, and conidiophore layers and its lacking shows hyperbranching. Rho GTPases, Cdc42p is function as a molecular on/off switch, relaying signals in its active GTP-bound form. Lack of fermentable carbon source causes the trigger for filamentous growth. Global nutrient regulatory GTPase Ras2 is required for filamentous growth regulation. There are four signaling pathways such as rat sarcoma/protein kinase A (RAS/ PKA), sucrose nonfermentable (SNF), target of rapamycin (TOR), and MAPK, are found to regulate filamentous growth.

Bacteria are found single, or in groups such as in small chains or clumps and could in coccoid or rod form. Bacterial cell size depends on the biomass synthesis via cell division capability. Megacells are larger bacterial cells more than normal size. Megacell could be formed after cell elongation process start without septum formation. It is abnormal growth, occurred in certain bacterial species, such as *E. coli, Helicobacter pylori, Caulobacter crescentus* and some species of *Bacillus* due to delayed cell division protein such as Fts Z expression. Megacell formation could be induced in bacteria via responding to various conditions such as stresses, (*i.e.* DNA damage or inhibition of replication), pH change, temperature change effect and metabolic alteration. Inhibition of Z-ring formation and nucleoid separation is occurred in elongated cells. MinC and MinD act an inhibitor is capable of blocking formation of the polar Z ring septums. Assembly of FtsZ is important parameter for bacterial cytokinesis. Z ring formation is an essential first step in bacterial cytokinesis. FtsZ is a GTP-binding protein with GTPase activity. In the early stage of bacterial cell division, FtsZ self-assembles to form an extremely dynamic cytokinetic Z-ring at the middle of the cell, which provides a platform for the recruitment of nine other proteins involved in the cytokinesis. Cell division inhibitors such as MinC and MinD act to form an inhibitor capable of blocking formation of the polar Z ring septums. Inhibition of Z-ring formation and nucleoid separation is occurred in elongated bacterial cells. Bacterial nucleoid was stained with 4', 6-diamidino-2-phenylindole (DAPI) and was visualized at excitation (359nm) and emission wavelengths (461 nm). Nine proteins (*i.e.* FtsA, FtsI, FtsK, FtsL, FtsN, FtsQ, FtsW, FtsZ, and ZipA) have now been identified that are required for cytokinesis of bacterial cells. Increased expression of ssd in *M. smegmatis* and *M. tuberculosis* has inhibited septum formation. Autocatalytic cleavage of LexA causes the SOS genes derepression and form normal cell where as complete LexA protein causes the SOS genes expressed for repair. Cell division is inhibited and delayed resulting filamentation to allow repair before cell division. Base Excision Repair (BER) lacking activity via mutating the genes (such as gene of glycosylase or AP-endonucleases enzymes) in *E. coli* could develop strong filamentation structure via elongation or megacell formation. Normally filamentous nature of bacteria is not found but due to external or internal stimuli response, non- filamentous bacteria develop the filamentous nature. This megacell nature of bacterial study is beneficial for developing the drug for cancer disease and some metabolites production enhancement. Author assumes that these types of study would be beneficial for industrial bioprocess development, drug design and treatment of cancer.

References

1. Addinall S and Holland B. The tubulin ancestor FtsZ draughtsman, designer and driving force for bacterial cytokinesis. J Mol. Biol, 2002, **318**: 219–236.

2. Addinall SG and Lutkenhaus J. FtsA is localized to the septum in an FtsZ–dependent manner. Journal of Bacteriology 1996, **178**(24): 7167–7172.

3. Allison, C., Coleman, N., Jones, PL., Hughes, C. Ability of *Proteus mirabilis* to invade human urothelial cells is coupled to motility and swarming differentiation Infect. Immun., 1992, **60**: 4740–4746.

4. Arai, T., Mikami, Y., Yokoyama, K., Kawata, T., Masuda, K. Morphological Changes in Yeasts as a Result of the Action of 5–Fluorocytosine. Antimicrobial Agents and Chemotherapy, 1977, **12** (2): 255–260.

5. Beuria, TK., Santra, MK., and Panda, D. Sanguinarine Blocks Cytokinesis in Bacteria by Inhibiting FtsZ Assembly and Bundling. Biochemistry 2005, 44, 16584–16593.

6. Bharucha, N., Ma, JC., Dobry, J., Lawson, SK., Yang Z. Analysis of the yeast kinome reveals a network of regulated protein localization during filamentous growth. Mol. Biol.Cell 2008, **19**: 2708–2717.

7. Bi E and Lutkenhaus J. Fts Z ring structure associated with division in *Escherichia coli*. Nature Biotechnology 1991, **354**: 161–164.

8. Bi, E and Lutkenhaus J. Cell division inhibitors SulA and MinCD prevent formation of the FtsZ ring. The Journal of Bacteriology, 1993, **175** (4): 4 1118–1125.

9. Bi, E and Pringle JR. ZDS1 and ZDS2, genes whose products may regulate Cdc42p in *Saccharomyces cerevisiae*. Mol. Cell. Biol., 1996, **16**: 5264–5275.

10. Bourett, TM and Howard, RJ. Ultrastructural immunolocalization of actin in a fungus. Protoplasma 1991, **163**: 199–202.

11. Boyce, KJ., Hynes, MJ., Andrianopoulos, A. Control of morphogenesis and actin localization by the *Penicillium marneffei* RAC homolog. J. Cell Sci., 2003, **116**: 1249–1260.

12. Bramhill D, Thompson C., 1994. GTP–dependent polymerization of *Escherichia coli* FtsZ protein to form tubules. Proc Natl Acad Sci USA 91: 5813–5817.

13. Bramhill D. Bacterial cell division. Annu Rev Cell Dev Biol 1997, **13**: 395–424.

14. Brown, JA., Sherlock, G., Myers, CL., Burrows, NM., Deng, C., Wu, HI., McCann, KE., Troyanskaya, OG., Brown JM. Global analysis of gene function in yeast by quantitative phenotypic proling. Mol. Syst. Biol 2006, **2**: 0001

15. Buijs, J., Dofferhoff, AS., Mouton, JW., van der Meer, JW. Pathophysiology of *in vitro* induced filaments, spheroplasts and rod–shaped bacteria in neutropenic mice. Clin Microbiol Infect 2006; **12**: 1105–1111.

16. Carson, M., Barondess, J., Beckwith M. The FtsQ protein of *Escherichia coli*: membrane topology, abundance, and cell division phenotypes due to overproduction and insertion mutations. J. Bacteriol 1991, **173**: 2187–95.

17. Chang, F., Peter, M. Yeasts make their mark. Nat. Cell. Biol., 2003, **5**: 294–299.

18. Chauhan, A., Madiraju, M., Fol, M., Lofton, H., Maloney, E., Reynolds, R., Rajagopalan, M. Mycobacterium tuberculosis cells growing in macrophages are filamentous and deficient in FtsZ rings. Journal of Bacteriology 2006, **188**(5): 1856–1865

19. Chung, Y., Byung, H., Cho C. *Flectobacillus lacus* sp. nov., isolated from a highly eutrophic pond in Korea. International Journal of Systematic and Evolutionary Microbiology 2006, **56**: 1197–1201.

20. Cook, JG., Bardwell, L., Thorner, J. Inhibitory and activating functions for MAPK Kss1 in the *S. cerevisiae* filamentous–growth signalling pathway. Nature 1997, **390**: 85–88.

21. Cutler, NS., Pan, X., Heitman, J., Cardenas, ME. The TOR signal transduction cascade controls cellular differentiation in response to nutrients. Mol. Biol. Cell, 2001, **12**: 4103–4113.

22. Cvrckova, F., Devirgilio, C., Manser, E., Pringle, R., Nasmyth, K. Ste.20–like protein–kinases are required for normal localization of cell–growth and for cytokinesis in budding yeast. Genes Dev., 1995, **9**: 1817–1830.

23. Dai, K., Mukherjee, A., Xu, YF., Lutkenhaus J. Mutations in Ftsz That Confer Resistance to Sula Affect the Interaction of Ftsz with Gtp. Journal of Bacteriology 1994, **176**(1): 130–136.

24. Dallo, SF and Weitao, T. Bacteria under SOS evolve anticancer phenotypes. Infectious Agents and Cancer 2010, **5**: 3.

25. Dari R, Huisman O. Novel Mechanism of Cell–Division Inhibition Associated with the Sos Response in *Escherichia coli*. Journal of Bacteriology 1983, 156(1): 243–250

26. de Boer, P., Crossley, R., Rothfield L. The essential bacterial cell–division protein FtsZ is a GTPase. Nature 1992, **359**: 254–56.

27. Dofferhoff, AS and Buijs, J. The influence of antibiotic–induced filament formation on the release of endotoxin from Gram negative bacteria. J Endotox Res, 1996, **3**: 187–194.

28. Du, L. X., Jia, S. J., Lu, F. P., Morphological changes of *Rhizopus chinesis* 12 in submerged antibiotic fermentations. Biotechnology Letters, 1980, **2**: 261–266.

29. El–Enshasy, HA., Farid, MA., El–Sayed, ESA. Influence of inoculum type and cultivation conditions on natamycin production by *Streptomyces natalensis*. Journal of Basic Microbiology, 2000, **40**: 333–342.

30. England, K., Crew, R., Slayden, RA. *Mycobacterium tuberculosis* septum site determining protein, Ssd encoded by rv3660c, promotes filamentation and elicits an alternative metabolic and dormancy stress response BMC Microbiology 2011, **11**: 79.

31. Errington, J., Daniel, RA., and Scheffers, DJ. Cytokinesis in bacteria. Microbiol. Mol. Biol. Rev., 2003, **67**: 52–65.

32. Everis, L., and Betts G. pH stress can cause cell elongation in *Bacillus* and *Clostridium* species: a research note". Food Control, 2001, **12**(1), 53–56.

33. Fortini, P., Pascucci, B., Parlanti, E., D'Errico, M., Simonelli, V., Dogliotti E. The base excision repair: mechanisms and its relevance for cancer susceptibility. Biochimie., 2003, **85**(11): 1053–71.

34. Gardner, TS., Cantor, CR., Collins, JJ. Construction of a genetic toggle switch in *Escherichia coli*. Nature 2000, **403**(6767): 339–42.

35. Ghojavand, H., Bonakdarpour, B., Heydarian, SM. and Hamedi J. The inter-relationship between inoculum concentration, morphology, rheology and erythromycin productivity in submerged cultivation of *Saccharopolyspora erythraea*. Brazilian Journal of Chemical Engineering 2011 **28**(04): 565 – 574.

36. Girneno, CJ., Ljungdahl, PO., Styles, CA., Fink, GR. Unipolar cell divisions in the yeast *Saccharomyces cerevisiae* lead to filamentous growth: regulation by starvation and RAS. Cell, 1992, **68** : 1077–1090.

37. Gorfer, M., Tarkka, MT., Hanif, M., Pardo, AG., Laitiainen, E., Raudaskoski, M. Characterization of small GTPases Cdc42 and Rac and the relationship between Cdc42 and actin cytoskeleton in vegetative and ectomycorrhizal hyphae of *Suillus bovinus*. Mol. Plant Microbe Interact., 2001, **14**: 135–144.

38. Gupta, K., Mishra PK., Srivastava, P. Correlative Evaluation of Morphology and Rheology of *Aspergillus terreus* during Lovastatin Fermentation. Biotechnol. Bioprocess Eng., 2007, **141**(2): 140–14.

39. Guzman, L., Barondess, J., Beckwith J. FtsL, an essential cytoplasmic membrane protein involved in cell division in *Escherichia coli*. J. Bacteriol 1992, **174**: 7716–28

40. Harris, SD and Momany, M. Polarity in filamentous fungi: moving beyond the yeast paradigm. Fungal Genetics and Biology, 2004, **41**: 391–400.

41. Harris, SD., Morrell, JL., Hamer, JE. Identication and characterization of *Aspergillus nidulans* mutants defective in cytokinesis. Genetics 1994, **136**: 517–532.

42. Heath, IB. The cytoskeleton in hyphal growth, organelle movements, and mitosis. In: Meinhardt, F., J.G.H, W. (Ed.), Growth, Differentiation and Sexuality. Springer–Verlag, Berlin Heidelberg, 1994, 43–65.

43. Heitman, J., Movva, NR., Hall, MN. Targets for cell cycle arrest by the immunosuppressant rapamycin in yeast. Science 1991, **253**: 905–909.

44. Higashitani, A., Ishii, Y., Kato, Y., Horiuchi K. Functional dissection of a cell–division inhibitor, SulA, of *Escherichia coli* and its negative regulation by Lon. Molecular and General Genetics 1997, **254**(4): 351–357.

45. Higashitani, N., Higashitani, A., Horiuchi, K. Sos Induction in *Escherichia coli* by Single–Stranded–DNA of Mutant Filamentous Phage – Monitoring by Cleavage of Lex a Repressor. Journal of Bacteriology 1995, **177**(12): 3610–3612.

46. Hu, Z and Lutkenhaus, J. Topological regulation of cell division in *Escherichia coli* involves rapid pole to pole oscillation of the division inhibitor MinC under the control of MinD and MinE. Mol. Microbiol., 1999, **34**: 82–90.

47. Hurtado, CA., Beckerich, JM., Gaillardin, C., Rachubinski, RA. Arac homolog is required for induction of hyphal growth in the dimorphic yeast Yarrowia lipolytica. J. Bacteriol., 2000, **182**: 2376–2386.

48. Jones, A., Beveridge, T., Woods, D. Intracellular survival of *Burkholderia pseudomallei*. Infection and Immunity 1996, **64**(3): 782–790.

49. Justice, S. Differentiation and developmental pathways of uropathogenic *Escherichia coli* in urinary tract pathogenesis. Proc. Natl Acad. Sci. USA 2004, **101**: 1333–1338

50. Justice, S., Hunstad, D., Cegelski, L., Hultgren, S. Morphological plasticity as a bacterial survival strategy. Nature Reviews Microbiology 2008, **6**(2): 162–168.

51. Justice, SS., Hunstad, DA., Seed, PC., Hultgren, SJ. Filamentation by Escherichia coli subverts innate defenses during urinary tract infection. Proceedings of the National Academy of Sciences of the United States of America 2006, **103**(52): 19884–19889.

52. Kataoka, T., Powers, S., McGill, C., Fasano, O., Strathern J. Genetic analysis of yeast RAS1 and RAS2 genes. Cell 1984, **37**: 437–445.

53. Knechtle, P., Dietrich, F., Philippsen, P. Maximal polar growth potential depends on the polarisome component AgSpa2 in the filamentous fungus *Ashbya gossypii*. Mol. Biol. Cell 2003, **14**: 4140–4154.

54. Kohler, JR., Fink, GR. *Candida albicans* strains heterozygous and homozygous for mutations in mitogen–activated protein kinase signaling components have defects in hyphal development. Proc.Natl. Acad. Sci. USA 1996, **93**: 13223–13228.

55. Lechler, T., Jonsdottir, GA., Klee, SK., Pellman, D., Li, R. A two–tiered mechanism by which Cdc42 controls the localization and activation of an Arp2/3–activating motor complex in yeast. J.Cell Biol., 2001, **155**: 261–270.

56. Lord, M., Yang, MC., Mischke, M., Chant, J. Cell cycle programs of gene expression control morphogenetic protein localization. J. Cell Biol., 2000, 151, 1501–1512.

57. Lu, C., Stricker, J., Erickson, H. FtsZ from *Escherichia coli, Azotobacter vinelandii,* and *Thermotoga maritima*–quantitation, GTP hydrolysis, and assembly. Cell Motil Cytoskeleton 1998, **40**: 71–86.

58. Lutkenhaus, J and Addinall, SG. Bacterial cell division and the Z ring. Annual Review of Biochemistry 1997, **66**: 93–116.

59. Lynn, RR and Magee PT. Development of the spore wall during ascospore formation in *Saccharomyces cerevisiae* . J. Cell Biol., 1970, **44**: 688–692.

60. Mattick, KL., Jørgensen, F., Legan, JD., Cole, MB., Porter, J., Lappin–Scott, HM., Humphrey, TJ. Survival and Filamentation of *Salmonella enterica* Serovar *Enteritidis* PT4 and *Salmonella enterica* Serovar *Typhimurium* DT104 at Low Water Activity. Applied and Environmental Microbiology, 66(4): 1274–1279.

61. Mosch, HU and Fink, GR. Dissection of filamentous growth by transposon mutagenesis in *Saccharomyces cerevisiae*. Genetics 2000, **145**: 671–684.

62. Mukherjee A, Dai K, Lutkenhaus J. *Escherichia coli* cell division protein FtsZ is a guanine nucleotide binding protein. Proc. Natl. Acad. Sci. USA 1993, **90**: 1053–57.

63. Mukherjee, A and Lutkenhaus J. Dynamic assembly of FtsZ regulated by GTP hydrolysis. EMBO Journal 1998, **17**(2): 462–469.

64. Myszka, K and Craczyk, K. Effect of starvation stress on morphology changes and production of adhesive exopolysaccharides (EPS) by *Proteus vulgaris. Acta Sci. Pol., Techol. Aliments.* 2011, **10**(3): *303–312.*

65. Norton, S., Lacroixa, C., Vuillemarda JC. Effect of pH on the morphology of Lactobacillus helveticus in free–cell batch and immobilized–cell continuous fermentation". Food Biotechnology, 1993, 7(3): 235 – 251.

66. Ohnuki, S., Nogami, S., Kanai, H., Hirata, D., Nakatani, Y., Morishita, S., Ohya Y. Diversity of Ca^{2+} –Induced Morphology Revealed by Morphological Phenotyping of Ca^{2+}–Sensitive Mutants of *Saccharomyces cerevisiae*. Eukaryotic cell, 2007, 6(5): 817.

67. Papagianni M and Mattey M. Morphological development of *Aspergillus niger* in submerged citric acid fermentation as a function of the spore inoculum level. Application of neural network and cluster analysis for characterization of mycelial morphology. Microbial Cell Factories, 2006, **5**: 3, 1–12.

68. Papagianni, M. Fungal morphology and metabolite production in submerged mycelial processes. Biotechnol Adv, 2004, **22**: 189–259.

69. Partridge, SR and Wake, RG. Ftsz and Nucleoid Segregation During Outgrowth of Bacillus–Subtilis Spores. Journal of Bacteriology 1995, **177**(9): 2560–2563.

70. Paul, JC., George, FS. Jr. The Regulation of Filamentous Growth in Yeast. Genetics, 2012, **190**: 23–49.

71. Pringle, J.R., Bi, E., Harkins, H.A., Zahner, J.E., DeVirglio, C., Chant, J., Corrado, K., Fares, H. Estabishment of cell polarity in yeast. Cold Spring Harbor Symposia on Quantitative Biology LX, 1995, 729–744.

72. Pruyne, D., Evangelista, M., Yang, C., Bi, E., Zigmond, S., Bretscher, A., Boone, C. Role of formins in actin assembly: nucleation and barbed–end association. Science 2002, **297**: 612–615.

73. Radcliffe, PA., Binley, KM., Trevethick, J., Hall M., Sudbery PE. Filamentous growth of the budding yeast *Saccharomyces cerevisiae* induced by overexpression of the WHl2 gene. Microbiology, 1997, **143**: 1867–1 876.

74. Rahman, DRJ., Sudbery, PE., Kelly, S., Marison, IW. The effect of dissolved oxygen concentration on the growth physiology of *Saccharomyces cerevisiae* whi2 mutants. J Gen Microbiol., 1988, **134**: 2241–2248.

75. Ram, AF., Arentshorst, M., Damveld, RA., Punt, PJ., van den Hondel, CAMJJ. Characterization of the racA gene in *Aspergillus niger*. Fungal Genet. Newslett., 2001, **48**, 443.

76. Rather, PN. Swarmer cell differentiation in *Proteus mirabilis*. Environ. Microbiol 2005, **7**: 1065–1073.

77. RayChaudhuri, D and Park J. *Escherichia coli* cell–division gene ftsZ encodes a novel GTP–binding protein. Nature 1992, **359**: 251–54.

78. Richman, TJ., Sawyer, MM., Johnson, DI. *Saccharomyces cerevisiae* Cdc42p localizes to cellular membranes and clusters at sites of polarized growth. Eukaryot. Cell 2002, **1**: 458–468.

79. Riquelme, M., Reynaga–Pena, CG., Gierz, G., Bartnicki–Garcia, S. What determines growth direction in fungal hyphae? Fungal Genet. Biol., 1998, **24**: 101–109.

80. Roberson, RW. The hyphal tip cell of *Sclerotium rolfsii*: cytological observations. In: Latge, J.P., Boucias, D. (Eds.), Fungal Cell Wall and Immune Response. Springer–Verlag, Berlin, 1991, 27–37.

81. Rothfield, L and Zhao C. How do bacteria decide where to divide? Cell 1996, **84**: 183–86

82. Rothfield, L., Justice S., Garcia–Lara, J. Bacterial cell division. Annual Review of Genetics 1999, **33**: 423–448.

83. Ryan, O., Shapiro, RS., Kurat, CF., Mayhew, D., Baryshnikova1, A,. Chin, B., Lin Z–Y, Cox1, MJ., Vizeacoumar, F., Cheung, D., Bahr, S., Tsui, K., Tebbji, F,. Sellam, A,. Istell, F., Schwarzmüller, T., Reynolds, TB., Kuchler, K., Gifford, DK., Whiteway, M., Giaever, G., Nislow, C., Costanzo, M., Gingras, A–C., Mitra, RD., Andrews, B., Fink GR., Cowen, LE., Boone, C. Global Gene Deletion Analysis Exploring Yeast Filamentous Growth. Science, 2012, **337** (6100): 1353–1356.

84. Sagot, I., Klee, SK., Pellman, D. Yeast formins regulate cell polarity by controlling the assembly of actin cables. Nat. Cell. Biol., 2002a, **4**: 42–50.

85. Sagot, I., Rodal, A.A., Moseley, J., Goode, B.L., Pellman, D. An actin nucleation mechanism mediated by Bni1 and prolin. Nat. Cell. Biol., 2002b, **4**: 626–631.

86. Sakai, N., Yao, M., Itou, H., Watanabe, N., Yumoto, F., Tanokura, M., Tanaka, I. The three–dimensional structure of septum site–determining protein MinD from *Pyrococcus horikoshii* OT3 in complex with Mg–ADP. Structure., 2001, **9**(9): 817–26.

87. Santra M and Panda D. Acid–Induced Loss of Functional Properties of Bacterial Cell Division Protein FtsZ: Evidence for an Alternative Conformation at Acidic pH. Proteins: Structure, Function, and Bioinformatics 2007, **67**: 177–188

88. Schenkman, LR., Caruso, C., Page, N., Pringle, JR. The role of cell cycle–regulated expression in the localization of spatial landmark proteins in yeast. J. Cell Biol., 2002, **156**: 829–841.

89. Schimmel, TG., Coffman AD., Parsons, SJ. Effect of Butyrolactone I on the Producing Fungus, *Aspergillus terreus*. Applied and Environmental Microbiology, 1998, **64** (10): 3707–3712.

90. Seiler, S., Plamann, M.,. Genetic basis of cellular morphogenesis in the filamentous fungus *Neurospora crassa*. Mol. Biol. Cell 2003, **14**: 4352–4364.

91. Sharpless, K.E., Harris, S.D. Functional characterization and localization of the *Aspergillus nidulans* formin SEPA. Mol. Biol. Cell 2002, **13**: 469–479.

92. Shi, X., Sha, Y., Kaminskyj, SG. *Aspergillus nidulans* hypA regulates morphogenesis through the secretion pathway. Fungal Genet. Biol., 2003, **41**: 75–88.

93. Shimada, Y., Gulli, M.P., Peter, M. Nuclear sequestration of the exchange factor Cdc24 by Far1 regulates cell polarity during yeast mating. Nat. Cell. Biol., 2000, **2**: 117–124.

94. Shively, CA., Eckwahl, MJ., Dobry, CJ., Mellacheruvu, D., Nesvizhskii, A., Kumar A. Genetic Networks Inducing Invasive Growth in *Saccharomyces cerevisiae* Identified Through Systematic Genome–Wide Overexpression. Genetics 2013, **193**: 1297–1310;

95. Smith, C.M., Calam, CT. Variations in inocula and their influence on the productivity of culture and its relationship with antibiotic production. Process Biochemistry, 2003, **38**: 1643–1646.

96. Specht, M., Schätzle, S., Peter, L. Graumann and Barbara Waidner. *Helicobacter pylori* possesses four coiled–coil–rich proteins that form extended filamentous structures and control cell shape and motility. Journal of Bacteriology, 2011, **193** (17): 4523–4530.

97. Srivastava, RK., Jaiswal, R., Panda, D., Wangikar, PP. Megacell Phenotype and Its Relation to Metabolic Alterations in Transketolase Deficient Strain of *Bacillus pumilus*. Biotechnology and Bioengineering, 2009, **102**(5): 1387–97.

98. Stricker, J., Maddox, P., Salmon, E., Erickson H. Rapid assembly dynamics of the *Escherichia coli* FtsZ–ring demonstrated by fluorescence recovery after photobleaching. Proc Natl Acad Sci USA 2002, **99**: 3171–3175.

99. Sudbery, PE. The germ tubes of *Candida albicans* hyphae and pseudohyphae show different patterns of septin ring localization. Mol. Microbiol., 2001, **41**: 19–31.

100. Teng, Y and Xu Y. Culture condition improvement for whole cell lipase production in submerged fermentation by *Rhizopus chinensis* using statistical method. Bioresour Technol 2008, **99**: 3900–3907.

101. Teng, Y., Xu, Y., Wang, D. Changes in morphology of *Rhizopus chinensis* in submerged fermentation and their effect on production of mycelium–bound lipase. Bioprocess Biosyst Eng 2009, **32**: 397–405.

102. Tetart, F., Albigot, R., Conter, A., Mulder, E., Bouche JP. Involvement of Ftsz in Coupling of Nucleoid Separation with Septation. Molecular Microbiology 1992, **6**(5): 621–627.

103. Toda, T., Cameron, S., Sass, P., Zoller, M., Wigler, M. Three different genes in *S. cerevisiae* encode the catalytic subunits of the cAMP–dependent protein kinase. Cell 1987, **50**: 277–287.

104. Toda, T., Uno, I., Ishikawa, T., Powers, S., Kataoka T. In yeast, RAS proteins are controlling elements of adenylate cyclase. Cell 1985, **40**: 27–36

105. Trueheart, J., Boeke, JD., Fink GR. Two genes required for cell fusion during yeast conjugation: evidence for a pheromone–induced surface protein. Mol. Cell. Biol., 1987, **7**: 2316–2328.

106. Ushinsky, SC., Harcus, D., Ash, J., Dignard, D., Marcil, A., Morchhauser, J., Thomas, DY., Whiteway, M., Leberer, E. CDC42 is required for polarized growth in human pathogen *Candida albicans*. Eukaryot. Cell., 2002, **1**: 95–104.

107. Van Dyk, D., Hansson, G., Pretorius IS., Bauer, FF. Cellular differentiation in response to nutrient availability: the repressor of meiosis, Rme1p, positively regulates invasivegrowth in *Saccharomyces cerevisiae*. Genetics 2003, **165**: 1045–1058.

108. Villena GK and Gutierrez–Correa M. Morphological patterns of *Aspergillus niger* biofilms and pellets related to lignocellulolytic enzyme productivities. Microbiology, Letters in Applied Microbiology 2007, **45**: 231–237.

109. Wang, F and Lee SY. Production of Poly (3–Hydroxybutyrate) by Fed–Batch Culture of Filamentation–Suppressed Recombinant *Escherichia coli*. Applied and Environmental Microbiology, 1997, **63**(12), 4765–4769.

110. Wang, H and Gayda R. Quantitative determination of FtsA at different growth rates in *Escherichia coli* using monoclonal antibodies. Mol. Microbiol 1992, **6**: 2517–24

111. Ward, MP., Gimeno, CJ., Fink, GR., Garrett, S. SOK2 may regulate cyclic AMP–dependent protein kinase–stimulated growth and pseudohyphal development by repressing transcription. Mol. Cell. Biol., 1995, **15**: 6854–6863.

112. Weinzierl, G., Leveleki, L., Hassel, A., Kost, G., Wanner, G., Bolker, M. Regulation of cell separation in the dimorphic fungus *Ustilago maydis*. Mol. Microbiol., 2002, **45**: 219–231.

113. Wendland, J and Philippsen, P. Cell polarity and hyphal morphogenesis are controlled by multiple rho–protein modules in the filamentous ascomycete *Ashbya gossypii*. Genetics 2001, **157**: 601–610.

114. Wendland, J., Philippsen, P. Determination of cell polarity in germinated spores and hyphal tips of the filamentous ascomycete *Ashbya gossypii* requires a rhoGAP homolog. J. Cell Sci., 2000, **113**: 1611–1621.

115. Westfall, PJ and Momany, M. *Aspergillus nidulans* septin AspB plays pre– and postmitotic roles in septum, branch, and conidiophore development. Mol. Biol. Cell., 2002, **13**: 110–118.

116. Wortinger, M., Quardokus, E., Brun, Y. Morphological adaptation and inhibition of cell division during stationary phase in *Caulobacter crescentus*. Molecular Microbiology 1998, **29**(4): 963–973.

117. Wright, RM., Repine, T., Repine, JE. Pseudohyphal growth in haploid *Saccharomyces cerevisiae* is a reversible process. Curr Genet 1993, **23**: 388–391.

118. Wu, LJ and Errington, J. Coordination of cell division and chromosome segregation by a nucleoid occlusion protein in *Bacillus subtilis*. Cell 2004, **117**: 915–925.

119. Wucherpfennig, T., Hestler, T., Krull R. Morphology engineering – Osmolality and its effect on *Aspergillus niger* morphology and productivity. Microbial Cell Factories 2011, **10**: 58, 1–15.

120. Yu, Y., Jiang, YW., Wellinger, RJ., Carlson, K., Roberts, JM., Stillman DJ. Mutations in the homologous ZDS1 and ZDS2 genes affect cell cycle progression. Mol. Cell. Biol., 1996, **16**: 5254–5263.

121. Zhang, Z., Smith, MM., Mymryk, JS. Interaction of the E1A oncoprotein with Yak1p, a novel regulator of yeast pseudohyphal differentiation, and related mammalian kinases. Mol.Biol. Cell 2001, **12**: 699–710.

122. Zhitao, X., Pilch, D., Srinivasan, A., Olson, W., Geacintov, N., Breslauer, K. Modulation of Nucleic Acid Structure by Ligand Binding: Induction of a DNA.RNA.DNA Hybrid Triplex by DAPI Intercalation. Bioorg Med Chemist 1997, **5**(6): 1137–1147.

123. Ziman, M., Preuss, D., Mulholland, J., O Brien, JM., Botstein, D., Johnson, DI. Subcellular localization of Cdc42p, a *Saccharomyces cerevisiae* GTP–binding protein involved in the control of cell polarity. Mol. Biol. Cell., 1993, **4**: 1307–1316.

Biotechnology: An Overview (2015)
Editors: Rajan Kumar Gupta, Nasim Akhtar and Deepak Vyas
Published by: DAYA PUBLISHING HOUSE, NEW DELHI

Pages 353–359

Chapter 21

Study on Bioremediation of Petroleum Hydrocarbons in Soil

Maryada Goyal[1] and Suneel Kumar Singh[2]*

[1]*Department of Zoology, H.N.B. Garhwal University Campus,
Pauri, Pauri Garhwal – 246 155, Uttarakhand*
[2]*Department of Biotechnology, Modern Institute of Technology,
Dhalwala, Rishikesh – 249 201, Uttarakhand*

ABSTRACT

Using engine oil as petroleum product study on bioremediation was performed. The micro-organisms capable of bioremediation were isolated from soil containing petroleum products mixed in it. Serial dilutions were prepared from the soil sample and labeled as 10^{-1}, 10^{-2}, 10^{-3}, 10^{-4}, 10^{-5}, 10^{-6}, 10^{-7} and 10^{-8}. Spread plate technique was used to isolate micro-organisms from soil. Then these were mixed with known amount of soil under sterile conditions. Twenty five gm samples were taken out from soil to check concentration of petroleum after every 3 days interval by means of Soxhlet extraction method using chloroform as solvent. The rate of biodegradation was found to be 19 per cent per week at initial and 25 per cent later on.

Keywords: Bioremediation, Petroleum hydrocarbons.

Introduction

World environment is contaminated due to huge amount of organic and inorganic compounds released every year by anthropogenic activities. These compounds effect not only the pollution but degrading the soil fertility also which is boon of agricultural industry. Various techniques are available to refine the contaminated soil but bioremediation based on the metabolic activity of micro-organisms has certain advantages (Werner, 1985; Barker *et al.*, 1987; Korda *et al.*, 1997; Kaviyarasan and

---

* *Corresponding Author:* E-mail: drsuneelkumarsingh@gmail.com

Karthikeyan, 2003; Schulze, 2009 and Sinha *et al.,* 2009). Environmental pollution especially, petroleum and petrochemical products like complex mixtures of Hydrocarbons, has been identified as one of most serious problems in today's scenario. Bioremediation has become an alternative way of remediation of oil contaminated sites with the help of specific micro-organisms (*e.g.,* Bacteria, Cyanobacteria, Algae, Fungi, Protozoa etc.) already present in soil can improve biodegradation efficiency through *in-situ* or *ex-situ* methods (Gilman, 1998; April *et al.,* 2012; Nocentini *et al.,* 2000; Salleh *et al.,* 2003; Hu *et al.,* 2006; Oboh *et al.,* 2006; Kishore and Ashisk, 2007; Tang, 2007; Wei *et al.,* 2007; Wang *et al.,* 2008; George *et al.,* 2009; Obire and Putheti, 2009; Xu, 2012; Maruthi *et al.,* 2013. Environmental factors, specially physical, chemical and biological, have complex effects on hydrocarbon biodegradation in soil (Baker, 1970; Calabrese and Kosteck, 1991; Banerji *et al.,* 1995; Bossert and Compeau, 1995). Workers like He *et al.* (1999), Chaudhary *et al.* (2012) and Lotfinasabasl *et al.* (2012) provided the assessment on bioremediation process while Wilson and Jones (1993) and Thapa *et al.* (2012) reviewed the work on the same subject.

Regular increasing vehicles and stationary engines have led to fast growth of automobile workshops in small cities and towns areas also. The washing and servicing of engines and vehicles generate a large volume of waste products which mixed with municipal drains and soil in and around service stations. Many compounds in waste oil and grease are known to be toxic. PHCs (polycyclic hydrocarbons) are very diverse mixture of chemicals; like *n*-alkanes, paraffins, olefins, aromatics, asphaltics etc. Oil spills cause severe damage to the ecosystems and pose threats of fire, ground water percolation and air pollution due to evaporation. These can also severely affect vegetation. A consortium of bacterial species has been developed hereby to combat pollution created by petroleum hydrocarbons in soil. Thus, the aim of this study was to find out whether or not biodegradation occurs in petroleum hydrocarbons and to what extent is this degradation effective.

Materials and Methods

All reagents and chemicals used in present investigation were of Analytical Grade (AR) and purchased from standard manufacturing companies (like Merk, Qualifier and Ranbaxy etc.). Various chemicals (chloroform, crystal violet, gram's iodine, safranin, NaCl, Peptone, Yeast extract, Agar etc), apparatus and instruments (Laminar Airflow, incubator, autoclave, Vortex mixer, hot air oven etc.) were used for the purpose. Soil sample (500 gm) was taken firstly and made it free of contamination and then engine oil of requisite quantity was added, along with growth media. These were left in incubator for growth of petroleum hydrocarbon feeding micro-organisms. After 4-5 days micro-organisms were found generated. Petroleum hydrocarbon feeding micro-organisms were isolated in the laboratory on nutrient agar plates. This was done by preparing serial dilutions of the above soil sample containing crude micro-organisms and labeled as 10^{-1}, 10^{-2}, 10^{-3}, 10^{-4}, 10^{-5}, 10^{-6}, 10^{-7} and 10^{-8}. Spread plate technique was applied for isolation. Sample of 10^{-7} and 10^{-8} dilution was spread on nutrient agar plate and observed for colonies of micro-organisms. These were further inoculated in liquid Lactose Broth media for obtaining culture of micro-organisms.

From 4-5 days old soil sample, 25 gm of soil was taken out and the amount of petroleum hydrocarbon present in it was calculated by Soxhlet extraction method using chloroform as solvent. Afterward, the grown micro-organisms were added to this soil sample and mixed properly before kept in incubator. After 3 days sample was taken out and the amount of petroleum hydrocarbon was calculated again using Soxhlet extraction method. The amount of petroleum hydrocarbon remains lower this time due to the microbial consortium added to the sample has consumed these PHCs. The process was repeated continuously till the amount of PHCs comes out to be too low. Finally, the rate of biodegradation was determined with respect to time, by using the given formulae:

Rate of biodegradation= {(A-B)/A}*100

where,

    A: Initial amount of PHCs in the 25 gm sample.

    B: Final amount of PHCs in the 25 gm sample.

Results and Discussion

Three hundred seventy two colonies of micro-organisms were obtained from 10^{-7} dilution and 40 from 10^{-8}. The result obtained from these colonies was counted through Colony Forming Units (CFU/ml) which came out to be 3720 CFU/ml for 10^{-7} dilution and 400 CFU/ml for 10^{-8} dilution, respectively. CFU count for un-autoclaved soil in 10^{-7} dilution was noted as 2760 CFU/ml. present results proves that hydrocarbon contaminated soils contain more micro-organisms than uncontaminated soils, but diversity of micro-organisms was reduced. This was proved further by studying the colony morphology of the bacterial colonies obtained in both cases. Morphology of the bacterial colonies in PHC containing soil, were medium in size with diameter ranging from 2.5 – 3 mm. Their shapes were round and margins were regular with convex elevation. Besides all this, the colonies were shiny white and glistening with opaque opacity. On the other hand, in case of normal soil, the colonies obtained were of varying shapes, sizes, opacity, margins and colour. While some colonies were found, round and oval shapes. The colour of colonies varies from blue, white, and green to reddish-yellow, indicating the diversity of micro-organisms in the soil sample, which was reduced, in the PHC containing sample.

The rate of biodegradation in the first 15 days came out to be 19.4 per cent which increased to 49 per cent in last 15 days. The overall rate of biodegradation in water, were noted 59 per cent (Table 21.1). It has been noted that the amount of extracted petroleum hydrocarbons decreased continuously from initial amount at all the rest stages. The decreasing amount was low in the beginning but increased later on. The day wise extracted amount of petroleum hydrocarbons indicate the decrease in amount of extracted petroleum hydrocarbons which was due to the increase in number of micro-organisms (Figure 21.1).

Many workers have noted that PHCs are degraded by means of micro-organisms at a remarkable rate (Werner, 1985; Lee *et al.*, 1988; Wilson and Jones, 1993; Banerji *et al.*, 1995; Bossert and Compeau, 1995; He *et al.*, 1999; April *et al.*, 2000; Jurgensen *et al.*,

Table 21.1. Weight of PHCs extracted from soil at regular intervals.

Sl.No.	Time (Days)	Weight of Sample (gm)	Weight of Extracted PHCs (gm)
1.	0 (initial)	25	2.9634
2.	3	25	2.8862
3.	6	25	2.7822
4.	9	25	2.6784
5.	12	25	2.5642
6.	15	25	2.3888
7.	18	25	2.1446
8.	21	25	1.7322
9.	24	25	1.6538
10.	27	25	1.4222
11.	30	25	1.2136

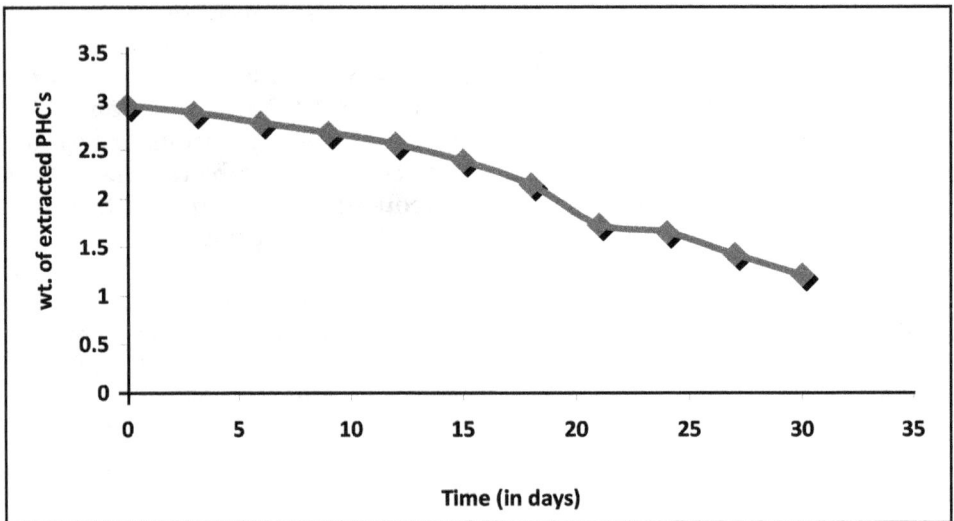

Figure 21.1. Day-wise extracted amount of petroleum hydrocarbons.

2000; Namkoong *et al.*, 2002; Salleh *et al.*, 2003; Wei *et al.*, 2007; Chaudhary *et al.*, 2012; Lotfinasabasl *et al.*, 2012; Thapa *et al.*, 2012 and Maruthi *et al.*, 2013). Present report also indicating that soil pollution caused by PHCs can be removed by generating micro-organisms capable of biodegradation. A number of micro-organisms present in soil, which can include variously hydrocarbon utilizing bacteria and fungi, representing 1 per cent of total population of some 10^4- 10^6 cells/g of soil. In addition, cyanobacteria and algae present in soil were reported to degrade hydrocarbons (Gilman, 1998; George *et al.*, 2000; Obire and Putheti, 2009 and Maruthi *et al.*, 2013).

Hydrocarbon contaminated soils have been found to contain more micro-organisms than uncontaminated soils, but diversity of micro-organisms was reduced. Thus, PHCs play significant role in soil pollution, which can be combat by means of bioremediation. Moreover, bioremediation method is easy, eco- friendly and cost effective.

Acknowledgement

The authors are thankful to the Campus Director, for providing laboratory facility, Director, Indian Institute of Petroleum and Forest Research Institute for providing literature on the subject.

References

1.   April,T.M., Foght, J.M., and Currah, R.S., 2000. Hydrocarbon degrading filamentous fungi isolated from flare pit soils in Northern and Western Canada. *Canadian Journal of Microbiology*, 46(1): 38–49.

2.   Atlas, R.M., 1996. Petroleum biodegradation and oil spill bioremediation. *Oceanographic Literature Review*, 43(6): 628.

3.   Baker, J.M., 1970. Studies on salt marsh communities. The effects of a single oil spillage. In: *The Ecological Effects of Oil Pollution on Littoral Communities*, (Ed.) E.B. Colwell. Institute of Petroleum, London, pp. 16–43.

4.   Banerji, S.K., Zappi, M.E., Teetar, C.L., Gunnison, D., Cullinane, M.J. and Morgan R.T., 1995. Bioremediation of soils contaminated with petroleum hydrocarbons using bioslurry reactors. *U.S. Army Corps Engineering–IRRP*, 95(2): 1–78.

5.   Barker, J.E., Patrick, G.C. and Major, D., 1987. Natural attenuation of aromatic hydrocarbons in a shake sand aquifer. *Gr. War. Mon. Rev. (Winter)* pp. 64–71.

6.   Bossert, I.D. and Compeau, G.C., 1995. Cleanup of petroleum hydrocarbon contamination in soil. In: *Microbial Transformation and Degradation of Toxic Organic Chemicals*, (Eds.) L.Y. Young and C.E. Cerniglia, pp. 77–125.

7.   Calabrese, J. and Kosteck, P.T., 1991. *Hydrocarbon Contaminated Soil*. Lewis Publishers, New York, p. 411–421.

8.   Chaudhary, S., Luhach, J., Sharma, V. and Sharma, C., 2012. Assessment of diesel degrading potential of fungal isolates from sludge contaminated soil of petroleum refinery, Haryana. *Research on Microbiology*, DOI: 10.3923/jm.2012.

9.   George-Okafor, U., Tasie, F. and Muotoe-Okafor, F., 2009. Hydrocarbon degradation potentials of indigenous fungal isolates from petroleum contaminated soils. *Journal of Physical Nature Science*, 3(1): 1–6.

10. Gilman, J.C., 1998. *A Manual of Soil Fungi*. Daya Publishing House, New Delhi, pp. 14–24.

11. He, L.J., Wei, D. and Zhang, W.Q., 1999. Research of microbial treatment of petroleum contaminated soil. *Advance in Environmental Science*, 7(3): 110–115.

12. Hu, X., Xia, F. and Zhu, N., 2006. Study on bioremediation of oil-contaminated soil. *Journal of Environmental Chemistry*, 25(6): 593–597.

13. Jurgensen, K.S., Puustinen, J. and Suortti, A.M. 2000. Bioremediation of petroleum hydrocarbon-contaminated soil by composting in biopiles. *Environmental Pollution*, 107: 245–254.

14. Kaviyarasan, M.V. and Karthikeyan, E., 2003. Biodegradation of Azo Dye by *Pleurotus ostreatus* and *Tricholoma lobaynese*. *Journal of Pollution Research*, 22(1): 77 – 80.

15. Kishore, D. and Ashisk, M., 2007. Crude petroleum-oil biodegradation efficiency of *Bacillus subtilis* and *Pseudomonas aeruginosa* strains isolated from a petroleum-oil contaminated soil from North-East India. *Bioresource Technology*, 98(7): 1339–1345.

16. Korda, A., Santas, P., Tenente, A. and Santas, K., 1997. Petroleum hydrocarbon remediation: Sampling and analytical techniques, *in situ* treatments and commercial micro-organisms currently used. *Applied Microbiology and Biotechnology*, 48: 677–686.

17. Lee, M.D., Thomson, J.M., Borden, R.C., Bedient, P.B., Ward, C.H. and Wilson, J.T., 1988. Bio-restoration of aquifers contaminated with organic compounds. *CRC Critical Reviews in Environmental Control*, 18(1): 29–89.

18. Lotfinasabasl, S., Gunale, V.R. and Rajurkar, N.S., 2012. Assessment of petroleum hydrocarbon degradation from soil and tarball by fungi. *Bioscience Discovery*, 3(2): 186–192.

19. Maruthi, Y.A., Hossain, K. and Thakre, S., 2013. *Aspergillus flavus*: A potential Bioremediator for oil contaminated soils. *European Journal of Sustainable Development*, 2: 1, 57–66.

20. Namkoong, W., Hwang, E.Y., Park, J.S. and Choi, J.Y., 2002. Bioremediation of diesel contaminated soil with composting. *Environmental Pollution*, 109: 23–31.

21. Nocentini, M., Pinelli, D. and Fava, F., 2000. Bioremediation of a soil contaminated by hydrocarbon mixtures: the residual concentration problem. *Chemosphere*, 41: 1115–1123.

22. Obire, O. and Putheti, R.R., 2009. Fungi in bioremediation of oil polluted environments, pp. 1–10. http: //energy .sigmaxi.org/wpcontent/uploads/ 2009/09/obire_putheti_bioremediation.pdf.

23. Oboh, B.O., Ilori, M.O., Akinyemi, J.O. and Adebusoye, S.A., 2006. Hydrocarbon degrading potentials of bacteria isolated from a Nigerian bitumen (Tarsand) deposit. *Nature and Science*, 4(3): 51–57.

24. Salleh, A.B., Ghazali, F.M., Rahman, R.N.Z.A. and Basri, M., 2003. Bioremediation of petroleum hydrocarbon pollution. *Indian Journal of Biotechnology*, 2: 411–425.

25. Scholze, R., 2009. Environmental parameter optimization for bioremediation of petroleum hydrocarbon contaminated soil. *Engineer Research and Development Centre*, 1–24.

26. Sinha, R.K., Valani, D., Sinha, S., Singh, S. and Heart, S., 2009. Bioremediation of contaminated sites: A low-cost nature's biotechnology for environmental clean

up by versatile microbes, plants and earthworms. In: *Solid Waste Management and Environmental Remediation*.

27. Tang, Z., 2007. The study of bioremediation of petroleum polluted soil in northwest loess area. *University of Architecture and Technology*, p. 19–33.

28. Thapa, B., Ajay Kumar, K.C. and Ghimire, A., 2012. A review on bioremediation of petroleum hydrocarbon contaminants in soil. *Kathmandu University Journal of Science, Engineering and Technology*, 8(1): 164–170.

29. Wang, J., Zhang, Z.Z., Su, Y.M., He, W., He, F. and Song, H.G., 2008. Phytoremediation of petroleum polluted soil. *Petroleum Science*, 5(2): 167.

30. Wei, X., Zhang, Z. and Guo, S., 2007. *Ex situ* bioremediation of petroleum contaminated soil with exogenous microbe. *Journal of Petrochemical Universities*, 20(2): 1–4.

31. Werner, J.J., 1985. A new way for the documentation of polluted aquifers by biodegradation. *Water Supply*, 3: 41–47.

32. Wilson, S.C. and Jones, K.C., 1993. Bioremediation of soil contaminated with polycyclic aromatic hydrocarbons: A review. *Environmental Pollution*, 81(3): 229–249.

33. Xu, J., 2012. Bioremediation of crude oil contaminated soil by petroleum–degrading active bacteria. Introduction to enhanced oil recovery (EOR) processes and bioremediation of oil-contaminated sites, p. 1–39.

Biotechnology: An Overview (2015)   *Pages 361–373*
Editors: Rajan Kumar Gupta, Nasim Akhtar and Deepak Vyas
Published by: DAYA PUBLISHING HOUSE, NEW DELHI

Chapter 22

Risks Associated with the Use of Transgenic Plants

Anil Kumar Dhiman*

Information Scientist, Gurukul Kangri University,
Haridwar – 249 404, Uttarakhand

ABSTRACT

Transgenic plants are getting attention worldwide due to their potential benefits for the human beings. The main objective of them is their increased resistance towards biotic and abiotic stresses, improved food quality with altered carbohydrate, protein and lipid composition, and production of polypeptides for medical, pharmaceutical or industrial use. Already, many plants like that soybean, cotton, corn, rice, rape and poplar have been genetically modified for maximum benefits. Other transgenic crops like potato, sugar beet, tomato, papaya, watermelon, zucchini, and squash are also being used throughout the world. But there are also various risks associated with the production of Transgenic Plants or Genetically Modified Plants, which not only affect the human beings but also the environment.

This chapter attempts to discuss some of the risks associated with transgenic or genetically modified plants and the remedies to overcome the problems.

Keywords: Genetically modified plants, Regulatory authorities, Risk assessment, Risk management, Transgenic plants.

Introduction

Since the ancient times, when man was a savage, he subsisted largely on fruits and roots of the plants. Later, he learnt about the use of plants and their parts including products to get rid off their diseases beside their use in food. Thus, the man is using plants for different purposes since the time immortal (Dhiman, 2006).

---

* *Corresponding Author:* E-mail: akvishvakarma@rediffmail.com

Transgenic plants (TPs) are those plants which contain a gene or genes that have been artificially inserted instead of the plant acquiring them through pollination. Two types of methods for plant transformation are used - the use of *Agrobacterium* as a biological vector for foreign gene transfer; and direct gene transfer techniques, in which DNA is introduced into cells by the use of physical, electrical or chemical means (Tsaftaris *et al.*, 2000). Transgenic plants provide improved nutritional quality and are insect resistant, disease resistant, herbicide resistant and salt tolerant with delaying fruit ripening properties and are used in biopharmaceuticals and vaccines.

The "first generation" of transgenic crops was aimed at improving traits involving single genes. Now we are on the verge of a new step in crop modification, fueled by the rate at which new genes that are important for plant growth and development metabolism and stress tolerance are characterized (Tsaftaris *et al.*, 2000). James (2001) mentions that the first transgenic crops were commercially planted in 1995. Since then, Transgenic or Genetically modified (GM) crops are becoming an increasingly common feature of agricultural landscapes. By 2000, a total of 44.2 million hectare of transgenic plants were grown in 13 countries. James (2003) further mentions that the total world's area planted to transgenic crops has increased dramatically, from 3 million hectares in 1996 to nearly 67.5 million hectares in 2003 and it is on further increase yet.

The plants that contain transgenes are often termed as Genetically Modified Plants. A variety of novel transgenic crops have been developed- some offering nutritional benefits and others that are tolerant of drought and other forms of stress, or higher yielding, are at the advanced stages of testing. Soybean, cotton, corn, rice, rape and poplar are some of the most used and well known transgenic plants. Others varieties in the market include potato, sugar beet, tomato, papaya, watermelon, zucchini, and squash.

Objectives of Forming Transgenic Plants

Genetically modified or transgenic, organisms are created through genetic engineering techniques that allow genetic material to be moved between similar or vastly different organisms with the aim of changing their characteristics for a purpose.

The main objectives of creating transgenic plants is to modify the metabolic pathways for the production of tailor-made low molecular weight compounds and polymers, increased resistance towards biotic and abiotic stresses, improved food quality with altered carbohydrate, protein and lipid composition, and production of polypeptides for medical, pharmaceutical or industrial use (Teli and Timoko, 2004). Thus, transgenic plants offer various advantages for the human beings.

However, transgenic plants need monitoring before they are grown for commercial purposes. They are to be field-tested for one or the several reasons. It is seen that the tests conducted for the risk assessment often are not state of the art. The molecular, physiological, metabolic and nutrition tests are regularly outmoded and the statistical methods used for significance analysis have a very low discrimination power (Moch, 2006). Therefore, their risk assessment is necessary.

Risk Assessment of Transgenic Plants

Risk assessment (RA) is a process that takes into account new developments and the progress of science. RA involves proceeding through several steps of assessment prior to obtaining results with an acceptable level of uncertainty. It evaluates and compares scientific evidence regarding the risks associated with alternative activities.

A formalized framework of science-based risk assessment and risk management measures usually governs the intentional introduction into the environment or market of Genetically Modified Organisms (GMOs). Therefore, the objective of an RA is, on a case-by-case basis, to identify and evaluate potential adverse effects of a GMO (Craig *et al.*, 2008). RA must consider the unintended consequences of the environmental release of a transgenic plant, particularly as this may impact on existing agricultural practices and the agro-ecosystem.

Garcia and Altieri (2005) have tabulated following benefits and impacts for transgenic or genetically modified plants.

Table 22.1. Possible benefits and impacts of genetically modified crops (After Garcia and Altieri, 2005).

Potential Benefits	Potential Impacts
Reduced pesticide use	Enhancement of "clean-crop" and monoculture paradigm
Scope for threshold driven herbicide use	Reduction of agro-ecosystem biodiversity
Simplification of farming practices	Increasing vulnerability of crops to environmental changes, new pests, and diseases
More efficient short-term production	Disruption of natural and biological control resources; Promotion of secondary pests; Impact on non-target arthropods, soil biota, and biogeochemical cycles; Selection of herbicide-and/or insect-resistant aggressive weeds; Contamination and erosion of genetic resources for agriculture; Contamination of natural flora and fauna (genetic pollution); Reduction of productivity due to yield drag effect on genetically modified crops and Taking over of natural area by agriculture reducing biodiversity

Potential Benefits of Transgenic Plants

The common benefits of transgenic plants include - pest and disease resistance; herbicide tolerance; cold, drought, and salinity tolerance; increased crop yield and nutritional values; pharmaceutical uses; phytoremediation and soil and water conservation; retention of genetic diversity; and reduced environmental impacts from pesticides and herbicides (ASLA, 2003, R2008). Lovei (2001) has mentioned following benefits of Transgenic Plants/Genetically Modified Crops.

Reduced Environmental Impact from Pesticides

Transgenic crops may decrease the use of environmentally harmful chemicals to control weeds and pests. Cultivation of transgenic plants is seen to help to reduce the use of chemical pesticides in cotton production, as well as in the production of many other crops, which could be engineered to contain the *Bacillus thuringiens* gene.

Herbicide tolerance has also been achieved in transgenic plants that are accomplished exploiting at least three different mechanisms – overexpression of the target enzyme, modification of the target enzyme, and herbicide detoxification (Tsaftaris, 1996). Glyphosate is an environmentally more benign, widely used broad-spectrum herbicide. It is easily degraded in the agricultural environment and works by interfering with the EPSPS enzyme system that is present only in plants. Unfortunately, the herbicide kills crop plants as well as weeds. Besides, the risk that weeds may become resistant to herbicide is well known. But Garcia and Altieri (2005) mentions that there are no studies which clearly support long-term reduction of pesticide use in GM plants/crops. This is because studies are tracking the use of all pesticides and herbicides.

Increased Yield

Less cultivated area would be needed to produce the total amount of food required by people, if crop yields increased. This could result in a lower pressure on land not yet under cultivation and could allow more land to be left under protection. The potential environmental benefits of this type may be greatest in developing countries where most of the agricultural production increase was due to new areas taken into cultivation.

Soil Conservation

Theoretically, pests and weeds could be managed more easily within GM crops than in conventional crops. Some argue that herbicide tolerant crops may offer options to bring more diversity to conventional agriculture. However, herbicide-tolerant crops may allow farmers to abandon the use of soil-incorporated pre-emergent herbicides. This shift to post-emergent weed control may increase the notill and conservation tillage practices, decreasing soil erosion, water loss, and increasing soil organic matter.

Phytoremediation

Genetically modified plants of transgenic plants and micro-organisms can be used for *in situ* remediation of soil and water pollution. Transgenic plants can sequester heavy metals from soils or detoxify the pollutants. But this has not yet been used widely, so its environmental impact has not been studied.

Potential Risks of Transgenic Plants

The potential risks associated with the use of transgenic plants may include - unintended harm to other organisms, including beneficial, non-native, and native species, through direct or indirect effects; reduced effectiveness of pesticides and herbicides; gene-transfer to and other impacts on non-target species; allergenicity and other unknown effects on animal, including human, health; invasiveness; new viral diseases; variability and unexpected results through cascade and cumulative effects; and reduced genetic diversity.

Lovei (2001) mentions that agricultural fields are also part of the "ecological theatre" in which the "evolutionary play" is continuously being played. When transgenic plants are planted in the field, they will inevitably come into contact with many other species that together perform several ecological processes operating in agricultural fields. Thus, there are the risks with transgenic plants/genetically modified crops.

Based on Tsaftaris *et al.* (2000), Lovei (2001), Garcia and Altieri (2005) and Lauer *et al.* (2012), following risks have been observed for TP/GM plants.

Gene Escape/Invasiveness

Gene escape is recognized as a potentially significant hazard. Outcrossing and hybridization with wild relatives is possible for many crops but the ecological consequences of this could be serious if the new trait changes fitness parameters or invasiveness of the modified plants.

Invasiveness is a global threat. Crawley *et al.* (2001) in a long-term study of survival in the wild state for invasiveness of herbicide-resistant crop plants in different area of British Isles found that no genetically modified plant line survived longer than 4 years when planted in natural habitats. However, invasion success is scale-related, and it is rather difficult to predict the consequences of wide-scale planting of transgenic crops from limited-scale studies.

Effects on Non-Target Organisms

Harmful effects on non-target species with the expression are also there, for example, of insecticide toxins that can kill beneficial as well as targeted insects. It is observed that phytophagous organisms which were not targeted may still be affected by insect-resistant plants. For example, transgenic maize pollen, deposited on milkweed leaves could cause larval mortality of the monarch butterfly (*Danais plexippus*), which is a species of important nature conservation.

Further, a plant may be considered a pest but not a weed. For example, a plant that produces an allelopathic substance may be considered a pest if the toxin produced has an undesirable environmental effect. Equally, there is also the possibility of spreading of transgenes to wild or weedy relatives that may cause problems.

Effects on Natural Enemies

Insect-resistant plants are aimed at reducing the densities of certain phytophagous insects, which, however, also serve as prey for a range of natural enemies. An important potential effect of transgenic plants is the consequences of changing the occurrence and density of prey for natural enemies. If the density of prey is reduced, a direct flow-on effect could be a reduced density of their natural enemies.

Predatory and parasitoid insects are also sensitive to prey quality, and prey quality can be influenced by host plants, giving rise to tri-trophic interactions. Parasitoids can also react at a behavioural level to a host originating on transgenic plants. Also the reduced weed diversity is seen that is leading to higher pest damage because of resource concentration effects or impoverished natural enemy communities.

Effects on Pollinators

Plants that are pollinated by animals provide more than 25 per cent of the world's food. Pollinating organisms in the temperate regions are mostly insects, namely bees and wasps. Thus, they can be agents of pollen spread and exposed to any transgenic product that is expressed in pollen or nectar but Bees and Bumble Bees can be affected by transgenic products.

Effects on Soil

Harmful effects on ecosystems is seen when transgenic plant products interfere with natural biochemical cycles. The root exudates of transgenic plants are detected at concentrations that can kill insects, however, their long-term consequences are not yet known. The study of soil organisms and processes is generally less advanced, reflecting the relative emphasis on above versus below-ground ecological processes but Griffiths *et al.* (2000) found transient effects and significant changes in soil protozoan populations in soil under genetically engineered potato lines. As soil fertility maintenance is a biological process, tests of the effects of TP/GM plants on soil processes are very important.

Besides, it is also seen that accumulation of the Bt toxins remain active in the soil after the crop is plowed under and bind tightly to clays and humic acids. Also the escalation of herbicide use in herbicide tolerant crops is seen with consequent environmental impacts including reduced weed populations and diversity.

Effects on Biodiversity

Harmful effects on biodiversity may be there, if a transgene offers an adaptive advantage in transgenic plants escaped in the area of cultivation or in wild relatives where it could be transferred by cross-fertilization. This is practically important if occurring at the centers of genetic variation of cultivated plants. In addition, biodiversity concerns have been raised for current cultivation systems including many locally adopted varieties if they will be substituted by a few new transgenics.

Besides, the widespread application of conventional agricultural technologies such as herbicides, pesticides, fertilizers and tillage has resulted in severe environmental damage in many parts of the world. Herbicide-resistant crops are expected to allow more efficient weed control. Consequently, changes in cultivated land due to transgenic crops are expected to create problems for biodiversity only if invasiveness is affected. Concerns have also been raised, especially in the United Kingdom that this will have negative consequences for countryside biological diversity, with fewer surviving flowering plants to provide resources for organisms ranging from invertebrates to birds. Watkinson *et al.* (2000) used a weed (*Chenopodium album*) and a songbird (*Alauda arvensis*) model in a landscape context to predict the effects of herbicide resistant sugarbeet on biological diversity in general. Their work points to potentially significant negative effects on seed-eating birds.

Similar concerns prompted the U.K. government to ban commercial growing of transgenic plants and initiate a 4-year farm-scale field trial to study what effect herbicide resistant transgenic plants will have on biodiversity (Firbank *et al.*, 1999).

However, studies published so far on the effects of transgenic plants on agricultural biodiversity are rather imperfect (Hilbeck *et al.*, 2000). Additionally, reduced weed populations has also been observed that leads to the declines in bird populations who feed on or shelter in weeds or feed on the arthropods supported by weeds.

Risks for Human Health

It is observed that though classical plant breeding techniques existed, but present day cultivated crops have become significantly different from their wild counterparts. Many of these crops were originally less productive and at times unsuitable for human consumption, but the advent of GM technology over the years has allowed further development. However, Tsaftaris *et al.* (2000) have listed following risks associated with human beings and environment.

☆ Formation of new allergens from the novel proteins expressed in the transgenic organism, which could trigger allergic reactions at some stage.

☆ Creation of new toxins through unexpected interactions between the product of the genetic modification and other endogenous constituents of the organism.

☆ Dispersion of antibiotic resistance genes used as markers from the genetically modified organism derived food to gut micro-organisms and intensification of problems with antibiotic resistant pathogens.

Risks for the Environment

The gene transfer from the transgenic plant to related species as a result of hybridization could also lead to new pests. The transgenic plant escapes its intended use and becomes an invader to the natural environment. Besides, the development of resistance from the continuous use of the same agent on the target organism is of great concern.

Therefore, the environmental risks of new TP/GM technologies need to be reconsidered in the light of the risks of continuing to use conventional technologies and other commonly used farming techniques.

Managing Risk associated with TP or GMP

Risk management is associated with developing strategies to prevent and control risks within acceptable limits and relies on risk assessment. Organization for Economic Co-operation and Development (OECD, 1993) has given guidelines on risk assessment. Some of them are as given below:

☆ Safety in biotechnology is achieved by the appropriate application of risk/safety analysis and risk management. Risk/safety analysis comprises of hazard identification and, if a hazard has been identified then risk assessment.

☆ Risk/safety analysis is based on the characteristics of the organism, the introduced trait, the environment into which the organism is introduced, the interaction between these, and the intended application.

☆ Risk/safety analysis is conducted prior to an intended action and is typically a routine component of research, development and testing of new organisms, whether performed in a laboratory or a field setting.

☆ Risk/safety analysis is a scientific procedure which does not imply or exclude regulatory oversight or imply that every case will necessarily be reviewed by a national or other authority.

We have seen there are negative impacts of TP/GM Plants, so attention is needed while going for commercialization of them. Meilan (2007) also opines that while working towards commercialization of genetically engineered or modified products, potential petitioners should be mindful that:

☆ Early consultation with the appropriate regulatory agencies can save a lot of time and needless actions and delays.

☆ A petition for non-regulated status can include more than one independent line (transformation event).

☆ Species-gene combinations are to be evaluated on a case-by-case basis.

☆ It is possible to use safety data submitted by previous petitioners with the appropriate permission.

☆ Gene flow is not considered a risk a priori, but it must be evaluated to determine whether a perceived risk is genuine.

☆ It may be possible to commercialize a product derived from a transgenic plant under rare circumstances. However, the developer must still abide by permit and notification requirements and other regulations.

☆ Pharmaceutical-producing plants are always to be regulated in some way.

Role of Regulatory Authorities in Managing Risks Associated with TP/GMP

Many countries have, or are currently putting into place, a framework for undertaking RAs of the cultivation or production of genetically modified foods, and as such, it represents a significant opportunity to work towards international harmonization on many levels. However, the criteria and factors that determine biosafety assessment of transgenic plants vary in different countries. In European Union all plants produced by genetic modification are to be assessed (technology based assessment) whereas in the USA and Canada only plants modified with particular genes are regulated (product based assessment).

Meilan (2007) has mentioned following major conclusions drawn in the report of the 'Biological Confinement of Genetically Engineered Organisms.'

☆ To evaluate efficacy, the Genetically Engineered Organism (GEO) should be compared with its progenitor before release. For trees, it may be necessary to begin field tests before such comparisons are possible.

☆ Various confinement methods should be tested separately and in combination, in a variety of appropriate environments, and in representative organisms.

☆ The need for bioconfinement should be considered early in the development of a genetically engineered organism.

☆ Non-food crops should be utilized for genes encoding products that need to be kept out of the food supply.

☆ An integrated confinement system should be based on risk. Therefore, it is unlikely that any single confinement method will be 100 per cent effective.

☆ The stringency of confinement should reflect the consequences of genetically engineered organism escape.

☆ International cooperation should be sought for managing genetically engineered organism confinement.

☆ Transparency and public participation are needed to develop and implement the most appropriate bioconfinement approach.

☆ Social and ethical values should be considered when assessing the stringency needed for confinement.

Dhiman (2010) has also supported the view that there is also a need to develop a code of conduct where simply academic discussion will not be enough rather there should be some ethical rules to carry out such types of the researches where human beings are involved (Dhiman, 2010).

Various sources of relevant information like, scientific literature and risk assessment dossiers previously submitted to, or authored by, competent authorities concerning applications for the deliberate release of TP/GM crop plants are available. Many of these latter documents can be found online at different sites. Craig *et al.* (2008) have listed some important of them, as given in Table 22.2.

These can be consulted while going to develop a TP/GM crops on commercial level and for in-depth study of them. Lastly, continued resistance management research should be conducted to evaluate the effectiveness of TP/GM crops for their commercial use.

Conclusion

Genetic modification of crops has provided the opportunity to plant breeders to modify plants in novel ways and has the potential to overcome the important problems of modern agriculture. Transgenic plants (TPs) display considerable potential to benefit both developed and developing countries. Carpenter and Gianessi (2000) mention that expressing insecticidal proteins or proteins providing tolerance to herbicides or resistance to environmental stresses TPs are revolutionizing agriculture.

Teli and Timko (2004) rightly state that within a very short time span the use of genetically modified plants for the production therapeutic compounds has moved from being an experimental system with significant potential to a commercially viable process poise to deliver products in useful in animal and human therapies. The rapid pace of development witnessed thus far is likely to accelerate in the very near future as additional, novel uses of transgenic plants as production systems for human therapeutics are explored.

Table 22.2. Relevant regulatory authority websites (After Craig *et al.*, 2008).

Country/Region	Region Authority	Website Address
Argentina	Comisión Nacional Asesora de Biotecnologýa Agropecuaria (CONABIA)	http://www.sagpya.mecon.gov.ar/new/0–0/programas/conabia/index.php
Australia	Office of the Gene Technology Regulator (OGTR)	http://www.ogtr.gov.au/
Australia and New Zealand	Food Standards Australia New Zealand (FSANZ)	http://www.foodstandards.gov.au/
Brazil	National Biosafety Technical Commission (CTNBio)	http://www.ctnbio.gov.br/
Cambodia	Biosafety Clearing Houses (BCH)	http://www.cambodiabiosafety.org/
Canada	Canadian Food Inspection Agency (CFIA) Health Canada	http://www.inspection.gc.ca/english/toce.shtmlhttp://www.hc-sc.gc.ca/index_e.html
China	Biosafety Clearing Houses (BCH)	http://english.biosafety.gov.cn/
Colombia	Biosafety Clearing Houses (BCH)	http://www.bch.org.co/bioseguridad/index.jsp
Costa Rica	Biosafety Clearing Houses (BCH)	http://cr.biosafetyclearinghouse.net/
Europe Union	Joint Research Council Biotechnology and GMOs Information Website (JRC) European Food Safety Authority (EFSA)	http://gmoinfo.jrc.ithttp://www.efsa.eu.int/
India	Biosafety Clearing Houses (BCH)	http://www.indbch.nic.in/
Iran	Biosafety Clearing Houses (BCH)	http://biosafety.irandoe.org/BCH/domestic.htm
Korea	Biosafety Clearing Houses (BCH)	http://www.biosafety.or.kr/
Mexico	Mexican Comisión Intersecretarial de Bioseguridad de los Organismos Geneticamente Modificados (CIBIOGEM)	http://www.cibiogem.gob.mx/
New Zealand	Environmental Risk Management Authority New Zealand (ERMA)	http://www.ermanz.govt.nz/
Norway	Norwegian Directorate for Nature Management (Dir. for Nat. Mgmt.)	http://english.dirnat.no/
Philippines	National Committee on Biosafety of the Philippines (NCBP)	http://www.ncbp.dost.gov.ph/
Japan	Biosafety Clearing Houses (BCH)	http://www.bch.biodic.go.jp/english/e_index.html
Switzerland	Swiss Expert Committee for Biosafety (SECB)	http://www.umwelt-schweiz.ch/buwal/eng/fachgebiete/fg_efbs/index.html
Thailand	Biosafety Clearing Houses (BCH)	http://www.biosafety.biotec.or.th/?idgroup=-1519023173
USA	US Regulatory Agencies Biotechnology Website (Unified)	http://usbiotechreg.nbii.gov/

However, the potential for genetically modified crops to benefit biodiversity conservation and sustainable agriculture is negligible or at least questionable, the potential for impacts or threats of genetically modification technology given the evidence so far appears substantial, particularly because genetically modified crops are truly biological novelties that would not exist via natural processes. The release of these new biological phenotypes into the environment has led to serious concerns about the unpredictable ecological and evolutionary responses GM species. Likewise, the next generation of transgenic crops, pose even more dangerous risks if released in the environment, especially as containment of transgenes is not assured.

Park *et al.* (2011) is of the opinion that it is difficult to predict that beyond 2015, what will come to market and when, or indeed what the nature and extent of international and national regulatory frameworks relating to transgenic crops will be. However, it is likely that new crop events related to nutritional benefits, nitrogen use efficiency, drought and salt tolerance and yield enhancement will be available by 2020. They further state that despite the growth and use of transgenic crops in many areas of the world, some governments, organizations and individuals still hesitate to acknowledge that transgenic crops provide economic and environmental benefits that are unobtainable in a timely manner via non-transgenic advances in plant breeding. Therefore, before transgenic plants can be grown for commercial purposes, a petition for non-regulated status needs to be submitted to respective regulatory authority in the country to their risks so that their maximum benefits can be obtained.

However, it is true- as our level of understanding of the factors that impact transgene expression in plants improves, we will see improvement in levels of production of target molecules (peptide, proteins, and antibodies), decreased costs of production, and greater overall exploitation of plant based production systems.

References

1. ASLA, 2003, R2008. *Transgenic Plants and the Environment*. American Society of Landscape Architects. Available at: http: //www.asla.org/uploadedFiles/ CMS/Government_ Affairs/Public_Policies/Transgenic_Plants.pdf.

2. Carpenter, J. and Gianessi, L., 2000. Herbicide use on roundup ready crops. *Science*, 287: 803–804.

3. Craig, Wendy, Tepfer Mark, Giuliano, Degrassi and Ripandelli, Decio, 2008. An overview of general features of risk assessments of genetically modified crops. *Euphytica*, 164: 853–880.

4. Crawley, M.J., Brown, S.L., Hails, R.S., Kohn, D.D. and Rees, M., 2001. Transgenic crops in natural habitats. *Nature*, 409: 682–683.

5. Dhiman, Anil Kumar, 2006. *Ayurvedic Drug Plants*. Daya Publishing House, New Delhi.

6. Dhiman, Anil Kumar, 2010. Developing trends and emerging needs of ethics in biotechnology in present environment. In: *Advances in Applied Biotechnology*, (Eds.) Pradeep Parihar and Leena Parihar. Agrobios India, Jodhpur, pp. 293–302.

7. Firbank, L.G., Dewar, A.M., Hill M.O., May, M.J., Perry, J.N., Rothery, P., Squire, G.R. and Woiwod, I.P., 1999. Farm-scale evaluation of GM crops explained. *Nature*, 399: 727–728.

8. Griffiths, B.S., Geoghegan, I.E. and Robertson, W.M., 2000. Testing genetically engineered potato, producing the lectins GNA and Con A, on non-target soil organisms and processes. *J. Applied Ecology*, 37: 159–170.

9. Garcia, Maria Alice and Altieri, Miguel A., 2005. Transgenic crops: Implications for biodiversity and sustainable agriculture. *Bulletin of Science, Technology and Society*, 25(4): 335–353.

10. Hilbeck, A., Meier, M.S. and Raps, A., 2000. Review on non-target organisms and *Bt* plants. Ecostrat Gmbh, Zurich, p. 77.

11. James, C., 2001. Preview: Global review of commercialized transgenic crops: 2000. ISAAA Briefs No. 21. *International Service for the Acquisition of Agri-Biotech Application*, Ithaca, NY.

12. James, C., 2003. Preview: Global status of commercialized transgenic crops: 2002. ISAAA Brief No. 30. *International Service for the Acquisition of Agri–Biotech Application*, Ithaca, NY.

13. Lovei, G.L., 2001. Ecological risks and benefits of transgenic plants. *New Zealand Plant Protection.* , 54: 93–100.

14. Lauer, Joe, Shi, Guanming, Chavas, Jean-Paul and LaForge, Matt, 2012. Managing risk using transgenic crops. *Proceeding of Wisconsin Crop Management Conference*, pp. 28–30. Available at : http://www.soils.wisc.edu/extension/wcmc/2012/pap/Lauer_2.pdf.

15. Meilan, Richard, 2007. Challenges to commercial use of transgenic plants. In: *Plant Biotechnology in Ornamental Use*, (Eds.) Pei, Yan and Li, Yi. CRC Press, USA. Available at: http://www.agriculture.purdue.edu/fnr/html/faculty/Meilan/documents/JCropImp_final.pdf.

16. Moch, Katja (Ed.), 2005. Epigenetics, transgenic plants and risk assessment. *Proceedings of the Conference* at Literaturhaus, Frankfurt am Main, Germany. Available at: http://www.oeko.de/oekodoc/277/2006–002–en.pdf.

17. OECD, 1993. *Safety Considerations for Biotechnology: Scale–up of Crop Plants.* Available at : http://www.oecd.org/science/biotrack/1958527.pdf.

18. Park, Julian Raymond, McFarlane, Ian, Phipps, Richard Hartley and Ceddia, Graziano, 2011. The role of transgenic crops in Ssustainable development. *Plant Biotechnology Journal*, 9: 2–21.

19. Teli, Nilesh P. and Timko, Michael P., 2004. Recent developments in the use of transgenic plants for the production of human therapeutics and biopharmaceuticals. *Plant Cell, Tissue and Organ Culture*, 79: 125–145.

20. Tsaftaris, A., 1996. The development of herbicide-tolerant transgenic crops. *Field Crops Research*, 45: 115–123.

21. Tsaftaris, A.S., Polidoros, A.N., Karavangeli, M., Nianiou-Obeidat, I., Madesis, P. and Goudoula, C., 2000. Transgenic crops: Recent developments and prospects. In: *Biological Resource Management: Connecting Science and Policy*, (Eds.) E. Balazs, E. Galante, J.M. Lynch, J.S. Schepers, J.-P. Toutant, D. Werner and P.A.T.J. Werry. Springer Verlag, Berlin Heidelberg, pp. 187–203.

22. Watkinson, A.R., Freckleton, R.P., Robinson, R.A. and Sutherland, W.J., 2000. Predictions of biodiversity response to genetically modified herbicide: Tolerant crops. *Science,* 289: 1554–1557.

Biotechnology: An Overview (2015)
Editors: Rajan Kumar Gupta, Nasim Akhtar and Deepak Vyas
Published by: DAYA PUBLISHING HOUSE, NEW DELHI

Pages 375–385

Chapter 23

Biosensors: A Commercial Aspect

Vibhu Sharma[1] and Ashish Bhardwaj[2]

[1]Department of Botany, Pt. L.M.S. Govt. P.G.College,
Rishikesh, Uttarakhand
[2]Department of Chemical Technology, I.I.T.-Roorkee, Uttarakhand

ABSTRACT

Biosensors are analytical tool that make use of some immobilized biological component to interact with analyte thereafter the produced biological signal is converted to a readable signal with the help of a transducer. Biosensors market is growing year after year with applications; today biosensors are used in various fields such as health care, process industries, environmental monitoring, food and beverages, defense and security etc. Biosensors are playing, has the potential to play more and more important role to improve our daily life. Biosensors industry now worth billions of US dollars, attracting the research groups throughout the world with thousands of publications every year for the last two decades. Home diagnostics and bio-defense is expected to grow in future with emergence of newer technologies in the field.

Introduction

Biosensor can be defined as an analytical tool, consist of an immobilized biological component which can be enzyme, or DNA, cell organelle, whole cells or tissue etc. to interact with analyte, in close proximity to a transducer with a read out device to convert the intricate biological interaction in an easily readable signal. (Figure 23.1). Thus biosensors make use of a biological sensing element to exploit the natural specificity and sensitivity of biological world.

History and Background

In 1916 first report on immobilization of an invertase came, as we could understand transducer is a vital for the development of a biosensor, the history of biosensor starts with the name Prof. Leland C Clarke. Clake explained about oxygen electrode that later used in the development of glucose biosensor.

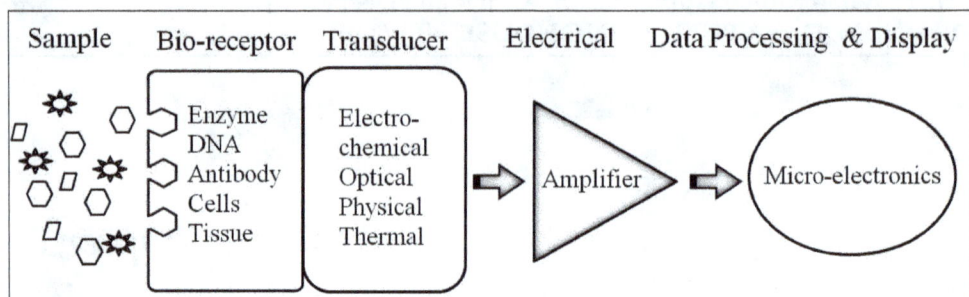

Figure 23.1. Basic working principle of biosensors.

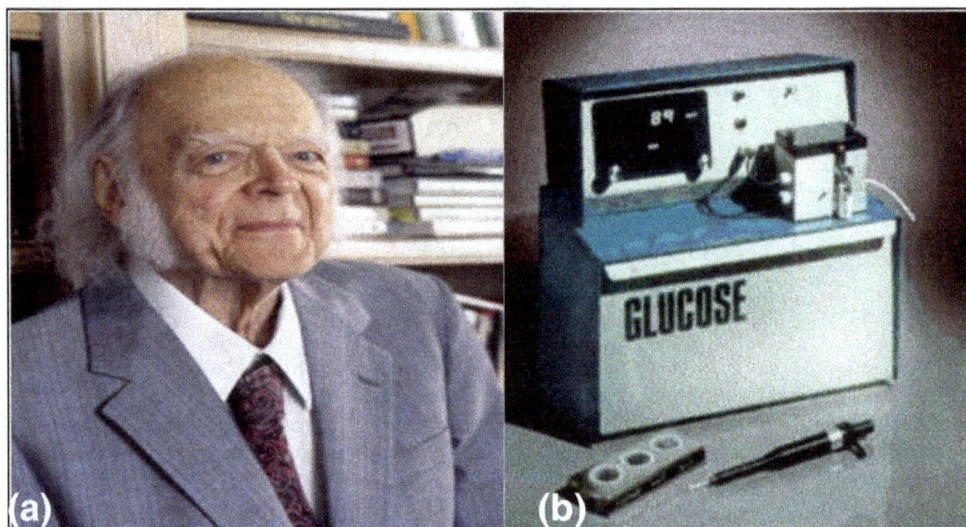

Figure 23.2. (a) Leland C. Clarke "Father of biosensor" (b) 1 st commercialized Glucose biosensor (reproduced from Newman and Turner, 2005).

If we look in to the history of biosensor, the main milestones* are:

★ 1916 – First report on the immobilisation of proteins: adsorption of invertase on activated charcoal

★ 1922 – 1st Glass pH Electrode

★ 1956 – Professor Leland C Clarke explained oxygen electrode in his paper

★ 1962 – Very 1st description of an amperometric glucose biosensor

★ 1969 – 1st Potentiometric biosensor for urea

★ 1974 – Thermal transducers such as thermal enzyme probes and enzyme thermistors were proposed.

☆ 1972/75 – 1st Commercial biosensor, idea of Clark came to reality (Yellow Springs Instrument, Ohio, USA)

☆ 1975 – Divis suggested that bacterial whole cells could be used as bio-component for measurement of alcohol and named as microbial electrode.

☆ 1975 – Lubbers and Opitz explained the concept of an optical biosensor for alcohol, immobilized indicator with a fiber optic sensor to measure carbon dioxide or oxygen, coined the term optode.

☆ 1982 – Shichiri *et al.*, reported in vivo application of glucose biosensors, which is the first needle-type enzyme electrode for subcutaneous implantation.

☆ 1984 – Ferrocene and its derivatives were immobilized, used as a mediator with oxidoreductases.

☆ 1982/86 – 1st Commercial biosensor in Europa (CSE Berlin, Germany/ PGW Medingen/ENH Hamburg, Germany)

☆ 1987 – 1st Commercial biosensor for Blood Glucose Home Monitoring (MediSense/ExacTec)

☆ 1990 – Launch of SPR based BIA core (Pharmacia/Sweden)

☆ 1990 – The sale of this home blood glucose monitoring reached 175 million dollars.

☆ 1996 — Glucocard launched

☆ 1997 – IUPAC committee did agree on the following definition of a biosensor: "An electrochemical biosensor is a self-contained integrated device, which is capable of providing specific quantitative or semi-quantitative analytical information using a biological recognition element (biochemical receptor), which is retained in direct spatial contact with an electrochemical transduction element. Because of their ability to be repeatedly calibrated, we recommend that a biosensor should be clearly distinguished from a bio analytical system, which requires additional processing steps, such as reagent addition. A device that is both disposable after one measurement *i.e.* single use and unable to monitor the analyte concentration continuously or after rapid and reproducible regeneration should be designated as a single-use biosensor."

☆ 1998 — Launch of LifeScan Fast Take blood glucose biosensor

☆ 1998 -- Merger of Roche and Boehringer Mannhein to form Roche Diagnostics

☆ 2001 — LifeScan purchases Inverness Medical's glucose testing business for $1.3 billion

☆ 2003 — i-STAT acquired by Abbott for $392 million.

☆ 2004-- Abbott acquires TheraSense for $1.2 billion Commercially available biosensors for different analytes

☆ 2010 – Biosensors International Launch its BioMatrix Flex™abluminal biodegradable polymer DES

☆ 2012 – Collaboration between Universal Biosensors Inc. and Siemens to commercialize a range of novel hand-held analyzers for the point-of-care.

* Source of information: Newman and Turner, 2005; Turner, 2013 and websites given below

http://www.biosensors.com/intl/about–biosensors–milestones–highlights; http://www.biospectrumasia.com/biospectrum/news/1758/universal–biosensors–achieves–research–milestone

Application of Biosensors

Since the inception of biosensor technology, it was expected to have immense applications in the fields of health, environment, food safety, security and industrial monitoring etc. (Monosik *et al.*, 2012 a). An application of biosensors is just limited to development of a bio-assay principle and its compatibility to a transducer. Figure 23.3 is depicting different areas of biosensors applications while Table 23.1 reviews the commercially available biosensors.

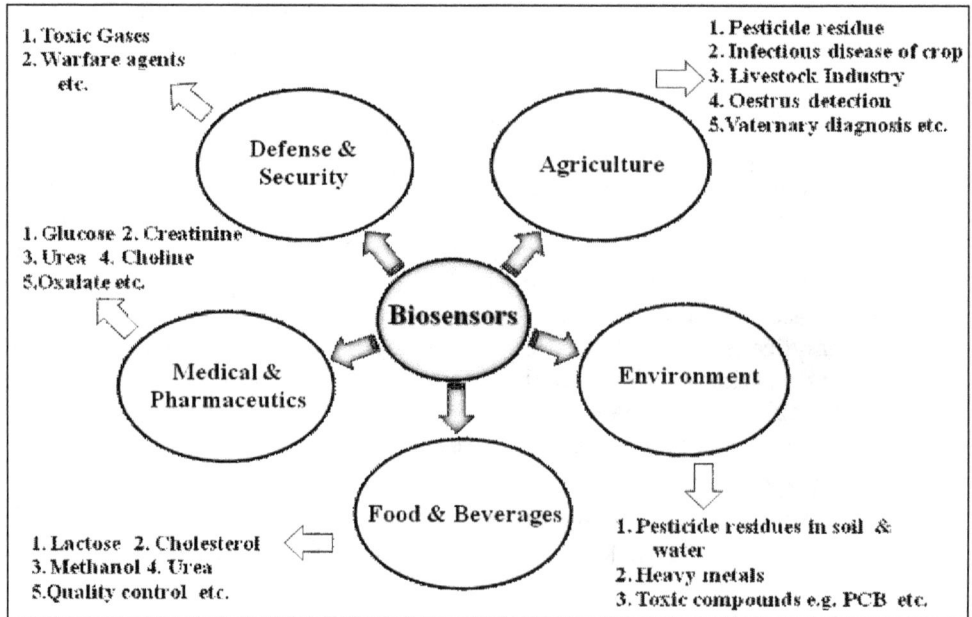

Figure 23.3. Applications of biosensors in various fields.

Medical and Pharmaceutics

Despite the enormous progress in the field of biosensors in the recent past, glucose biosensors leads the global biosensor market acquiring about 90 per cent of that, application for other analytes is not very common. Though biosensor has been developed for various analytes *e.g.* Urea in urine or blood samples (Verma *et al.*, 2012). Blood glucose has been the most common analyte with amperometric measurement (Malhotra and Chaubey 2003; Belluzo *et al.*, 2008; Her *et al.*, 2013). Apart from glucose commercial biosensors are available for lactate, triglycerides, blood gases and creatinine etc. (Monosik *et al.*, 2012b).

Food and Beverages

Food and beverages industry is another important are for the applications of biosensors but not as common compared to medical and pharmaceutics (Dzyadevych *et al.*, 2008). Hall, 2002 focused on biosensor technologies for detecting microbiological food borne hazards. Various biosensors have been developed by the scientist for the detection different heavy metals (Verma *et al.*, 2010; Verma *et al.*, 2011)

Environment

Usually for the monitoring of environmental pollution we use chemical analysis which sometime doesn't provide the sufficient information on the ecological risk (Castillo *et al.*, 2001). With an inherent quality of providing the bioavailable concentration of the analyte, biosensors can play an important role in environmental monitoring. Biosensors for different analytes *e.g.* pesticide residues, BOD, nitrates, phosphates, heavy metal ions, PCBs, surfactants has already been developed (Rodriguez *et al.*, 2006; Somerset 2011).

Agriculture

In the dairy industry, with importance in monitoring of progesterone and thereby fertility management through artificial insemination biosensors can play a big role because the successful prediction of ovulation leads to considerable cost saving (up to £10 000 pa in a dairy farm with a 100-cow herd) (Velasco-Garcia and Mottram, 2001). Biosensors have their application in agriculture industry addressing the problems *e.g.* pesticides residues in soil and crop; in process control *e.g.* bacteriological food safety, quality control of products; diagnosis of infectious disease in crops; in animal production *e.g.* oestrous detection, veterinary drug residue screening and veterinary diagnosis etc. (Velasco-Garcia and Mottram, 2003).

Defence and Security

After the event of September 11, 2001 detection of biohazards has become an important issue, researchers of different universities *e.g.* New Mexico State University, has a focus on the development of biosensors for the detection of biowarfare agents (Gooding, 2006; Luong *et al.*, 2008; Mongra and Kaur, 2012). Mascini and Palchetti, 2005 reported the development of a biosensor based on acetylene choline esterase and DNA for the detection of neurotoxic and genotoxic compounds respectively. Among the biowarfare agents biosensors has been developed for Anthrax, Botulinum toxin (BoTN), Cholera toxin, Staphylococcal enterotoxin B, Ricin, *Bacillus globigii* and *F. tularensis*, a few of them has already been commercialized by Response Biomedicals, Biowarfare Agent Detection Devices (BADD) and BioVeris etc. (Gooding, 2006). As far global market for biosensors in the field of defence and security is concerned, it is uncertain besides USA and a few other countries as there is no space for a false negative or too many false positive results in this case (Luong *et al.*, 2008).

Table 23.1. List of commercially available biosensors for different analytes from different companies.

Company	Biosensor	Analyte (Target Compound)
Danvers (USA)	Apec glucose analyzer	Glucose
Biometra Biomedizinische	Biometra biosensors for HPLC	Glucose, ethanol and methanol
Eppendorf (Germany)	ESAT 6660 Glucose Analyzer	Glucose
Scola-Tacussel (France)	Glucoprocesseur	Glucose and lactose
Universal Sensors (USA)	Amperometric Biosensor Detector	Glucose, galactose, I-amino acids, ascorbate and ethanol
Yellow Spring Instruments (USA)	ISI Analyzer	Glucose, lactate, L-amino acids, cholesterol
Toyo Jozo Biosensors (Japan)	Models: PM-1000 and PM-1000 DC (on line), M-100, AS-200 and PM-1000 DC	Glucose, lactate, L-amino acids, cholesterol Triglycerides, glycerine, ascorbic acid, alcohol
Orinetal Electric (Japan)	Orientel Freshness Meter KV101	Fish Fressnesh
Swedish BIACORE AB (Sweden)	BIACORE	Bacteria
Malthus Instruments (UK)	Malthus 2000	Bacteria
Biosensori SpA (Italy)	Midas Pro	Bacteria
Biotrace (UK)	Unilite	Bacteria
Nova Biomedicals (Waltham MA)	Bio profile Chemistry Analyzer	Glucose, Lactate, Glutamate, Glutamine
TRACE Biotech AG (Germany)	Process TRACE 1.2	Glucose, Gtamate, Glutamine
SensAlyse Ltd (UK) Malic Acid	Alcohol Sensor	Alcohol, Sugars, Ascorbic acid,
Gwent Sensors Ltd. (UK)	The Answer 8000	Glucose
Applied Enzyme Technology Ltd. (UK)	-	Glutamate, Alcohol
Bioanalytical Systems Inc (USA)	Peroxidase Redox Polymer kit	Glucose
Biotech Products (USA) Distributor	Micro Dialysis Biosensor by Scypogel	Glucose Lactate, Glycerol, Ascorbate, (D/L Amino Acids) Adenosine, Xanthene
Flownamics Analytical Instruments (USA)	FAIZA 110-P	Glucose, Lactose, Sucrose, Galactose, Lactate, Ethanol/ Methanol, L-Glutamate, L-amino acids
Analox Istruments Ltd (UK)	LM5 Lactate Analyzer, AM2 Industrial Alcohol Analyser, GM10 Industrial Glucose Analyser, GM7 Micro-Stat Multiassay Analyser	Lactate, Alcohol, Glucose, Sucrose, Glutamine, Ammonia, Methanol

Contd...

Table 23.1–Contd...

Company	Biosensor	Analyte (Target Compound)
IBA GmbH (Germany)	OLGA On-Line General Analyzer	Sucrose, Glucose, Alcohol
Biosensor Technology GmbH (Germany)	Enzyme Membranes, Thick film biosensors Glukometer 3000	Glucose, Lactate, Ascorbic Acid Lactose
EKF Diagnostics (Germany)	Biosen 5040	Glucose, Lactate
BioFutura S.r.1 (Italy)	PerBaco 2000, PerBaco2002	Glucose, Fructose, Lactate, Malate
Ismatec S.A. (Zuric)	ASIA Flow Injection Analyzer	Glucose, Alcohol, Xanthine, Galactose, Choline
Roche Diagnostics AG	Accu-Chek Plus Glucose meter	Glucose
Carewell Biotech Pvt. Ltd. (India)	EUKARE	Glucose
Modern Water	Microtox®	Toxicitiy
– (Wang *et al.*, 2008)	Cellsense®	Metals and Organic compounds
LifeScan Inc.	Onetouch®	Glucose, cholesterol
DuPont	RiboPrinter®	Bacteria
Oxford Biosensors Ltd.	Multisense	Different analytes in Blood, Urine, Serum etc.

Source: Edited and upgraded from Mello and Kubota, 2002; Prodromidis and Karayannis 2002; Luong *et al.*, 2008 and Wang *et al.*, 2008

Table 23.2. A list of commercially available nucleic acid-based biosensors for pathogen detection with their mode of detection and the sample source.

Organism	Biosensor	Company
Candida sp.	BD Affirm™ APIII	Beckton Dickinson, Inc.
Chlamydia trachomatis	HC2 CT-ID	Qiagen
	APTIMA® CT	Gen-probe
	PACE2 CT	Gen-probe
	BD ProbeTec™ CT	Beckton Dickinson, Inc.
	COBASAMPLICOR CT	Roche
Escherichia coli O157:H7	BAX system	Qualicon, Inc.
Gardnerella	BD Affirm™ APIII	Beckton Dickinson, Inc.
Mycobacterium avium	Accuprobe®	Gen-probe
Mycobacterium gordonae	Accuprobe®	Gen-probe
Mycobacterium intracellulare	Accuprobe®	Gen-probe
Mycobacterium kansasii	Accuprobe®	Gen-probe

Contd...

Table 23.2–Contd...

Organism	Biosensor	Company
Mycobacterium tuberculosis	Accuprobe® MTD	Gen-probe
	BD ProbeTec™ ET	BD ProbeTec™ Inc.
	COBAS AMPLICOR MTB	Roche
Neisseria gonorrhoeae	HC2 GC-ID	Qiagen Chem
	APTIMA® GC	Gen-probe
	PACE2 GC	Gen-probe
	BD ProbeTec™ GC	Beckton Dickinson, Inc.
	COBAS AMPLICOR NG	Roche
Streptococci Group A	GASDirect®	Gen-probe
Streptococci Group B	IDI-StrepB	Infectio Diagnostic Inc.
Trichomonas vaginalis	APTIMA®	Gen-probe
	Beckton BD Affirm™ APIII	Dickinson, Inc.

Source: Singh *et al.*, 2013.

Table 23.3. A list of nucleic acid and protein-based commercial products for foodborne pathogen detection with their method and limit of detection.

Organism	Biosensor	Company	Limit of Detection (cfu/ml)
E. coli O157:H7	BAX®	Dupont	10^4
	Lateral Flow System	Dupont	1 (per 25 g food)
	Reveal®	Neogen	10^4
	GeneQuence®	Neogen	1 (per 25 g food)
	VIDAS	Biomerieux	–
Campylobacter	BAX®	Dupont	10^4
	VIDAS	Biomerieux	–
	ACCUPROBE	Biomerieux	–
Listeria	BAX®	Dupont	10^4
	Lateral Flow System	Dupont	1 (per 25 g food)
	Reveal®	Neogen	10^6
	ANSR™	Neogen	10^4
	VIDAS	Biomerieux	–
Salmonella	ANSR™	Neogen	10^4
	GeneQuence®	Neogen	1 (per 25 g food)
	Reveal®	Neogen	10^6
	BAX®	Dupont	10^4
	Lateral Flow System	Dupont	1-4 (per 25 g food)
Enterobacter	BAX®	Dupont	–
	VIDAS	Biomerieux	–
Vibrio	BAX®	Dupont	10^4

Source: Singh *et al.*, 2013.

Market Push and Pulls

As the research paper published year after year in the field of biosensor shows, the research is upward and scientist are being attracted but even after the huge research only a few have been commercialized, till date development of biosensor is somewhat incremental with a low success rate except glucose. There are various technical hurdle *e.g.* biosensor must function continuously over a long period say for one month or so, response time should be as low as possible in minutes only, matrix interference etc. With 500 companies worldwide working in the field of biosensors (Mongra and Kaur, 2012) market potential is there to accept biosensor if the said hurdles are covered (Figure 23.4).

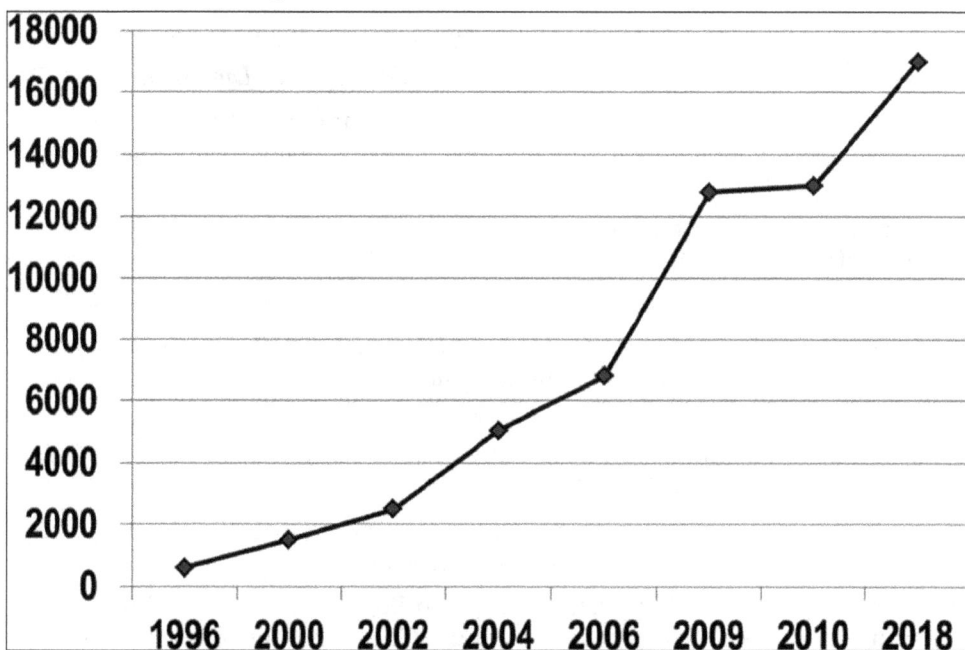

Figure 23.4. Graph of the world market for biosensors estimated from various commercial sources and predicted for the future in US$ millions (Source: Turner, 2013).

Conclusion

Biosensor technology has achieved huge success in the field of research as well as commercial level, though till date the glucose biosensors have outshined the biosensor market but it should be taken as model for the commercial success in other fields *e.g.* agriculture sector and more importantly in the field of biodefence and security. As far the technology is concerned it is ever changing, with the new findings in nanotechnology offer new possibilities for biosensor technology and thereby the commercial aspects get improved.

References

1. Castillo, J. *et al.*, 2004. *Sensors and Actuators B*, 102: 179–194.

2. Rodriguez-Mozaz, S., Lopez de Alda, M. and Barcelo, D., 2006. Biosensors as useful tools for environmental analysis and monitoring. *Anal Bioanal Chem*, 386: 1025–1041.

3. Mascini, M. and Palchetti, I. (Chapter author).

4. Morrison, D., *et al.* (Eds.), 2005. *Defense against Bioterror: Detection Technologies, Implementation Strategies and Commercialization Opportunities*, Springer, p. 245–259.

5. Somerset, 2011. In: *Environmental Biosensors*. InTech, Rijeka p. 486–493.

6. Monosik, R., Stredansky, M. and Sturdik, E., 2012. *J. Clin. Lab. Anal.*, 26: 22–34..

7. Monosik, Rastislav, Stredanskyb, Miroslav and Sturdika, Ernest, 2012. Biosensors: Classification, characterization and new trends. *Acta Chimica Slovaca*, 5(1): 109–120.

8. Dzyadevych, S.V., Arkhypova, V.N., Soldatkin, A.P., Elskaya, A.V., Martelet, C. and Jaffrezic-Renault, N., 2008. *IRBM* 29: 171–180.

9. Malhotra, B.D. and Chaubey, A., 2003. Sensor Actuat. *B–Chem*. 91: 117–127.

10. Maria, N., Velasco-Garcia and Toby Mottram, 2003, Biosensor technology addressing agricultural problems. *Biosystems Engineering*, 84(1): 1–12.

11. Maria, N., Velasco-Garcia and Toby Mottram, 2001. Biosensors in the livestock industry: An automated ovulation prediction system for dairy cows. *Trends in Biotechnology*, 19(11): 433–434.

12. Hall, Robert H., 2002. Biosensor technologies for detecting microbiological food borne hazards. *Microbes and Infection*, 4: 425–432.

13. Her, Jim-Long, Wu, Min-Hsien, Peng, Yen-Bo, Pan, Tung-Ming, Weng, Wen-Hui, Pang, See-Tong and Chi, Lifeng, 2013. High performance GdTixOy electrolyte-insulator-semiconductor pH sensor and biosensor. *Int. J. Electrochem. Sci.*, 8: 606–620.

14. Mongra, A.C. and Kaur, A., 2012. Biosensors activties around the globe digest. *Journal of Nanomaterials and Biostructures*, 7(4): 1457–1471.

15. Luong John, H.T., Male Keith, B. and Glennon Jeremy, D., 2008. Biosensor technology: Technology push versus market pull. *Biotechnology Advances*, 26: 492–500.

16. Amit, Singh, Somayyeh, Poshtiban and Stephane, Evoy, 2013. Recent advances in bacteriophage based biosensors for food-borne pathogen detection. *Sensors* 13: 1763–1786.

17. Gooding, J. Justin, 2006, Biosensor technology for detecting biological warfare agents: Recent progress and future trends. *Analytica Chimica Acta*, 559: 137–151.

18. Prodromidis Mamas, I. and Karayannis Miltiades, I., 2002. Enzyme Based Amperometric biosensors for food analysis. *Electroanalysis*, 14(4): 241–261.

19. Wang, Hong, Jiang, Xue, Jian, Wang, Zhao, Fu and Chen, Ling, 2008, Toxicity assessment of heavy metals and organic compounds using CellSense biosensor with *E.coli*. *Chinese Chemical Letters*, 19: 211–214.

20. Verma, N., Kumar, S. and Kaur, H., 2010. Fiber optic biosensor for the detection of Cd in milk. *J. Biosens. Bioelectron*, 1: 102. doi: 10.4172/2155–6210.1000102.

21. Verma, Neelam, Kumar, Sachin, and Kaur, Hardeep, 2011. Whole cell based disposable biosensor for Cadmium detection in milk. *Advances in Applied Science Research*, 2 (6): 354–363.

22. Verma, Neelam, Kumar, Rajiv and Kumar, Minhas Sachin, 2012, Simple, qualitative cum quantitative, user friendly biosensor for analysis of urea. *Advances in Applied Science Research*, 3(1): 135–141.

Biotechnology: An Overview (2015)
Editors: Rajan Kumar Gupta, Nasim Akhtar and Deepak Vyas
Published by: DAYA PUBLISHING HOUSE, NEW DELHI

Pages 387–398

Chapter 24

Biochemical and Molecular Mechanisms that Decipher Drought Stress Tolerance in Plants

D. Easwar Rao, K. Divya, G. Venkata Ramana
and K.V. Chaitanya*

Department of Biotechnology, GITAM Institute of Technology,
GITAM University, Viskhapatnam – 530 045

ABSTRACT

Drought stress is a major environmental factor which shows adverse affects on plant growth and productivity and is considered as a major threat for the crop production under sustainable conditions. Drought stress triggers a plethora of responses in plants resulting severe alterations in physiological, biochemical and molecular processes. Details of the drought stress-induced biochemical and molecular responses in plants with special emphasis to the enhancement of secondary metabolites in plants are discussed in this chapter, considering the progress made in identifying the drought stress tolerance mechanisms with enhanced tolerance to drought stress by over expression of the enzymes, hormones and transcription factors.

Introduction

Drought Stress in plants is an abiotic stress caused by the changes in the environment with temperature fluctuations and with less rainfall leading to the reduction of available water in the soil and prolonged loss of water leads to the condition of desiccation either by transpiration or evaporation of water [1, 2]. Plants experience a problem of stress due to various factors like Drought, salt stress, Heavy metals, etc, Which ultimately affects the growth of plant by decreased photosynthetic activity, low respiration rates, altered metabolisms and uptake of ions and finally leads to the death of plants under severe conditions [4]. Being sessile, plants are

---

* *Corresponding Author:* E-mail: viswanatha.chaitanya@gmail.com

frequently exposed to the environmental stresses like drought, which triggers a plethora of responses in plants resulting severe alterations in physiological, biochemical and molecular processes. Plants experience drought stress for two major reasons, firstly when the water supply to the roots becomes deficit and secondly, when the transpiration rates are increased due to rise in atmospheric temperature. These conditions mostly prevail under arid and semi-arid climates. In order to overcome the low levels of water availability, plants adapt stress avoidance and stress tolerance mechanisms [5]. Plants undergo various morphological changes like increased development of root hairs, deepening of roots, rolling of leaves to physiological adaptations such as alterations in carbon partitioning and isotope discrimination, osmotic adjustment and decreased efficiency of photosynthesis [6]. Drought causes of ABA levels in plant leaves, triggering major changes in gene expression leading to physiological responses [7]. Root-to-shoot biomass ratio is governed by a functional balance between water uptake by the root and photosynthesis by the shoot. Stomatal closure is considered as a main line of defence against drought. In response to phytohormone – Abscisic acid, the uptake and loss of water is regulated in guard cells resulting in stomatal closure by hydropassive and hydroactive closure mechanisms. Drought stress decreases the rate of photosynthesis by affecting chlorophyll components by damaging the photosynthetic apparatus and reduction in the activity of the Calvin cycle enzymes [8].

Antioxidant Defence Mechanisms Under Drought Stress

In plants photosynthesis acts as a well-established source of Reactive Oxygen Species (ROS), where the photosynthetic electron transport chain (PET) functions in aerobic environment [9]. Molecular oxygen has been introduced into the environment approximately 2.7 billion years ago by the photosynthetic organisms which evolve oxygen along with which ROS also became a component of aerobic life [10, 11]. Oxygen molecule with two impaired electrons possessing same quantum number is a free radical. This characteristic of oxygen persuades it to accept one electron at a time which generates ROS. In the daily metabolism, production of ROS will takes place continuously from cell components like mitochondria, chloroplast and peroxisomes [12]. Photosynthesis takes place in chloroplast for higher plants and algae where a well-organized thylakoid membrane system will group all the photosynthetic apparatus which captures light. During the process of photosynthesis oxygen which is generated in the chloroplasts, will accept electrons while passing through the photosystem, ultimately leading to the formation of ROS. Inspite of these hostilities, plants will survive the toxicity imposed by the abiotic stress-induced ROS by the help of antioxidative defence mechanism comprising of enzymatic and non-enzymatic antioxidants [13].

During the drought stress, plant will exhibit methodologies to counteract the hostilities at the molecular level apart from senescence and rolling of leaves. Enzymatic anti oxidants such as Superoxide dismutase (SOD), Ascorbate peroxidase (APX), Glutathione peroxidase (GPX), Glutathione S Transferase (GST) and Catalase (CAT) will have very good ROS scavenging abilities, which act as a strong defence against the damage of oxidation. Along with enhancement of these antioxidant enzymes, plant tolerance against the stress will also improve the cellular compartments substantiating the cellular survival of plants [14]. Scavenging of O_2^- will be done by the SOD which will catalyse the dismutation of superoxide radical to H_2O_2. SOD enzyme is present in all sub cellular components and in all aerobic organisms where the risk of oxidation is more [15]. SOD is classified into four types basing on their metal cofactors as Mn SOD, Fe SOD, Cu/Zn SOD and Ni SOD. Cellular membranes will be protected by the GPX under drought stress which catalyses hyperoxides using GSH. GPX can also react with H_2O_2 but this is a bit slow process. Both Catalase and Peroxidases will regulate intracellular H_2O_2 were Catalase work effectively via catalase-H_2O_2 complex which is Compound I. This will produce dioxygen and water or it will decay as Compound II which is inactive [16].

Non Enzymatic low molecular metabolites such as ascorbic acid, tocopherol, carotenoids etc., also has a prominent role in conservation of plants under drought stress [17]. Ascorbic acid is found to be most powerful non enzymatic antioxidant. Apart from the apoplast it was also detected in the plant cell organells. In aqueous phase AA is found to be the main ROS detoxifying compound since its caliber to donate the electrons is high in both enzymatic and non enzymatic reactions [18]. It can scavenge most of the ROS forms such as superoxide, hydroxyl radicals, singlet oxygen and reduced H_2O_2 to water by ascorbate peroxidase reaction. Tocopherols were infamous for their enzymatic and non enzymatic functions during the drought stress conditions. Vitamin E is an antioxidant with the chain breaking function which can repair the oxidizing radicals directly. In addition to this tocopherols act as scavengers of singlet oxygen by charge transfer mechanism which evolved them as chemical scavengers of singlet oxygen [19].

Role of Secondary Metabolites in Enhancing the Drought Stress Tolerance

Secondary metabolites are the bioactive compounds synthesized by the process of metabolism from primary metabolic compounds that gets accumulated in plants and plays a major role in providing protection by overcoming stress with defence mechanisms against abiotic stress like drought [20]. Response of plants to drought stress varies from species to other depending upon the degree of intensity, period of exposure and production of secondary metabolites [21]. Secondary metabolites are produced as a response to stress, which can be derived from various parts of plants like leaves, bark, stem, root etc. possessing anti-oxidant properties, which can helps in preventing cancer, diabetes and heart diseases in humans [22]

Secondary metabolites are produced by plants under the control of genes, which helps in the growth and development of plants without supporting the survival rate of plants. Their main role involves in the adaptation of plants to different climatic changes in the environment and also protection against herbivores, pathogens, dispersal of seeds, as well as aids in the protection against various abiotic stress factors caused with the exposure of plants to water limited conditions [23]. Plant secondary metabolites conducts electrostatic interactions between polyamines which carry out positive charge and macromolecule loci possessing negative charge resulting in cell structure stabilization. Structural and functional integrity in plants is brought about by the interactions between polyamines with phosphoric acid residues of DNA and uronic acid residues of cell wall matrix [21]. The properties of plant secondary metabolites against various abiotic stress factors are due to the interactions of quinoid groups with microbial proteins and plant proteins [24]. During photosynthesis, plant tissues will be protected from photoinhibition by secondary metabolites with the presence of conjugated double bonds which acts as UV protectors [25].

In plants, classification of secondary metabolites depends on the chemical structure, composition, solubility, and the pathways from which they are synthesized. Terpenes, phenolics and Alkaloids are three main groups of plant secondary metabolites, which are Aromatic, hydro aromatic, aliphatic and heterocyclic. Phenolic acids, phenolic alcohols, unsaturated aromatic hydrocarbons belongs to the class of aromatic secondary metabolites. Polyamines, ethylene, isoprene comes under the aliphatic type; whereas jasmonic acid under hydro aromatic class and indole derivatives, flavonoids comes under the class of hydrophilic secondary metabolites [26]. Plant phenols also called as phenolics are white crystalline compounds derived from benzene containing aromatic ring with one or more hydroxyl groups, also called as carbolic acid. Based on the number of phenolic units present within the molecule phenols are classified as simple phenols or polyphenols which includes phenolic acids, stilbenes, anthocyanins, flavonoids, tannins, phytoalexins, and furanocoumarins[27]. Among them flavonoids are the most abundant phenols containing a three-ring structure at the center with 15 carbon atoms synthesized by the phenylpropanoid pathway, which are used to provide protection against free radicals that are produced during the process of photosynthesis under drought stress [28]. Flavonoids also called as Vitamin P or citrin found mostly in plants producing yellow and other colored pigments. The backbones of flavonoids are called chalcones, aromatic ketones with two phenyl rings produced by the combination of Malonyl CoA and 4-coumaryol CoA with the help of amino acid phenylalanine. The subclasses of flavonoids include isoflavones, aurones, dihydrocharcones, flavans, flavanols, flavanones, flavanolols, anthocyanidins, catechins, and leucoanthocyanidins possessing anti-oxidative properties [29]. Phenolic compounds are synthesized by plants as a response caused due to stress or pressure from pathogens, insects, Drought or salt stress conditions, photosynthetic stress, inter-specific competition [30]. As these plant phenolic components rich in anti-microbial, anti-oxidant and anti-septic properties they are widely used in the preparations of ointments, disinfectants and bio-pesticides. Terpenes are hydrocarbons with isoprene precursor units, largest group of plant secondary metabolites and are classified as Monoterpenes, Diterpenoids, Triterpenes, tetraterpenes and polyterpenoids depending on the number of precursor units present [31]. Ascorbic acid comes under the group of triterpenoids. Carotenoids and xanthophylls under the group tetraterpenoids and Rubber included under the group of polyterpenoids with complex isoprene precursor units [32].

Alkaloids are water soluble nitrogen or sulphur containing plant secondary metabolite compounds with one or two nitrogen, carbon, hydrogen and oxygen atoms occur as N-oxides derived from various amino acids such as tyrosine, tryptophan, lysine and aspartate[33]. Morphine, nicotine, cocaine and caffeine are some of the examples of alkaloids which are bitter in taste found in plants

Metabolisms Involved in the Synthesis of Plant Secondary Metabolites

Secondary plant metabolites are often used in the cell signalling pathways and in the regulation of primary metabolic pathways. Metabolic pathways involves series of reactions catalyzed by enzymes synthesizing specific products nothing but a substrate or metabolite. The pathways that are responsible for the synthesis of secondary plant metabolites includes

1. Pentose phosphate pathway
2. Shikimic acid pathway
3. Malonyl coA pathway and
4. Mevalonate pathway.

Pentose phosphate pathway also called as hexose monophosphate pathway or phosphogluconate pathway. The end-products produced by this pathway includes NADPH, Ribose-5-phosphate, fructose 6-phosphate and glyceraldehyde 3-phosphate which gives rise to Glycosides.HMP shunt considered as an alternate pathway for the oxidation of glucose and is anabolic. Oxidative and Non-oxidative are the two phases of the pathway [34]. Shikimic Acid pathway also called as Chorismic acid pathway from which most of the aromatic compounds namely phenylalanine, tyrosine and tryptophan found in plants are derived by the conversion of carbohydrate precursors synthesized by the glycolysis and pentose phosphate pathways with the formation of intermediate metabolite chorismate. These aromatic amino acids not only serve as precursors for secondary metabolite synthesis but also used for the growth of plants and for the synthesis of plant hormones such as salicylate and Auxins. Alkaloids, Quinones, Lignins, Phenols, Coumarins, Tannins, Flavones, and Vanillin which are plant secondary metabolites are obtained by this pathway. The aromatic amino acids phenyl alanine, tyrosine and tryptophan act as precursors for the synthesis of various plant secondary metabolites via Shikimic acid pathway. Metabolites namely salicylate, auxins and various other pigments are synthesized by this pathway. Alkaloids, plastoquinones and some phenylpropanoids are synthesized with the help of precursor tryptophan. From the precursor phenyl alanine, phenylpropanoids, glucosinolates and 2-phenylethanol are synthesized and from the other aromatic amino acid tryptophan, secondary metabolites namely glucosinolates, alkaloids and camalexin compounds can be produced [35]. Malonyl CoA Pathway also called as long chain acyl CoA pathway or polyketide pathway. The main regulatory step in the pathway of fatty acid synthesis is the conversion of acetyl Co A to malonyl CoA by the enzyme acetyl CoA carboxylase. Plant secondary metabolites such as fatty acids, tetracyclines, anthraquinones, flavonoids, isoflavonoids, terpenoids are derived by this pathway utilizing malonyl CoA as a precursor. [36]. Mevalonate pathway also called as HMG CoA reductase pathway or isoprenoid pathway. Terpenes, terpenoids, sterols, and steroids are produced in this pathway by an intermediate mevalonic acid and the key enzyme mevalonate kinase [37].

Up Regulation of Secondary Metabolites during Drought Stress

Drought stress in plants is brought about by loss of available water in tissues. Plants adopt themselves to different climatic changes by overcoming stress is brought about by the presence of secondary metabolites in plants. The abiotic stress factors include temperature, salinity, drought, radiation, chemical stress and mechanical stress shows more or less impact on the growth of plants and the production of secondary metabolites. Plants exposed to drought stress carry out several reactions in common. The level of phenolic compounds concentration increases with the lack of potassium, sulfur and magnesium. Presence of lower levels of iron also causes increase in the level of phenolic compounds in plants. Phenylpropanoids gets accumulated due to the deficiency of nitrogen and phosphate. Formation of plant secondary metabolites and biomass production occurs with the exchange of carbon compounds when the plants were stressed. Accumulation or the deposition of anthocyanin pigments are also seen in drought stressed plants . In response to several Abiotic stress factors, secondary metabolites are found to play a major role in defence mechanisms thereby providing protection to plants. The level of secondary metabolites likes flavonoids, phenolic compounds, carotenoids, terpenes, and anthocyanins increases with the drought stress in plants. Saponins are the class of plant secondary metabolites which occur in roots, leaves, stems, fruits and in flowers get accumulated mostly in the reproductive organs of plants in response to drought Stress. Secondary plant metabolites can be produced in higher concentrations with the help of invitro plant cell culture techniques and with the use of genetic engineering technology by regulating the genes involved in the synthesis of bioactive compounds of plants which will be an additional advantage in agriculture for their medicinal and aromatic properties.

Molecular Responses of Drought Stress in Plants

The molecular mechanisms that sense environmental stress like drought, consists of a number of classes of cell surface receptors such as receptor-like kinases (RLKs), ion channel–linked receptors, G-protein-coupled receptors (GPCRs) and two-component histidine kinase receptors. These receptors initiate a cascade responds to transmit the information through signal-transducing pathway. The signal transduction cascade involves protein phosphorylation and dephosphorylation mediated by several protein kinases and phosphatases whose genes are upregulated during drought stress. The increase in cytoplasmic Ca^{2+} concentration mediates the integration of different signalling pathways. The most abundant regulatory protein kinases that mediate signalling during drought stress include Ca^{2+} dependent protein kinases (CDPK) and mitogen activated protein kinases (MAPK). These protein kinases transfer the dehydration signals from plasma membrane to the nucleus. Similarly, in tobacco cells, a SIMK-like MAP kinase named SIPK (salicylic acid-induced protein kinase) is activated by hyperosmotic stress resulting in the accumulation of osmolytes that helps re-establish the osmotic balance, protection from stress damage or repair mechanisms by induction of LEA/dehydrin-type stress genes. The translocation of the MAPK into the nucleus brings about the activation of transcription factors through phosphorylation [38]. Phospholipids also generates signalling molecules like inositol 1,4,5-trisphosphate (IP3), diacylglycerol (DAG) and PA, which play a role in the transmission of the signal across plasma membrane and in intracellular signalling that results in the activation of Transcription Factors (TFs) that help in drought tolerance. These transcriptional factors include Dehydration Responsive Transcription Factors (DREBs) and c-Repeat Binding Factors (CBFa) that respectively bind to the Dehydration Response Element (DRE) and c-repeat terminal (CRT) *cis* acting elements, ethylene responsible element binding factor (ERF), zinc-finger family, WRKY family, basic helix-loop-helix (bHLH) family, basic-domain leucine zipper (bZIP) family, NAC family and homeodomain transcription family. Drought stress– induced gene expression was seen to be regulated by TFs belonging to bZIP, AP2/ERF, HD-ZIP, MYB, bHLH, NAC, NF-Y, EAR and ZPT2 families [39].

Drought responsive genes are divided into two groups, ABA-dependent and ABA-independent genes, according to their dependency on ABA for induction [40] (Figure 24.1). The products of drought-inducible genes identified through microarray analysis in *Arabidopsis* are classified into two groups. The first group includes proteins-LEA, osmotin, chaperones, antifreeze proteins, m-RNA binding proteins, water channel proteins, sugar and proline transporters, detoxification enzymes and other proteases. The second group includes regulatory proteins- proteins that regulate signal transduction, stress-responsive gene expression, transcription factors, protein kinases, protein phosphatases, enzymes of phospholipid metabolism, and signalling molecules [41].

ABA-Independent Regulatory System

Drought stress results in the synthesis of ABA in shoots and roots of plants. ABA biosynthesis can be assumed to start at the epoxidation of zeaxanthinepoxidase (ZEP) to form epoxyxanthophyll precursor [42]. ABA synthesis in roots and leaves is catalysed by 9-cis-epoxycarotenoid dioxygenase (NCED), an enzyme that converts the epoxycarotenoid precursor to xanthoxin in plastid. The produced xanthoxin is then converted to ABA by cytosolic enzymes. Abscisic aldehyde oxidase is the final enzyme that catalyses the ABA synthesis. Overexpression of the gene encoding 9-cisepoxycarotenoiddioxygenase (NCED) improves drought stress tolerance in transgenic Arabidopsis plants [43]. The ABA-independent expression of drought inducible genes consists of a *cis*-element DRE/CRT, which is characterised with the promoter region of RD29A gene in Arabidopsis [40]. As the plants sense environmental stress the *cis*-element trans-factors CBF/DREB1 and DREB2 are expressed which help in the up regulation of the target genes involved in drought tolerance. The ABA-activated SnRK2 protein kinase (OST1/SRK2E) functions in the ABA signal transduction pathway controlling stomatal closure. The ERD1 gene is also induced by drought and is also up-regulated during natural senescence and dark-induced senescence.

ABA-Dependent Regulatory System

The major *cis*-acting element in ABA-responsive gene expression is ABRE. In *Arabidopsis*, RD29B gene consists two ABRE motifs that control in ABA-responsive

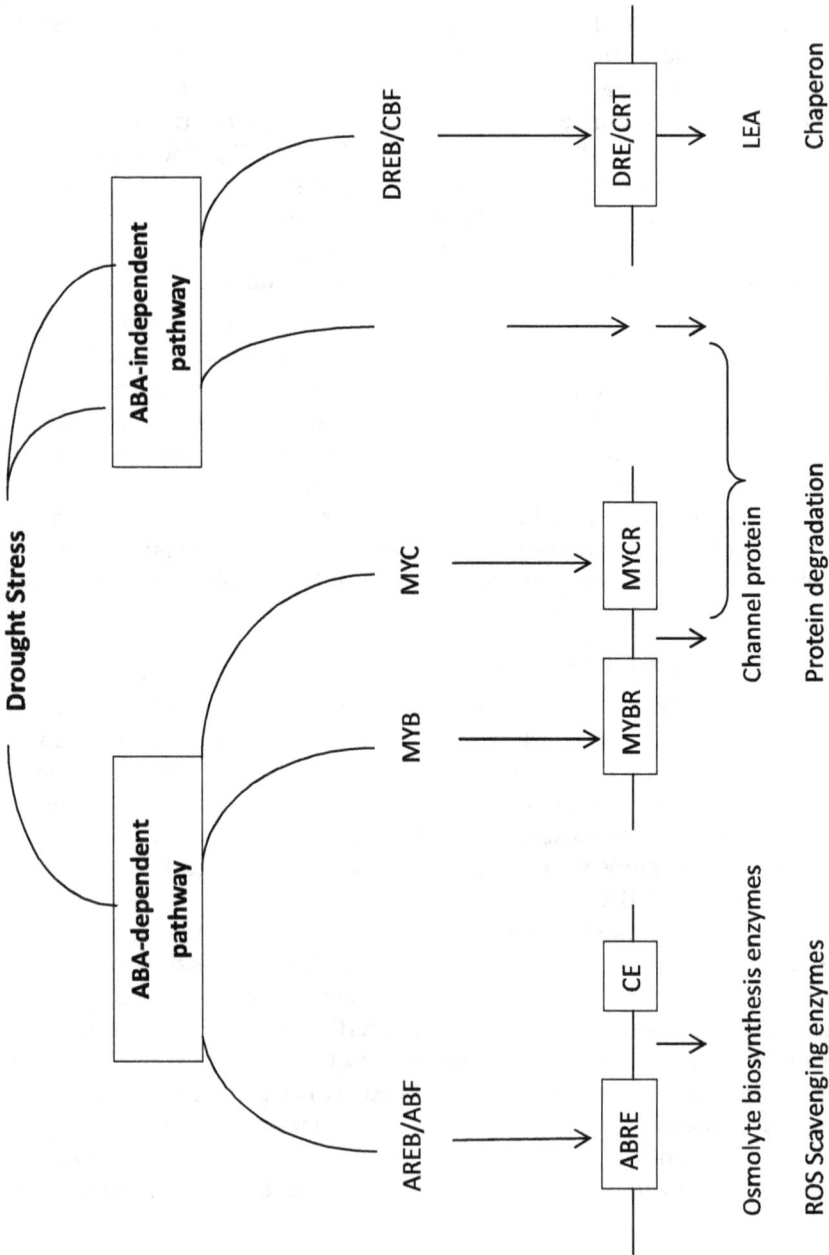

Figure 24.1. Transcriptional factors and *cis*-acting elements involved in drought stress responsive gene expression.
ABA: Abscisic acid; AREB: ABREbinding proteins;ABRE: ABA-responsive element;CE: Coupling element; MYBR: MYB recognition site; MYC: MYC recognition site; DREB: DRE-binding protein;CBF: Crepeat-binding factor; DRE/CRT: Dehydration-responsive element/C-repeat; LEA: Late embryogenesis abundant protein.

gene expression. Basic leucine zipper (bZIP) transcription factors like AREB/ABF, can bind to ABRE and activate ABA dependent gene expression. The ABA-mediated signal required for the activation of these AREB/ABF proteins as indicated by their reduced activity in the ABA-deficient *aba2* and ABA-insensitive *abi1* mutants and their enhanced activity in the ABA hypersensitive *era1* mutant of Arabidopsis. Induction of the drought-inducible RD22 gene is mediated by ABA and requires protein biosynthesis for its ABA-dependent expression. A MYC transcription factor, AtMYC2 (RD22BP1) and a MYB transcription factor, AtMYB2 were shown to bind *cis*-elements in the RD22 promoter and co-operatively activate RD22. The accumulation of endogenous ABA results in the synthesis of MYC and MYB proteins which play a major role in plants stress responses. Overexpression of both *AtMYC2* and *AtMYB2* not only resulted in an ABA-hypersensitive phenotype but also improved drought stress tolerance of the transgenic plants. RD26 gene is a drought inducible gene with a NAC transcription factor, the expression of which is induced by drought, high salinity, ABA and JA treatments. It was observed that ABA and stress-inducible genes were up-regulated in the RD26-overexpressing transgenics and repressed in the RD26 repressor lines.

Acknowledgements

Research lab of K.V.Chaitanya is funded by DBT, DST and UGC, Government of India. All authors acknowledge the agencies for the Junior Research Fellowships.

References

1. Jaleel, C.A., Manivannan, P., Wahid, A., Farooq, M. and Al-Juburi, H.J., Somasundaram, R. and Rajam, P. 2009. Drought stress in plants: A review on morphological characteristics and pigments composition. *Int. J. Agric. Biol.*, 11: 100–105.

2. Daniel, A., Jacob-Velazquez and Luis, C.-Z., 2012. An alternative use of horticultural crops: Stressed plants as biofactories of bioactive phenolic compounds. *Agriculture*, 2: 259–271.

3. Beatriz, X.-C., Francisco, A.R.-O., Leonardo, F.-E. and Roberto, R.-M., 2011. Drought tolerance in crop plants. *American Journal of Plant Physiology*, 5: 241–256.

4. Julia, K. and Claudia, J., 2012. Drought, salt and temperature stress: Induced metabolic rearrangements and regulatory networks. *Journal of Experimental Botany*, p. 1–16 doi: 10.1093.

5. Lawlor, D.W., 2013. Genetic engineering to improve plant performance under drought: Physiological evaluation of achievements, limitations, and possibilities. *J. Exp. Bot.*, 64: 83–108.

6. Raghavendra, A.S., Gonugunta, V.K., Christmann, A. and Grill, E., 2010. ABA perception and signalling. *Trends Plant Sci.*, 15: 395–401.

7. Kawamitsu, Y., Driscoll, T. and Boyer, J.S., 2000. Photosynthesis during desiccation in an intertidal alga and a land plant. *Plant Cell Physiol*, 41: 344–353.

8. Wang, W.-H., Yi, X.-Q., Han, A.-D., Liu, T.-W., Chen, J., Wu, F.-H., Dong, X.-J., He, J.-X., Pei, Z.-M. and Zheng, H.-L., 2012. Calcium-sensing receptor regulates stomatal closure through hydrogen peroxide and nitric oxide in response to extracellular calcium in Arabidopsis. *J. Exp. Bot.*, 63: 177–190.

9. Foyer, C.H. and Shigeru, S., 2011. Understanding oxidative stress and antioxidant functions to enhance photosynthesis. *Plant Physiology*, 155: 93–100.

10. Gill, S.S. and Tuteja, N., 2010. Reactive oxygen species and antioxidant machinery in abiotic stress tolerance in crop plants. *Plant Physiology and Biochemistry*, 48: 909-930.

11. Halliwell, B., 2006. Reactive species and antioxidants. redox biology is a fundamental theme of aerobic life. *Plant Physiol*, 141: 312-322.

12. Navrot, N., Rouhier, N., Gelhaye, E. and Jaquot, J.P., 2007. Reactive oxygen species generation and antioxidant systems in plant mitochondria. *Physiol. Plant*, 129: 185-195.

13. Foyer, C.H. and Noctor, G., 2005. Redox homeostis and antioxidant signaling: a metabolic interface between stress perception and physiological responses. *Plant Cell*, 17: 1866-1875.

14. Del Rio, L.A., Sandalio, L.M., Corpas, F.J., Palma, J.M. and Barroso, J.B., 2006. Reactive oxygen species and reactive nitrogen species in peroxisomes. Production, scavenging, and role in cell signalling. *Plant Physiol*, 141: 330-335.

15. Blokhina, O., Eija, V. and Kurt, V.F., 2003. Antioxidants, oxidative damage and oxygen deprivation stress: A review. *Anals of Botany*, 91: 179–194.

16. Paravaiz, A., Cheruth, A.J., Azooz, M.M. and Gowher, N., 2009. Generation of ROS and non enzymatic antioxidants during abiotic stress in plants. *Botany Research International*, 2: 11–20.

17. Bhattachrjee, S., 2005. Reactive oxygen species and oxidative burst: roles in stress, senescence and signal transduction in plant. *Curr. Sci.*, 89: 1113-1121.

18. Ahmad, P., Sarwat, M. and Sharma, S., 2008. Reactive oxygen species, antioxidants and signaling in plants. *J. Plant Biol.* 51: 167–173.

19. Arora, A., Sairam, R.K. and Srivastava, G.C., 2002. Oxidative stress and antioxidative systems in plants. *Current Science*, 82: 1227–1239.

20. Ramakrishna, A. and Ravishankar, G.A., 2011. Influence of abiotic stress signals on secondary metabolites in plants. *Plant Signal Behav.*, 6: 1720–1731.

21. Edreva, A., Velikova, V., Tsonev, T., Dagnon, S., Gurel, A., Aktas, L. and Gesheva, E., 2008. Stress protective role of secondary metabolites: Diversity of functions and mechanisms. *Gen. Appl. Plant Physiol.*, 34: 67–78.

22. Jaafar, H.Z., Ibrahim, M.H. and Mohamad Fakri, N.F., 2012. Impact of soil field water capacity on secondary metabolites, phenylalanine ammonia–lyase (PAL), maliondialdehyde (MDA) and photosynthetic responses of Malaysian Kacip Fatimah (*Labisia pumila* Benth). *Molecules*, 17: 7305–22.

23. David, O.K. and Emma, L.W., 2011. Herbal extracts and phytochemicals: Plant secondary metabolites and the enhancement of Human brain function. *Adv. Nutr.*, 2: 32–50.

24. Vincenzo, L., Veronica, M.T.L., and Angela, C., 2006. Role of phenolics in the resistance mechanisms of plants against fungal pathogens and insects. *Phytochemistry: Advances in Research*, p. 23–67.

25. Michalak, A., 2006. Phenolic compounds and their anti-oxidant activity in plants growing under heavy metal stress. *Polish J. of Environ. Stud.*, 15: 523–530.

26. Jyothi, N., Shashank, B., Suresh, D. and Nupur, J., 2013. Phytochemical screening of secondary metabolites of *Dhatura stramonium* L. *Int. J. Curr. Pharm. Res.*, 5: 151–153.

27. Khoddami, A., Wilkes, M.A. and Roberts, T.H., 2013. Techniques for analysis of plant phenolic compounds. *Molecules*, 19: 2328–75.

28. Brunetti, C., Ferdinando, M.D., Fini, A., Pollastri, S. and Tattini, M., 2013. Flavonoids as antioxidants and developmental regulators: Relative significance in plants and humans. *Int. J. Mol. Sci.*, 14: 3540–3555.

29. Amina, Abd and El-Hamid, A.L.Y., 2010. Biosynthesis of phenolic compounds and water soluble vitamins in CULANTRO Plantlets as affected by low doses of gamma radiation. *Tom.*, 17(2): 356–361.

30. Gun-Ae, Y., Kyung-Jin, Y. and Yang Cha, L.-K., 2012. Carotenoids and total phenol contents in plant foods commonly consumed in Korea. *Nutr. Res. Pract.*, 6: 481–490.

31. Sam, Z. and Chhandak, B., 2008. Plant terpenoids: Applications and future potentials. *Biotechnology and Molecular Biology Reviews*, 3: 1–7.

32. Adeyemi, M.M.H., 2011. A review of secondary metabolites from plant materials for post harvest storage. *International Journal of Pure and Applied Sciences and Technology*, 6: 94–102.

33. Toni, M.K., 1995. Alkaloid biosynthesis: The basis for metabolic engineering of medicinal plants. *The Plant Cell*, 7: 1059–1070.

34. Asaph, A. and Gad, G., 2011. Metabolic engineering of the plant primary-secondary metabolism interface. *Curr. Opin. Biotechnol.*, 22: 239–44

35. Vered, T. and Gad, G., 2010. New insights into the shikimate and aromatic amino acids biosynthesis pathways in plants. *Molecular Plant*, 3: 956–972.

36. Klaus, M.H., 1995. The shikimate pathway: Early steps in the biosynthesis of aromatic compounds. *The Plant Cell*, 7: 907–919.

37. Hui, C., Hyun, U.K. and John, B., 2011. Malonyl CoA synthetase, encoded by acyl activating enzyme 13, is essential for growth and development of Arabidopsis. *Plant Cell*, 23: 2247–2262.

38. Taylor, I.B., Burbidge, A. and Thompson, A.J., 2000. Control of abscisic acid synthesis. *Journal of Experimental Botany*, 51: 1563–1575.

39. Iuchi, S., Kobayshi, M., Taji, T., Naramoto, M., Seki, M., Kato, T., Tabata, S., Kakubari, Y., Yamaguchi-Shinozaki, K. and Shinozaki, K., 2001. Regulation of drought tolerance by gene manipulation of 9-cis-epoxycarotenoid, a key enzyme in abscisic acid biosynthesis in Arabidopsis. *The Plant Journal*, 27: 325–333.

40. Doubuzet, J., Sakuma, Y., Kasuga, M., Doubouzet, E., Miura, S., Seki, M., Shinozaki, K. and Yamaguchi Shinozaki, K., 2003. OsDREB genes in rice, *Oryza sativa L.*, encode transcription factors that function in drought, high salt and cold responsive gene expression. *Plant J.*, 33: 751–763.

41. Yoshida, R., Hobo, T., Ichimura, K., Mizoguchi, T., Takahashi, F., Alonso, J., Ecker, J.R. and Shinozaki, K., 2002. ABA-activated SnRK2 protein kinase is required for dehydration stress signaling in Arabidopsis. *Plant Cell Physiology*, 43: 1473–1483.

42. Uno, Y., Furihata, T., Abe, H., Yoshida, R., Shinozaki, K. and Yamaguchi-Shinozaki, K., 2000. Arabidopsis basic leucine zipper transcriptional transcription factors involved in an abscisic acid-dependent signal transduction pathway under drought and high-salinity conditions. *Proceedings of the National Academy of Sciences*, 97: 11632–11637.

43. Abe, H., Urao, T., Ito, T., Seki, M., Shinozaki, K. and Yamaguchi-Shinozaki, K. Arabidopsis AtMYC2 (bHLH) and AtMYB2 (MYB) function as transcriptional activators in abscisic acid signaling. *The Plant Cell*, 15: 63–78.

Biotechnology: An Overview (2015)
Editors: Rajan Kumar Gupta, Nasim Akhtar and Deepak Vyas
Published by: DAYA PUBLISHING HOUSE, NEW DELHI

Pages 399–403

Chapter 25

Biotechnology for Discovery of New Drugs from Medicinal and Aromatic Plants

Nivedita Srivastava

Managing Director, Navjeevanam Kayakalp and Medical Research Centre,
Shiva Enclave, Rishikesh

ABSTRACT

Medicinal plants represent rich chemical diversity, which will continue to be important source of lead molecule for doing development. In recent year sensitive biological testing system coupled with biotechnology, automation and robotics. Application have been developed that will permit rapid screening of large number of samples. Development of Bio activity guided fractionation or use of receptors and chromatographic separation techniques and spectroscopy for structure determination will further add to existing knowledge of chemical diversity of the natural world.

Several thousands of plants are used as alternative medicine and is needed to evaluate their clinical efficacy on scientific protocols. It is therefore expected that plant will continue to be source of navel drug compounds and source of lead molecules for drug development.

Introduction

About 80 per cent if the World Population primarily in developing countries depend on traditional system of medicine for their primary health care needs. India has rich traditional system of medicine. Ayurveda, Unani, Homeopathy, Sidha and Amchi system are popular make extensive use of herbs in therapeutic treatments. The historical usage of traditional medicine of over 5000 years provides some level of confidence about its safety and efficacy. There are over 7000 manufacturing units in Indian system of medicine (ISM & H). About 5,64,000 ISM & H practitioners with an addition of 8000 new chemical equities for development of modern drugs, also use of natural products as dietary supplements (nutraceuticals), ingredients of food and beverages, phytocosmetics and other herbal products. Global market for herbal products which includes medicine, health supplement, herbal cosmeceuticals is around US $ 65 billion. According to world bank report, the international herbal medicine is expanded to reach us 5 trillion in 2050 with an annual growth rate of between 10 to 20 percent. According to Ayurvedic drug manufactures Association (ADMA) estimates the current value of Indian trade in ISM & H Medicine is about Rs. 4200 Crores (or US $ 1 billion).

Medicinal Plants as a Template for New Drug Development

Pure chemical constituents from plants can be used even in unmodified form as a source of useful drugs. Many drug substances of plant origin are new produced commercially through synthetic routes and those include : Caffeine, theophylline, theobromine, ephedrine, pseudoepedrine, emetin, papaverine, L-dopa, salicylic acid and 9- tetrahydrocannabinol. Apart from synthesis, in various cases plant constituents have general as important precursors for synthetic modifications. For example, diosgenin from Dioscorea spp and hecogenin from Agave sislana are used as starting material for manufacture or oral contraceptive and other steriods hormons.

Plant constituents and natural product are better choice as "templets" or "lead" molecules than synthetic chemicals because of their known biochemical or biological activity. They have also been utilized as chemical models or templetes for the drug designs and synthesis of many important drugs (Loche, 1977); (Assady and Douros, 1980); Gund *et al.*, 1980; De souza *et al.*, 1982; Swader, 1985; Baldwin, 1987; Midgley, 1988; Tyler *et al.*, 1988; Foye, 1989; Buss and Waigh, 1995. White willow bark (Salix alba) and certain flowers, meadow sweet (Spiraea ulmaria) were found to contain Salicin, which is metabolized in the human-intestine to salicylic acid. Salicyclic acid then become head molecule for synthesis of acety/salicylate drug under trade name "Aspirin" in year 1989.

Another opium alkaloides, papaverine (smoth muscle relaxant), which is now synthesized, led to the synthesis or the open chain analog varapamil. It is used to treat cardio vascular diseases and hypertension occasionally. The analog of the lead compound may exhibit unexpected prological activity.

A series of analgesics were discovered initially in attempt to obtain smoth muscle, relaxant based on tropane alkaloid, atropine (Midgley, 1988).

The application of bitechnology for conservation, characterization, Micropropagation, cell culture production of secondary metabolites isolation and characterization of novel. Bioactive agents, development of standardized and safe herbal formulation and genetic improvement of selected medicinal and aromatic plants.

Conservation and Characterization

A network of four national gene banks on medicinal and aromatic plants at TBGRI, Thiruvananthapuram; CIMAP, Lucknow; NBPGR, New Delhi and RRL, Jammu have been further strengthened. A total of about 8500 accessions of prioritized species are conserved in different forms such as in field bank, seed bank, in vitro repository, cryobank and DNA bank. Set up of germplasm bank for medicinal plants used in Ayurveda.

Micropropagation

☆ *In vitro* protocols have been developed for multiplication of selected medicinal and aromatic plants such as *Garcinia indica, Holarrhena antidysenterica, Lavendula officinalis, Pterocarpus marsupium, Chlorophytum borivilianum.*

☆ Field evaluation of the performance of tissue-culture raised elite varieties of large cardamom (*Amomum subulatum*) over a total area of 50 acres in Uttarakhand initiated in association with the Spices Board.

Cell-Culture Production of Therapeutic Agents

☆ Efforts initiated towards protocol development of production of important therapeutic agents through cell culture methods such podophyllotoxin from *Podophyllum hexandrum*, hyoscyamine from *Hyoscyamus muticus*, guggulsterones Z and E from *Commiphora wightii* and comptothecin from *Ophiorrhyza* spp.

☆ Four fast growing cell-lines of *P.hexandrum* capable of synthesizing podophyllotoxin devoid of a peltatins established.

☆ Cell suspension cultures raised from leaf and hairy root derived callus of *Hyoscyamus muticus* have been scaled-up in bioreactor (15-litre capacity) toward production of hyoscyamine.

Up-Scaling of Process for Extraction of Lead Therapeutic Compounds from Plants

☆ A process for extraction of 10-DAB has been scaled-up in pilot-plant with 100 kg fresh *Taxus Wallichiana* needles. An improved process has been scaled-up to 30 kg/batch raw material (twigs and stems of *Nothapodytes foetida*) for extraction of comptothecin. A pilot-scale process for isolation of silymarine from the seeds of *Silybum marianum* has been developed (40 kg seeds/batch level).

Cell-Based Screening System

☆ *In vitro* bioscreens developed for screening plant extracts having anti-cancer and anti-diabetic properties. Multi-institutional project implemented on using these bioscreens alongwith modern cell signal targets to identify anti-diabetic, anti-cancer and immunomodulatory agents from plants that have been used in Indian traditional system.

☆ Screening system has been developed for screening of extracts from medicinal plants (used for amoebiosis in Indian traditional system of medicine) against *Entamoeba histolytica* trophozytes. Lead extracts have been identified for further standardization and product development.

Isolation and Characterization of New Bioactives/Therapeutic Agents

☆ Under multi-institutional project, after bio-activity based *in vitro* screening of 60 medicinal plants (used in Indian traditional system of medicine), a total of 35 lead molecules identified so far.

☆ Anti-cancer - 15

☆ Anti - diabetic - 5

☆ Immunomodulatory - 15

☆ Two anti-cancer lead molecules (from *Aegle marmelos* and *Phyllanthus urinaria*) have been patented.

☆ A lead medicinal plant extract exhibiting promising osteotenic (bone forming activity) using several in vitro and in vivo test systems have been identified. The patent for the above is being filed.

☆ The active principle (saturated fatty acid) isolated from *Oxalis corniculata* showing significant anti-proliferative activity against *Entamoeba histolytica* has been further characterized.

☆ A lead fraction from *Piper nigrum* having anti-tubercular activity (effective against both sensitive and resistant strains of *Mycobacterium tuberculosis*) has been identified.

☆ Efforts are in progress for isolation and characterization of anti-cancer, anti-tubercular, anti-viral, hepatoprotective, and immunomodulatory agents from medicinal plants used in Indian traditional system of medicine.

Development of Standardized Herbal Formulation

☆ Based on leads already available, a collaborative project has been initiated to develop a standardized and safe herbal product from *Terminalia arjuna* for left ventricular dysfunction. Clinical trials is an integral component of this project. Animal models for left ventricular dysfunction have been successfully developed. Efficacy studies (ED50) with standardized aqueous extract of *T. arjuna* bank are in progress. Seven marker compounds have been isolated Acute toxicity studies of standardized aqueous extract of *T. arjuna* have been completed.

☆ Projects for developing standardized herbal products for hepatoproteciton, diabetes - type 2, amoebiosis and atherosclerosis are in progress.

☆ A multi-institutional network project has been recently initiated for developing a standardized herbal product for bovine mastitis.

Genomic Resources and Metabolic Pathways

☆ The capsaicin synthase enzyme (key regulatory enzyme for capsaicin biosynthesis) and its gene (csy 1) have been characterized from placental tissues of *Capsicum* sp. Patent has been filed for csy 1 gene.

☆ Five lines of *Catharanthus roseus* that hyperproduce serpentine and its product ajmalcine have been developed through DNA marker assisted pyramidation of the concerned gene loci.

☆ Projects have been recently initiated to develop ESTs data base, understand the biosynthetic pathway/regulatory genes involved in production of artemisinin in *Artemisia annua* : morphine alkaloids in *Papaver somniferum;* santalol in *Santalum album;* picrosides in *Picrorhyza kurrooa* and podophyllotoxin in *Podophyllum hexandrum*.

Acknowledgement

The author wish to thank Dr. R.K. Gupta, Department of Botany, Govt. P.G. College, Rishikesh and Dr. D.K. Srivastava, M.D., Gold medilist, Renowned Ayurvedic Physician Rishikesh.

Biotechnology: An Overview (2015)   *Pages 405–415*
Editors: **Rajan Kumar Gupta, Nasim Akhtar and Deepak Vyas**
Published by: **DAYA PUBLISHING HOUSE, NEW DELHI**

Chapter 26

Role of Nanotechnology in Materials Science

Vijendra Lingwal

Pt. L.M.S. Government P.G. College (An Autonomous College),
Rishikesh, Dehradun

ABSTRACT

Materials science is an interdisciplinary field involving the properties of matter and its applications to various areas of science and engineering. This science investigates the relationship between the structure of materials and their properties. It includes elements of applied physics and chemistry, as well as chemical, mechanical, civil and electrical engineering. The material of choice of a given era is often its defining point; the Stone Age, Bronze Age, and Steel Age are examples of this. Materials science is one of the oldest forms of engineering and applied science, deriving from the manufacture of ceramics. Modern materials science evolved directly from metallurgy, which itself evolved from mining. A major breakthrough in the understanding of materials occurred in the late 19th century, when Willard Gibbs demonstrated that thermodynamic properties relating to atomic structure in various phases are related to the physical properties of a material. Important elements of modern materials science are a product of the space race: the understanding and engineering of the metallic alloys, and silica and carbon materials, used in the construction of space vehicles enabling the exploration of space. Materials science has driven, and been driven by, the development of revolutionary technologies such as plastics, semiconductors, and biomaterials.

Before the 1960s (and in some cases decades after), many materials science departments were named *metallurgy* departments, from a 19th and early 20th century emphasis on metals. The field has since broadened to include every class of materials, including: ceramics, polymers, semiconductors, magnetic materials, medical implant materials and biological materials.

With significant media attention to nanoscience and nanotechnology in recent years, materials science has been propelled to the forefront at many universities. It is also an important part of forensic engineering and forensic materials engineering, the study of failed products and components.

In this chapter a brief introduction of materials and their importance has been given. The field of nanotechnology is introduced with historical background, common nano-materials, assemblers, few important applications, and lastly conclude with visions of good and visions of harm.

Fundamentals of Materials Science

In materials science, rather than haphazardly looking for and discovering materials and exploiting their properties, one instead aims to understand materials fundamentally so that new materials with the desired properties can be created.

The basis of all materials science involves relating the desired properties and relative performance of a material in a certain application to the structure of the atoms and phases in that material through characterization. The major determinants of the structure of a material and thus of its properties are its constituent chemical elements and the way in which it has been processed into its final form. These, taken together and related through the laws of thermodynamics, govern a material's microstructure, and thus its properties.

An old adage in materials science says: *"materials are like people; it is the defects that make them interesting"*. The manufacture of a perfect crystal of a material is currently physically impossible. Instead materials scientists manipulate the defects in crystalline materials such as precipitates, grain boundaries, interstitial atoms, vacancies or substitutional atoms, to create materials with the desired properties.

In addition to industrial interest, materials science has gradually developed into a field which provides tests for condensed matter or solid state theories. New physics emerge because of the diverse new material properties which need to be explained.

On the basis of atomic arrangement materials are divided broadly into two category:

Crystalline

In chemistry, mineralogy, and materials science, a *crystal* is a solid in which the constituent atoms, molecules, or ions are packed in a regularly ordered, repeating pattern extending in all three spatial dimensions and has a definite melting point. The word *crystal* is a loan from the ancient Greek word (*krustallos*), which had the same meaning, but according to the ancient understanding of crystal. At root it means anything congealed by freezing, such as *ice*. The word once referred particularly to quartz, or "rock crystal". If the regalarity of the atoms break at some extent and than again shows regular arrangement to other direction it is called polycrystalline materials. Most metals encountered in everyday life are polycrystals. The process of forming a crystalline structure from a fluid or from materials dissolved in the fluid is often referred to as *crystallization*. In the ancient example referenced by the root meaning of the word crystal, water being cooled undergoes a phase change from liquid to solid beginning with small ice crystals that grow until they fuse, forming a polycrystalline structure. The physical properties of the ice depend on the size and arrangement of the individual crystals, or grains, and the same may be said of metals solidifying from a molten state. Again crystals are devided into seven system and subsystem (Figure 26.1).

Glassy or Amorphous

In the common sense refers to a hard, brittle, transparent super-cool liquid that appears solid, having no definite melting temperature, such as used for windows,

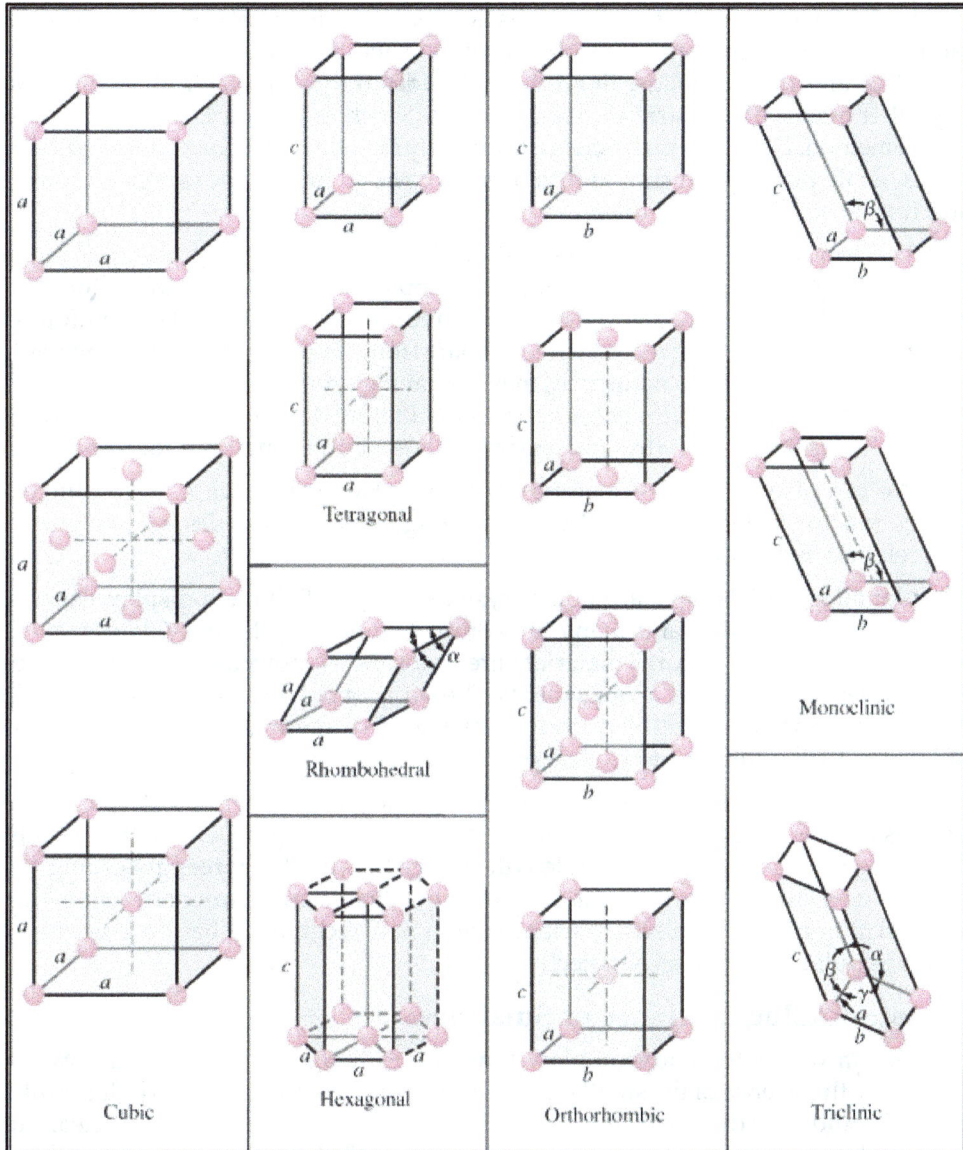

Figure 26.1: Different crystal system.

many bottles, or eyewear, including soda-lime glass, acrylic glass, sugar glass, isinglass (Muscovy-glass), or aluminium oxynitride. It is said to be a liquid due to it's arrangement of molecules, wich are in a rigid but random arrangement. In the technical sense, glass is an inorganic product of fusion which has been cooled to a rigid condition without crystallizing. Many glasses contain silica as their main component and glass former.

In the scientific sense the term glass is often extended to all amorphous solids (and melts that easily form amorphous solids), including plastics, resins, or other silica-free amorphous solids. In addition, besides traditional melting techniques, any other means of preparation are considered, such as ion implantation, and the sol-gel method. However, glass science commonly includes only inorganic amorphous solids, while plastics and similar organics are covered by polymer science, biology and further scientific disciplines.

The optical and physical properties of glass make it suitable for applications such as flat glass, container glass, optics and optoelectronics material, laboratory equipment, thermal insulator (glass wool), reinforcement fiber (glass-reinforced plastic, glass fiber reinforced concrete). Apart from random atomic arrangement, glasses are symmetrical in nature, i.e., they are having same properties in all direction whereas crystals shows different properties in different direction, i.e., asymmetric in nature. This is the reason why studying crystals is a interesting topic for researcher.

We can also catagorize material in terms of their bandgap, as conductor, semiconductor and insulator and also according to their properties, e.g., Magnetic, Piezoelectric, Ferroelectric etc.

Not all materials have a regular crystal structure. Polymers display varying degrees of crystallinity, and many are completely non-crystalline. Glasses, some ceramics, and many natural materials are amorphous, not possessing any long-range order in their atomic arrangements. The study of polymers combines elements of chemical and statistical thermodynamics to give thermodynamic, as well as mechanical, descriptions of physical properties.

Whatever is the material, if we keep on decreasing it's dimensions, at a particular dimension (100 nm down to 0.2 nm) its properties start changing with that of its bulk form. The physics of this small scale will change because the forces governing the material now changes. On the basis of its attractive characteristics modern technology starts exploiting this small scale science and a new technology has been emerged and is popularly known as nanotechnology.

Nanotechnology: Science of Smallness

Dream big, elder's advice. Now how about dreaming small? Scientists are exploiting the science of the small – Nanotechnology – to promise us wonder results that we could only dream of until now. While there's the promise of self –cleaning glasses, clothes that never get dirty are very much on the horizon. Intelligent clothing that measure pulse and respiration rate are also on the anvil. And so are corrosionless paints and paints that changes colour on command. There is also the likelihood of tiny nanorobots, armed with lasers, patrolling your blood stream for any menacing-looking foreign matter.

Nanotechnology aims at manipulating materials at the molecular and atomic levels and come up with totally new products with wonder properties. Also called the science of making things small, nanotechnology gives to the scientist the power of isolating and playing with individual atoms. Individually these atoms behave much differently than when they are in clustered form and this is what nanotechnology hopes to exploit.

For instance, gold, although inert in the bulk form, when broken into particles in the 10-100 nanometer range, becomes a highly effective catalyst. Aluminium foils come in handy to wrap sandwiches and keep them warm, but as a nanoparticle aluminium becomes an integral ingredient of explosive mixtures. There are other materials that turn into superconductors of electricity. Similarly, carbon atoms can be arranged into nanotube structures that are stronger than steel, conduct electricity better than copper and are virtually impervious to heat. There are lots of wondrous things that happen to materials at the nano scale.

And what are the products one could expect from nanotechnology? There are lots to come if scientists are to be believed. Nanotechnology promises to revolutionize the way we detect and treat diseases; increasing miniaturization of gadgets is very much on its agenda; electronic circuits could become tinier and yet more powerful leading to miniature supercomputers; protecting the environment could become easier; energy production and storage could be revolutionized, and so and so forth, the list is endless.

Some of the outcomes of nanotechnology are especially desirable. For instance, tiny robots are on the anvil that would scour the insides of the human body seeking out troublesome germs and destroying them with lasers. Such nanorobots could also deliver drugs at specified targets thus multiplying the effectiveness of drugs manifold and also leaving the surrounding healthy tissues untouched. Diseases like cancer would be the primary beneficiaries where a technique like chemotherapy kills even healthy cells in the vicinity of cancerous tumours. Especially relevant to Indian conditions is a technology that involves pumping of nanomaterials into the ground where they can convert hazardous chemicals in the groundwater into benign products. In fact, this technology has already been developed at the University of Western Ontario, USA. Nobel laureate Harry Kroto also added his voice in favour of the technology when he said nanotechnology has the potential to reduce costs with its multiple applications and the inherent ability to produce new materials like non-corroding and flexible iron.

Nanotechnology is emerging as an industrial force worldwide. The NanoBusiness Alliance trade group estimates that in 2004, $13 billion worth of products incorporated nanotechnology, which is less than 0.1 per cent of global output. But by 2014, that figure is expected to rise to nearly $3 trillion, or 15 per cent of manufacturing output.

India too needs to put ample resources into nanotechnology to be able to harvest ots fruits in times to come. In fact, one of the most eminent scientists of the country Prof. C.N.R. Rao, has this to say: *"India cannot afford to miss the revolution in nanotechnology. We should not be at the receiving end when the world is driven by nanotechnology'.* Indeed, for once the country should not let go of the opportunity to be in the driver's seat.

Early Development

Scientists have now given a clear picture about the nature of matter and atoms, showing how atoms combine. Research by chemists in the 1950s showed the workings of natural molecular machines. The concept of nanotechnology had its genesis in a lecture by physicist and former Cornell professor Richard P. Feynman (1918-1988) to the American Physical Society in 1959 titled "There's Plenty of Room at the Bottom". Though he never used the term nanotechnology, Feynman envisioned the direct manipulation of atoms by developing small-scale machine tools that would evolve into future generations of smaller and smaller tools for the task.

The term nanotechnology was coined in 1974 by Tokyo Science University professor Norio Taniguch; to describe precision manufacturing of materials at the nanometer level. In 1986, K. Eric Drexler wrote Engines of Creation: The coming Era of Nanotechnology, a work that introduced the public to the wider possibilities in the field. He talked about building machines on the scale of molecules, a few nanometers wide-motors, robot arms, and even whole computers, far smaller than a cell. However, his ideas seemed too good to be true, and many scientists pronounced the whole thing impossible. While Drexler remained firm in his conviction, spending the next ten years describing and analyzing these incredible devices, he also often had to respond to accusations of science fiction. However, laws of physics care little for hope and fear. Subsequent analysis kept returning the same answer: it will take time, but it is not only possible but also almost unavoidable. As technology developed the ability to build simple structures on a molecular scale, nanotechnology became an accepted concept. Soon the U.S. National Nanotechnology Initiative was created to fund this kind of nanotech. Drexler taught the first course on the subject at Stanford University in 1988. Progress in nanotechnology is being made in many laboratories around the world, notable in the U.S., Japan, and Europe. Three fields of work have been seen as most relevant: protein design, biomimetic chemistry, and atomic imaging and positioning. Major advances in protein design have been made in the last two years.

Introduction to Nanomaterials

A key driver in the development of new and improved materials, from the steels of the 19th century to the advanced materials of today, has been the ability to control their structure at smaller and smaller scales. The overall properties of materials as diverse as paints and silicon chips are determined by their structure at the micro- and nanoscales. As our understanding of materials at the nanoscale and our ability to control their structure improves, there will be great potential to create a range of materials with novel characteristics, functions and applications.

Although a broad definition, we categorize nanomaterials as those which have structured components with at least one dimension less than 100 nm. Materials that have one dimension in the nanoscale (and are extended in the other two dimensions) are layers, such as a thin films or surface coatings. Some of the features on computer chips come in this category. Materials that are nanoscale in two dimensions (and extended in one dimension) include nanowires and nanotubes (Figure 26.2). Materials that are nanoscale in three dimensions are particles, for example precipitates, colloids and quantum dots (tiny particles of semiconductor materials). Nanocrystalline materials, made up of nanometer sized grains, also fall into this category. Some of these materials have been available for some time; others are genuinely new. We will look an overview of the properties, and the significant expectable applications of some key nanomaterials.

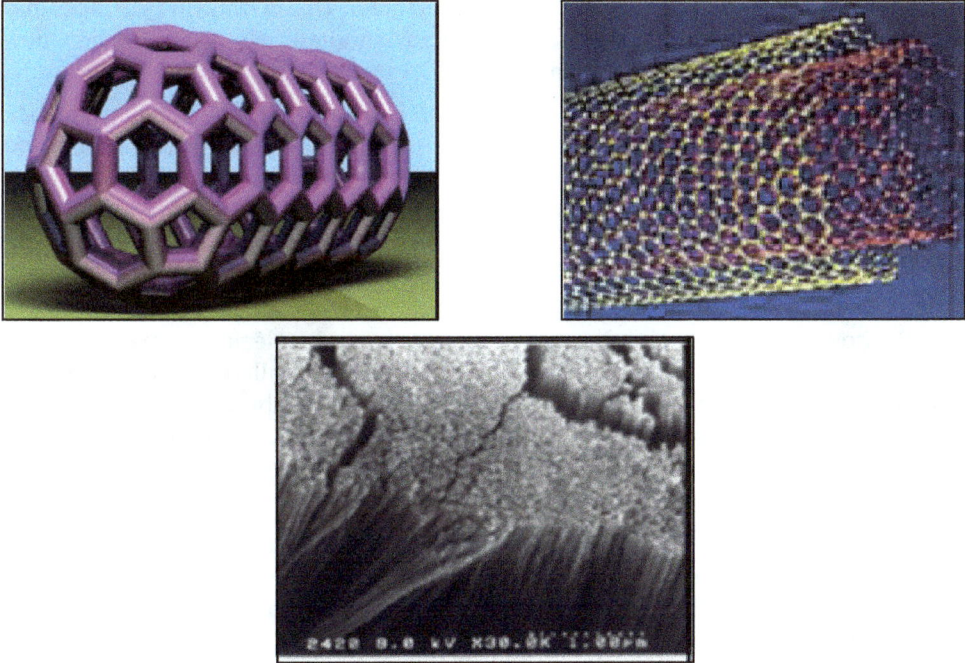

Figure 26.2: Different types of nanotubes.

Two principal factors cause the properties of nanomaterials to differ significantly from other materials: increased relative surface area, and quantum effects. These factors can change or enhance properties such as reactivity, strength and electrical characteristics. As a particle decreases in size, a greater proportion of atoms are found at the surface compare to those inside. For example, a particle of size 30 nm has 5 per cent of its atoms on its surface, at 10 nm 20 per cent of it's atoms, and at 3 nm 50 per cent of its atoms. Thus nanoparticles have a much greater surface area per unit mass compare with larger particles. As growth and catalytic chemical reactions occur at surfaces, this means that a given mass of material in nanoparticulate form will be much more reactive than the same mass of material made up of larger particles.

In tandem with the surface-area effects, quantum effects can begin to dominate the properties of matter as size is reduced to the nanoscale. These can affect the optical, electrical and magnetic behavior of materials, particularly as the structure or particle size approaches the smaller end of the nanoscale. Materials that exploit these effects include quantum dots, and quantum well lasers for optoelectronics.

For other materials such as crystalline solids, as the size of their structural components decreases, there is much greater interface area within the material; this can greatly affect both mechanical and electrical properties. For example, most metals are made up of small crystalline grains; the boundaries between the grain slow down or arrest the propagation of defects when the material is stressed, thus giving it strength. If these grains can be made very small, or even nanoscale in size, the interface area within the material greatly increases, which enhances its strength. For example, nano crystalline nickel is as strong as hardened steel. Understanding surfaces and interfaces is a key challenge for those working on nanomaterials, and one where new imaging and analysis instruments are vital.

Nanomaterials are not simply another step in the miniaturization of materials. They often require very different production approaches. There are several processes to create nanomaterials, classified as 'top-down' and bottom-up'. Although many nanomaterials are currently at the laboratory stage of manufacture, a few of them are being commercialized.

The Assembler

But how do scientists visualize assembling atoms and molecules to make devices? To get every atom in the right place, we needed a machine called the assembler. The assembler can force site-specific chemical reactions. To find the structures consistent with the laws of chemistry and physics, we must use molecular modeling software. To reduce manufacturing costs, we may need to develop some type of replication for the assemblers.

A self-replicating assembler would work by using its ability to make site-specific chemical reaction to make copies of itself. These copies can then make copies of themselves, and so on. Eventually, the assembler multitude can then working parallel to build molecular structures.

This massive parallelism leads to great economics of scale. These assemblers can be compared to the molecular machinery evident in cells today. For example, a seed can be seen as the instructions to create a vast structure of cellulose such as a redwood tree or in something as simple as a potato plant. With the right instructions, an assembler could make products in an analogous way.

Wings to Manufacturing Process

According to Eric Drexler, nanotechnology will render the traditional manufacturing process obsolete. Because the techniques of nanotechnology will be able to copy themselves, assemblers will be inexpensive. We can see this by recalling that many other products of molecular machine-firewood, hay, and potatoes-cost very little.

In its advanced form, nanotechnology is expected to offer better built, longer lasting, cleaner, safer, and smarter products for the home, for communications, for agriculture, and for industry in general.

Apart from new products, nanotechnology will also usher in an era of vastly improved manufacturing processes. For instance, how would it be if the building of products becomes as cheap as the copying of files in a computer? You would have a desktop machine that will make for you products as simply. That's the kind of possibilities that nanotechnology has to offer and that is the reason why it is sometimes seen as _"the next industrial revolution"_.

The first products made from nanomachines will be stronger fibers. Eventually, we will be able to replicate anything, including diamonds, water and food. For instance, machines that fabricate foods to feed the hungry could eradicate famine.

By working in large teams assemblers and more specialized nanomachines will be able to build objects cheaply. By insuring that each atom is properly placed, they will manufacture products of high quality and reliability. Leftover molecules would be subject to this strict control as well, making the manufacturing process extremely clean.

Precautions

But there will have some precautions as well. Radiation can break bonds and misarrange atoms within a device. Such defects can be dealt with in two ways:1) by using designs in which when one part fails, another takes over; engineers call this redundancy, 2) by using repair devices left within the object to make molecular repairs when needed. Without such precautions, molecular machines would eventually break down and stop working.

New Computation Methodology

In the computer industry, the ability to shrink the size of transistors on silicon microprocessors will soon reach its limits. Nanotechnology will be needed to create a new generation of computers components. Molecular computers could contain storage devices capable of storing trillions of bytes of information in a structure the size of a sugar cube.

Assembler based manufacturing will enable the construction of extremely small computers. The equivalent of a modern mainframe computer could fit into a cubic micron, a volume far smaller than that of a single human cell. Once such nanocomputers have been designed and the technology is in hand, building them will be inexpensive, enabling use many of them at once. A laptop computer could then have more power than all the computers in the world today put together.

Judicious Use of Natural Resources

Another promising scenario is with relation to using natural resources judiciously. Rather than clear-cutting forests to make paper, it is possible to have assemblers for synthesizing paper. Rather than using oil for energy, it will be better to have molecule-sized solar cells mixed into road pavement. With such solar nanocells, a sunny patch of pavement of a few hundred square miles could generate enough energy for various uses.

Famine could be synthesized easily and cheaply with a microwave-sized nanobox that pulls the raw materials, mostly carbon, from the air or the soil. By using nanorobots as cleaning machines that break down pollutants, we would be able to counteract the damage we've done to the earth since the industrial revolution.

Balancing the Environment

Nanotechnology could have a positive effect on the environment as well. Building devices at the molecular level means products would be smaller minimizing waste and also leading to less waste in the production process and in the trash, when nano-devices are discarded at the end of their lives. The overall trend would be towards sustainable development.

Nanotechnology could even help repair some of the damage that human have inflicted on the environment. Foe instance, airborne nanorobots could be programmed to rebuild the thinning ozone layer. Contaminants could be automatically removed from water sources, and oil spills could be cleaned up instantly.

Our dependence on non-renewable resources would diminish with nanotechnology. Many resources could be constructed by nanomachine. Cutting down trees, mining coal or drilling for oil may no longer be necessary. Resources could simply be constructed by nanomachines.

Nanomedicine

Nanotechnology may have its biggest impact on the medical industry. The technology is widely recognized as a great opportunity for diseases prevention (*e.g.*, improved food safety), early disease detection (*e.g.*, sensors for cancer detection) or medical treatment (*e.g.*, controlled drug delivery by nanocapsules).

It could also mean the end of disease. For instance, if somebody suffers from cold or fever, he/she would just have to drink a teaspoon of liquid containing an army of molecule-sized nanorobots programmed to enter into the body's cells and fight viruses. In case of genetic diseases, the patient would have to ingest nanorobots that would burrow into the DNA and repair the defective gene, the technique could even eliminate traditional plastic surgery, as medical nanorobots could change one's colour, alter the shape of nose, or even give a complete sex change without surgery.

There is even speculation that nanorobots could slow or reverse the aging process, and life expectancy could increase significantly.

Boosting the Economy

Nanotechnology will accelerate the national economies, revolutionizing most industries, and has been compared in importance to humanity's taming of fire. Because assemblers will be able to build copies of themselves quickly using inexpensive materials, little energy and no human labour, a single assembler could be used to make billions of items. Once we have software to program assemblers to make consumer goods, each household could use an assembler system to produce goods cheaply and quickly. Manufacturing, mining, transportation, and other industries will change radically. Individuals will be able to make at home much of what they need, reducing the need to transport goods.

Conclusion

This technology is one of the fastest growing research field which will have impact not only on all sectors of science and industries but will change the way we live, communicate, entertain etc., in a manner which is not possible to predict. Could there be a flip side to nanotechnology as well? without any doubt, yes. A technology that can build sophisticated products quickly and inexpensively could be used to quickly build a vast arsenal of powerful weapons. Further, new types of weapons might be developed, combining features of today's chemical and biological weaponry and greater control and hence greater military usefulness.

Ideally, the race for early breakthroughs would be won in a country or group of countries firmly under democratic control, where a free press and public scrutiny could help prevent abuse. Broadly based international cooperation seems clearly desirable if we are to minimize the chance of friendly competition turning into hostile competition and then an unstated arms race.

According to some scientists, discussions on nanotechnology should be restricted to the research community until they have been actually developed, since premature exposure might lead to confusion, and perhaps to inappropriate and premature regulation. However, any such new and powerful technologies deserve early and through consideration, to help us maximize benefits and minimize problems.

The promises of nanotechnology sound great, may be even unbelievable. But researchers say that we will achieve these capabilities within the next century. And if nanotechnology is, in fact, realized, it might be the human race's greatest scientific achievement yet, completely changing every aspect of the way we live.

Nanotechnology has so much to offer the world and if used with proper caution and careful planning, it could be the best thing humankind has ever done for the environment. There is no doubt that nanotechnology will be incorporated into every facet of our lives, making things faster, easier and longer lasting than one can imagine today. On the other hand, this technology could also be abused for warfare purposes. The most appropriate words that come to mind are from the movie Spider Man, '*With great power comes great responsibility*'.

References

1. Narottam Sahoo, '*Nanotechnology: Small Science-Big Dreams*', Science Reporter, pp. 10-15, April 2007.

2. C. Kittel, 'Introduction to Solid State Physics', 8th Ed., Wiley Pub.

3. Vijendra Lingwal, '*Developments in Materials Science Research*', A project Report Submitted to the UGC-ASC, University of Kerala in Partial Fulfillment of the 97th Orientation Course, 2008.

Biotechnology: An Overview (2015)
Editors: Rajan Kumar Gupta, Nasim Akhtar and Deepak Vyas
Published by: DAYA PUBLISHING HOUSE, NEW DELHI

Pages 417–423

Chapter 27

Applications of Nanoparticles in the Field of Medicine

Ramna Tripathi[1]* and Akhilesh Kumar[2]

[1]Department of Physics, THDC-Institute of Hydropower Engineering and Technology, Tehri, Uttarakhand
[2]Department of Physics, Govt. Girls P. G. College, Rajajipuram, Lucknow, U.P.

ABSTRACT

Nanomaterials are at the leading edge of the rapidly developing field of nanotechnology. Their unique size-dependent properties make these materials superior and indispensable in many areas of human activity. This brief review tries to summarise the most recent developments in the field of applied nanomaterials, in particular their application in the field of medicine.

Keywords: Cancer therapy, Quantum dots, Medicine, Nanomaterials, Tissue engineering.

Introduction

Nanotechnology is derived from the combination of two words Nano and Technology. Nano means very small or "miniature". So, Nanotechnology is the technology in miniature form. It is the combination of Bio- technology, Chemistry, Physics and Bio-informatics, etc.

The three chief divisions of Nanotech are Nano-electronics, Nano-materials, and Nano-Biotechnology. The implications of Nanotechnology in India can be found in the field of telecommunications, computing, aerospace, solar energy, and environment. However, Nanotech's major contribution can be seen in the computing, communication and medical field.

---

* *Corresponding Author:* E-mail: ramna_tripathi@yahoo.co.in

Nano-medicine is the most important field of Nanotechnology. The nano level gadgets and materials are used for diagnosing and treatment of diseases. Nano-Pharmacology has generated a specific category of smart drugs that affect negligible side effects. The Council of Scientific and Industrial Research, also known as CSIR has set up 38 laboratories in India dedicated to research in Nanotechnology. This technology will be used in diagnostic kits, improved water filters and sensors and drug delivery. The research is being conducted on using it to reduce pollution emitted by the vehicles.

Looking at the progressive prospects of Nanotechnology in India, Nanobiosym Inc., a US-based leading nanotechnology firm is planning to set up India's first integrated nanotechnology and biomedicine technology park in Himachal Pradesh. Nanotechnology has certainly acquired an essential position in the Indian Economy and Scientific Research Department and it is expected to reach the pinnacle of Development thereby making India a role model for the countries of the world.

Nanotechnology [1] is enabling technology that deals with nano-meter sized objects. It is expected that nanotechnology will be developed at several levels: materials, devices and systems. The nanomaterials level is the most advanced at present, both in scientific knowledge and in commercial applications. A decade ago, nanoparticles were studied because of their size-dependent physical and chemical properties [2]. Now they have entered a commercial exploration period [3,4].

Living organisms are built of cells that are typically 10 ìm across. However, the cell parts are much smaller and are in the sub-micron size domain. Even smaller are the proteins with a typical size of just 5 nm, which is comparable with the dimensions of smallest manmade nanoparticles. This simple size comparison gives an idea of using nanoparticles as very small probes that would allow us to spy at the cellular machinery without introducing too much interference [5]. Understanding of biological processes on the nanoscale level is a strong driving force behind development of nanotechnology [6].Out of plethora of size-dependant physical properties available to someone who is interested in the practical side of nanomaterials, optical [7] and magnetic [8] effects are the most used for biological applications.

Applications of nanomaterials in the field of medicine are fluorescent biological labels [11-13], drug and gene delivery [14,15], bio detection of pathogens [16], detection of proteins [17], probing of DNA structure [18], tissue engineering [19,20], tumour destruction via heating (hyperthermia), separation and purification of biological molecules and cells, MRI contrast enhancement etc.

As mentioned above, the fact that nanoparticles exist in the same size domain as proteins makes nanomaterials suitable for bio tagging or labelling. However, size is just one of many characteristics of nanoparticles that it is rarely sufficient if one is to use nanoparticles as biological tags. In order to interact with biological target, a biological or molecular coating or layer acting as a bioinorganic interface should be attached to the nanoparticle. Examples of biological coatings may include antibodies, biopolymers like collagen, or monolayers of small molecules that make the nanoparticles biocompatible. In addition, as optical detection techniques are wide spread in biological research, nanoparticles should either fluoresce or change their optical properties.

Nano-particle usually forms the core of nano-biomaterial. It can be used as a convenient surface for molecular assembly, and may be composed of inorganic or polymeric materials. It can also be in the form of nano-vesicle surrounded by a membrane or a layer. The shape is more often spherical but cylindrical, plate-like and other shapes are possible. The size and size distribution might be important in some cases, for example if penetration through a pore structure of a cellular membrane is required. The size and size distribution are becoming extremely critical when quantum-sized effects are used to control material properties. A tight control of the average particle size and a narrow distribution of sizes allow creating very efficient fluorescent probes that emit narrow light in a very wide range of wavelengths. This helps with creating biomarkers with many and well distinguished colours. The core itself might have several layers and be multifunctional. For example, combining magnetic and luminescent layers one can both detect and manipulate the particles.

The core particle is often protected by several monolayers of inert material, for example silica. Organic molecules that are adsorbed or chemisorbed on the surface of the particle are also used for this purpose. The same layer might act as a biocompatible material. However, more often an additional layer of linker molecules is required to proceed with further functionalisation. This linear linker molecule has reactive groups at both ends. One group is aimed at attaching the linker to the nanoparticle surface and the other is used to bind various moieties like biocompatibles (dextran), antibodies, fluorophores etc., depending on the function required by the application.

Tissue Engineering

Natural bone surface is quite often contains features that are about 100 nm across. If the surface of an artificial bone implant were left smooth, the body would try to reject it. Because of that smooth surface is likely to cause production of a fibrous tissue covering the surface of the implant. This layer reduces the bone-implant contact, which may result in loosening of the implant and further inflammation. It was demonstrated that by creating nano-sized features on the surface of the hip or knee prosthesis one could reduce the chances of rejection as well as to stimulate the production of osteoblasts. The osteoblasts are the cells responsible for the growth of the bone matrix and are found on the advancing surface of the developing bone.

The effect was demonstrated with polymeric, ceramic and, more recently, metal materials. More than 90 per cent of the human bone cells from suspension adhered to the nanostructured metal surface, but only 50 per cent in the control sample. In the end this findings would allow to design a more durable and longer lasting hip or knee replacements and to reduce the chances of the implant getting loose.

Titanium is a well-known bone repairing material widely used in orthopaedics and dentistry. It has a high fracture resistance, ductility and weight to strength ratio [19].

A real bone is a nanocomposite material, composed of hydroxyapatite crystallites in the organic matrix, which is mainly composed of collagen. Thanks to that, the bone is mechanically tough and, at the same time, plastic, so it can recover from a mechanical damage. The actual nanoscale mechanism leading to this useful combination of properties is still debated.

An artificial hybrid material was prepared from 15–18 nm ceramic nanoparticles and poly (methyl methacrylate) copolymer [20]. Using tribology approach, a viscoelastic behaviour (healing) of the human teeth was demonstrated. An investigated hybrid material, deposited as a coating on the tooth surface, improved scratch resistance as well as possessed a healing behaviour similar to that of the tooth.

Cancer Therapy

The small size of nanoparticles endows them with properties that can be very useful in oncology, particularly in imaging. Quantum dots (nanoparticles with quantum confinement properties, such as size-tunable light emission), when used in conjunction with MRI (magnetic resonance imaging), can produce exceptional images of tumour sites. These nanoparticles are much brighter than organic dyes and only need one light source for excitation. This means that the use of fluorescent quantum dots could produce a higher contrast image and at a lower cost than today's organic dyes used as contrast media. The downside, however, is that quantum dots are usually made of quite toxic elements.

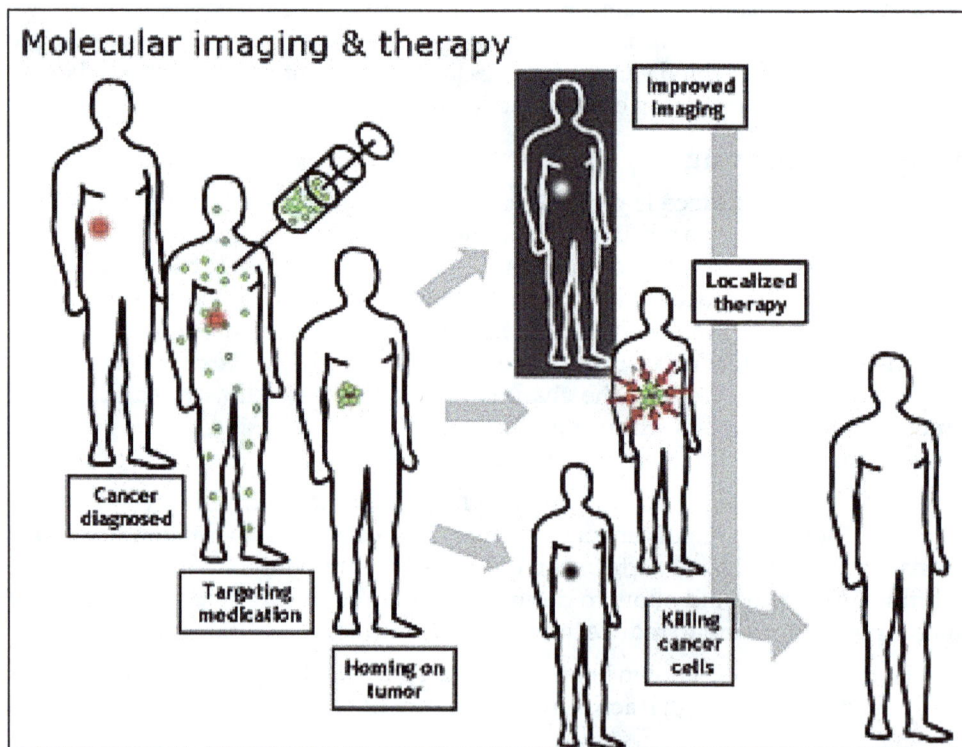

Figure 27.1: A schematic illustration showing how nanoparticles or other cancer drugs might be used to treat cancer.

Another nanoproperty, high surface area to volume ratio, allows many functional groups to be attached to a nanoparticle, which can seek out and bind to certain tumour cells. Additionally, the small size of nanoparticles (10 to 100 nanometers), allows them to preferentially accumulate at tumour sites (because tumours lack an effective lymphatic drainage system). A very exciting research question is how to make these imaging nanoparticles do more things for cancer. Vigorous investigation is going on to manufacture multifunctional nanoparticles that would detect, image, and then proceed to treat a tumour. A promising new cancer treatment that may one day replace radiation and chemotherapy is edging closer to human trials. Kanzius RF therapy attaches microscopic nanoparticles to cancer cells and then "cooks" tumours inside the body with radio waves that heat only the nanoparticles and the adjacent (cancerous) cells.

Sensor test chips containing thousands of nanowires, able to detect proteins and other biomarkers left behind by cancer cells, could enable the detection and diagnosis of cancer in the early stages from a few drops of a patient's blood.

The basic point to use drug delivery is based upon three facts:

1. Efficient encapsulation of the drugs,
2. Successful delivery of said drugs to the targeted region of the body, and
3. Successful release of that drug there.

Nanoparticles of cadmium selenite (CdSe) quantum dots glow when exposed to ultraviolet light [21]. When injected, they seep into cancer tumours. The surgeon can see the glowing tumour, and use it as a guide for more accurate tumour removal.

Protein Detection

Proteins are the important part of the cell's language, machinery and structure, and understanding their functionalities is extremely important for further progress in human well-being. Gold nanoparticles are widely used in immunohistochemistry to identify protein-protein interaction. However, the multiple simultaneous detection capabilities of this technique are fairly limited. Surface-enhanced Raman scattering spectroscopy is a well-established technique for detection and identification of single dye molecules. By combining both methods in a single nanoparticle probe one can drastically improve the multiplexing capabilities of protein probes. The group of Prof. Mirkin has designed a sophisticated multifunctional probe that is built around a 13 nm gold nanoparticle. The nanoparticles are coated with hydrophilic oligonucleotides containing a Raman dye at one end and terminally capped with a small molecule recognition element (*e.g.* biotin). Moreover, this molecule is catalytically active and will be coated with silver in the solution of Ag(I) and hydroquinone. After the probe is attached to a small molecule or an antigen it is designed to detect, the substrate is exposed to silver and hydroquinone solution. A silver-plating is happening close to the Raman dye, which allows for dye signature detection with a standard Raman microscope. Apart from being able to recognise small molecules this probe can be modified to contain antibodies on the surface to recognise proteins. When tested in the protein array format against both small molecules and proteins, the probe has shown no cross-reactivity.

Future Directions

As it stands now, the majority of commercial nanoparticle applications in medicine are geared towards drug delivery. In biosciences, nanoparticles are replacing organic dyes in the applications that require high photo-stability as well as high multiplexing capabilities. There are some developments in directing and remotely controlling the functions of nano-probes, for example driving magnetic nanoparticles to the tumour and then making them either to release the drug load or just heating them in order to destroy the surrounding tissue. The major trend in further development of nanomaterials is to make them multifunctional and controllable by external signals or by local environment thus essentially turning them into nano-devices.

References

1.	Feynman R (1991). There's plenty of room at the bottom. Science (254): 1300–1301.

2.	Murray CB, Kagan CR, Bawendi MG (2000). Synthesis and characterisation of monodispersenanocrystals and close-packed nanocrystal assemblies. Rev Mater Sci. (30): 545–610.

3.	Mazzola L (2003). Commercializing nanotechnology.Nature Biotechnology (21): 1137–1143.

4.	Paull R, Wolfe J, Hebert P, Sinkula M (2003). Investing in nanotechnology. Nature Biotechnology (21): 1134–1147.

5.	Taton TA (2002). Nanostructures as tailored biological probes.Trends Biotechnology. (20): 277–279.

6.	Whitesides GM (2003). The 'right' size in Nanobiotechnology.Nature Biotechnology (21): 1161–1165.

7.	Parak WJ, Gerion D, Pellegrino T, Zanchet D, Micheel C, Williams CS, Boudreau R, Le Gros MA, Larabell CA, Alivisatos AP (2003). Biological applications of colloidal nanocrystals. Nanotechnology (14): 15–27.

8.	Pankhurst QA, Connolly J, Jones SK, Dobson J (2003). Applications of magnetic nanoparticles in biomedicine. J Phys D: Appl Phys. (36): 167–181.

9.	Yan H, Park SH, Finkelstein G, Reif JH, LaBean TH (2003). DNA-templated self-assembly of protein arrays and highly conductive nanowires. Science (301): 1882–1884.

10.	Keren K, Berman RS, Buchstab E, Sivan U, Braun E (2003). DNA-templated carbon nanotube field-effect transistor. Science (302): 1380–1382.

11.	Bruchez M, Moronne M, Gin P, Weiss S, Alivisatos AP (1998). Semiconductor nanocrystals as fluorescent biological labels. Science (281): 2013–2016.

12.	Chan WCW, Nie SM (1998). Quantum dot bio-conjugates for ultrasensitive non-isotopic detection. Science (281): 2016–2018.

13.	Wang S, Mamedova N, Kotov NA, Chen W, Studer J (2002). Antigen/antibody immunocomplex from CdTe nanoparticle bio-conjugates. Nano Letters (2): 817–822.

14. Mah C, Zolotukhin I, Fraites TJ, Dobson J, Batich C, Byrne BJ (2000)Microsphere-mediated delivery of recombinant AAV vectors in vitro and in vivo. Mol Therapy (1): S239.

15. Panatarotto D, Prtidos CD, Hoebeke J, Brown F, Kramer E, Briand JP, Muller S, Prato M, Bianco A (2003)Immunization with peptide-functionalized carbon nanotubes enhances virus-specific neutralizing antibody responses. Chemistry and Biology (10): 961–966.

16. Edelstein RL, Tamanaha CR, Sheehan PE, Miller MM, Baselt DR, Whitman LJ, Colton RJ (2000). The BARC biosensor applied to the detection of biological warfare agents. Biosensors Bioelectron (14): 805–813.

17. Nam JM, Thaxton CC, Mirkin CA (2003). Nanoparticles-based bio-bar codes for the ultrasensitive detection of proteins. Science (301): 1884–1886.

18. Mahtab R, Rogers JP, Murphy CJ (1995). Protein-sized quantum dot luminescence can distinguish between "straight", "bent", and "kinked" oligonucleotides. J Am Chem Soc. (117): 9099–9100.

19. Ma J, Wong H, Kong LB, Peng KW (2003). Biomimetic processing of nanocrystallite bioactive apatite coating on titanium.Nanotechnology (14): 619–623.

20. de la Isla A, Brostow W, Bujard B, Estevez M, Rodriguez JR, Vargas S, Castano VM (2003)Nanohybrid scratch resistant coating for teeth and bone viscoelasticity manifested in tribology. Mat ResrInnovat (7): 110–114.

21. Tripathi R, Dutta A, Kumar A, SinghR S, SinhaT P(2010)Characterization and Optical Spectra of CdSe Nanostructure. National Conference on Nanomaterials and Nanotechnology ISSN 2229-4872, (1): 115-119.

Biotechnology: An Overview (2015)
Editors: Rajan Kumar Gupta, Nasim Akhtar and Deepak Vyas
Published by: DAYA PUBLISHING HOUSE, NEW DELHI

Pages 425–439

Chapter 28

Nanotechnology as Clean Energy and Resources for the Future

Akhilesh Kumar [1]* and Ramna Tripathi [2]

[1]*Department of Physics, Govt. Girls P.G. College,
Rajajipuram, Lucknow, U.P, India*
[2]*Department of Physics, THDC-Institute of Hydropower Engineering and
Technology, Tehri, Uttarakhand, India*

ABSTRACT

Nanotechnology is the technology of 21st century. When size of any material is less than 100 nm, then it is called nanomaterials. By reducing size of any material, when we reach in the nanometer regime, due to increase in surface to volume ratio and quantum confinement, physical, chemical, thermal, electrical, optical, structural and biological etc. properties of material changes. Due to changes in these properties of nanomaterials; Nanotechnology can be used as clean energy and resources for the future. In fact, conventional technology is exhausting its resource base at an accelerating rate. Nanotechnology is the only way to resolve the prospect of providing something like a sustainable standard of living for the entire world. Possible resource related applications of nanotechnology in energy sectors are *to fossil fuels, fission energy, geothermal energy, Solar energy indirect sources (hydropower, wind, wave energy, ocean thermal energy conversion), Solar energy direct sources (Photovoltaics, Artificial Photosynthesis, Solar thermal, Nanofabrication, Dispersed collection), tidal energy and nuclear fusion* etc. In the present paper we have discussed, how nanotechnology is applicable in these energy resources. Conventional technology is depleting its resource base at an accelerating rate and creating an increasingly unlivable mess in doing so, even as rising expectations around the world heighten demand still further. We simply cannot maintain the standard of living in the industrialized world much longer with conventional technology.

* *Corresponding Author:* E-mail: singhakhilesh26@rediffmail.com

Introduction

Nanoscale materials (or nanomaterials) contain nanoparticles or are developed using nanotechnology. Nanoparticles are commonly considered to be materials that have at least one dimension that is less than 100 nm[1]. Nanoparticles can be distinguished according to their origin. They can occur naturally (*e.g.*, from volcanic eruptions); they can be produced incidentally during other processes (*e.g.*, from fuel combustion); and they can be manufactured intentionally. Manufactured nanomaterials can be classified according to their method of production. Some are produced "from the top down," as when a bulk material (*e.g.*, gold, silicate) is reduced to a mass of nanoscale particles[2,3,4]. Because of their very small size, these nanoscale metals, metal oxides, powders, and dusts have physical, chemical, magnetic, electrical, mechanical, and other properties that differ from those of the bulk materials from which they are derived. They are found today in sunscreen products (titanium dioxide nanoparticles), solar cells (aluminum oxide nanoparticles), and several other applications in research and commerce.

The second type of manufactured nanoparticles are built "from the bottom up," atom-by atom or molecule-by-molecule. Engineered nanoparticles of this type are still relatively difficult and expensive to manufacture, but they have the potential to impact energy development and use, transportation, electronics, manufacturing, and other disciplines. Specific examples of nanoparticles with the potential to impact energy transmission system development are highlighted in the next section.

Nanoparticles with the Potential to Impact Energy Transmission System Development

Examples of nanoparticles with the potential to impact energy transmission system development include the following:

1. A carbon nanotube (CNT) is a type of fullerene (carbon-only) molecule that is formed when atoms of carbon link together into tubular shapes. CNTs are generally extremely light, strong, and resilient, and some CNTs can be many times more electrically conductive than steel or copper. CNTs are available for industrial applications in bulk quantities [5].
2. Carbon atoms may also link to form spherical nanostructures that can be coated or filled with atoms; these "buckyball" fullerenes are used in mechanical and semiconductor operations.
3. Nanodots, or quantum dots, are nanoscale semiconductor crystals having electrical and optical properties that may make for more efficient lighting and solar collection.

A specific type of CNT, the "armchair" CNT, has the potential to greatly impact electrical conductivity and transmission. The armchair CNT is 30 to 100 times stronger than steel, conducts heat better than diamond, and conducts electricity better than any other molecule discovered to date. Some researchers like CNTs to a new miracle polymer. Richard Smalley, the Nobel Prize winner who discovered fullerenes in 1985, described the potential of the armchair nanotube as follows:

Electrons move down this tube as a coherent quantum particle, much like a photon of light travels down a single-mode optic fiber. Individual armchair tubes can conduct as much as 20 micro amps of current. This doesn't sound like much until you realize that this little molecular wire is only 1 nanometer in diameter.

A half-inch-thick cable made of these tubes aligned parallel to each other along the cable would have over 100 trillion conductors packed side-by-side like pipes in a hardware store.

Energy Applications of Nanotechnology

Nanoparticles and nano manufacturing techniques may impact energy transmission system development and use for many years to come. For example, nanotechnologies may make more efficient use of transportation fuels, possibly slowing the increase in demand for long distance shipment of liquid fuels. Construction materials made from nanoparticles may be stronger but occupy less volume than today's materials, which may reduce the footprints required for the construction and maintenance of pipelines and electricity transmission lines.

General Energy Applications

Nanotechnology is being used or considered for use in many applications targeted to provide cleaner, more efficient energy supplies and uses. While many of these applications may not affect energy transmission directly, each has the potential to reduce the need for the electricity, petroleum distillate fuel, or natural gas that would otherwise be moved through energy transmission ROWs (rights of ways). More efficient energy generation and use (and the consequent reduced need to transmit energy over long distances) may decrease the amount of construction, maintenance, repair, and decommissioning activities along the ROWs that would otherwise be needed to meet increased energy demands. Energy-related technologies in which nanotechnology may play a role include [6,7,8]:

☆ Lighting,
☆ Heating,
☆ Transportation,
☆ Renewable energy,
☆ Energy storage,
☆ Fuel cells,
☆ Hydrogen generation and storage, and
☆ Power Chips.

Examples of how nanotechnology may be integrated into each of these technology areas are explained in the following sections.

Lighting

Roughly 20 per cent of all electricity is consumed in providing incandescent and fluorescent lighting. Because of their compactness, durability, low heat generation, and electrical efficiency, light-emitting diodes (LEDs) now rival incandescent light sources in many parts of the visible spectrum and are being used in displays, automobile lights, and traffic lights. Semiconductors used in the preparation of LEDs for lighting are increasingly being built at nanoscale dimensions, and projections indicate that nanotechnology-based lighting advances have the potential to reduce worldwide consumption of energy by more than 10 per cent. Nanocrystals, also known as quantum dots, are known primarily for their ability to produce distinct colors of light as the size of the individual crystals is varied. The discovery has implications for using nanotechnology to produce light for residential, commercial, and industrial applications without the heat that accounts for a large portion of the incandescent light bulb's poor energy efficiency.

Heating

Nanotechnology may help accelerate the development of energy-efficient central heating. When added to water, CNTs disperse to form a nanofluid. Researchers have developed nanofluids whose rates of forced convective heat transfer are four times better than the norm by using CNTs. When added to a home's commercial water boiler, such nanofluids could make the central heating device 10 per cent more efficient. The researchers say that the technology is 3 to 5 years away from commercial home use.

Transportation

Nanotechnology may enable more efficient transportation via catalysts in fuels; lighter, stronger materials; and more efficient batteries.

Diesel Fuel Additives for more Efficient Combustion

The Envirox™Fuel Borne Catalyst, developed by Oxonica, Ltd., is an example of a commercially proven product that improves diesel fuel combustion, reducing fuel consumption and harmful exhaust emissions. The additive uses nanoscale (10 nm across) particles of cerium oxide to catalyze the combustion reactions between diesel fuel and air. The small particle size creates a greater surface area for catalyzing the reactions, causes the particles to remain more evenly suspended in the fuel, and allows the additive to be used at concentrations as low as five parts per million, or one-tenth the concentration of previous additives [9]. Fuel economy benefits of up to 10 per cent have been demonstrated in independently assessed field trials under commercial operating conditions.

Stronger Materials

More energy-efficient transportation resulting from the use of high-strength, low-weight materials developed with nanotechnology may reduce the need for transportation fuels that would be shipped via pipeline along a ROW. Nanoparticle-reinforced materials that are as strong as or stronger than today's materials but weigh less will help provide better fuel economy. By using high-strength nanomaterials, parts for automobiles and other modes of transportation could be more than 50 per cent lighter than conventional alternatives. The reduction in weight could cut fuel requirements, thereby potentially reducing the demand for petroleum fuel (and its attendant pipeline transportation in ROWs). Similarly, new materials developed through nanotechnology will permit the miniaturization of systems and equipment, which may further improve fuel economy.

Batteries and Capacitors

More efficient batteries developed by using better electrolytes (composed of nanomaterials) may also reduce the need for transportation fuels. Nanotechnology is being used in lithium-ion and other batteries that are expected to increase the efficiencies of hybrid and electric vehicles. Nanoscale capacitors made from multi walled CNTs dramatically boost the amount of surface area, and thus the electrical charge, that each metal electrode in the capacitor can possess. Smaller and more powerful capacitors may facilitate the development of microchips having greatly increased circuit density. Such nanoscale capacitors may also impact the development of compact and cost-effective super capacitors, which could help reduce the amount of weight in hybrid-electric vehicles, thus improving fuel consumption.

Renewable Energy

Nanotechnologies may also facilitate the generation of electricity directly from solar, wind, and geothermal sources. Using such energy at or near the source could enable distributed energy production of electricity, thereby minimizing transmission losses and reducing the need for ROW-based transmission of electricity, oil, and gas. Practical energy collectors that are simple and automated may result from cheap nanofabrication. Also, more efficient electricity transmission may enable the generation of increased amounts of electricity in remote locations (*e.g.*, nonpopulated areas with abundant renewable energy) to be sent to high-energy demand areas via nanoenhanced transmission. Solar photovoltaic technology, which at present relies on crystalline-silicon wafers that are costly to produce, is deployed economically only in limited settings. Less costly quantum dot (nanocrystal) technologies could make important contributions to improving the efficiency of solar energy systems. Examples of some of these potential nanotechnology-enabled improvements are highlighted below:

☆ High-performance semiconductor nanocrystals (nanodots) that are active over the entire visible spectrum and into the near-infrared have been combined with conductive polymers to create ultrahigh-performance solar cells. The solar cells have improved efficiencies because the nanocrystals harvest a greater portion of the energy spectrum. Solar roofing tiles using quantum dots that are based on metal nanoparticles are expected to be commercialized within the next several years.

☆ Highly ordered nanotube arrays have demonstrated remarkable properties, when used in solar cells. Researchers explain that the nanotube arrays provide excellent pathways for electron percolation, acting as "electron highways" for directing the photo-generated electrons to locations where they can do useful work. Research results suggest that highly efficient solar cells could be made simply by increasing the length of the nanotube arrays.

☆ One solar energy company (Konarka Technologies) creates a photoactive nanoscale material that can be printed on a variety of surfaces, including flexible plastics that can be manufactured in rolls. The material can be cut up and used for such applications as roofing and interior wall material, and can even be stitched onto or woven into a soldier's backpack. The product's costs are one-third those of conventional photovoltaics, and the projected capital cost for manufacturing equipment and facilities is about one-fifth that of the prevailing cost for conventional solar cells.

☆ Nanoadditives, including nanoparticles and nanopowders, could be used to enhance the transfer of heat from solar collectors to storage tanks. When added to heat-transfer fluids, the solid nanoparticles conduct heat better than the fluids alone, and they stay suspended longer than larger particles

Energy Storage

The ability to store energy locally can reduce the amount of electricity that needs to be transmitted over power lines to meet peak demands. Energy storage could allow downsizing of baseload capacity and is a prerequisite for increasing the penetration of renewable and distributed generation technologies such as wind turbines at reasonable economic and environmental costs. Suitable energy storage is critical to the increased use of renewable energies, particularly solar and wind, because these are inconsistent resources. Nanotechnology may play a role in distributed generation through the development of cost-effective energy storage in batteries, capacitors, and fuel cells. The next generation of storage devices may be optimized by nanoengineered advances and the use of nanoscale catalyst particles (Foster 2006). Richard Smalley has described a model for storage using nanotechnology (Foster 2006) in which he suggests that by 2050 every house, business, and building would have its own local electrical storage device # an uninterruptible power supply capable of handling all of the needs of the owner for 24 hours. Because such devices would be small and relatively inexpensive, they could be replaced with new models every 5 years or so as technological innovation continues. Today, such a unit, using lead-acid storage batteries and storing 100-kilowatt-hours (kWh) of electrical energy for a typical house, would occupy a small room and cost more than $10,000. Through advances in nanotechnology, it may be possible to shrink an equivalent unit to the size of a washing machine and drop the cost to less than $1,000. With these advances, the electrical grid could become exceedingly robust. Such advances could also permit some or all of the primary electrical power on the grid to come from solar and wind energy [10-11].

Batteries

CNTs have extraordinarily high surface areas and good electrical conductivity, and their linear geometry makes their surface areas highly accessible to a battery's electrolyte. These properties could enable CNT-based electrodes in batteries to generate increased electricity output as compared to traditional electrodes. This ability to increase the energy output from a given amount of material means not only that batteries could become more powerful, but also that smaller and lighter batteries could be developed for a wider range of applications. Commercial firms are actually developing such next-generation batteries today. For example, in April 2006, a nanotechnology company (Altairnano) owner testified before Congress about a new nanotechnology battery with potentially broad applications. The battery technology utilizes 25-nm nanostructured lithium titanate spinel (a hard, glassy mineral) as the electrode material in the anode of a rechargeable lithium-ion battery, replacing the graphite electrode typically used in such batteries and contributing to performance and safety issues. The new battery offers vastly faster discharge and charge rates, meaning that the time to recharge the battery can be measured in minutes rather than in hours. The nanostructured materials also increase the useful lifetime of the battery by 10 to 20 times over current lithium batteries and provide battery performance over a broader range of temperatures than currently achievable; over 75 per cent of normal power would be available at temperatures between "40°F and +152°F. These types of batteries may enable the U.S. auto industry to "leapfrog" the next generation of hybrid-electric vehicles, thereby accelerating the reduction of the need for petroleum (and for the pipeline transmission of petroleum).

Other commercial applications for these batteries are for uninterruptible power supplies (UPSs) and mergency backup power (EBP). Present-day UPS and EBP systems typically use lead-acid batteries because of their reliability and low initial cost. However, lead-acid batteries must be replaced every 2 to 3 years, and hazardous materials issues surround their manufacture, handling, and maintenance. Lead-acid batteries also lose charge quickly in extreme temperatures (<32°F and >112°F) and suffer from power declines and a decreased ability to accept a recharge over time. By comparison, prototype batteries using the nanostructured lithium titanate electrode material show promising improvements. The advanced lithium-ion battery is virtually unaffected by temperature extremes, its charge is fully available immediately, and it can accept a full recharge in a few minutes. It also has a much longer lifetime with no decline in performance, and there are no hazardous materials issues. With these kinds of advantages, UPS and EBP systems could become reliable components of distributed mini-grids.

Capacitors

While batteries, which derive electrical energy from chemical reactions, are effective in storing large amounts of energy, they must be discarded after many charges and discharges. Capacitors, however, store electricity between a pair of metal electrodes. They charge faster and longer than normal batteries, but because their storage capacity is proportional to the surface area of their electrodes, even today's most powerful capacitors hold 25 times less energy than similarly sized chemical batteries. Researchers, however, have covered capacitor electrodes with millions of nano tubes to increase electrode surface area and thus the amount of energy that they can hold. The researchers claim that the new technology "combines the strength of today's batteries with the longevity and speed of capacitors and has broad practical possibilities, affecting any device that requires a battery" (Limjoco 2006). Because conventional lithium-ion batteries cannot charge at temperatures below 32°F and explode at temperatures higher than 208°F, this characteristic would permit the new batteries to be used in physical environments that today cannot be served by lithium-ion batteries because of safety concerns or because they require complex, expensive electronic control circuitry and temperature maintenance.

Fuel Cells

A fuel cell is a device used for electricity generation that is composed of electrodes that convert the energy of a chemical reaction directly into electrical energy, heat, and water. It is similar to a battery, except that it is designed for continuous replenishment of the reactants that become consumed, thereby requiring no recharging. It produces electricity from an external supply of fuel and oxygen, rather than the limited internal energy storage capacity of the battery. Fuel cells come in various sizes and provide useful power in remote locations such as spacecraft and weather stations. Fuel cells are often considered in the context of hydrogen, because they change hydrogen and oxygen into water, producing electricity and heat in the process but no other by-products. A fuel-cell system running on hydrogen has no major moving parts and can be compact and lightweight. Many believe that in the future, fuel cells will be used to power everything from handheld electronic devices to cars, buildings, and utility power plants. Fuel cells are not new, but the materials' costs and complex manufacturing processes have limited their development. Nanoengineered materials may help improve fuel cells' efficiency in several ways; some examples are highlighted below:

☆ Fuel cells operate by catalyzing the conversion of hydrogen into energy as the hydrogen passes through a catalytic medium. Advanced designs for next generation fuel cells involve the use of a polymer membrane as the structure through which the hydrogen passes and on which the catalysis occurs. The use of nano engineered membrane materials may increase the volume of hydrogen conversion and thus result in more energy.

☆ Precious metal nanoparticles of various compositions have been optimized to act as effective electro- catalysts in polymer electrolyte fuel cells and direct methanol fuel cells at both the anode and the cathode sides.

☆ A materials design concept used to control and manipulate the structure of a new material on the nanoscale could lead to more powerful fuel cells than currently available and to devices that enable more efficient energy extraction from fossil fuels and carbon-neutral fuels. The new electrode material allows more efficient direct utilization of natural gas or biogas (produced from waste) in fuel cells.

☆ CNTs' high strength and toughness-to-weight characteristics may be important for composite components in fuel cells that are deployed in transport applications where durability is important.

Hydrogen Generation and Storage

The hydrogen economy is a hypothetical future economy in which hydrogen is the primary form of stored energy for mobile applications and load balancing. It is typically discussed as an alternative to today's fossil-fuel economy. Many barriers need to be overcome for the hydrogen economy to become a reality. These include producing the hydrogen (for which adequate sources of electricity are needed), transporting it (including the possible need for additional ROWs), and storing it. Nanotechnology may play a role in helping to meet these challenges. As discussed previously, nanotechnology may help accelerate the use of solar and other renewables to generate electricity. Also, nanowires may increase the efficiency of long-distance electrical energy transmission. Nanotechnology may help address hydrogen storage problems. Because hydrogen is the smallest element, it can escape from tanks and pipes more easily than conventional fuels. There are two ways to store hydrogen in materials. One way involves absorption of the hydrogen within the material, and the other is to store the hydrogen in a container. The challenge for absorption is to control the diameter of the nano tube so that the absorption energy of hydrogen on the outside and inside of the tube is high enough to provide the desired storage capacity at an acceptable pressure. If the absorption behavior of the optimized tube is acceptable, the challenge then is to develop a process capable of producing the material at a reasonable cost. Single-walled CNTs are a leading candidate for solving the storage problem for hydrogen-fueled cars and trucks. However, if this approach cannot be carried out, CNTs could still facilatae storage in a container. Small hydrogen-fueled vehicles would have a pressurized tank, and large hydrogen-fueled vehicles, ships, and planes would have a cryogenic liquid hydrogen tank. In this case, CNTs may be used in super-strong composites in the bodies of the vehicles to make them lighter. Additional options might include metal nanoclusters which have been shown to be some of the best catalysts available for reversible hydrogen storage. Core-shell cobalt nanoparticles with tailored chemical compositions can provide protection against corrosive environments. The nanoparticles can easily be removed, leaving a graphite carbon shell. The low-density shell's excellent permeability and electron conductivity make it a candidate for hydrogen storage applications.

Power Chips

Numerous nanotechnology applications will likely be developed within the next 20 years, but one example serves to illustrate the potential impact that nanotechnology may have on energy generation, and, in turn, on the reduced need for energy ROWs in the future. "Power Chips" are nanotechnology devices that use thermionics to convert heat directly into electricity. If successful, these small solid-state devices could improve current power generation and waste heat recovery techniques. They are estimated to deliver up to 70 to 80 per cent of the maximum (Carnot) theoretical efficiency for heat pumps (conventional power-generation equipment operates at up to 40 per cent Carnot efficiency). Currently under development, Power Chips contain no moving parts or motors and can be either miniaturized or scaled to very large sizes for use in a variety of applications.

Nanotechnology Applications having Particular Relevance to Energy Transmission Technologies

Numerous nanomaterials and other nano-related applications relevant to electricity transmission and petroleum distillate fuel and gas pipeline transport are in various stages of research, development, and deployment. These applications have the potential to directly or indirectly reduce the environmental impact associated with the construction, operation, and dismantlement of energy transmission technologies. The remainder of this section highlights examples of nanotechnology applications relevant to transmission of electricity via cables and of fossil fuels (*i.e.*, petroleum distillate fuel and natural gas) through pipelines. Potential pitfalls and timeframes have been identified in the literature. In general, however, the potential for practical scale-up of most of the techniques in use today for nanoparticle production is limited by high capital costs, low production rates, the need for exotic and expensive precursor materials, and limited control over nanoparticle physical and chemical homogeneity. Breakthroughs in nanotechnology research may accelerate the development and implementation of these technologies.

Nanotechnology Applications Relevant to Electricity Transmission

Wires and Cables

Nanotechnology may help improve the efficiency of electricity transmission wires. Today, aluminum conductor steel reinforced (ACSR) wire is the standard overhead conductor against which alternatives are compared. By 2010, the development of new overhead conductors is expected to increase the capacity of existing ROWs by five times that of ACSR wire at current costs (DOE 2006). The 3M Corporation has developed a nano material-based metal-matrix overhead conductor known as the aluminum conductor composite reinforced (ACCR) wire, which is designed to resist heat sag and provide more than twice the transmission capacity of conventional conductors of similar size. This ACCR wire is currently in use, or has been selected for use, by six major utilities across the country. According to 3M (2006): "Aluminum has been a key ingredient in bare overhead conductors for decades. The difference is that ACCR wire is based on the use of aluminum processed in new ways to create high-performance and reliable overhead conductors that retain strength at high temperatures and are not adversely affected by environmental conditions." The ACCR wire's strength and durability derive from its nano crystalline aluminum oxide fibers, which are embedded in the high-purity 3M aluminum matrix core wires using a patented process. The constituent materials are chemically inert with respect to each other and can withstand extreme temperatures without chemical reactions or any appreciable loss in strength. The material used in the core of the cable replaces the steel used in conventional cables (3M 2006).

Other Electrical Transmission Infrastructure

Nanotechnology applications may help improve other components of the electric transmission infrastructure, thereby potentially reducing environmental impacts. The examples below pertain to transformers, substations, and sensors [12-15].

Transformers

Fluids containing nanomaterials could provide more efficient coolants in transformers, possibly reducing the footprints, or even the number, of transformers. Nanoparticles may increase heat transfer, and solid nanoparticles conduct heat better than liquid. Nanoparticles stay suspended in liquids longer than larger particles, and they have a much greater surface area, which is where heat transfer takes place (Strem 2006). Using nanoparticles in the development of HTS transformers could result in compact units with no flammable liquids, which could help increase siting flexibility.

Substations

Substation batteries are important for load-leveling peak shaving, providing uninterruptible supplies of electricity to power substation switchgear, and for starting backup power systems. Smaller, more efficient batteries could reduce the footprints of substations and possibly the number of substations within a ROW.

Sensors

Nanoelectronics have the potential to revolutionize sensors and power-control devices. Nanotechnology-enabled sensors would be self-calibrating and self-diagnosing. They could place trouble calls to technicians whenever problems were predicted or encountered. Such sensors could also allow for the remote monitoring of infrastructure on a real-time basis. Miniature sensors deployed throughout an entire transmission network could provide access to data and information previously unavailable.

Other Materials

Advanced materials using nanomaterials or nanotechnology may extend service life, lower failure rates, and reduce the potential for environmental damage. Two examples of such advanced materials follow.

Smart Materials

The Electric Power Research Institute, Inc. (EPRI) has described how nanotechnology may accelerate the development of "smart materials and structures" (SMSs) (EPRI 2003). According to EPRI, SMSs have the unique capability to sense and physically respond to changes in their environment (*e.g.*, temperature, acidity levels, magnetic field). Generally consisting of a sensor, an actuator, and a processor, an SMS device can function autonomously in an almost biological manner. On a transmission line, SMSs could monitor and assess the condition of conductors, breakers, and transformers in real time to avoid outages.

Ceramics

Ceramics are hard and resist heat and chemical attack, but they are also very brittle. Researchers at the University of California at Davis have mixed aluminum oxide with 5 per cent to 10 per cent CNTs and 5 per cent finely milled niobium and processed it to consolidate ceramic powders at lower temperatures than conventional processes. The resulting material has up to five times the fracture toughness (resistance to cracking under stress) of conventional alumina. The material also shows electrical conductivity seven times that of previous ceramics made with nanotubes [16].

Nanotechnology Applications Relevant to Pipeline Transmission of Petroleum Distillate Fuel and Natural Gas

Today, most of the identified nanotechnology applications for pipelines involve material coatings (insulation, corrosion, and multipurpose). Other potential applications include nanosensors, which have the potential to minimize environmental damage by identifying potential leaks before they spread, and oil spill remediation with nanomaterials, which may minimize damage should a leak occur.

Because the current and expected future applications of nanotechnology for petroleum distillate fuel pipelines are basically the same as those for natural gas pipelines [17-18].

Materials

Advanced materials using nanotechnology may extend service life, reduce failure rates, and limit the potential for environmental damage. Nanocoating metallic surfaces can help achieve super hardening, low friction, and enhanced corrosion protection. Stronger materials may reduce wear, corrosion, and the chances of puncturing associated with third-party damage. Also, because nanomaterials can be stronger per unit volume than conventional materials, the use of pipe materials that contain or coated with nanomaterials may mean fewer disturbances to the environment during installation, maintenance, and dismantlement.

Nanosensors

Nanosensors, or sensors made of nanomaterials, can be extremely sensitive, selective, and responsive. As such, they could be smaller and cheaper, and consume less power than conventional sensors. Sensors and controls that are small in size; work safely in the presence of electromagnetic fields, high temperatures, and high pressures; and can be changed cost-effectively may provide the ability to monitor conditions in the infrastructure and monitor for pollutants [19-20].

Potential Environmental, Safety and Health Risks

Due to their extremely small size and relatively large surface areas, nanomaterials may interact with the environment in ways that differ from more conventional materials. Potentially harmful effects of nanotechnology could result from the nature of the nanoparticles themselves and from products made with them. Environmental, safety, and health (ES&H) risks could occur during research, development, production, use, and end-of-life processes. The 21st Century Nanotechnology Research and Development Act, signed into law in December 2003, calls for addressing potential environmental and societal concerns associated with nanotechnology. About 10 per cent of the Federal nanotechnology budget is characterized as environmental, but much of this amount is for developing nanotechnologies to address existing environmental problems rather than for investigating potential ES&H effects. Some research has been conducted on the toxicology of nanomaterials, but research on the fate, transport, and transformation; risk characterization, mitigation, and communication; and exposure, bioaccumulation, and personal protection has yet to come. All of the departments and agencies with Federal funding from the National Nanotechnology Initiative have environmental research planned or underway, but industry, nongovernmental organizations, and others have questioned whether the amount of research on the potential ES&H impacts of nanotechnologies is sufficient, given the number of unknowns and the size of the potential nanotechnology market. There have been relatively few studies on the ES&H risks of engineered nanoparticles [21-25]. There are also no regulatory requirements to conduct such studies, and little funding is allocated to them. The limited results to date are neither conclusive nor consistent. For example, evidence indicates that nanoparticles in the lungs may cause more severe damage than conventional toxic dusts, but few if any inhalation or exposure studies have been conducted. Nanotechnologies may speed cleanup of soil and water contamination, but in the process may harm local soil ecology. The impacts of large quantities of nanomaterials on the environment or human health have not been studied, and there are no studies on accumulation or other long-term impacts.

References

1. Anderson, R., P. Chu, R. Oligney, R. Smalley, *et al.*, 2006, *White Paper, Smart Grid of the Future: A National Test Bed*, Lamont-Doherty Earth Observatory, Columbia University. Available at http://www.ldeo.columbia.edu/res/pi/4d4/testbeds/Smart-Grid-White-Paper.pdf. Accessed July 7, 2006.

2. Aspen, 2006, Aspen Aerogels. Available at http://www.aerogel.com/. Accessed July 7, 2006.

3. Davis, K., 2006, "Tiny Dreams for the Future of Transmission Capacity," *Utility Automation & Engineering T&D Magazine,* April issue.

4. EPRI (Electric Power Research Institute, Inc.), 2003, *Electricity Technology Roadmap: 2003*

   Summary and Synthesis – Power Delivery and Markets, Nov. Available at http://www.hoffmanmarcom.com/docs/pd&m_roadmap_2003.pdf. Accessed June 27, 2006.

5. Foley, M., 2003, *How Do Carbon Nanotubes Work: Carbon Nanotubes 101*. Available at http://www.nanovip.com/node/2077. Accessed February 19, 2007.

6. Foster, L., 2006, *Nanotechnology: Science, Innovation, and Opportunity*, Prentice Hall, Upper Saddle River, NJ.

7. Fox, B., 2006, *Efficiency Trials for Oxonica Nano Fuel Additive, Envirox*. Available at http://www.azonano.com/details.asp?ArticleID=31. Accessed June 26, 2006.

8. Gillett, S.L., 2002, *Nanotechnology: Clean Energy and Resources for the Future*, White Paper for Foresight Institute, Oct. Available at http://www.foresight.org/impact/whitepaper_illos_rev3.PDF. Accessed July 13, 2006.

9. Gotcher, A., 2006, Written statement of Alan Gotcher, Ph.D., President and CEO, Altair Nanotechnologies, Inc., to the U.S. Senate Commerce Committee, June 14, 2006. Available at http://commerce.senate.gov/public/_files/Gotcher061406.pdf. Accessed July 7, 2006. 24

10. Hoffert, M., 2004, "Renewable Energy Options – An Overview," in *The 10-50 Solution: Technologies and Policies for a Low-Carbon Future*, workshop cosponsored by the Pew Center on Global Climate Change and the National Commission on Energy Policy, Washington, D.C., March 25–26. Available at http://www.pewclimate.org/docUploads/10 per cent 2D50 per cent 5FHoffertper cent 2Epdf. Accessed February 19, 2007.

11. IBM (IBM Business Consulting Services), 2004, "Revitalizing the Utilities Network," prepared for Montgomery Research, March. Available at http://www-03.ibm.com/industries/utilities/doc/content/bin/EU_revitalize_network.pdf. Accessed July 7, 2006.

12. Industrial Nanotech, Inc., 2006, Homepage. Available at http://www.voyle.net/Nano per cent 20Biz per cent 2005-100+/NanoBiz-05-100+-0027.htm. Accessed February 19, 2007.

13. Lamba, B., 2005, "Nanotechnology for Recovery and Reuse of Spilled Oil," PhysOrg.com, Sept. 9. Available at http://www.physorg.com/news6358.html. Accessed July 11, 2006.

14. Limjoco, V., 2006, "Super Battery," ScienCentralNews. Available at http://www.sciencentral. com/articles/view.php3?type=article&article_id=218392803. Accessed June 29, 2006.

15. McGahn, D.P., 2006, "Nanotechnology and Its Impact on Industry." Available at http://www.mtpc.org/institute/research/nano_report_04/energy.pdf. Accessed July 12, 2006.

16. Nanotechwire, 2004a, "American Superconductor's Nanotechnology Breakthrough Significantly Increases Performance of Superconductor Wire." Available at http://www.nanotechwire.com/news.asp?nid=666&ntid=117&pg=7. Accessed July 7, 2006.

17. Nanotechwire, 2004b, "NanoDynamics Acquires 12 Patents and Patent Applications Covering New Category of Nanocomposite Thin Film Technology." Available at http://nanotechwire.com/news.asp?nid=1169. Accessed March 12, 2007.

18. National Energy Policy Development Group, 2001, *National Energy Policy*, U.S. Government Printing Office, Washington, D.C., May, pp. 7-5 to 7-6.

19. NNI (National Nanotechnology Initiative), 2000, *The Initiative and Its Implementation Plan*, National Science and Technology Council Committee on Technology Subcommittee on Nanoscale Science, Engineering and Technology, July.

20. Oxonica, 2006, *ENVIROX™ Fuel Borne Catalyst*. Available at http://www.oxonica.com/energy/energy_envirox_intro.php. Accessed June 27, 2006.

21. Penn State, 2006, "Titania Nanotube Arrays Harness Solar Energy," PhysOrg.com, Jan. 25. Available at http://www.physorg.com/news10244.html. Accessed June 29, 2006.

22. Pollitt, M., 2006, "Tiny Tubes Could Bring Big Savings on Fuel Bills," *The Guardian*, April 13. Available at http://technology.guardian.co.uk/weekly/story/0,,1752275,00.html. Accessed June 27, 2006. 25

23. Ruiz-Morales, J.C., J. Canales-Vázquez, C. Savaniu, D. Marrero-López, W. Zhou, and J.T.S. Irvine, 2006, "Disruption of Extended Defects in Solid Oxide Fuel Cell Anodes for Methane Oxidation," *Nature* 439:568–571, Feb. 2.

24. Salisbury, D.F., 2005, "Quantum Dots That Produce White Light Could Be the Light Bulb's Successor," *Exploration*, Oct. 20. Available at http://www.vanderbilt.edu/exploration/stories/quantumdotled.html. Accessed July 10, 2006.

25. Science Daily, 2004, "Titania Nanotube Hydrogen Sensors Clean Themselves," March 29. Available at http://www.sciencedaily.com/releases/2004/03/040325073047.htm. Accessed June 27, 2006.

Index

Rhizodeposition 132

Rhizofilteration 159, 162

Rhizosphere 129, 157

RhoGTPase modules 342

Ribosomal intergenic spacer analysis (RISA) 39

Risk assessment 361, 363

Risk management 361

RNA 32

RNAi 105

RNAs 109

Rodenticides 307

RRNA 40

S

Saccharomyces 298

Saccharomyces cerevisiae 187, 298, 317, 322, 326

Saccharopolyspora erythraea 330

Safranin 354

Saltwater ecosystems 27

Saving ecosystems 26

Scenedesmus 14

Scenedesmus obliquus 12, 15

Schizochytrium sp. 14

Secondary metabolites 389

Security 379

Semiconductor 68

Shikimic acid pathway 391

Signature tagged mutagenesis (STM) 188

Silybum marianum 401

Single strand conformation polymorphism (SSCP) 39

SMD 56

Snf1 pathway 327

Soil aggregation 132

Soil conditioning 281

Soil conservation 364

Soil fertility 353

Soil management 44

Soil microbial diversity 43

Soil remediation 279

Soil type 43

Solar 10

Solar energy 425

Soybean 11

Spirogyra sp. 12

Spirulina platensis 12

Sports 77

Streptomyces hygrosporicus 208

Superoxide dismutase 388

Symbiotic association 90

Symbiotic bioengineers 129

Synechococcus sp. 12

Synthetic biology 101, 109

System biology 101, 107

T

TAG 106

TAIR 61

Taxol 179

Taxus baccata 179, 181

Taxus brevifolia 179, 181

Taxus canadensis 179, 181

Taxus chinensis 181

Taxus cuspidata 179, 181

Taxus mairel 181

Taxus media 181

Taxus wallichiana 181

Taxus yew 179

Taxus yunnanensis 181

Temperature 45

Temperature gradient gel electrophoresis (TGGE) 38

Terminal restriction fragment length polymorphism 40